AF587392

CURRENT ADVANCES IN MOLECULAR MYCOLOGY

Current Advances in Molecular Mycology

Youssuf Gherbawy,
Robert Ludwig Mach
and
Mahendra Rai
Editors

Nova Science Publishers, Inc.
New York

For permission to use material from this book please contact us:
Telephone 631-231-7269; Fax 631-231-8175
Web Site: http://www.novapublishers.com

LIBRARY OF CONGRESS CATALOGING-IN-PUBLICATION DATA

Current advances in molecular mycology / Youssuf Gherbawy, Robert Ludwig Mach, Mahendra Rai (editors).
p. ; cm.
Includes bibliographical references and index.
ISBN 978-1-60456-909-4 (hardcover)
1. Fungal molecular biology. 2. Phytopathogenic fungi. 3. Fungal diseases of plants--Molecular aspects. 4. Fungi--Genetics. I. Gherbawy, Youssuf. II. Mach, Robert Ludwig. III. Rai, Mahendra.
[DNLM: 1. Fungi--genetics. 2. Molecular Biology--methods. 3. Mycology--methods. 4. Mycoses--genetics. QK 604.2.M64 C976 2008]
QK604.2.M64C87 2008
572.8'295--dc22
200802783

Published by Nova Science Publishers, Inc. ✝ New York

CONTENTS

Contributing Authors		**vii**
Preface		**xiii**
Chapter I	Where are the Sequences that Control Multicellular Development in Filamentous Fungi? *D. Moore and A. Meškauskas*	**1**
Chapter II	Journey of Phytopathogenic Fungi from Genetics to Genomics *A. Pain, A. K. Dhar and C. Chattopadhyay*	**39**
Chapter III	In Silico Tools to Study Phytofungi *K. A. Abd-Elsalam, J-R. Guo and J.-A. Verreet*	**63**
Chapter IV	Fusaria: Mycotoxin Production and Species Differentiation by Molecular Markers *A. P. Ingle, A. S. Karwa, M. K. Rai and Y. Gherbawy*	**85**
Chapter V	Fusarium DNA Levels in Finnish Cereal Grains *T. Yli-Mattila, P. Parikka, T. Lahtinen, S. Rämö, M. Kokkonen, A. Rizzo, M. Jestoi and V. Hietaniemi*	**107**
Chapter VI	Molecular Markers for Fusarium Oxysporum Formae Speciales Characterization *Y. Gherbawy, B. El-Deeb, S. Hassan , A. Altalhi and M. Rai*	**139**
Chapter VII	Morphology and Molecular Biology of Phoma *L. Irinyi, A. K. Gade, A. P. Ingle, G. J. Kövics, M. K. Rai and E. Sándor*	**171**
Chapter VIII	Gene Regulation of Hydrolase Expression: Lessons from the Industrially Important Fungus Trichoderma Reesei (Hypocrea Jecorina) *A. R. Stricker and R. L. Mach*	**205**

Chapter IX pH-Related Transcriptional Regulation of Lignocellulolytic Enzymes and Virulence Factors Genes in Fungi **223**
Marcio José Poças-Fonseca, Thiago Machado Mello-de-Sousa and Sérgio Martin Aguiar

Chapter X Exploring Ascochyta Rabiei on Chickpea as a Model to Study Pathogenicity Factors of Ascochyta Blight of cool Season Grain Legumes **249**
W. Chen and D. White

Chapter XI Diversity, Genotypic Identification, Ultrastructural and Phylogenetic Characterization of Zygomycetes from Different Ecological Habitats and Climatic Regions: Limitations and Utility of Nuclear Ribosomal DNA Barcode Markers **263**
Kerstin Hoffmann, Sabine Telle, Grit Walther, Martin Eckart, Martin Kirchmair, Hansjörg Prillinger, Adam Prazenica, George Newcombe, Franziska Dölz, Tamás Papp, Csaba Vágvölgyi, G. Sybren deHoog, Lennart Olsson and Kerstin Voigt

Chapter XII Revision of the Family Structure of the Mucorales (Mucoromycotina, Zygomycetes) based on Multigene-Genealogies: Phylogenetic Analyses suggest a Bigeneric Phycomycetaceae with Spinellus as Sister Group to Phycomyces **313**
K. Voigt, K. Hoffmann, E. Einax, M. Eckart, T. Papp, C. Vágvölgyi and L. Olsson

Chapter XIII Molecular Identification and Characterization of Indoor Wood Decay Fungi **333**
O. Schmidt

Chapter XIV Molecular Tools for Identification and Differentiation of Different Human Pathogenic Candida Species **349**
V. V. Tiwari, M. N. Dudhane and M. K. Rai

Index **371**

CONTRIBUTING AUTHORS

Abdulla Altalhi
Biological Sciences department, Faculty of Science, Taif University, Saudia Arabia

Adam Prazenica
University of Idaho, College of Natural Resources, Department of Forest Resources, 975 W. 6th Street, Moscow, Idaho 83844-1133, U.S.A.

Aldo Rizzo
Finnish Food Safety Authority (Evira), Research Department, Chemistry and Toxicology Unit, Helsinki , Finland .

Alka Karwa
Department of Biotechnology, SGB Amravati University, Amravati 444 602, Maharashtra, India

Arnab Pain
Senior Computer Biologist, The Pathogen Sequencing Unit, The Wellcome Trust Sanger Institute, Genome Campus, Hinxton, Cambridge CB 10 1SA, UK.

Arun Kumar Dhar
Adjunct Associate Professor, San Diego State University, Biology Department, 5500 Campanile Drive, San Diego, California 92182, USA; Senior Scientist, Advanced Bionutrition Corporation, 7155 Gateway Drive, Columbia, Maryland 21046, USA.

Astrid R. Stricker
Gene Technology, Gene Technology and Applied Biochemistry, Institute of Chemical Engineering, TU Wien, Getreidemarkt 9/166/5/2, A-1060 Wien, Austria.

Audrius Meškauskas
Faculty of Life Sciences, The University of Manchester, 1.800 Stopford Building, Oxford Road, Manchester M13 9PT, UK.

Avinash Ingle
Department of Biotechnology, SGB Amravati University, Amravati 444 602, Maharashtra, India.

Bahig El-Deeb
Biological Sciences department, Faculty of Science, Taif University, Saudia Arabia.

Csaba Vágvölgyi
University of Szeged, Faculty of Sciences, Department of Microbiology, Közép fasor 52, 6726 Szeged, Hungary.

Chirantan Chattopadhyay
Principal Scientist (Plant Pathology), National Research Centre on Rapeseed-Mustard (ICAR), Sewar, Bharatpur 321303, India (chirantan_cha@hotmail.com).

David Moore
Faculty of Life Sciences, The University of Manchester, 1.800 Stopford Building, Oxford Road, Manchester M13 9PT, UK.

David White
Formerly Department of Crop and Soil Sciences, Washington State University, Pullman, WA 99164 (USA).

Esra Einax
University of Jena, Institute of Microbiology, Fungal Reference Centre, Neugasse 24, 07743 Jena, Germany.
Thüringer Tierseuchenkasse, Victor-Goerttler-Str. 4, 07745 Jena, Germany Thüringer Tierseuchenkasse, Victor-Goerttler-Str. 4, 07745 Jena, Germany.

Franziska Dölz
University of Jena, Institute of Mikrobiology, Fungal Reference Centre, Neugasse 24, 07743 Jena, Germany.

Gade Aniket Krishnarao
Department of Biotechnology, Amravati University, Amravati-444 602, Maharashtra, India.

George Newcombe
University of Idaho, College of Natural Resources, Department of Forest Resources, 975 W. 6th Street, Moscow, Idaho 83844-1133, U.S.A.

Sybren deHoog
Centraalbureau voor Schimmelcultures, Uppsalalaan 8, 3584 Utrecht, The Netherlands.

Grit Walther
Centraalbureau voor Schimmelcultures, Uppsalalaan 8, 3584 Utrecht, The Netherlands.

Hansjörg Prillinger
University of Natural Resources and Applied Life Sciences Vienna, Department of Biotechnology, Institute of Applied Microbiology, Muthgasse 18, 07743 Vienna, Austria.

Ingle P. Avinash
Department of Biotechnology, Amravati University, Amravati-444 602, Maharashtra, India.

Irinyi László
Debrecen University, Faculty of Agriculture, Department of Plant Protection H-4015 Debrecen, P.O.Box 36, Hungary.

Jian-Rong Guo
Institute of Phytopathology, University of Kiel, Hermann-Rodewald-Str. 9, D-24118, Kiel, Germany.

Joseph-Alexander Verreet
Institute of Phytopathology, University of Kiel, Hermann-Rodewald-Str. 9, D-24118, Kiel, Germany.

Kamel A. Abd-Elsalam
Agricultural Research Centre, Plant Pathology Research Institute, Cotton Diseases Section, 9-Gamaa St., Giza, Egypt (abdelsalamka@gmail.com).

Kerstin Hoffmann
University of Jena, Institute of Microbiology, Fungal Reference Centre, Neugasse 24, 07743 Jena, Germany.

Kerstin Voigt
University of Jena, Institute of Microbiology, Fungal Reference Centre, Neugasse 24, 07743 Jena, Germany kerstin.Voigt@uni-jena.de.

Kövics György János
Debrecen University, Faculty of Agriculture, Department of Plant Protection H-4015 Debrecen, P.O.Box 36, Hungary.

Lennart Olsson
University of Jena, Institute of Systematic Zoology and Evolutionary Biology, Erbertstrasse 1, 07743 Jena, Germany.

Mahendra Rai
Department of Biotechnology, SGB Amravati University, Amravati-444 602, Maharashtra state, India (mkrai123@rediffmail.com).

Martin Eckart
University of Jena, Institute of Microbiology, Fungal Reference Centre, Neugasse 24, 07743 Jena, Germany.

Martin Kirchmair
University of Innsbruck, Faculty of Biology, Institute of Microbiology, Technikerstrasse 25, 6020 Innsbruck, Austria.

Marika Jestoi
Finnish Food Safety Authority (Evira), Research Department, Chemistry and Toxicology Unit, Helsinki , Finland .

Marcio José Poças-Fonseca
Vienna University of Technology (TU Wien), Institute of Chemical Engineering, Gene Technology Group. Getreidemarkt 9/166/5/2 A-1060. Vienna, Austria

Meri Kokkonen
Finnish Food Safety Authority (Evira), Research Department, Chemistry and Toxicology Unit, Helsinki , Finland .

Milind Dudhane
Department of Biochemistry, PDMMC, Amravati i-444 602, Maharashtra state, India.

Olaf Schmidt
Department of Wood Biology, University of Hamburg, Leuschnerstr. 91d, 21031 Hamburg, Germany (o.schmidt@holz.uni-hamburg.de).

Päivi Parikka
Plant Production/Plant Protection, MTT Agrifood Research Finland, Jokioinen, Finland.

Robert L. Mach
Gene Technology, Gene Technology and Applied Biochemistry, Institute of Chemical Engineering, TU Wien, Getreidemarkt 9/166/5/2, A-1060 Wien, Austria (rmach@mail.zserv.tuwien.ac.at).

Sabine Telle
University of Jena, Institute of Mikrobiology, Fungal Reference Centre, Neugasse 24, 07743 Jena, Germany.
Universität Hohenheim, Institut für Botanik (210), FG Spezielle Botanik, Garbenstrasse 30, D-70599 Stuttgart, Germany.

Sabry Hassan
Biological Sciences department, Faculty of Science, Taif University, Saudia Arabia

Sándor Erzsébet
Debrecen University, Faculty of Agriculture, Department of Plant Protection H-4015 Debrecen, P.O.Box 36, Hungary.

Sari Rämö
MTT Agrifood Research Finland, Laboratories, Finland, Jokioinen, Finland.

Sérgio Martin Aguiar
Department of Cellular Biology, Institute of Biological Sciences, University of Brasilia. 70.910-900, Brasilia-DF. Brazil (mpossas@unb.br).

Taina Lahtinen
Lab. of Plant Physiology and Molecular Biology, University of Turku, Turku, Finland.

Tamás Papp
University of Szeged, Faculty of Sciences, Department of Microbiology, Közép fasor 52, 6726 Szeged, Hungary.

Tapani Yli-Mattila
Lab. of Plant Physiology and Molecular Biology, University of Turku, Turku, Finland.

Thiago Machado Mello-de-Sousa
Department of Cellular Biology, Institute of Biological Sciences, University of Brasilia. 70.910-900, Brasilia-DF. Brazil.

Vaibhav Tiwari
Department of Biotechnology, SGB Amravati University, Amravati-444 602, Maharashtra state, India Department of Biotechnology, SGB Amravati University, Amravati-444 602, Maharashtra state, India.

Veli Hietaniemi
MTT Agrifood Research Finland, Laboratories, Finland, Jokioinen, Finland.

Weidong Chen

USDA-ARS, Grain Legume Genetics and Physiology Research Unit, 303 Johnson Hall, Washington State University, Pullman, WA 99164 (USA), 509-335-9178 (w-chen@wsu.edu).

Youssuf Gherbawy

Biological Sciences department, Faculty of Science, Taif University, Saudia Arabia. (Youssufgherbawy@yahoo.com).

PREFACE

Recent advances in the molecular genetics of filamentous fungi are finding increased application in the pharmaceutical, agricultural, and enzyme industries, and this trend promises to continue as the genomics of fungi is explored and new techniques to speed genetic manipulation become available. Molecular techniques, particularly polymerase chain reaction technology, have revolutionized biology. The incorporation of molecular techniques into what has traditionally been a morphologically-based taxonomy of fungi has yielded surprising changes and new species. Fungal genomics is also progressing at a fast pace. Databases of fungi are being created and the fungal world is being unveiled with the help of bioinformatic tools. Comparative genomics has shown new interrelationships and phylogeny among the fungi. In this volume, we approach problems of fungus identification by increasing the number of fungi characterized with molecular techniques. Identification on the basis of morphology alone can create confusion, and this becomes worse when mycelium is obtained from a soil or plant environment. Moreover, the book focuses on the filamentous fungi and highlights the advances of the past decade, both in methodology and in the understanding of genomic organization and regulation of gene and pathway expression. The work offers an update of progress in the development of fungal molecular techniques, and draws attention to potential and associated problems, as well as integrating theory and practice.

The book will be essential reading to the researchers, microbiologists, scientists, post graduate students of mycology, agriculture, biotechnology and medical microbiology also.

MARKETING BLURB

Molecular mycology has been playing a pivotal role in 21st century. It is emerging with full impact. It is multi-disciplinary and includes molecular markers, recombinant DNA techniques, cloning, phylogeny and bioinformatics. Varying in application of concepts, practice, scale, style and substance, molecular mycology is amongst the latest globalizing frontiers of the corporate world. This branch is being regarded as a core subject in colleges and universities. In the present book, various topics on molecular mycology are uniquely combined to provide a complete overview of the subject.

The book attempts to address the role of molecular and bioinformatics tools in solving the problems of identification of fungi. The contributors are from UK, Hungary, India, USA, Germany, Austria, Finland and Egypt.

Special Features

- Discusses current trends in Molecular Mycology
- Includes functional genomics
- Covers application of *in silico* tools in mycology
- Incorporates revision of zygomycetes based on multigene-genealogies
- Focuses on gene regulation of hydrolase expression in *Trichoderma*
- Addresses phylogeny in species of *Fusarium oxysporum*

The book will cater the need of researchers, microbiologists, scientists, post graduate students of mycology, agriculture, biotechnology and medical microbiology.

In: Current Advances in Molecular Mycology
Eds: Y. Gherbawy, R.L. Mach and M.K. Rai
ISBN 978-1-60456-909-4

Chapter I

WHERE ARE THE SEQUENCES THAT CONTROL MULTICELLULAR DEVELOPMENT IN FILAMENTOUS FUNGI?

D. Moore and A. Meškauskas
Faculty of Life Sciences, The University of Manchester, 1.800 Stopford Building, Oxford Road, Manchester M13 9PT, UK

ABSTRACT

We describe a wide-ranging data-mining exercise to search for homologies to gene sequences assigned to the category 'development' in the Gene Ontology Consortium AmiGo database (www.godatabase.org) across all three of the crown group of eukaryote kingdoms. Internet web-agents were used to automate the similarity searches for 552 such developmental sequences in 1069 genomes. It emerged that only 78 of the sequences are shared between all three Kingdoms, 72 are shared only between fungi and animals, 58 sequences are shared between plants and fungi, and 4 sequences were common only to *Dictyostelium* and fungi. No sequences were strictly fungus specific, but 68 occurred only in plants (Viridiplantae) and 239 occurred only in animals (Metazoa). Although some homology was indicated for a total of 219 fungal sequences, 143 (65%) of the matches returned were assigned *E*-values of 0.05 and were not considered significant. These findings show that there are no resemblances between the crown group of eukaryotic Kingdoms in the ways they control and regulate their developmental processes. Current understanding of phylogenetic relationships is that the major kingdoms of eukaryotes separated from one another at a stage prior to the multicellular grade of organization. Consequently, in the course of their evolutionary history these very different organisms may have found different strategies to solve the same sorts of morphogenetic control problems. Finally, this means, in particular, that we are currently totally ignorant of the way fungi regulate their multicellular development.

Key words: automated, sequence analysis, homology, web-agents, eukaryotes.

INTRODUCTION

Filamentous fungi inhabit every environment and habitat on planet Earth. Their success is achieved because of the apical growth of filamentous hyphae, which enables them to populate the available substratum and make efficient use of nutrient resources. Investigation of the habitat and capture of resources depends on sub-apical branching and, especially, negative autotropism which together are crucial to proliferation of the growing mycelium and, above all, to its spread through the habitat. Consequently, ***the*** characteristic feature of the fungal mycelium is the aggressive exploration and control of new substrate (Pringle and Taylor, 2002; Trinci *et al.*, 1994).

But this is not all that mycelia accomplish, because fungal mycelia of Ascomycota and Basidiomycota produce a range of structures that distribute spores or other propagules, including ascomata and basidiomata – the structures that release sexual spores (meiospores) in Ascomycota and Basidiomycota respectively, as well as a range of structures that produce asexual spores (mitospores) and some somatic (vegetative) structures, such as stromata and sclerotia, that can survive adverse conditions. Obviously, the phrase 'fruit body', which is usually used, encompasses a very wide range of organs but their common feature is that they are multicellular, they are constructed of organised tissues that have specific functions, and their shape and form emerge as a result of a sequence of developmental adjustments. That is, they exhibit a characteristic pattern of cell and tissue morphogenesis (Moore, 1998).

As we will illustrate later, there is suggestive evidence that fungal cells within the developing tissues of a fruit body embark on their particular course of differentiation in response to the interaction of their inherent genetic programme with external physical signals (light, temperature, gravity, humidity, physical contact), and/or chemical signals from other regions of the developing structure. These chemicals may be termed organisers, inducers or morphogens, and may inhibit or stimulate entry to particular states of determination. There is also evidence for morphogenetic fields around fungal structures (cell or organ), which permits continued development of that structure but inhibits formation of another structure of the same type within the field. All of these phenomena contribute to the pattern formation that characterises the 'body plan' created by the particular distribution of differentiated tissues in the multicellular structure. Pattern formation depends on positional information, which prompts or allows the cell to differentiate in a way appropriate to its position in the structure and may be conveyed by concentration gradients of one or more morphogens emitted from one or more spatially distinct organisers. Pattern formation thus involves a process that provides positional information, and a second process, in which the receiving cell or tissue responds to that information.

The cells that make up a mature fungal fruit body are generally found to be totipotent (that is, able to dedifferentiate and subsequently follow any pathway of differentiation), because a mycelial culture can be produced *in vitro* from a fragment of a mature, fully differentiated structure, e.g. a mushroom stem or the inner tissues of a mushroom cap. This feature results in a morphogenetic plasticity which surpasses that of other organisms and provides an intellectual challenge in terms of developmental biology, taxonomy and genetics (Watling and Moore, 1994). The only exceptions to totipotency are the meiocytes (the cells within which meiosis occurs), which are committed to sporulation, but only when they have progressed beyond meiotic prophase (Chiu, 1996; Chiu and Moore, 1988a, b, 1990, 1993). On the other hand, even meiocytes can serve non-sporulation functions: the hymenium of *Agaricus bisporus* is packed with basidia held in an arrested meiosis and serving a purely structural function (Allen *et al.*, 1992).

It is tempting to draw parallels with morphogenesis in animals and the vocabulary established to describe morphogenetic events in animals can be used without making presumptions about the mechanisms which may be involved (Moore, 1998, 2005). For example, during the progress of the developmental process the intermediate stages represent successive reduction in developmental potential in comparison with the previous stage. Each such adjustment (or, effectively, each developmental 'decision') is made by cells already specified by earlier adjustments that drew them into a particular branch of the developmental pathway. Consequently, developmental decisions are made from progressively smaller numbers of alternatives until the particular structure to which the cell will contribute is finally determined. It is those decisions and that sequence of developmental adjustments that, in animals and plants, are controlled by genetic regulators of the sort that we seek in fungi.

The mating type factors are the only major morphogenetic control elements that are presently known in fungi. Mating type factors are complex genetic elements (most of which specify transcription factors) that regulate pheromone production and pheromone receptors involved in mating, ranging from recognition between sexually competent cells in yeast to regulating growth of clamp connections, internuclear recognition, and the distance between the two nuclei in Basidiomycota (Casselton, 2002). They reach their highest expression in the basidiomycetes, where compatibility of the mating type factors permits the development of complex fruit bodies containing several different interacting tissues. However, not all fungi possess mating type factors, and, indeed, even in species that have a well-developed mating type system, haploid (that is, unmated) cultures can form apparently normal fruit bodies and fruit body formation can usually be separated from other parts of the sexual pathway by mutation (see chapter 5 in Moore, 1998). The occurrence of fruiting bodies outside the influence of mating type factors makes the real influence of the latter on events beyond the initial mating reaction difficult to judge.

Even though knowledge of major developmental gene sequences in fungi is lacking, we can at least make a comparative analysis to determine whether developmental sequences important in plants and animals can be found in fungi. There is now a sufficient number of filamentous fungal genomes, covering a representative range of fungal diversity, in the public sequence databases to make direct sequence comparisons with animal and plant genomes meaningful. A recent search of a few selected filamentous fungal genomes with a small selection of gene sequences generally considered as being essential and highly conserved components of normal animal and/or plant development failed to reveal any homologies (Moore *et al.*, 2005). This was taken to suggest that fungal and animal lineages may have diverged from their common opisthokont line (Cavalier-Smith and Chao, 1995) at the unicellular level. The unique cell biology of filamentous fungi could be presumed to cause control of multicellular development as it arose in fungi to evolve in a completely different way from that emerging concurrently in animals and plants. This line of argument was based on the current phylogenetic understanding that the major kingdoms of eukaryotes separated from one another at a very early stage in their evolution. If this is the case, these very different organisms may have needed to solve the same sorts of morphogenetic control problems in the course of their evolutionary history and may have found some common strategies, but there is no logical reason to expect that the Kingdoms will share features that contribute to multicellular developmental biology unless arrived at by convergent evolution. The fungal hypha differs from animal and plant cells in many important respects and significant differences in the way cells interact in the construction of organised tissues must be expected (Moore, 2005).

Broad conclusions like this need comprehensive comparisons and in the study described here we expanded our sequence comparisons to include all sequences assigned to the biological process 'development' in the Gene Ontology Consortium's AmiGO database (http://amigo.geneontology.org/cgi-bin/amigo/go.cgi) (Harris *et al.*, 2004). In this database the term 'development' is currently defined as the biological process "whose specific outcome is the progression of the [*specific process*] over time, from its formation to the mature structure." It is a node in the ontology and incorporates terms like morphogenesis, formation, differentiation, specification, metamorphosis, maturation, etc. Using this source of sequences allowed us to collect any and all sequences which the authors of the database entry had assigned or identified with any developmental process.

All of these 'developmental sequences' were then used in similarity searches against all genomes of cellular organisms included as Metazoa, Fungi or Viridiplantae in the NCBI Taxonomy database (www.ncbi.nlm.nih.gov/Taxonomy). This represents an estimated total of 590,000 similarity searches. Even at 100 manual searches a day that works out to 16 years, doing 1 search every 15 minutes, 24/7, so to make such a job possible we used web agents (also known as web-robots), which are reusable programming modules that interact with the Internet seeking user-defined goals, for example 'get the sequence data', 'get the taxonomy information' or 'get the similarity search results', etc. We created the agents, as described below, using an application called *Sight*, which is a Java-based package that provides a user-friendly interface to generate and connect web agents for automatic genomic data mining (visit http://bioinformatics.org/jSight/) (Meškauskas *et al.*, 2004).

AUTOMATED GENOMIC DATA MINING

To create the web agents employed in this report we used an application called *Sight*, which is a package of Java™-based programs that offers a user-friendly interface for the assembly and interconnection of agents for automatic genomic data mining (Meškauskas *et al.*, 2004).

Sight web agents are effectively scripts of active flow charts in which each element is a preprogrammed working routine. *Sight* enables the user to assemble a flow chart tailored to the task to be performed. The application was originally developed for automated analysis of the human genome but has been modified to include loops, convergences and other features that suit it to servers carrying fungal databases (version 3.2.0 beta available for free download from http://bioinformatics.org/jSight/).

The web agent comprises two data structures: one defines the query submitted to the selected database and the other processes the response(s) received from the database. The user provides all the necessary input information, but the program writes the code. The *Sight* application interface provides a web form appropriate to the chosen database comprising the fields, checkboxes and other controls needed by the user for entry of the initial data that generates the agent. The application itself converts these data into string values for the various named items (data fields) that represent the automated query. Single queries often generate multiple responses because, for example, several homologies may be found to the query sequence, or there may be multiple genes in a sequence, or multiple motifs in a sequence, etc.

Consequently, the *Sight* agent response data structure needs to be programmed as an array of records of multiple named fields. Because the query and response formats differ for each agent, the agents also contain explanations of the formats, defining the type, name and identifying comment for each query or response field. Default values for query fields may also be included.

The initial query in this analysis was to the Gene Ontology Consortium's AmiGO database to extract information on gene sequences involved in development (database ID GO:0032502),which belongs to the larger group 'biological process' (database ID GO:0007275). The query was sent to the AmiGo server and the responses received were stored locally as an HTML document.

Search hits from the AmiGo server contained two hyperlink references, one to the entry within the GO database itself, the other to the external server from which the original data was derived; 552 paired references were returned when the survey was first completed in January 2005 (there were 603 in July 2007). Where the GO database page contained the protein sequence, the sequence was taken from there; otherwise, the domain of the link to the original data source was checked and one of the specialized sequence retrievers was called. Sequence retrievers were written for www.pir.uniprot.org (Apweiler *et al.* 2004); www.tigr.org (Venter *et al.* 1992); www.arabidopsis.org (Huala *et al.* 2001); www.flybase.org (Ashburner and Drysdale, 1994); www.informatics.jax.org (MGI) (Blake *et al.*, 2003); dictybase.org (Kreppel *et al.*, 2004). DNA similarity search web agents were written for genome.jgi-psf.org/whiterot1/whiterot1.home.html (sequence data produced by the US Department of Energy Joint Genome Institute at http://www.jgi.doe.gov/); and protein, DNA and RNA similarity search web agents were written for tigrblast.tigr.org (Venter *et al.* 1992) and www.ncbi.nlm.nih.gov/BLAST (Wheeler *et al.*, 2005); and a taxonomy search web agent for www.ncbi.nlm.nih.gov/Taxonomy (Wheeler *et al.*, 2005). The number of links to Gramene, the database for genomes of rice, maize and other grasses (Wheeler *et al.*, 2003), the Rat Genome Database (Twigger *et al.*, 2002) and the Zebrafish Information Network, ZFIN, were too few to justify writing a tailored web agent, so these sequences were retrieved manually.

Similarity searches (Altschul *et al.*, 1990) were performed in both protein (using BLASTP) and nucleic acid (using TBLAST) sequence databases. Low complexity filters, which remove short, very widespread sequence fragments, were turned on. All other search options were left with the default values proposed by the research groups that administer the search servers. Only hits with *E*-values less than or equal to 0.05 were accepted (the significance of the *E*-value is discussed briefly below; for further information we recommend the page dealing with the statistics of sequence similarity scores at this URL: http://www.ncbi.nlm.nih.gov/BLAST/tutorial/Altschul-1.html).

Each and every 'developmental' sequence retrieved from the AmiGo server was then used in similarity searches against all genomes of cellular organisms included as Metazoa (875 genome sequences), Fungi (141 genome sequences) or Viridiplantae (53 genome sequences) in the NCBI Taxonomy database (www.ncbi.nlm.nih.gov/Taxonomy). The initial query to the Gene Ontology database retrieved a total of 552 sequences so this represents an estimated total of 590,000 similarity searches.

Some dedicated databases contain sequence data either from a single species or from several related species, so the taxonomic position was immediately evident. The taxonomic position of the organism corresponding to a sequence retrieved from a more general database was identified in several ways. Some sequences contained the Latin binomial of the organism in the sequence header; in this case the web agent submitted the organism name to the NCBI taxonomy search service referred to above, and automatically extracted the taxonomy of the organism from the web page received in response. For the NCBI nucleic acid database, the search was performed by explicitly limiting the scope of the web agent's search to a specified Kingdom.

For processing the search data we used the *E*-value to identify the most similar sequence found in any organism from each of the Metazoa, Viridiplantae and Fungi groups for each sequence entry retrieved from the AmiGO database. These contributed to an overall comparison table for all 552 searches (in HTML format) that preserved all appropriate hyperlinks. Useful descriptive annotations for each sequence were subsequently retrieved manually from the databases using those hyperlinks. This complete table, containing live hyperlinks, is available by e-mail from the authors.

Advantages of Web Agents over Manual Searching

We have suggested above that the number of similarity searches completed during this survey would take about 16 years to complete manually, assuming it would be possible to complete 4 searches every hour of every day of the week. Apart from speeding this process sufficiently to make the operation feasible, the web agents we used have other advantages.

The *Sight* application program permits inclusion of routines into the agent algorithm effectively to 'anticipate' a variety of interactions with the search server. For example, the ability to follow multiple links, conditional behaviour and loops which, for example, enable the agent to make a positive ordered response to a transient server error, or to react to server delays (for example, NCBI BLAST can tell the user to wait for a given duration a number of times before returning the results). So when constructed by an experienced bioinformatician web agents can be at least as effective as a human investigator.

As we used several independent database services (and different web servers, therefore), requests could be submitted in parallel, for which the agents ran in separate execution threads and task queues. This effectively created a kind of distributed computing, although arranging the work of ordinary web servers in parallel like this significantly differs from distributed computing as normally understood. In particular, the concept of load balancing is not applicable because each web server is specialised to its own group of tasks. Also, users have no ability to adapt software on the server(s) to suit distributed computing. Nevertheless, the possibility of reliably mimicking parallel computing gives the web agents a significant advantage over manual management of this sort of work. While it is certainly possible for an individual worker to submit tasks in parallel from several running instances of their ordinary web browser, frequent switching of concentration between the parallel searches demands extreme skill and minor loss of attention can generate multiple errors. Web agents are greatly superior to their human equivalents in this.

As we have implied in the descriptions given above, web agents can use each other in organised workflows. For example, a sequence retrieval web agent can pass the sequence it recovers from one server to a similarity search web agent for submission to another server, and take part of the header of each similarity returned hit for submission to a third server to identify the taxonomy of the organism. Manual working of this process requires multiple copy/paste operations and switching between several browser windows. The time required for these clerical operations is comparable with the waiting time for server response, but during hours and days of such work a human operator is likely to make mistakes. Humans can use their knowledge to speed such analysis, of course. For example the Latin binomials of frequently-used organisms, such as *Homo sapiens*, *Drosophila melanogaster*, *Arabidopsis thaliana* and many other popular research organisms, were retrieved frequently in many thousands of sequence headers. A human operator recognising these repetitive features can avoid the time penalty of approximately ten seconds that it takes to execute a query to the NCBI taxonomy search server. However, this can also be mimicked in the web agents by adding retrieved names and taxonomy to a locally-stored cache. The agent can then be written to search the cache of previously-retrieved names, a matter of milliseconds only, before issuing a query for any newly-encountered name.

Researcher time is needed for creating each web agent and building the workflow system, though this is significantly reduced by using specialised development platforms like *Sight* (Meškauskas *et al.*, 2004), as used in this project. An overwhelming advantage, of course, is that while running, the agent system needs no researcher attention and web agents can run around the clock; using the night hours when servers are less loaded and respond more quickly. We believe the analysis reported here demonstrates that the web agents generated using tools like *Sight* can be extremely useful for an extensive range of highly repetitive tasks.

Comparison of the Occurrence of Developmental Gene Sequences in the Genomes of Eukaryotes

The overall summary of the results of this survey (Table 1) shows that of the 552 developmental sequences retrieved by the initial query to the AmiGO database only 78 are shared between all three Kingdoms, 72 are shared only between fungi and animals, 58 sequences are shared between plants and fungi, and 4 sequences were common only to *Dictyostelium* and fungi. No sequences were strictly fungus specific, but 68 occurred only in Viridiplantae and 239 occurred only in Metazoa.

In many respects these latter two results constitute 'control' searches by representing positive hits within the Kingdom from which the original reference sequence was obtained. It is significant, therefore, that all of these similarities have *E*-values markedly less than our arbitrary cut-off value of 0.05. Broadly speaking, the lower the *E*-value, the better the match; an *E*-value of zero indicates identity of the compared sequences. *E*-values indicate the likelihood of the observed similarity between the sequences being found by chance. *E*-values less than 0.01 are numerically very similar to probability statements. *E*-values of 0.05 mean that there is ***more*** than one chance in twenty of the similarity being found by chance – ***and we do not assign any significance to these*** (because, even if not due entirely to chance, they most likely indicate possession of similar functional motifs – like shared DNA binding sites, membrane spanning regions, etc.). Consequently, ***very*** low *E*-values observed for hits within the Kingdom from which the original reference sequence was obtained validate the process by showing that the reference sequences can be shown to retrieve highly similar sequences from within their own Kingdom. Unfortunately, there are no fungal sequences that are categorised as being involved in developmental processes. This is not a fault in the AmiGO database; rather the deficiency accurately reflects the paucity of research interest in the multicellular developmental biology of Kingdom Fungi.

Table 1. Overall summary of similarities returned

Kingdom	Hits	Remarks*
Animal only	239	all *E*-values well below 0.05
Plant only	68	all *E*-values well below 0.05
Common to fungi and *Dictyostelium*	4	of which three had *E*-values of 0.05, and the fourth an *E*-values of 0.03.
Common to animal and plant	33	13 had *E*-values of 0.05
Common to fungi and animal	72	64 had *E*-values of 0.05
Common to fungi and plant	58	55 had *E*-values of 0.05
Common to all three kingdoms	78	14 plant homologies had *E*-values of 0.05 20 fungal homologies had *E*-values of 0.05
Total	552	219 showed some homology with fungal sequences, though 143 of these had *E*-values of 0.05

*Broadly speaking, the lower the *E*-value, the better the match; an *E*-value of zero indicates identity of the compared sequences. *E*-values indicate the likelihood of the observed similarity between the sequences being found by chance. *E*-values less than 0.01 are numerically very similar to probability statements. *E*-values of 0.05 mean that there is ***more*** than one chance in twenty of the similarity being found by chance – ***and we do not assign any significance to these*** (because, even if not due entirely to chance, they most likely indicate possession of similar functional motifs – like shared DNA binding sites, metal ion binding domains, membrane spanning regions, etc.).

Cross-kingdom comparisons are interesting, although for Kingdom Fungi they are mostly negative. Searches with 44 'no apical meristem' (NAM) family proteins failed to detect any similarities with animals or fungi (Table 2); a further 42 NAM family protein sequences showed weak (not significant) similarities (*E*-value = 0.05) with fungal genomes, but still with no similarity in Metazoa (Table 3). NAM sequences have a role in determining positions of meristems and are required for pattern formation in embryos and flowers, so there is no great surprise that the developmental functions represented by the family of NAM proteins are restricted to plants. The one exception we discovered in this protein family is NAM locus AT4G28500 (a predicted protein of *Arabidopsis thaliana* with transcription factor activity) for which the search revealed homology (E-value = 5×10^{-5}) with clone RP11-26F2 of the *Homo sapiens* chromosome 15, and only an insignificant similarity (E-value = 0.05) with the fungal *Phanerochaete* genome (which was not annotated when the survey was carried out) (Table 4).

Table 2. Plant-only similarities found

Entries under 'genome hit' show E-value as returned by the sequence comparison software: a reference to the specific database entry and brief descriptive annotation. No hits were returned to these plant sequences by searches of Metazoan or Fungal genomes. ***NOTE*** that small E-values are shown as exponential functions, i.e. 1.217E-29 = 1.217×10^{-29}

AmiGO description and hyperlinks	Viridiplantae genome hit	AmiGO description and hyperlinks	Viridiplantae genome hit
49199.m00050: *Arabidopsis* no apical meristem (NAM) family protein.	1.217E-29: 4337200 NAM (no apical meristem)-like protein [*Arabidopsis thaliana*].	60023.m00237: *Arabidopsis* no apical meristem (NAM) family protein.	2.44843E-30: 4325286 NAC domain protein NAM [*Arabidopsis thaliana*].
49203.m00039: *Arabidopsis* no apical meristem (NAM) family protein.	2.53372E-27: 5306267 NAM (no apical meristem)-like protein [*Arabidopsis thaliana*].	60208.m00069: *Arabidopsis* no apical meristem (NAM) family protein.	2.07314E-29: 6016718 hypothetical protein [*Arabidopsis thaliana*].
49299.m00025: *Arabidopsis* no apical meristem (NAM) family protein.	3.53623E-29: 4544462 NAM (no apical meristem)-like protein [*Arabidopsis thaliana*].	60250.m00126: *Arabidopsis* no apical meristem (NAM) family protein.	2.99757E-28: 6223650 NAM-like protein (no apical meristem) [*Arabidopsis thaliana*].
51050.m00108: *Arabidopsis* no apical meristem (NAM) family protein.	2.07588E-29: 15217678 no apical meristem (NAM) family protein [*Arabidopsis thaliana*].	60460.m00049: *Arabidopsis* no apical meristem (NAM) family protein (NAC2).	5.45563E-30: 7021736 putative jasmonic acid regulatory protein [*Arabidopsis thaliana*].
51050.m00231: *Arabidopsis* no apical meristem (NAM) family protein.	8.42797E-31: 15217677 *Arabidopsis* no apical meristem (NAM) family protein.	60460.m00230: *Arabidopsis* no apical meristem (NAM) family protein.	3.53623E-29: 7021735 putative jasmonic acid regulatory protein [*Arabidopsis thaliana*].
51050.m00232: *Arabidopsis* no apical meristem (NAM) family protein.	2.07588E-29: 15217699 no apical meristem (NAM) family protein [*Arabidopsis thaliana*].	6548.m00385: *Arabidopsis* no apical meristem (NAM) family protein.	7.87791E-29: 2459430 putative NAM (no apical meristem)-like protein [*Arabidopsis thaliana*].
51104.m00153: *Arabidopsis* no apical meristem (NAM) family protein.	1.09927E-30: 20857250 product At2g17040/At2g17040 [Arabidopsis thaliana].	67041.m00008: *Arabidopsis* no apical meristem (NAM) family protein.	2.29166E-28: 7268195 putative NAM-like protein [*Arabidopsis thaliana*].

Table 2. (Continued)

AmiGO description and hyperlinks	Viridiplantae genome hit	AmiGO description and hyperlinks	Viridiplantae genome hit
51241.m00131: *Arabidopsis* no apical meristem (NAM) family protein.	5.45452E-30: 34222060 product At1g54330 [*Arabidopsis thaliana*].	67041.m00010: *Arabidopsis* no apical meristem (NAM) family protein.	6.6677E-28: 7268197 putative NAM-like protein [*Arabidopsis thaliana*].
51442.m00188: *Arabidopsis* no apical meristem (NAM) family protein.	7.87791E-29: 5091626 Similar to gb|X92204 NAM gene product from Petunia hybrida [*Arabidopsis thaliana*].	67099.m00015: *Arabidopsis* no apical meristem (NAM) family protein.	1.21539E-29: 22136592 putative NAM/NAP [*Arabidopsis thaliana*].
51476.m00248: *Arabidopsis* no apical meristem (NAM) family protein.	8.71005E-28: 6692113 product F22C12.13 [*Arabidopsis thaliana*].	67170.m00140: *Arabidopsis* no apical meristem (NAM) family protein.	9.30401E-30: 30984582 protein product At4g28530 [*Arabidopsis thaliana*].
51641.m00076: *Arabidopsis* no apical meristem (NAM) family protein.	1.58702E-29: 6227016 Contains similarity to gb|AF123310 NAC domain protein NAM gene from *Arabidopsis thaliana.*	67175.m00005: *Arabidopsis* no apical meristem (NAM) family protein.	2.44843E-30: 7269821 putative protein [*Arabidopsis thaliana*].
51784.m00174: *Arabidopsis* no apical meristem (NAM) family protein.	2.71119E-29: 16612277 protein product At1g01010/T25K16_1 [*Arabidopsis thaliana*].	67241.m00008: *Arabidopsis* no apical meristem (NAM) family protein.	1.02889E-28: 7649380 putative protein [*Arabidopsis thaliana*].
51864.m00069: *Arabidopsis* no apical meristem (NAM) family protein.	1.3435E-28: 7715611 protein product F20B17.1 [*Arabidopsis thaliana*].	67267.m00207: *Arabidopsis* no apical meristem (NAM) family protein.	4.18276E-30: 24030450 putative NAC2 protein [*Arabidopsis thaliana*].
67299.m00017: *Arabidopsis* no apical meristem (NAM) family protein.	1.43759E-30: 7594530 putative protein [*Arabidopsis thaliana*].	68090.m00129: *Arabidopsis* no apical meristem (NAM) family protein.	1.217E-29: 23506087 protein product At5g13180/T19L5_140 [*Arabidopsis thaliana*].
67299.m00021: *Arabidopsis* no apical meristem (NAM) family protein.	7.38326E-27: 7594534 putative protein [*Arabidopsis thaliana*].	68097.m00013: *Arabidopsis* no apical meristem (NAM) family protein.	5.11307E-28: 9955520 putative protein [*Arabidopsis thaliana*].
67622.m00135: *Arabidopsis* no apical meristem (NAM) family protein.	2.9936E-28: 30023658 protein product At5g14000 [*Arabidopsis thaliana*].	68151.m00018: *Arabidopsis* no apical meristem (NAM) family protein.	5.11307E-28: 15237698 no apical meristem (NAM) family protein [*Arabidopsis thaliana*].
67644.m00021: *Arabidopsis* no apical meristem (NAM) family protein.	6.66905E-28: 29824189 putative NAM (no apical meristem) protein [*Arabidopsis thaliana*].	60615.m00165: *Arabidopsis* seven in absentia (SINA) protein.	1.09927E-30: 16323494 putative seven in absentia protein [*Arabidopsis thaliana*].
67644.m00144: *Arabidopsis* no apical meristem (NAM) family protein.	3.19839E-30: 9758912 unnamed protein product [*Arabidopsis thaliana*].	60735.m00029: *Arabidopsis* putative seven in absentia (SINA) protein.	1.48571E-27: 25404638 hypothetical protein T12I7.6 of *Arabidopsis thaliana.*
67651.m00128: *Arabidopsis* no apical meristem (NAM) family protein.	2.9936E-28: 21592559 NAC-domain protein-like [*Arabidopsis thaliana*].	60735.m00031: *Arabidopsis* seven in absentia (SINA) protein.	1.13757E-27: 25404642 hypothetical protein T12I7.8 of *Arabidopsis thaliana.*

Table 2. (Continued)

AmiGO description and hyperlinks	Viridiplantae genome hit	AmiGO description and hyperlinks	Viridiplantae genome hit
67651.m00132: *Arabidopsis* no apical meristem (NAM) family protein.	9.64284E-27: 18700194 protein product AT5g22290/MWD9_7 [*Arabidopsis thaliana*].	1342.m00052: *Arabidopsis* seven in absentia (SINA) protein.	1.13907E-27: 1871185 putative RING zinc finger protein; tRNA-Ser [*Arabidopsis thaliana*].
67745.m00110: *Arabidopsis* no apical meristem (NAM) family protein.	1.34377E-28: 31322199 no apical meristem-like protein [*Arabidopsis thaliana*].	67304.m00152: *Arabidopsis* seven in absentia (SINA) protein.	2.29212E-28: 21593355 putative RING zinc finger protein [*Arabidopsis thaliana*].
67746.m00011: *Arabidopsis* no apical meristem (NAM) family protein.	2.9936E-28: 10177980 NAM (no apical meristem)-like protein [*Arabidopsis thaliana*].	67319.m00137: *Arabidopsis* seven in absentia (SINA) protein.	7.87632E-29: 21536945 seven in absentia-like protein [*Arabidopsis thaliana*].
67757.m00015: *Arabidopsis* no apical meristem (NAM) family protein.	6.0319E-29: 9759155 unnamed protein product [*Arabidopsis thaliana*].	67733.m00003: *Arabidopsis* seven in absentia (SINA) protein.	2.9936E-28: 9758487 unnamed protein product [*Arabidopsis thaliana*].
67798.m00005: *Arabidopsis* no apical meristem (NAM) family protein.	1.21514E-29: 8885600 NAM-like [*Arabidopsis thaliana*].	67733.m00006: *Arabidopsis* seven in absentia (SINA) protein.	8.71005E-28: 9758490 unnamed protein product [*Arabidopsis thaliana*].
67828.m00003: *Arabidopsis* no apical meristem (NAM) family protein.	4.17723E-30: 51972128 protein product At5g50820 [*Arabidopsis thaliana*].	67908.m00010: *Arabidopsis* seven in absentia (SINA) protein.	1.58734E-29: 15241972 *Arabidopsis* seven in absentia (SINA) protein.
67905.m00020: *Arabidopsis* no apical meristem (NAM) family protein.	4.17723E-30: 8809651 NAM (no apical meristem)-like protein [*Arabidopsis thaliana*].	67308.m00012: *Arabidopsis* putative auxin-responsive protein.	1.48541E-27: 7529750 putative protein [*Arabidopsis thaliana*].
67913.m00149: *Arabidopsis* no apical meristem (NAM) family protein.	2.9936E-28: 24030239 unknown protein [*Arabidopsis thaliana*].	68169.m00337: *Arabidopsis* auxin-responsive family protein.	5.10631E-28: 27363428 product At4g17280/dl4675c [*Arabidopsis thaliana*].
67919.m00003: *Arabidopsis* no apical meristem (NAM) family protein.	2.61645E-32: 10178056 unnamed protein product [*Arabidopsis thaliana*].	67113.m00113: *Arabidopsis* putative auxin-responsive protein.	1.13757E-27: 21593814 unknown protein [*Arabidopsis thaliana*].
60500.m00168: *Arabidopsis* auxin-responsive family protein.	3.30981E-27: 9294186 unnamed protein product [*Arabidopsis thaliana*].	Gramene gi|7227890|sp|O24175|FL_O: rice RYSA putative transcription factor FL (RFL).	0.0: 7489570 protein RFL of rice.
67804.m00010: *Arabidopsis* putative auxin-responsive protein.	1.9404E-27: 9758781 unnamed protein product [*Arabidopsis thaliana*].	29427.m00038: *Arabidopsis* rcd1-like cell differentiation family protein.	3.91494E-28: 21805678 hypothetical protein [*Arabidopsis thaliana*].
52129.m00067: *Arabidopsis* SEUSS transcriptional co-regulator.	4.61846E-29: 18033922 SEUSS transcriptional co-regulator [*Arabidopsis thaliana*].	67670.m00010: *Arabidopsis* turnip crinkle virus-interacting protein / TCV-interacting protein (TIP).	2.07314E-29: 32441252 protein product At5g24590 [*Arabidopsis thaliana*].

Table 2. (Continued)

AmiGO description and hyperlinks	Viridiplantae genome hit	AmiGO description and hyperlinks	Viridiplantae genome hit
52277.m00207: *Arabidopsis* transcription activator NAC1.	4.62458E-29: 6056383 Similar to NAM protein [*Arabidopsis thaliana*].	67936.m00109: *Arabidopsis* putative AP2 domain-containing transcription factor.	4.93441E-31: 21593812 floral homeotic protein apetala2-like [*Arabidopsis thaliana*].
67292.m00019: *Arabidopsis* putative AP2 domain-containing transcription factor.	4.77936E-26: 4678294 APETALA2-like protein [*Arabidopsis thaliana*].	67845.m00006: *Arabidopsis* seed maturation family protein.	4.04597E-25: 9759173 unnamed protein product [*Arabidopsis thaliana*].
68000.m00048: *Arabidopsis* root cap 1 (RCP1).	4.32275E-27: 20466029 putative root cap protein RCP1 [*Arabidopsis thaliana*].	UniProt P83139 Antifungal protein 5 [fragment] from cheeseweed (*Malva parviflora*).	2.27601E-4: 31879432 lipid transfer protein [*Atriplex nummularia*].

Table 3. Plant – Fungal similarities found

Entries under 'genome hit' show *E*-value as returned by the sequence comparison software: a reference to the specific database entry and brief descriptive annotation. No hits were returned to these plant sequences by searches of genomes of Metazoa. ***NOTE*** that small *E*-values are shown as exponential functions, i.e. 1.217E-29 = 1.217×10^{-29}

AmiGO description and hyperlinks	Viridiplantae genome hit	Fungi genome hit
Gene 1944691: Phosphoribosyl-anthranilate isomerase of *Arabidopsis thaliana.*	2.34934E-142: 28058927 putative phosphoribosylanthranilate isomerase of *Arabidopsis thaliana.*	1.20724E-21: 50285849 TRPF of *Candida glabrata.*
67844.m00001: *Arabidopsis* putative oxidoreductase.	9.30589E-30: 53828609 unknown protein of *Arabidopsis thaliana.*	7E-18: Putative short-chain dehydrogenase/reductase (Afu6g11650) of *Aspergillus fumigatus.*
DDB0214816 |Protein| locus: mybC: *Dictyostelium myb* transcription factor.	6.27106E-10: 28829358 hypothetical protein At1g08840.1 of *Arabidopsis thaliana.*	4.80158E-10: 46433415 hypothetical protein CaO19.10173 of *Candida albicans.*
60278.m00050: *Arabidopsis* NAM (no apical meristem) family protein (transcription factor?).	4.18276E-30: 7547102 no apical meristem hypothetical protein [*Arabidopsis thaliana*].	0.05: *Cryptococcus* hypothetical protein.
67041.m00011: *Arabidopsis* NAM (no apical meristem) family protein.	1.02868E-28: 7268198 putative NAM-like protein [*Arabidopsis thaliana*].	0.05: *Cryptococcus* putative glutathione transferase.
51799.m00289: *Arabidopsis* no apical meristem (NAM) family protein.	4.17723E-30: 18396807 *Arabidopsis* no apical meristem (NAM) family protein.	0.05: *Cryptococcus* putative cytoplasm protein.
60723.m00078 *Arabidopsis* no apical meristem (NAM) family protein.	9.30589E-30: 25403180 unknown protein of *Arabidopsis thaliana.*	0.05: *Cryptococcus* expressed protein.
51786.m00193: *Arabidopsis* no apical meristem (NAM) family protein.	1.217E-29: 8671840 *Arabidopsis* sequence with Strong similarity to OsNAC6 protein from *Oryza sativa.*	0.05: *Cryptococcus* conserved hypothetical protein.

Table 3. (Continued)

AmiGO description and hyperlinks	Viridiplantae genome hit	Fungi genome hit
51781.m00034: *Arabidopsis* no apical meristem (NAM) family protein.	3.3142E-27: 42562475 *Arabidopsis* no apical meristem (NAM) family protein.	<0.05: *Phanerochaete* genomic homology (not annotated).
67320.m00183 *Arabidopsis* no apical meristem (NAM) family protein.	4.61846E-29: 21536577 *Arabidopsis* no apical meristem (NAM) family protein.	0.05: *Cryptococcus* conserved hypothetical protein.
36000.m00043 *Arabidopsis* no apical meristem (NAM) family protein.	5.10528E-28: 27754598 putative *Arabidopsis* no apical meristem (NAM) family protein.	0.05: *Cryptococcus* conserved hypothetical protein (same as above).
60052.m00190 *Arabidopsis* no apical meristem (NAM) family protein.	8.70829E-28: 32452837 cup-shaped cotyledon 3 of *Arabidopsis*.	0.05: *Cryptococcus* conserved hypothetical protein (same as above).
60085.m00279 *Arabidopsis* no apical meristem (NAM) family protein.	2.45216E-30: 30793825 putative GRAB1 protein of *Arabidopsis thaliana*.	0.05: *Cryptococcus* conserved hypothetical protein (same as above).
67625.m00138 *Arabidopsis* no apical meristem (NAM) family protein.	8.41682E-31: 7573474 putative protein of *Arabidopsis thaliana*.	0.05: *Cryptococcus* conserved hypothetical protein.
68152.m00941 *Arabidopsis* no apical meristem (NAM) family protein.	2.71119E-29: 19424091 unknown protein of *Arabidopsis thaliana*.	<0.05: *Phanerochaete* genomic homology (not annotated).
67871.m00006 *Arabidopsis* no apical meristem (NAM) family protein	2.89282E-31: 10176766 unnamed protein product of *Arabidopsis thaliana*.	<0.05: *Phanerochaete* genomic homology (not annotated).
50828.m00139: *Arabidopsis* no apical meristem (NAM) family protein.	1.87469E-30: 42561659 *Arabidopsis* no apical meristem (NAM) family protein	0.05: *Cryptococcus* putative vacuolar membrane protein.
60250.m00041: *Arabidopsis* no apical meristem (NAM) family protein.	1.03025E-28: 6223651: NAM-like protein of *Arabidopsis*.	0.05: *Cryptococcus* conserved hypothetical protein.
67165.m00132: *Arabidopsis* no apical meristem (NAM) family protein.	7.12527E-30: 24417372 unknown protein of *Arabidopsis thaliana*.	0.05: *Cryptococcus* conserved hypothetical protein.
67119.m00016: *Arabidopsis* no apical meristem (NAM) family protein.	9.31821E-30: 7268550 *Arabidopsis* no apical meristem (NAM) family protein	0.05: *Cryptococcus* hypothetical protein.
60242.m00298: *Arabidopsis* no apical meristem (NAM) family protein.	2.07272E-29: 8567779 putative NAC (nascent polypeptide-associated complex)(chaperone) protein of *Arabidopsis*.	0.05: *Cryptococcus* expressed protein.
67583.m00136: *Arabidopsis* no apical meristem (NAM) family protein.	2.07588E-29: 22136362 putative protein of *Arabidopsis*.	0.05: *Cryptococcus* expressed protein.
60242.m00304: *Arabidopsis* no apical meristem (NAM) family protein.	9.62814E-27: 8567777 unknown protein of *Arabidopsis thaliana*.	0.05: *Cryptococcus* conserved hypothetical protein.
60242.m00303: *Arabidopsis* no apical meristem (NAM) family protein.	4.32188E-27: 18398893 *Arabidopsis* no apical meristem (NAM) family protein	0.05: *Cryptococcus* conserved hypothetical protein (as above).

Table 3. (Continued)

AmiGO description and hyperlinks	**Viridiplantae genome hit**	**Fungi genome hit**
67915.m00045: *Arabidopsis* no apical meristem (NAM) family protein.	3.53623E-29: 10176954 unnamed protein product of *Arabidopsis thaliana.*	0.05: *Cryptococcus* conserved hypothetical protein (as above).
60023.m00007: *Arabidopsis* no apical meristem (NAM) family protein.	2.70705E-29: 21436105 putative NAM protein of *Arabidopsis.*	0.05: *Cryptococcus* conserved hypothetical protein (as above).
67242.m00003: *Arabidopsis* no apical meristem (NAM) family protein.	6.67788E-28: 7529769 NAC domain-like protein [*Arabidopsis thaliana*].	0.05: *Cryptococcus* putative oxidoreductase.
60743.m00243: *Arabidopsis* no apical meristem (NAM) family protein.	7.12383E-30: 15293163 unknown protein of *Arabidopsis thaliana.*	0.05: *Cryptococcus* conserved expressed protein.
67899.m00120: *Arabidopsis* no apical meristem (NAM) family protein.	7.12527E-30: 9757865 NAM-like protein of *Arabidopsis.*	0.05: *Cryptococcus* conserved hypothetical protein.
50821.m00244: *Arabidopsis* no apical meristem (NAM) family protein.	2.44843E-30: 30725366 unknown protein of *Arabidopsis thaliana.*	0.05: *Cryptococcus* putative peptide-binding protein.
60278.m00142: *Arabidopsis* no apical meristem (NAM) family protein.	1.03025E-28: 23507759 unknown protein of *Arabidopsis thaliana.*	0.05: *Cryptococcus* putative myo-inositol transporter 2.
60507.m00052: *Arabidopsis* no apical meristem (NAM) family protein.	4.17723E-30: 11994103 unnamed protein product of *Arabidopsis thaliana.*	0.05: *Cryptococcus* conserved hypothetical protein.
60482.m00108: *Arabidopsis* no apical meristem (NAM) family protein.	6.66905E-28: 30984532 unknown protein of *Arabidopsis thaliana.*	0.05: *Cryptococcus* conserved hypothetical protein.
60025.m00139: *Arabidopsis* no apical meristem (NAM) family protein.	7.87632E-29: 21553558 NAM-like protein of *Arabidopsis.*	0.05: *Cryptococcus* conserved hypothetical protein.
51903.m00352: *Arabidopsis* no apical meristem (NAM) family protein.	6.66905E-28: 14334572 putative NAM protein of *Arabidopsis.*	0.05: *Cryptococcus* conserved hypothetical protein.
67601.m00151: *Arabidopsis* no apical meristem (NAM) family protein.	2.44892E-30: 30102618 NAM-like protein of *Arabidopsis.*	0.05: *Cryptococcus* conserved hypothetical protein.
67850.m00011: *Arabidopsis* no apical meristem (NAM) family protein.	9.94249E-32: 10177257 CUC2 (NAM-family) protein of *Arabidopsis.*	0.05: *Cryptococcus* conserved hypothetical protein.
60499.m00015: *Arabidopsis* no apical meristem (NAM) family protein.	9.30589E-30: 9294586 unnamed protein product of *Arabidopsis thaliana.*	0.05: *Cryptococcus* hypothetical protein.
67637.m00010: *Arabidopsis* no apical meristem (NAM) family protein.	1.58734E-29: 10177071 NAM-like protein of *Arabidopsis.*	0.05: *Cryptococcus* conserved hypothetical protein.
60247.m00053: *Arabidopsis* no apical meristem (NAM) family protein.	2.45216E-30: 6714418 NAM-like protein of *Arabidopsis.*	0.05: *Cryptococcus* conserved hypothetical protein.

Table 3. (Continued)

AmiGO description and hyperlinks	**Viridiplantae genome hit**	**Fungi genome hit**
43133.m00053: *Arabidopsis* no apical meristem (NAM) family protein.	1.58734E-29: 24030186 putative NAM protein of *Arabidopsis*.	0.05: *Cryptococcus* hypothetical protein.
67026.m00194: *Arabidopsis* no apical meristem (NAM) family protein.	5.45563E-30: 34222068 similar to putative NAM protein of *Arabidopsis*.	0.05: *Cryptococcus* conserved hypothetical protein.
51766.m00048: *Arabidopsis* no apical meristem (NAM) family protein.	7.87791E-29: 6714280 putative NAM-like protein of *Arabidopsis*.	0.05: *Cryptococcus* conserved hypothetical protein.
51442.m00191: *Arabidopsis* no apical meristem (NAM) family protein.	3.53623E-29: 23397178 putative NAM protein of *Arabidopsis*.	0.05: *Cryptococcus* conserved hypothetical protein.
51079.m00082: *Arabidopsis* no apical meristem (NAM) family protein.	2.99757E-28: 15223376 NAM-family protein of *Arabidopsis*.	0.05: *Cryptococcus* conserved hypothetical protein.
67733.m00005: *Arabidopsis* putative seven in absentia (SINA) protein.	2.80193E-26: 9758489 unnamed protein product [*Arabidopsis thaliana*].	0.05: *Cryptococcus* expressed protein.
60737.m00068 putative seven in absentia (SINA) protein of *Arabidopsis thaliana*.	1.13757E-27: 12322287 hypothetical protein of *Arabidopsis thaliana*.	<0.05: *Phanerochaete* genomic homology (not annotated).
60735.m00030: putative seven in absentia (SINA) protein of *Arabidopsis thaliana*.	2.29212E-28: 25404640 hypothetical protein of *Arabidopsis thaliana*.	0.05: *Cryptococcus* conserved hypothetical protein.
67167.m00011: seven in absentia (SINA) family protein.	6.67788E-28: 7269641 putative zinc finger protein [*Arabidopsis thaliana*].	0.05: *Cryptococcus* putative beta-fructofuranosidase.
67733.m00009: seven in absentia (SINA) family protein.	1.02889E-28: 28827312 unknown protein of *Arabidopsis thaliana*.	<0.05: *Phanerochaete* genomic homology (not annotated).
51205.m00083: *Arabidopsis* seed maturation family protein.	7.87632E-29: 4587565 *Arabidopsis* protein similar to rab28 protein gb|X59138 from *Zea mays*.	0.05: *Cryptococcus* putative ligand-regulated transcription factor.
67845.m00007 *Arabidopsis* seed maturation family protein.	3.09789E-25: 9759174 unnamed protein product of *Arabidopsis thaliana*.	0.05: *Cryptococcus* hypothetical protein.
67962.m00005: *Arabidopsis* seed maturation family protein.	3.65869E-26: 26452310 putative embryonic abundant protein of *Arabidopsis*.	0.05: *Cryptococcus* hypothetical protein.
68169.m00284: putative DRE-binding transcription factor of *Arabidopsis*.	1.34377E-28: 7268425 apetala2 domain TINY like protein of *Arabidopsis*.	0.05: *Cryptococcus* putative vacuolar membrane protein.
67824.m00008: transducin family protein.	2.80193E-26: 9759025 unnamed protein product of *Arabidopsis thaliana*.	0.05: *Cryptococcus* hypothetical protein.
67813.m00007: putative auxin-responsive protein.	1.21539E-29: 9758874 unnamed protein product of *Arabidopsis thaliana*.	0.05: *Cryptococcus* putative protein-S-isoprenylcysteine O-methyltransferase.

Table 3. (Continued)

AmiGO description and hyperlinks	Viridiplantae genome hit	Fungi genome hit
67617.m00015: putative rcd1-like cell differentiation protein of *Arabidopsis* (similar to protein involved in sexual development in *Schizosaccharomyces pombe*).	1.81856E-25: 7630054 putative protein of *Arabidopsis*.	0.05: *Cryptococcus* putative regulation of transcription from Pol II promoter-related protein.
60485.m00171: putative rcd1-like cell differentiation protein of *Arabidopsis*.	1.64264E-26: 21689729 putative cell differentiation protein of *Arabidopsis*.	0.05: *Cryptococcus* putative regulation of transcription from Pol II promoter-related protein (same as above).

Table 4. Similarities found to be common to all three eukaryotic Kingdoms

Entries under 'genome hit' show *E*-value as returned by the sequence comparison software: a reference to the specific database entry and brief descriptive annotation. ***NOTE*** that small *E*-values are shown as exponential functions, i.e. 1.217E-29 = 1.217×10^{-29}

AmiGO description and hyperlinks	Metazoan genome hit	Viridiplantae genome hit	Fungi genome hit
67170.m00018: *Arabidopsis* no apical meristem (NAM) family protein.	5.06164E-5: *Homo sapiens* chromosome 15: clone RP11-26F2: complete sequence.	3.77738E-31: 7269704 predicted protein [*Arabidopsis thaliana*].	<0.05: *Phanerochaete* genomic homology (not annotated).
MGI:108062 symbol:Siah2: mouse seven in absentia 2 protein.	6.41984E-23: 40254613 seven in absentia 2 [*Mus musculus*].	0.00963779: 4584086 p210 protein [*Spermatozopsis similis*] (protein is located in a membrane-microtubule-linker at the distal end of basal bodies)	0.00565021: 32422095 predicted protein [*Neurospora crassa*].
P13481 monkey (*Cercopithecus aethiops*) cellular tumour antigen p53.	0.0: 129367 (*Cercopithecus aethiops*) cellular tumour antigen p53.	3.22101E-123: 48374980 putative tumour protein p53 [*Zea mays*].	0.0: *Ustilago maydis* 521: UM04579.1 predicted mRNA.
Chinese hamster cellular tumour antigen p53.	0.0: 1890325 Chinese hamster ellular tumour antigen p53.	2.83486E-103: 48374980 same as above.	0.0: same as above.
Rhesus monkey cellular tumour antigen p53.	0.0: 47117801 rhesus monkey cellular tumour antigen p53.	7.17567E-123: 48374980 same as above.	0.0: same as above.
Tree shrew cellular tumour antigen p53.	0.0: 10720194 tree shrew cellular tumour antigen p53.	8.22908E-111: 48374980 same as above.	0.0: same as above.
Woodchuck cellular tumour antigen p53.	0.0: 2440123 tumor suppressor [*Marmota monax*].	3.91678E-113: 48374980 same as above.	0.0: same as above.
Mongolian gerbil P53.	0.0: 16266760 p53 [*Meriones unguiculatus*].	8.46452E-100: 48374980 same as above.	<0.05: *Phanerochaete* genomic homology (not annotated).

Table 4. (Continued)

AmiGO description and hyperlinks	**Metazoan genome hit**	**Viridiplantae genome hit**	**Fungi genome hit**
Porcine cellular tumour antigen p53.	0.0: 47523088 tumour suppressor p53 [*Sus scrofa*].	3.72158E-108: 48374980 same as above.	<0.05: *Phanerochaete* genomic homology (not annotated).
Guinea pig cellular tumour antigen p53.	0.0: 4884046 p53 protein [*Cavia porcellus*].	8.74593E-105: 48374980 same as above.	<0.05: *Phanerochaete* genomic homology (not annotated).
Human putative p150 protein.	0.0: 2072958 putative p150 [*Homo sapiens*].	2.52883E-64: *Arabidopsis thaliana* DNA chromosome 4: ESSA I FCA contig fragment No. 4.	0.0: *Aspergillus nidulans* FGSC A4: AN2724.2 predicted mRNA.
Mouse Homeobox protein A10	0.0: 30046954 Homeobox protein A10 [*Mus musculus*].	0.0: *Solanum demissum* chromosome 5 BAC PGEC446O19 genomic sequence: complete sequence.	0.0: *S. cerevisiae* chromosome XIII cosmid 8337.
Mouse latent transforming growth factor beta binding protein 4.	0.0: 32189330 latent transforming growth factor beta binding protein 4 [*Mus musculus*].	2.46129E-14: *Oryza sativa* (japonica cultivar-group): mRNA.	0.0: *Eremothecium gossypii* ADR210Cp (ADR210C): mRNA.
Human latent transforming growth factor-beta binding protein 4S.	0.0: 3327808 latent transforming growth factor-beta binding protein 4S [*Homo sapiens*].	1e-14: 42407754: rice: putative wall-associated serine/threonine kinase [*Oryza sativa*].	0.0: *Ashbya gossypii* (= *Eremothecium gossypii*) ATCC 10895 chromosome II: complete sequence.
UNIPROT|Q9JM52 symbol:M4K6: mouse mitogen-activated protein kinase kinase kinase kinase 6.	7.08858E-22: 24850117 misshapen/NIK-related kinase isoform 2 [*Homo sapiens*].	0.05: *Arabidopsis* putative cyclin-dependent kinase.	0.0: *Saccharomyces cerevisiae* NRK1 gene for N-rich kinase 1.
gi|11131838|sp|Q9SLY8|CRTC_ORYSA: rice calreticulin precursor (but no longer on godatabase.org page).	1.07201E-127: 18858381 calreticulin [Danio rerio].	0.0: 6682833 calcium-binding protein [*Oryza sativa*].	0.0: chromosome F of strain CLIB99 of *Yarrowia lipolytica*.
UNIPROT|O15347 symbol:HMG4: human high mobility group protein 4.	7.5735E-24: 4885421 high-mobility group box 3 [*Homo sapiens*].	0.0: *Triticum aestivum* mRNA for high mobility group protein (HMGW).	0.0: *Saccharomyces cerevisiae* clone FLH113809.01X YPR052C gene: complete cds.
Human dentin sialophosphoprotein precursor.	0.0: 11036632 dentin sialophosphoprotein preproprotein [*Homo sapiens*].	6.0953E-64: 11994784 unnamed protein product [*Arabidopsis thaliana*].	7.8507E-112: 50546453 hypothetical protein [*Yarrowia lipolytica*].
MMHC188A7: Mouse casein kinase.	6.07491E-128: 46237616 casein kinase 2: beta subunit [*Rattus norvegicus*].	7.06435E-68: 37536920 putative casein kinase II beta subunit [*Oryza sativa* (japonica cultivar-group)].	3.506E-67: 452290 casein kinase II beta subunit [*Schizosaccharomyces pombe*].

Table 4. (Continued)

AmiGO description and hyperlinks	Metazoan genome hit	Viridiplantae genome hit	Fungi genome hit
gi|18056667|gb|AAL58107.1|AF395064_1 CSN complex subunit 6B [*Arabidopsis thaliana*] (related to COP9 signalosome).	1.5475E-58: 55741990 COP9 constitutive photomorphogenic homolog subunit 6 [*Xenopus tropicalis*].	0.0: 21593149 transcription factor-like [*Arabidopsis thaliana*].	1.11417E-40: 46095839 hypothetical protein UM00643.1 [*Ustilago maydis* 521].
Arabidopsis CSN6A gene - one of two genes encoding subunit 6 of COP9 signalosome complex.	1.07168E-59: 47226158 unnamed protein product [*Tetraodon nigroviridis*].	0.0: 18056665 CSN complex subunit 6A [*Arabidopsis thaliana*].	6.11364E-39: 50257625 hypothetical protein CNBF1410 [*Cryptococcus neoformans*].
Human pProtein-tyrosine phosphatase: non-receptor type 22.	0.0: 48928054 protein tyrosine phosphatase: non-receptor type 22 (lymphoid) isoform 1 [*Homo sapiens*].	2.50413E-35: *Oryza sativa* (japonica cultivar-group) cDNA clone:002-166-D08: full insert sequence.	1.1738E-34: *Yarrowia lipolytica* CLIB99: YALI0F24585g predicted mRNA.
gi|18426814|ref|NP_569084.1| dihydrofolate reductase [*Rattus norvegicus*].	1.52823E-104: 18426814 dihydrofolate reductase [*Rattus norvegicus*].	1.15727E-27: 21702230 dihydrofolate reductase-thymidylate synthase [*Pisum sativum*].	1.61483E-29: 42554965 hypothetical protein FG07210.1 [*Gibberella zeae* PH-1].
Human putative tyrosine phosphatase	1.37967E-164: 6650693 putative tyrosine phosphatase [*Homo sapiens*].	3.08241E-39: 30023688 At5g10480 [*Arabidopsis thaliana*].	3.77673E-29: 32421867 hypothetical protein [*Neurospora crassa*].
Mouse peroxisome proliferator-activated receptor.	0.0: 25990188 thyroid hormone receptor associated protein 220 [*Mus musculus*].	0.015399: *Oryza sativa* (japonica cultivar-group) cDNA clone:002-161-H03: full insert sequence.	1.3692E-28: 40645462 cell wall protein Awa1p [*Saccharomyces cerevisiae*].
Peroxisome proliferator-activated receptor binding protein.	0.0: 14193715 peroxisome proliferator-activated receptor binding protein [*Mus musculus*].	0.015399: *Oryza sativa* (japonica cultivar-group) cDNA clone:002-161-H03: full insert sequence.	3.99601E-28: 40645462 cell wall protein Awa1p [*Saccharomyces cerevisiae*].
Human Jagged 2 precursor.	0.0: 2605945 Jagged-2 [*Homo sapiens*].	5.29485E-51: *Oryza sativa* (japonica cultivar-group) cDNA clone:002-167-G01: full insert sequence.	3e-23: 3250920: Putative wall protein. [*Hypocrea lixii*]
tr|Q80ZV1 1110033J19: mouse RIKEN cDNA 1110033J19 gene.	3.53552E-29: 55715979 LOC495812 protein [*Xenopus laevis*].	1.75873E-20: 488739 ribosomal protein: small subunit 4e (RS4e) [*Gossypium hirsutum*].	2.21965E-23: 50426545 unnamed protein product [*Debaryomyces hansenii*].
Human male-specific lethal 3-like 1.	0.0: 21411116 Male-specific lethal 3-like 1: isoform a [*Homo sapiens*].	1.83009E-18: 4006854 putative protein [*Arabidopsis thaliana*].	7.93833E-22: 40745985 hypothetical protein AN1976.2 [*Aspergillus nidulans* FGSC A4].
MGI|MGI:98158 symbol:Rps4x: mouse ribosomal protein S4: X-linked.	2.20647E-23: 46048780 ribosomal protein S4 [*Gallus gallus*].	1.8092E-17: 22138108 40S ribosomal S4 protein [*Glycine max*].	2.52453E-19: 50426545 unnamed protein product [*Debaryomyces hansenii*].

Table 4. (Continued)

AmiGO description and hyperlinks	**Metazoan genome hit**	**Viridiplantae genome hit**	**Fungi genome hit**
Mouse Fliih: flightless I protein homolog.	0.0: 21595485 Fliih protein [*Mus musculus*].	2.25262E-55: 22136974 putative villin 2 protein [*Arabidopsis thaliana*].	1.47563E-15: *Aspergillus nidulans* FGSC A4: AN1306.2 predicted mRNA.
UNIPROT|Q9Y295 symbol:DRG1: human developmentally regulated GTP-binding protein 1.	3.88963E-20: 4758796 developmentally regulated GTP binding protein 1 [*Homo sapiens*].	2.44197E-14: 50939391 putative GTP-binding protein DRG [*Oryza sativa* (japonica cultivar-group)].	1.09615E-14: 32403724 hypothetical protein [*Neurospora crassa*].
gi|30580409|sp|O42182|FBLN1_BRARE: Zebrafish fibulin-1 precursor.	0.0: 18858663 fibulin 1 [*Danio rerio*].	0.0: *Oryza sativa* (japonica cultivar-group): predicted mRNA.	9e-14: 42549555: Hypothetical protein FG02898.1 (*Gibberella zeae*: anamorph *Fusarium graminearum*).
UNIPROT|Q16576 symbol:RBB7: human histone acetyltransferase type B subunit 2.	2.20647E-23: 31982059 retinoblastoma binding protein 7 [*Mus musculus*].	1.06065E-17: 50881441 putative MSI type nucleosome/chromatin assembly factor C [*Oryza sativa* (japonica cultivar-group)].	2.70349E-13: 46100907 hypothetical protein UM04760.1 [*Ustilago maydis* 521].
UNIPROT|O00429 symbol:O00429: human Dynamin 1-like protein.	1.93041E-19: 19352981 Dynamin 1-like protein: isoform 2 [*Homo sapiens*].	3.64904E-10: 50902394 putative dynamin-like protein ADL2 [Oryza sativa (japonica cultivar-group)].	1.75003E-12: 50556172 hypothetical protein [*Yarrowia lipolytica*].
DDB0215363 |Protein| locus: *Dictyostelium* alrA aldehyde reductase.	1.22079E-13: 39591260 Hypothetical protein CBG20740 [*Caenorhabditis briggsae*].	1.03346E-12: 53749361 putative aldose reductase [*Oryza sativa* (japonica cultivar-group)].	1.76282E-12: 40745643 hypothetical protein AN1679.2 [*Aspergillus nidulans* FGSC A4].
Human negative elongation factor A.	0.0: 11527781 Wolf-Hirshhorn syndrome candidate 2 protein [*Homo sapiens*].	1.49452E-10: 41400384 plus agglutinin [*Chlamydomonas reinhardtii*].	8.76169E-11: 46442651 hypothetical protein CaO19.11809 [*Candida albicans* SC5314].
DDB0215356 |Protein| locus: *Dictyostelium* putative myb transcription factor.	6.27106E-10: 38787935 BMP-2 inducible kinase isoform a [*Homo sapiens*].	1.54462E-8: 4914452 putative protein [*Arabidopsis thaliana*].	2.15533E-10: 32420087 predicted protein [*Neurospora crassa*].
UNIPROT|O14807 symbol:RASM: human Ras-related protein M-Ras.	9.27024E-22: 54696976 muscle RAS oncogene homolog [*Homo sapiens*].	1.54474E-6: *Oryza sativa* (japonica cultivar-group) cDNA clone:002-166-D02: full insert sequence.	2.79767E-10: 19114491 hypothetical protein SPAC17H9.09c [*Schizosaccharomyces pombe*].
DDB0191102 |Protein| locus: Dictyostelium apm1: clathrin-adaptor medium chain.	2.1549E-10: 48097723 similar to ENSANGP00000020532 [*Apis mellifera*].	7.6643E-8: 20466372 clathrin adaptor medium chain protein MU1B: putative [*Arabidopsis thaliana*].	3.11167E-9: 28949965 probable clathrin assembly protein AP47 [*Neurospora crassa*].

Table 4. (Continued)

AmiGO description and hyperlinks	Metazoan genome hit	Viridiplantae genome hit	Fungi genome hit
UNIPROT|Q94899 symbol:CSN2: *Drosophila* COP9 signalosome complex subunit 2.	2.87734E-23: 7297479 CG9556-PB: isoform B [*Drosophila melanogaster*].	2.21335E-7: 21593214 putative PCI domain protein [*Arabidopsis thaliana*].	9.93527E-8: 19571748 csn2 [*Schizosaccharomyces pombe*].
Human death effector domain-associated factor (RING1 and YY1 binding protein).	6.35467E-126: 15928993 RING1 and YY1 binding protein [*Homo sapiens*].	2.1166E-4: 50428710 putative FHA domain protein [*Oryza sativa* (japonica cultivar-group)].	1.01509E-6: 46435639 hypothetical protein CaO19.11553 [*Candida albicans* SC5314].
UNIPROT|Q92499 symbol:DDX1: human ATP-dependent helicase DDX1.	1.02359E-20: 34863163 DEAD (Asp-Glu-Ala-Asp) box polypeptide 1 [*Rattus norvegicus*].	2.07205E-5: 50917625 putative RNA helicase [*Oryza sativa* (japonica cultivar-group)].	2.70618E-5: 40740142 hypothetical protein AN4233.2 [*Aspergillus nidulans* FGSC A4].
UNIPROT|P31276 symbol:HXCD: human homeobox protein Hox-C13.	1.6872E-23: 24497536 homeo box C13 [*Homo sapiens*].	1.34306E-4: 50919335 putative prohibitin [*Oryza sativa* (japonica cultivar-group)].	3.53438E-5: 45190453 unnamed protein of *Eremothecium gossypii* (*Ashbya gossypii*).
Mouse C330013B04 product: peroxisome proliferator-activated receptor binding protein homolog [fragment].	0.0: 26339888 unnamed protein product [*Mus musculus*].	0.0131176: 55978791 hypothetical protein AT1G79480 [*Arabidopsis thaliana*].	3.68814E-5: 6322209 GPI-anchored cell surface glycoprotein required for diploid pseudohyphal formation and haploid invasive growth [*Saccharomyces cerevisiae*].
gi|50054384|ref|NP_076471.3| colony stimulating factor 1 (macrophage) [*Rattus norvegicus*].	0.0: 50054384 colony stimulating factor 1 (macrophage) [*Rattus norvegicus*].	8.26683E-4: 3063699 putative protein [*Arabidopsis thaliana*].	4.38345E-5: 50556110 hypothetical protein [*Yarrowia lipolytica*].
UNIPROT|Q8NG53 symbol:Q8NG53: human diacylglycerol kinase: delta.	4.9155E-23: 25777598 diacylglycerol kinase: delta 130kDa isoform 2 [*Homo sapiens*].	1.34484E-4: 51557999 chloroplast DnaJ-like protein 2 [*Chlamydomonas reinhardtii*].	1.02971E-4: 38100339 predicted protein [*Magnaporthe grisea* 70-15].
UNIPROT|Q9Y255 symbol:PX19: human Px19-like protein (may be important for the development of vital and immunocompetent organs).	2.87793E-23: 41190437 predicted protein similar to Px19-like protein [*Homo sapiens*].	0.05: *Arabidopsis* MSF1-like family protein similar to px19 of chicken.	1.7541E-4: 50551063 hypothetical protein [*Yarrowia lipolytica*].
DDB0216392 |Protein| locus: *Dictyostelium* CRTF transcription factor required for expression of aggregation genes.	1.5981E-5: 23093054 CG32223-PA [*Drosophila melanogaster*].	1.03586E-4: 28829358 *Dictyostelium* protein similar to hypothetical protein; protein id: At1g08840.1 of *Arabidopsis thaliana*.	1.76691E-4: 32416082 predicted protein [*Neurospora crassa*].

Table 4. (Continued)

AmiGO description and hyperlinks	Metazoan genome hit	Viridiplantae genome hit	Fungi genome hit
UNIPROT|O14511 symbol:NRG2: human pro-neuregulin-2 precursor.	2.28032E-20: 7669536 neuregulin 2 isoform 6 [*Homo sapiens*].	0.00564273: 15230121 hypothetical protein [*Arabidopsis thaliana*].	2.29092E-4: 34809539 adhesin of *Candida glabrata* mediating adherence to human epithelial cells.
UNIPROT|Q8IN81 symbol:FRU: Drosophila sex determination protein fruitless.	1.05679E-25: 23171647 CG14307-PE: isoform E [*Drosophila melanogaster*].	0.0164178: 16550925 zinc transporter [*Eucalyptus grandis*].	2.99203E-4: 50260923 hypothetical protein CNBA2200 [*Cryptococcus neoformans*].
DDB0214900: *Dictyostelium* amiB novel protein required for aggregation.	3.93627E-4: 23093054 CG32223-PA [*Drosophila melanogaster*].	5.14092E-4: 28829358 hypothetical protein; protein id: At1g08840.1 [*Arabidopsis thaliana*].	3.01389E-4: 6323816 Transcription factor involved in regulation of invasive growth and starch degradation [*Saccharomyces cerevisiae*].
DDB0214819 |Protein| locus: *Dictyostelium* autophagy protein 7 (homologue of yeast atg7; E1-like).	1.22363E-5: 12652685 APG7L protein [*Homo sapiens*].	0.05: *Arabidopsis* (APG7) nearly identical to autophagy 7 [*Arabidopsis thaliana*].	6.71425E-4: 46100947 hypothetical protein UM04880.1 [*Ustilago maydis* 521].
Human LIM homeobox protein cofactor	0.0: 3372807 LIM homeobox protein cofactor [*Homo sapiens*].	1.65327E-4: 7523675 Hypothetical protein [*Arabidopsis thaliana*].	8.20512E-4: 38102578 hypothetical protein MG01057.4 [*Magnaporthe grisea* 70-15].
Human nuclear LIM interactor [fragment].	0.0: 5123791 Nuclear LIM interactor [*Homo sapiens*].	1.81309E-4: 7523675 *Arabidopsis* hypothetical protein (same as above).	8.99825E-4: 38102578 *Magnaporthe* hypothetical protein (same as above).
UNIPROT|Q9TVM2 symbol:XPO1: Drosophila Exportin-1.	2.52453E-19: 28380309 CG13387-PA [*Drosophila melanogaster*].	0.05: *Arabidopsis* expressed protein.	0.00194195: 50285273 unnamed protein product [*Candida glabrata*].
UNIPROT|P31260 symbol:HXAA: human Homeobox protein Hox-A10.	2.87793E-23: 24497549 homeobox protein A10 isoform a [*Homo sapiens*].	1.02835E-4: 50909875 hypothetical protein [*Oryza sativa* (japonica cultivar-group)].	0.00330808: 32415003 hypothetical protein [*Neurospora crassa*].
Human transcription factor-like 5 protein.	0.0: 12314002 TCFL5 [*Homo sapiens*].	1.24779E-4: 7671199 flagellar autotomy protein Fa1p [*Chlamydomonas reinhardtii*].	0.00684683: 11877204 putative centromere binding factor 1 [*Candida albicans*].
UNIPROT|P25800 symbol:RHM1: human rhombotin-1.	7.5735E-24: 4505005 LIM domain only 1 [*Homo sapiens*].	0.05: *Arabidopsis* LIM domain-containing protein weak similarity to LIM-homeobox protein.	0.0103: *S. cerevisiae* (DBY874) LRG1 gene.
Human Carbohydrate sulfotransferase 2.	0.0: 27369497 carbohydrate (N-acetylglucosamine-6-O) sulfotransferase 2 [*Homo sapiens*].	7.63685E-4: 50934045 hypothetical protein [*Oryza sativa* (japonica cultivar-group)].	0.0110276: 38106495 hypothetical protein MG05921.4 [*Magnaporthe grisea* 70-15].

Table 4. (Continued)

AmiGO description and hyperlinks	Metazoan genome hit	Viridiplantae genome hit	Fungi genome hit
UNIPROT|P39880 symbol:CUT1: human CCAAT displacement protein.	1.74598E-20: 31652240 CCAAT displacement protein isoform a [*Homo sapiens*].	0.05: *Arabidopsis* CCAAT displacement protein-related.	0.0477686: 50256507 hypothetical protein CNBH3310 [*Cryptococcus neoformans*].
UNIPROT|O60869 symbol:O60869: human endothelial differentiation-related factor 1.	3.51803E-21: 15930118 Endothelial differentiation-related factor 1: isoform alpha [*Homo sapiens*].	7.12154E-6: 50944921 putative ethylene-responsive transcriptional coactivator [*Oryza sativa* (japonica cultivar-group)].	0.0477686: 40744391 hypothetical protein AN2996.2 [*Aspergillus nidulans* FGSC A4].
UNIPROT|P52566 symbol:GDIS: human Rho GDP-dissociation inhibitor 2.	1.4283E-22: 14327952 Rho GDP dissociation inhibitor (GDI) beta [*Homo sapiens*].	0.0280046: 34906916 putative Rho GDP-dissociation inhibitor [*Oryza sativa* (japonica cultivar-group)].	0.05: *Cryptococcus* putative Rho GDP-dissociation inhibitor 1.
DDB0191390 |Protein| locus: *Dictyostelium* atg5 autophagy protein 5.	1.00099E-7: 31201617 ENSANGP00000012467 [*Anopheles gambiae*].	9.36897E-6: 26450228 APG5 (autophagy 5) like protein [*Arabidopsis thaliana*].	0.05: *Cryptococcus* putative prostatic steroid 5-alpha-reductase type I: of *Cryptococcus neoformans*.
UNIPROT|Q15717 symbol:ELV1: human ELAV-like protein 1.	5.42754E-22: 38201714 ELAV-like 1 [*Homo sapiens*].	0.05: *Arabidopsis* RNA and export factor-binding protein: putative transcriptional coactivator.	<0.05: *Phanerochaete* genomic homology (not annotated).
67166.m00025: *Arabidopsis* oxidoreductase: forever young (FEY3).	3.93493E-11: Human DNA sequence from clone XXyac-65C7_A	1.217E-29: 7269630 forever young gene (FEY) (fragment) [*Arabidopsis thaliana*].	0.05: *Cryptococcus* putative transposable elements-Tcn5.
tr|Q6P6K5: mouse nuclear factor: erythroid derived 2.	1.21539E-29: 40254626 nuclear factor: erythroid derived 2 [*Mus musculus*].	0.05: *Arabidopsis* putative chloroplast division protein.	0.05: *Cryptococcus* hypothetical protein.
UNIPROT|Q12951 symbol:FXI1: human forkhead box protein I1.	2.12937E-26: 1911185 forkhead box L1 [*Homo sapiens*].	0.05: *Arabidopsis* expressed protein.	0.05: *Cryptococcus* conserved hypothetical protein.
UNIPROT|Q02223 symbol:TR17: human tumour necrosis factor receptor superfamily member 17.	5.80651E-24: 23238192 tumor necrosis factor receptor superfamily: member 17 [*Homo sapiens*].	0.05: *Arabidopsis* gibberellin-regulated protein 3 (GASA3).	0.05: *Cryptococcus* conserved hypothetical protein.
31052.m00046: *Arabidopsis* putative AP2 domain-containing transcription factor.	0.0: *Mus musculus* BAC clone RP24-121M11 from 13: complete sequence.	1.75501E-28: 20260076 putative AP2 domain transcription factor [*Arabidopsis thaliana*].	0.05: *Cryptococcus* conserved hypothetical protein.
67844.m00002: *Arabidopsis* putative oxidoreductase.	4.95871E-6: *Homo sapiens* chromosome 13q34 schizophrenia region contig 1 section 7 of 11 of the complete sequence.	7.12527E-30: 9757991 protochlorophyllide reductase; oxidoreductase required for shoot apex development [*Arabidopsis thaliana*].	0.05: *Cryptococcus* putative ubiquitin-protein ligase.

Table 4. (Continued)

AmiGO description and hyperlinks	Metazoan genome hit	Viridiplantae genome hit	Fungi genome hit
UNIPROT|Q9NRW4 symbol:Q9NRW4: human mitogen-activated protein kinase phosphatase x.	5.43472E-22: 9910432 dual specificity phosphatase 22 [*Homo sapiens*].	0.05: *Arabidopsis* dual specificity protein phosphatase family protein.	<0.05: *Phanerochaete* genomic homology (not annotated).
UNIPROT|P15976 symbol:GAT1: human erythroid transcription factor.	1.20913E-21: 14602571 GATA1 protein [*Homo sapiens*].	0.05: *Arabidopsis* GDSL-motif lipase.	0.05: *Cryptococcus* expressed protein.
tr|Q8VI44: mouse flightless I homolog (*Drosophila*) Cytoskeletal actin-modulating.	1.3435E-28: 4503743 flightless I homolog [*Homo sapiens*].	0.00114528: 37783214 resistance candidate RPP8-like protein [*Arabidopsis lyrata*].	0.05: *Cryptococcus* putative adenylate cyclase.
UNIPROT|P14653 symbol:HXB1: human Homeobox protein Hox-B1.	3.75869E-23: 72239 homeotic protein Hox B1 - human.	0.05: *Arabidopsis* zinc finger (CCCH-type) family protein.	0.05: *Cryptococcus* hypothetical protein.
UNIPROT|P50221 symbol:MOX1: human Homeobox protein MOX-1.	1.38022E-25: 7710150 mesenchyme homeo box 1 isoform 2 [*Homo sapiens*].	0.05: *Arabidopsis* hydroxyproline-rich glycoprotein family protein.	0.05: *Cryptococcus* hypothetical protein.
Arabidopsis abnormal inflorescence meristem 1 / fatty acid multifunctional protein (AIM1).	0.00568511: 50752176 predicted protein similar to enoyl-Coenzyme A: hydratase/3-hydroxyacyl Coenzyme A dehydrogenase [*Gallus gallus*].	2.80193E-26: 20465649 putative AIM1 protein [*Arabidopsis thaliana*].	0.05: *Cryptococcus* putative enoyl-CoA hydratase.
UNIPROT|Q9UGM4 symbol:Q9UGM4: human nuclear LIM interactor [fragment].	4.02113E-25: 34863521 similar to Ldb1a [*Rattus norvegicus*].	0.05: *Arabidopsis* hydroxyproline-rich glycoprotein family protein similar to extension.	0.05: *Cryptococcus* hypothetical protein.

The 'seven in absentia' (SINA) protein family has a more mixed distribution. SINA was originally discovered as a RING zinc finger-containing protein that is critically involved in development of the R7 photoreceptor cell in the *Drosophila* eye. The RING zinc finger domain (originally named for the acronym *really interesting new gene*) is a protein interaction domain consisting of two pairs of zinc ligands co-ordinately binding two zinc ions, which is implicated in a range of processes from transcriptional regulation to targeted proteolysis. Mammalian SINA homologues can act in the ubiquitin/proteasome pathway and plant (*Arabidopsis*) homologues are essential to seed (and embryo) development. The RING-finger domain is one of the most frequently detected domains in the *Arabidopsis* proteome, and is more abundant in *Arabidopsis* than in other eukaryotic genomes (Kosarev *et al.*, 2002).

Table 5. Animal-Plant similarities found

Entries under 'genome hit' show *E*-value as returned by the sequence comparison software: a reference to the specific database entry and brief descriptive annotation. No hits were returned to these animal sequences by searches of genomes of Fungi. ***NOTE*** that small *E*-values are shown as exponential functions, i.e. 1.217E-29 = 1.217×10^{-29}

AmiGO description and hyperlinks	Metazoan genome hit	Viridiplantae genome hit
Bovine beta-galactoside alpha-2:6-sialyltransferase (integral membrane protein).	0.0: 29135323 Bovine beta-galactoside alpha-2:6-sialyltransferase (integral membrane protein).	0.0: predicted mRNA of rice.
Chick CMP-N-acetylneuraminate-beta-galactosamide-alpha-2:6-sialyltransferase.	0.0: 45382551 Chick CMP-N-acetylneuraminate-beta-galactosamide-alpha-2:6-sialyltransferase.	0.0: APG5 (autophagy 5)-like protein of *Oryza sativa*.
AT1G66650.1 "seven in absentia" (SINA) family protein: located in nucleus.	1.06813E-17: 54641564 unnamed gene product of *Drosophila pseudoobscura*.	0.0: 12597767 hypothetical protein of *Arabidopsis*.
Rabbit cellular tumour antigen p53 (cell cycle regulator).	0.0: 2842741 Rabbit cellular tumour antigen p53.	7.13805E-115: 48374980 putative tumour protein p53 of *Zea mays*.
Beluga whale P53 protein.	0.0: 18997097 Beluga whale P53 protein.	5.58669E-112: 48374980 same as above.
Feline cellular tumour antigen p53.	0.0: 538225 feline p53 tumour-suppressor gene.	2.10835E-111: 48374980 same as above.
Canine cellular tumour antigen p53.	0.0: 50978974 Canine cellular tumour antigen p53.	4.60588E-111: 48374980 same as above.
Pig P53 protein.	0.0: 50979299 Pig P53 protein.	2.05624E-106: 48374980 same as above.
Sheep cellular tumour antigen p53.	0.0: 1709531 Sheep cellular tumour antigen p53.	5.84948E-106: 48374980 same as above.
Rodent P53 [fragment]	0.0: 56829 rat unnamed protein product.	1.2865E-105: 48374980 same as above.
Mouse tumour suppressor p53.	0.0: 2961247 mouse tumour suppressor p53.	1.64813E-103: 48374980 same as above.
67845.m00016: manatee seven in absentia protein.	7.66585E-8: 41054792 Zebrafish seven in absentia protein.	1.09927E-30: 26449935 putative ring finger E3 ligase SINAT5 [*Arabidopsis thaliana*].
60737.m00067: putative seven in absentia (SINA) protein of *Arabidopsis*.	0.0282149: 29293702 SINA protein of *Schistosoma mansoni*.	3.53623E-29: 33589720 product of ORF At1g66660/F4N21_20 of *Arabidopsis thaliana*.
67733.m00007 : seven in absentia (SINA) family protein of *Arabidopsis*.	0.00435293: 31207365 protein of *Anopheles gambiae*.	6.66905E-28: 9758491 unnamed protein product of *Arabidopsis thaliana*.
UNIPROT|Q9Y6A4: human transcription factor IIB.	1.58126E-21: 55643953 predicted hypothetical protein XP_511001 of *Pan troglodytes*.	9.9274E-16: 15795151 unnamed protein product [*Arabidopsis thaliana*].
Human DNA (cytosine-5)-methyltransferase 3B	0.0: 5901940 DNA cytosine-5 methyltransferase 3 beta isoform 1 [*Homo sapiens*].	4.78271E-9: 29467228 DNA methyltransferase [*Nicotiana tabacum*].
UNIPROT|Q8VI44 mouse flightless I protein homolog (possible coactivator in transcriptional activation).	4.9155E-23: 4503743 human flightless I protein homolog.	0.00253626: 37783214 resistance candidate RPP8-like protein [*Arabidopsis lyrata*].

Table 5. (Continued)

AmiGO description and hyperlinks	Metazoan genome hit	Viridiplantae genome hit
UNIPROT|P25100 symbol:A1AD: human alpha-1D adrenergic receptor.	7.08714E-22: 4501957 human alpha-1D adrenergic receptor.	0.00564159: 15081245 glycine-rich protein GRP16 of *Arabidopsis*.
UNIPROT|P31310 symbol:HXAA: Homeobox protein Hox-A10 of mouse.	1.4283E-22: 6680243 Homeobox protein Hox-A10 of mouse.	0.0280046: 42391853 cold shock domain protein 2 of *Triticum aestivum*.
UNIPROT|P52926 symbol: human high mobility group protein HMGI-C.	1.09361E-22: 4504431 high mobility group AT-hook 2 isoform a [*Homo sapiens*].	0.0477686: 50933035 AT1 protein of rice.
UNIPROT|Q9NQZ9 symbol: human TCP11 protein (receptor of fertilization promoting peptide).	1.09506E-22: 54887320 human TCP11 protein.	0.05: T-complex protein 11 of *Arabidopsis*.
Zebrafish Wnt-4a protein precursor.	0.0: 18859563 Zebrafish Wnt-4a protein	0.05: putative ubiquitin-specific protease 1 of *Arabidopsis*.
tr|Q8R002 C1qtnf5 protein: mouse C1q and tumor necrosis factor related protein 5 (transmembrane receptor).	4.17638E-30: 26024327 mouse C1q and tumor necrosis factor related protein 5.	0.05: *Arabidopsis* proline-rich family protein with proline rich extensin domains.
UNIPROT|P17483 symbol:HXB4: human homeobox protein Hox-B4.	9.8913E-24: 13273315 homeo box B4 [*Homo sapiens*].	0.05: mitochondrial transcription termination factor-related protein of *Arabidopsis*.
tr|Q8K5B8 HOXC12: Mouse homeo box C12.	8.41682E-31: 33859568 Mouse homeo box C12.	0.05: expressed protein identical to ORF1 [*Arabidopsis thaliana*].
UNIPROT|Q96PN7 symbol:Q96PN7 human Zinc finger transcription factor TReP-132.	5.80651E-24: 15812226 transcriptional regulating factor 1 isoform 1 [*Homo sapiens*].	0.05: Rho GDP-dissociation inhibitor family protein [*Arabidopsis thaliana*].
UNIPROT|O00192 symbol:ARVC: human armadillo repeat protein deleted in velo-cardio-facial syndrome.	2.27986E-20: 4502247 human armadillo repeat protein.	0.05: expressed protein of *Arabidopsis*.
UNIPROT|P48281 symbol:VDR: mouse Vitamin D3 receptor.	1.99567E-24: 1352836 mouse Vitamin D3 receptor.	0.05: zinc finger (C3HC4-type RING finger) family protein contains Pfam domain: of *Arabidopsis*.
tr|Q8C145 mouse solute carrier family 39 (metal ion transporter) protein.	1.48571E-27: 32822909 mouse Slc39a6 protein.	0.05: expressed protein of *Arabidopsis*.
UNIPROT|P30968 symbol:GRHR: human gonadotropin-releasing hormone receptor.	4.59468E-21: 1628390 human gonadotropin-releasing hormone receptor.	0.05: expressed protein of *Arabidopsis*.
Progonadoliberin I precursor of H*aplochromis (=Astatotilapia) burtoni*.	1.2395E-50: 6226866 Progonadoliberin I precursor of *Astatotilapia burtoni*.	0.05: hypothetical protein of *Arabidopsis*.
UNIPROT|P13562 symbol:GON1: mouse progonadoliberin I precursor.	1.86542E-22: 51093849 gonadotropin releasing hormone [*Mus musculus*].	0.05: *Arabidopsis* putative plastid developmental protein DAG (required for chloroplast differentiation).
Progonadoliberin I precursor of Japanese rice fish.	1.92893E-48: 34098705 Progonadoliberin I precursor of Japanese rice fish.	0.05: *Arabidopsis* expressed protein with weak similarity to a bacterial urease accessory protein.

No similarity in metazoan or fungal genomes can be detected for eight of the *Arabidopsis* SINA proteins (Table 2), but high levels of similarity (E-values less than 10^{-28}) were returned for four other *Arabidopsis* SINA proteins (Table 5), and the mouse siah2 protein showed moderate homology with a protein from the green alga *Spermatozopsis* (E-value = 9.6×10^{-3}) and a predicted protein of *Neurospora crassa* (E-value = 5.6×10^{-3})(Table 5). It is significant that 16 mammalian p53 protein sequences (Table 5) showed very high similarity (E-values less than 10^{-100}) with a protein from the *Zea mays* genome in view of the suggestion that SINA proteins may mediate p53-dependent cell-cycle arrest in man (Matsuzawa *et al.*, 1998). Eight of the mammalian p53 sequences failed to detect similarity with fungal genomes, although the sequences from African green monkey, Chinese hamster, rhesus monkey, tree shrew and woodchuck all showed complete homology (E-value = 0) with a predicted mRNA reported from the *Ustilago maydis* genome, whilst gerbil, porcine and guinea pig sequences were weakly similar (E-value reported as <0.05) to sequences in the *Phanerochaete* genome (Table 6).

Table 6. Animal – fungus similarities found

Entries under 'genome hit' show E-value as returned by the sequence comparison software: a reference to the specific database entry and brief descriptive annotation. No hits were returned to these animal sequences by searches of genomes of Viridiplantae. ***NOTE*** that small E-values are shown as exponential functions, i.e. 1.217E-29 = 1.217×10^{-29}

AmiGO description and hyperlinks	Metazoan genome hit	Fungi genome hit
MGI|MGI:101791 symbol: ISL1: mouse transcription factor: LIM/homeodomain (islet 1); may regulate insulin gene expression or islet cell development.	6.84994E-25: 8393633 ISL1 transcription factor: LIM/homeodomain 1 (*Rattus norvegicus*).	0.0: *Ustilago maydis* 521: UM05343.1 predicted mRNA.
DDB0185218 |Protein| locus: ifkA: *Dictyostelium* initiation factor 2 alpha (eIF2alpha) kinase.	0.00969732: 42733663 *Dictyostelium* initiation factor 2 alpha (eIF2alpha) kinase.	6.66905E-28: 28828088 *Dictyostelium* protein similar to Gcn2p of *Saccharomyces cerevisiae.*
UNIPROT|Q9Y3R5 symbol:CU05: human Protein C21orf5 ortholog of a *Caenorhabditis elegans* gene (pad-1) required for embryonic patterning.	7.83735E-21: 45827701 pad-1-like [*Homo sapiens*].	1.58651E-5: 46100006 hypothetical protein UM04150.1 of *Ustilago maydis.*
Zebrafish nuclear respiratory factor 1.	0.0: 16200181 nuclear respiratory factor 1 [*Danio rerio*].	0.00121777: 46444519 hypothetical protein: potential cell surface flocculin of *Candida albicans.*
DDB0191136 |Protein| locus: abpD: *Dictyostelium* actin binding protein; developmentally and cAMP-regulated; associates with intracellular membranes.	1.35288E-4: 39592126 Hypothetical protein CBG23326 [*Caenorhabditis briggsae*].	0.00195355: 38110842 hypothetical protein MG06475.4 [*Magnaporthe grisea*].
UNIPROT|O94761 symbol:RCQ4: human ATP-dependent DNA helicase Q4.	1.21073E-21: 4759030 RecQ protein-like 4 [*Homo sapiens*].	0.00565021: 45185185 hypothetical protein from *Eremothecium* (*Ashbya*) *gossypii.*
UNIPROT|O95376 symbol:ARI2: human Ariadne-2 protein homolog; might act as an ubiquitin-protein ligase.	4.908E-23: 5453557 ariadne homolog 2 [*Homo sapiens*].	0.00736814: 42546808 hypothetical protein FG00241.1 of *Gibberella zeae* PH-1 [anamorph = *Fusarium graminearum*].

Table 6. (Continued)

AmiGO description and hyperlinks	**Metazoan genome hit**	**Fungi genome hit**
Nrfl protein: responsible for the mutation: named "Not really finished" which is crucial for development of the zebrafish outer retina.	0.0: 27881974 nuclear respiratory factor 1 protein of Zebra fish; transcription factor.	0.0180559: 46444519 hypothetical protein: potential cell surface flocculin of *Candida albicans.*
UNIPROT|P29762 symbol:RET3: human cellular retinoic acid-binding protein.	1.58126E-21: 4758052 Human cellular retinoic acid-binding protein.	0.05: *Cryptococcus* putative glutathione transferase
MGI|MGI:97712 symbol:Prrx1: mouse paired related homeobox 1.	2.69366E-21: 5902024 paired mesoderm homeobox 1 isoform pmx-1a [*Homo sapiens*].	0.05: *Cryptococcus* conserved hypothetical protein
UNIPROT|Q9VB08 symbol:RNG1: *Drosophila melanogaster* polycomb complex protein Sce.	3.18614E-22: 7301619 *D. melanogaster* Polycomb group (PcG) protein.	0.05: *Cryptococcus* conserved hypothetical protein
UNIPROT|O95285 symbol:O95285: human erythroblast macrophage protein EMP.	2.69366E-21: 5031685 human macrophage erythroblast attacher.	<0.05: *Phanerochaete* genomic homology (not annotated).
UNIPROT|Q02386 symbol:ZN45: human zinc finger protein 45.	3.89479E-20: 4508029 zinc finger protein 45 [*Homo sapiens*].	0.05: *Cryptococcus* conserved hypothetical protein
Mouse transcription factor TFEC.	8.41682E-31: 13654264 transcription factor EC [*Mus musculus*].	<0.05: *Phanerochaete* genomic homology (not annotated).
UNIPROT|P80370 symbol:DLK: human Delta-like protein precursor (type I membrane protein).	8.9463E-25: 21361080 delta-like homolog (*Drosophila*): EGF-like: type I membrane protein.	0.05: *Cryptococcus* expressed protein
UNIPROT|O14682 symbol:ENC1: human ectoderm-neural cortex-1 protein.	2.28032E-20: 4505461 ectodermal-neural cortex 1 protein; nuclear matrix-associated: actin binding protein [*Homo sapiens*].	<0.05: *Phanerochaete* genomic homology (not annotated).
UNIPROT|P92189 symbol:STIL: *Drosophila* stand still protein.	1.29184E-23: 7303433 stand still protein (*Drosophila*) nuclear: participates in transcriptional activation.	0.05: *Cryptococcus* putative oxidoreductase:
tr|Q925K0: mouse peroxisome proliferator-activated receptor binding protein...	5.64569E-27: 14193715 mouse peroxisome proliferator-activated receptor binding protein.	0.05: *Cryptococcus* putative nuclear mRNA splicing: spliceosome-related protein.
tr|Q6AXE6: mouse LIM domain binding 2. LIM-homeodomain gene expressed in the developing forebrain.	2.21496E-31: 4504971 LIM domain binding 2 [*Homo sapiens*].	0.05: *Cryptococcus* putative phospholipase.
UNIPROT|P20719 symbol:HXA5: human hmeobox protein Hox-A5. DNA-binding transcription factor which provides cells with specific positional identities on the anterior-posterior axis.	8.9463E-25: 24497517 homeobox protein A5 - HOXA5 [*Homo sapiens*].	<0.05: *Phanerochaete* genomic homology (not annotated).

Table 6. (Continued)

AmiGO description and hyperlinks	**Metazoan genome hit**	**Fungi genome hit**
UNIPROT|P09629 symbol:HXB7: human HOXB7.	7.08858E-22: 25121963 homeo box B7 [*Homo sapiens*].	<0.05: *Phanerochaete* genomic homology (not annotated).
UNIPROT|P17481 symbol:HXB8: human HOXB8.	1.6872E-23: 13273317 homeo box B8 [*Homo sapiens*].	0.05: *Cryptococcus* expressed protein.
UNIPROT|Q8VDQ7 symbol:Q8VDQ7: mouse Ppar binding protein: isoform 2.	1.57917E-21: 14193715 peroxisome proliferator-activated receptor binding protein [*Mus musculus*].	0.05: *Cryptococcus* putative nuclear mRNA splicing: via spliceosome-related protein.
MGI|MGI:108063 symbol:Siah1b: mouse seven in absentia 1B. The sina protein contains a putative zinc finger domain and localises to the cell nucleus in *Drosophila*.	3.18614E-22: 6677949 seven in absentia 1B [*Mus musculus*]	0.05: *Cryptococcus* conserved hypothetical protein
MGI|MGI:108064 symbol:Siah1a: mouse seven in absentia homolog 1: isoform a.	1.09506E-22: 23274142 Seven in absentia homolog 1: isoform a [*Homo sapiens*].	0.05: *Cryptococcus* same as above.
UNIPROT|Q04900 symbol:MG24: human putative mucin core protein 24 precursor. Membrane glycoprotein with peanut agglutinin binding sites.	4.59468E-21: 219925 MGC-24 precursor [*Homo sapiens*].	0.05: *Cryptococcus* expressed protein.
UNIPROT|P10244 symbol:MYBB: human MYB-related protein B: transcription factor.	2.20355E-23: 4505293 MYB-related protein B [*Homo sapiens*].	0.05: *Cryptococcus* putative (NADP+) glutamate dehydrogenase.
UNIPROT|Q16621 symbol:NFE2: human Transcription factor NF-E2 45 kDa subunit.	5.42754E-22: 13477165 NFE2 protein [*Homo sapiens*].	<0.05: *Phanerochaete* genomic homology (not annotated).
UNIPROT|P91660 symbol:L259: *Drosophila* Probable multidrug resistance-associated protein lethal(2)03659.	1.4283E-22: 45445626 Integral membrane protein vital for development of *Drosophila melanogaster*	0.05: *Cryptococcus* putative branched-chain alpha-keto acid dehydrogenase E1-alpha subunit.
UNIPROT|Q9W1A4 symbol:TAMO: *Drosophila* tamozhennic protein (modulates the nuclear import of other proteins).	1.68943E-23: 21626728 CG4057-PB: isoform B [*Drosophila melanogaster*]	0.05: *Cryptococcus* putative adaptation to pheromone during conjugation with cellular fusion-related protein.
zfin: tr|Q9DE50 Zebra fish Platelet-derived growth factor alpha polypeptide. Growth factor activity. Membrane protein.	1.08506E-111: 35903201 platelet-derived growth factor alpha polypeptide of Zebra fish.	<0.05: *Phanerochaete* genomic homology (not annotated).
UNIPROT|Q03014 symbol:HMPH: human Homeobox protein PRH.	3.39959E-24: 4506049 hematopoietically expressed homeobox transcription factor [*Homo sapiens*].	0.05: *Cryptococcus* hypothetical protein.
UNIPROT|Q9Y5L5 symbol:L503: human lens epithelial cell protein LEP503.	1.86542E-22: 8923830 lens epithelial protein [*Homo sapiens*].	0.05: *Cryptococcus* expressed protein.
UNIPROT|Q15116 symbol:PCD1: human programmed cell death protein 1 precursor.	8.09159E-26: 4826890 programmed cell death 1 precursor [*Homo sapiens*].	0.05: *Cryptococcus* hypothetical protein.

Table 6. (Continued)

AmiGO description and hyperlinks	Metazoan genome hit	Fungi genome hit
UNIPROT|Q00587 symbol:MSE5: human CDC42 effector protein 1.	6.41135E-23: 23238226 CDC42 effector protein 1 isoform a [*Homo sapiens*].	0.05: *Cryptococcus* putative vesicular-fusion protein.
UNIPROT|P14652 symbol:HXB2: human homeobox protein Hox-B2.	5.42754E-22: 4504465 homeo box B2 transcription factor [*Homo sapiens*].	0.05: *Cryptococcus* putative Rho small monomeric GTPase.
FB|FBgn0010340 symbol:140up: *D. melanogaster* gene 'upstream of RpII140'.	4.15487E-22: 7299846 CG9852 [*Drosophila melanogaster*].	0.05: *Cryptococcus* conserved hypothetical protein.
UNIPROT|P32242 symbol:OTX1: human homeobox protein OTX1; encodes a member of the bicoid sub-family of transcription factors.	1.09361E-22: 20070107 orthodenticle 1 [*Homo sapiens*].	0.05: *Cryptococcus* putative polyadenylate-binding protein.
MGI|MGI:1194883 symbol:Crx: cone-rod homeobox containing gene (transcription factor).	2.43582E-22: 6681029 cone-rod homeobox containing gene [*Mus musculus*].	0.05: *Cryptococcus* conserved hypothetical protein.
UNIPROT|P05549 symbol:AP2A: human transcription factor AP-2 alpha.	4.4391E-24: 31981462 transcription factor AP-2: alpha [*Mus musculus*].	0.05: *Cryptococcus* putative chromatin modification-related protein.
UNIPROT|P11309 symbol:PIM1: human Proto-oncogene serine/threonine-protein kinase Pim-1.	5.42754E-22: 4505811 pim-1 oncogene [*Homo sapiens*].	<0.05: *Phanerochaete* genomic homology (not annotated).
MGI|MGI:1888519 symbol:Lmx1a: mouse LIM homeobox transcription factor 1 alpha.	2.28032E-20: 587461 hamster LIM homeobox transcription factor 1 alpha [*Mesocricetus auratus*].	<0.05: *Phanerochaete* genomic homology (not annotated).
UNIPROT|P48357 symbol:LEPR: human leptin receptor precursor (receptor for obesity factor (leptin); on ligand binding: mediates signaling).	5.24482E-25: 1589772 leptin receptor [*Homo sapiens*].	0.05: *Cryptococcus* conserved hypothetical protein.
MGI|MGI:88005 symbol:Amelx: mouse amelogenin X chromosome (amelogenin is the major enamel protein).	6.41005E-23: 9506381 amelogenin X chromosome [*Rattus norvegicus*].	0.05: *Cryptococcus* hypothetical protein.
UNIPROT|P07333 symbol:KFMS: human macrophage colony stimulating factor I receptor precursor (CSF-1-R).	1.86542E-22: 27262659 colony stimulating factor 1 receptor precursor [*Homo sapiens*].	0.05: *Cryptococcus* conserved hypothetical protein.
UNIPROT|P25116 symbol:PAR1: human proteinase activated receptor 1 precursor.	1.20913E-21: 30354672 Coagulation factor II receptor: precursor [*Homo sapiens*].	<0.05: *Phanerochaete* genomic homology (not annotated).
UNIPROT|P07996 symbol:TSP1: human thrombospondin-1 precursor.	2.43954E-22: 40317626 thrombospondin 1 precursor [*Homo sapiens*].	0.05: *Cryptococcus* putative chromatin assembly complex protein.
tr|Q9R0A6: mouse T-box 21 transcription factor.	1.09927E-30: 9507179 T-box 21 transcription factor [*Mus musculus*].	0.05: *Cryptococcus* expressed protein.
UNIPROT|P01344 symbol:IGF2: human insulin-like growth factor II precursor.	1.4283E-22: 30582865 insulin-like growth factor 2 (IGF2) [*Homo sapiens*].	<0.05: *Phanerochaete* genomic homology (not annotated).

Table 6. (Continued)

AmiGO description and hyperlinks	**Metazoan genome hit**	**Fungi genome hit**
UNIPROT|Q04743 symbol:EMX2: human homeobox protein EMX2 (critical for central nervous system and urogenital development).	5.42754E-22: 14149611 homeodomain transcription factor EMX2 [*Homo sapiens*].	0.05: *Cryptococcus* conserved hypothetical protein.
UNIPROT|P53539 symbol:FOSB: human protein fosB; interacts with Jun proteins enhancing their DNA binding activity.	5.42754E-22: 54673701 Protein fosB [*Homo sapiens*].	0.05: putative bZIP transcription factor (AtfA) of *Aspergillus fumigatus*.
UNIPROT|Q04724 symbol:TLE1: human transducin-like enhancer protein 1.	4.9155E-23: 34869030 similar to groucho protein GRG1-LDLZ2; Grg1-LDLZ2 [*Rattus norvegicus*] (record removed from further distribution at submitter's request).	0.05: *Cryptococcus* putative kinesin family member 21A.
UNIPROT|P08151 symbol:GLI1: human Zinc finger protein GLI1; may regulate transcription.	1.52601E-24: 4885279 glioma-associated oncogene homolog 1 [*Homo sapiens*].	<0.05: *Phanerochaete* genomic homology (not annotated).
UNIPROT|P17097 symbol:ZN07: human Zinc finger protein 7; possible transcription factor.	3.18614E-22: 37590636 Zinc finger protein 7.	<0.05: *Phanerochaete* genomic homology (not annotated).
UNIPROT|P50222 symbol:MOX2: human homeobox protein MOX-2 with a role in mesoderm induction and its earliest regional specification.	8.37349E-23: 8393773 mesenchyme homeo box 2 [*Rattus norvegicus*].	0.05: *Cryptococcus* putative tRNA binding protein.
UNIPROT|Q15699 symbol:CRT1: human cartilage homeoprotein 1; possible transcription repressor.	3.18128E-22: 55638599 predicted protein similar to cartilage paired-class homeoprotein 1 [*Pan troglodytes*].	0.05: *Cryptococcus* hypothetical protein.
UNIPROT|P04637 symbol:P53: human cellular tumour antigen p53; involved in cell cycle regulation..	3.18192E-22: 35214 protein p53 [*Homo sapiens*].	0.05: *Cryptococcus* putative sterol metabolism-related protein.
UNIPROT|P10361 symbol:P53: rat cellular tumour antigen p53.	2.43631E-22: 56829 unnamed protein product [*Rattus norvegicus*].	0.05: *Cryptococcus* putative F-actin capping protein.
Flounder gonadotropin releasing hormone precursor.	6.88528E-46: 18253176 salmon-type gonadotropin-releasing hormone precursor [*Verasper moseri*].	<0.05: *Phanerochaete* genomic homology (not annotated).
UNIPROT|P10071 symbol:GLI3: human Zinc finger protein GLI3.	1.33685E-20: 51094755 GLI-Kruppel family member GLI3 (Greig cephalopolysyndactyly syndrome) [*Homo sapiens*].	0.05: *Cryptococcus* conserved hypothetical protein.
Japanese rice fish progonadoliberin III precursor.	4.46291E-45: 34098704 gonadotropin releasing hormone precursor (Japanese rice fish).	0.05: *Cryptococcus* hypothetical protein.
tr|O88728: mouse interferon induced transmembrane protein 5.	2.07314E-29: 33504579 haemopoiesis related membrane protein 1 [*Mus musculus*].	<0.05: *Phanerochaete* genomic homology (not annotated).

Table 6. (Continued)

AmiGO description and hyperlinks	Metazoan genome hit	Fungi genome hit
tr|Q8BM71: mouse integrin beta 8.	2.9936E-28: 26330087 unnamed protein product [*Mus musculus*].	<0.05 *Phanerochaete* genomic homology (not annotated).
UNIPROT|P10072 symbol:HKR1: human Krueppel-related zinc finger protein 1 (HKR1): transcription factor.	1.57917E-21: 6177785 HKR1 [*Homo sapiens*].	0.05: *Cryptococcus* hypothetical protein.
UNIPROT|Q9HD85 symbol:Q9HD85: Human haematopoietic PBX-interacting protein (transcription corepressor).	2.43954E-22: 19923830 pre-B-cell leukemia transcription factor interacting protein 1 [Homo sapiens].	0.05: *Cryptococcus* hypothetical protein.
UNIPROT|Q9Y5Y0 symbol:FVR1: human feline leukemia virus subgroup C receptor-related protein 1.	8.37349E-23: 7661708 feline leukemia virus subgroup C cellular receptor [*Homo sapiens*].	0.05: *Cryptococcus* hypothetical protein.
UNIPROT|Q9Y458 symbol:TX22: Human T-box transcription factor TBX22.	5.08675E-20: 18375603 Human T-box transcription factor TBX22.	0.05: *Cryptococcus* putative 30S ribosomal protein S17.
UNIPROT|O75093 symbol:SLT1: Human Slit-1; cue for cellular migration.	3.07887E-25: 55634425 chimpanzee homolog of Slit-1.	0.05: *Cryptococcus* hypothetical protein.
UNIPROT|P00734 symbol:THRB: human prothrombin precursor (coagulation factor II).	4.60077E-21: 4503635 human prothrombin precursor (coagulation factor II).	<0.05: *Phanerochaete* genomic homology (not annotated).
UNIPROT|P09919 symbol:CSF3: human granulocyte colony-stimulating factor precursor.	7.08858E-22: 27437049 human CSF1.	0.05: *Cryptococcus* expressed protein.
tr|Q6PCS0 Itgb4 protein (Fragment): mouse integrin beta 4.	3.30981E-27: 6981108 rat integrin beta 4.	0.05: *Cryptococcus* expressed protein.
UNIPROT|Q04741 symbol:EMX1: human homeobox protein EMX1.	4.01582E-25: 45598369 human homeobox protein EMX1.	<0.05: *Phanerochaete* genomic homology (not annotated).

Only three of the other plant development sequences from AmiGO retrieved highly similar sequences from the fungal genomes (Table 3); namely, the phosphoribosylanthranilate isomerase of *Arabidopsis thaliana* which is similar to the TRP-F sequence of *Candida glabrata* (E-value = 1.2×10^{-21}); a putative oxidoreductase of *Arabidopsis* that is highly similar to a putative dehydrogenase/reductase of *Aspergillus fumigatus* (E-value = 7×10^{-18}); and a hypothetical protein of *Arabidopsis* highly similar (with an E-value = 6.3×10^{-10}) to the mybC transcription factor of *Dictyostelium* and to a hypothetical protein of *Candida albicans* (E-value = 4.8×10^{-10}). All other plant-fungus similarities were returned with E-value reported as 0.05, which corresponds to a weak similarity that is probably not significant.

Apart from the SINA and p53 homologies already noted above, very low E-values in other animal-plant similarities (Table 5) were limited to two sialyltransferases (E-value reported as zero), a cytosine methyl transferase (E-value = 4.8×10^{-9}), a transcription factor (E-value = 9.9×10^{-16}), a transcriptional co-activator (E-value = 2.5×10^{-3}), a receptor protein (E-value = 5.6×10^{-3}) and a homeobox domain protein (E-value = 2.8×10^{-2}). All other animal-plant comparisons returned E-values of 0.05 (Table 5).

Generally weak similarities were encountered in comparisons between animal developmental sequences and fungus genomes (Table 6); only six were sufficiently similar to be notable. A predicted mRNA from the *Ustilago maydis* genome proved to be homologous (*E*-value = 0) to the ISL1 mouse transcription factor, and a hypothetical protein of *U. maydis* was very similar (*E*-value = 1.6×10^{-5}) to the human ortholog of the *pad-1* gene of *Caenorhabditis elegans* which is required for embryonic patterning. *E*-values in the region of 10^{-3} were returned to a Zebrafish nuclear respiratory factor (compared with a potential cell surface flocculin of *Candida albicans*), a *Dictyostelium* actin binding protein (compared with a hypothetical protein of *Magnaporthe grisea*), a human ATP-dependent DNA helicase (compared with a hypothetical protein from *Eremothecium* (*Ashbya*) *gossypii*), and a human Ariadne-2 protein homolog (compared with a hypothetical protein of *Gibberella zeae* [anamorph = *Fusarium graminearum*]). *E*-values of 0.05 were returned for all other fungus-animal comparisons, and these are not considered significant (Table 6). The same applies to four *Dictyostelium* sequences which failed to retrieve any similarities in either Metazoa or Viridiplantae, but were just detectable in fungal genomes (Table 7). One, a putative GATA-binding transcription factor of *Dictyostelium* was marginally similar to a hypothetical protein of *Gibberella zeae* (*E*-value = 2.8×10^{-2}), but the other three (two transcription regulators and an adhesion modulator) returned similarities with *E*-values of 0.05 in *Cryptococcus* and *Phanerochaete* respectively.

Table 7. Similarities found between Dictyostelium sequences and fungal genomes

Entries under 'genome hit' show E-value as returned by the sequence comparison software: a reference to the specific database entry and brief descriptive annotation. No hits were returned to these Dictyostelium sequences by searches of genomes of Metazoa or Viridiplantae. **NOTE** thatsmall E-values are shown as exponential functions, i.e. 1.217E-29 = 1.217×10^{-29}

AmiGO description and hyperlinks	Fungi genome hit
Locus comH: putative GATA-binding transcription factor of Dictyostelium.	0.0282092: hypothetical protein FG03968.1 from Gibberella zeae.
Locus dstA: signal transducer and activator of transcription (STAT) family protein of Dictyostelium.	0.05: putative heat shock transcription factor 2 of Cryptococcus.
Locus gbfA: G-box binding transcription factor of Dictyostelium.	0.05: expressed protein of Cryptococcus.
Locus ampA: adhesion modulation protein A of Dictyostelium.	<0.05: Phanerochaete genomic homology (not annotated).

Overall Conclusions

Our purpose was to establish whether sequences reported to be involved in development in animals or plants could be found in fungal genomes. Overall, some similarity was indicated by the comparison software for a total of 219 sequences from fungal genomes, but

143 (65%) of these returned matches were assigned *E*-values of 0.05. This level of similarity corresponds approximately to a probability of one in fewer than 20 of finding the match purely by chance, and we believe this to be too low a level of similarity for much significance to be assigned to it. The highly similar matches found in this survey between sequences labelled as being concerned in animal or plant development and fungal genomes proved to be involved in basic cell metabolism or essential eukaryotic cell processes: enzymes in common metabolic pathways, many transcription regulators, binding proteins, receptors and membrane proteins.

These findings demonstrate that there is no strong resemblance between the crown group of eukaryotic Kingdoms in the 'higher-management' functions that integrate and regulate their developmental processes. In particular, NAM sequences are essentially limited to plants, and *Notch*, *TGF*, *Hedgehog* and *Wnt* sequences (all widely considered as essential, highly conserved, components of normal development in animals) are limited to animals. None of the sequences most closely involved in animal or plant multicellular development can be found in the genomes of fungi. Generally, there are no *Wnt*, *Hedgehog*, *Notch*, *TGF*, *p53*, *SINA*, or *NAM* sequences in fungi.

Is there any reason to believe that there should be; if not homologous, then analogous sequences at least? We would have to say 'yes' to this question. Fungi, like animals and plants, have a basic 'body plan' which is established very early on in development. The lack of NAM (no apical meristem) sequences in fungi is easy to correlate with the fact that fungi do not have apical meristems; but fungi do have organised growth centres that need to be controlled, so where are their regulators? Tissues are demarcated in even the earliest fungal fruit body initials, so they certainly exhibit regional specification, cell differentiation, and cell co-ordination essential to establishing pattern formation (Rosin and Moore, 1985; Horner and Moore, 1987; Allen *et al.*, 1992; Chiu and Moore, 1993; Moore, 1998; Moore *et al.*, 1998). The non-random distributions of cells and tissues in the (mushroom) fruit bodies of *Coprinopsis cinerea* have been interpreted as being dependent on interplay between activating and inhibiting 'morphogen' factors (Horner and Moore, 1987; Moore, 1988) in a pattern-forming process similar to the model developed by Meinhardt and Gierer (1974, 1980; Meinhardt, 1984, 1998). Successful application of this morphogenetic field model to fungi as well as to plants and animals indicates that the general rules of pattern formation apply similarly to all multicellular systems.

Many other similarities have emerged from observations of fungal, particularly mushroom, development (reviewed in Moore, 2005). These include commitment (Bastouill-Descollonges and Manachère (1984) and Chiu and Moore (1988a, b) demonstrated that basidia are specified irreversibly as meiocytes during meiotic prophase I, their maturation being an autonomous, endotrophic process that is able to proceed *in vitro*); regeneration (Chiu and Moore, 1988a, b; Brunt and Moore, 1989; Bourne *et al.*, 1996; Chiu *et al.*, 1998). Programmed cell death in fungi is used to sculpture the shape of the fruit body from the raw medium provided by the hyphal mass of the fruit body initial and primordium (Umar and Van Griensven, 1997a, b; 1998).

Thus, we reach the conclusion that fungal morphogenesis must be totally different from animals, because fungal cells have walls, and from plants (whose cells also have walls) because hyphae grow only at their tips and hyphal cross-walls form only at right angles to the long axis of the hypha, which together make fungal morphogenesis dependent on the placement of hyphal branches. On the other hand, there is no doubt that development of fungal multicellular structures involves a whole suite of cellular processes and interactions that are analogous or homologous to those that occur during animal or plant development.

Yet our findings show that there are no resemblances between the crown group of eukaryotic Kingdoms in the ways they control and regulate their developmental processes. Current understanding of phylogenetic relationships is that the major kingdoms of eukaryotes separated from one another at a stage prior to the multicellular grade of organization. Consequently, in the course of their evolutionary history these very different organisms may have found different strategies to solve the same sorts of morphogenetic control problems. Finally, this means, in particular, that we are currently totally ignorant of the way fungi regulate their multicellular development.

REFERENCEES

Allen, J.J., Moore, D. and Elliott, T.J. (1992). Persistent meiotic arrest in basidia of *Agaricus bisporus*. *Mycol. Res. 96*:125-127.

Altschul, S.F., Gish, W., Miller, W., Myers, E.W. and Lipman, D.J. (1990). Basic local alignment search tool. *J. Mol. Biol. 215*:403-410.

Apweiler, R., Bairoch, A., Wu, C.H., Barker, W.C., Boeckmann, B., Ferro, S., Gasteiger, E., Huang, H., Lopez, R., Magrane, M., Martin, M.J., Natale, D.A., O'Donovan, C., Redaschi, N. and Yeh, L.S. (2004). UniProt: the Universal Protein knowledgebase. *Nucleic Acids Res. 32*: D115-D119.

Ashburner, M. and Drysdale, R. (1994). FlyBase-the *Drosophila* genetic database. *Development 120*:2077-2079.

Bastouill-Descollonges, Y. and Manachère, G. (1984). Photosporogenesis of *Coprinus congregatus*: correlations between the physiological age of lamellae and the development of their potential for renewed fruiting. *Physiol. Plant. 61*:607-610.

Blake, J.A., Richardson, J.E., Bult, C.J., Kadin, J.A. and Eppig, J.T. (2003). MGD: the Mouse Genome Database. *Nucleic Acids Res. 31*:193-195.

Brunt, I.C. and Moore, D. (1989). Intracellular glycogen stimulates fruiting in *Coprinus cinereus*. *Mycol. Res. 93*: 543-546.

Bourne, A.M., Chiu, S.W. and Moore, D. (1996). Experimental approaches to the study of pattern formation in *Coprinus cinereus*. In Chiu, S.W. and Moore, D. Ed. *Patterns in Fungal Development*. Cambridge University Press, Cambridge, UK, pp. 126-155.

Casselton, L.A. (2002). Mate recognition in fungi. *Heredity* 88, 142-147.

Cavalier-Smith, T. and Chao, E.E. (1995). The opalozoan *Apusomonas* is related to the common ancestor of animals, fungi and choanoflagellates. *Proc. Roy. Soc., London Ser. B, 261*:1-6.

Chiu, S.W. (1996). Nuclear changes during fungal development. In: *Patterns in Fungal Development* (S.W. Chiu and D. Moore, eds). Pp. 105-125. Cambridge: Cambridge University Press.

Chiu, S.W. and Moore, D. (1988a). Evidence for developmental commitment in the differentiating fruit body of *Coprinus cinereus*. *Trans. Br. Mycol. Soc. 90*:247-253.

Chiu, S.W. and Moore, D. (1988b). Ammonium ions and glutamine inhibit sporulation of *Coprinus cinereus* basidia assayed *in vitro*. *Cell Biol. Int. Repts 12*:519-526.

Chiu, S.W. and Moore, D. (1990). Sporulation in *Coprinus cinereus*: use of an *in vitro* assay to establish the major landmarks in differentiation. *Mycological Research 94*, 249-253.

Chiu, S.W. and Moore, D. (1993). Cell form, function and lineage in the hymenia of *Coprinus cinereus* and *Volvariella bombycina*. *Mycol. Res. 97*:221-226.

Chiu, S.W., Chan, Y.H., Law, S.C., Cheung, K.T. and Moore, D. (1998). Cadmium and manganese in contrast to calcium reduce yield and nutritional value of the edible mushroom *Pleurotus pulmonarius*. *Mycol. Res. 102*:449-457.

Harris, M.A., Clark, J., Ireland, A., Lomax, J., Ashburner, M., Foulger, R., Eilbeck, K., Lewis, S., Marshall, B., Mungall, C., Richter, J., Rubin, G.M., Blake, J.A., Bult, C., Dolan, M., Drabkin, H., Eppig, J.T., Hill, D.P., Ni, L., Ringwald, M., Balakrishnan, R., Cherry, J.M., Christie, K.R., Costanzo, M.C., Dwight, S.S., Engel, S., Fisk, D.G., Hirschman, J.E., Hong, E.L., Nash, R.S., Sethuraman, A., Theesfeld, C.L., Botstein, D., Dolinski, K., Feierbach, B., Berardini, T., Mundodi, S., Rhee, S.Y., Apweiler, R., Barrell, D., Camon, E., Dimmer, E., Lee, V., Chisholm, R., Gaudet, P., Kibbe, W., Kishore, R., Schwarz, E.M., Sternberg, P., Gwinn, M., Hannick, L., Wortman, J., Berriman, M., Wood, V., de la, Cruz N., Tonellato, P., Jaiswal, P., Seigfried, T. and White, R. (2004). The Gene Ontology (GO) database and informatics resource. *Nucleic Acids Res. 32*:D258-D261.

Horner, J. and Moore, D. (1987). Cystidial morphogenetic field in the hymenium of *Coprinus cinereus*. *Trans. Br. Mycol. Soc. 88*:479- 488.

Huala, E., Dickerman, A.W., Garcia-Hernandez, M., Weems, D., Reiser, L., LaFond, F., Hanley, D., Kiphart, D., Zhuang, M., Huang, W., Mueller, L.A., Bhattacharyya, D., Bhaya, D., Sobral, B.W., Beavis, W., Meinke, D.W., Town, C.D., Somerville, C. and Rhee, S.Y. (2001). The *Arabidopsis* Information Resource (TAIR): a comprehensive database and web-based information retrieval, analysis, and visualization system for a model plant. *Nucleic Acids Res. 29*:102-105.

Kosarev, P., Mayer, K.F.X. and Hardtke, C. (2002). Evaluation and classification of RING-finger domains encoded by the *Arabidopsis* genome. *Genome Biol.* 3 (4): research0016.1–0016.12 (online at http://genomebiology.com/2002/3/4/research/0016).

Kreppel, L., Fey, P., Gaudet, P., Just, E., Kibbe, W.A., Chisholm, R.L. and Kimmel, A.R. (2004). dictyBase: a new *Dictyostelium discoideum* genome database. *Nucleic Acids Res. 32*:D332-D333.

Matsuzawa, S., Takayama, S., Froesch, B.A., Zapata, J.M. and Reed, J.C. (1998). p53-inducible human homologue of *Drosophila* seven in absentia (Siah) inhibits cell growth: suppression by BAG-1. *EMBO J. 17*:2736–2747.

Meinhardt, H. and Gierer, A. (1974). Applications of a theory of biological pattern formation based on lateral inhibition. *J. Cell Sci. 15*:321-346.

Meinhardt, H. and Gierer, A. (1980). Generation and regeneration of sequence of structures during morphogenesis. *J. Theor. Biol. 85*:429-450.

Meinhardt, H. (1984). Models of pattern formation and their application to plant development. In Barlow, P.W. and Carr, D.J. Ed. *Positional Controls in Plant Development*. Cambridge University Press, Cambridge, UK, pp. 1-32.

Meinhardt, H. (1998). *The Algorithmic Beauty of Sea Shells*. Springer-Verlag, New York, Heidelberg, Berlin, pp. 236.

Meškauskas, A., Lehmann-Horn, F. and Jurkat-Rott, K. (2004). Sight: automating genomic data-mining without programming skills. *Bioinformatics 20*:1718-1720.

Moore, D. (1988). Recent developments in morphogenetic studies of higher fungi. *Mushroom J. Tropics 8*:109-128.

Moore, D. (1998). *Fungal Morphogenesis*. Cambridge University Press, New York. Pp. 469.

Moore, D. (2005) Principles of mushroom developmental biology. *Int. J. Med. Mushrooms 7*:79-102.

Moore, D., Chiu, S.W., Umar, M.H. and Sánchez, C. (1998). In the midst of death we are in life: further advances in the study of higher fungi. *Bot. J. Scotland 50*:121-135.

Moore, D., Walsh, C. and Robson, G.D. (2005). A search for developmental gene sequences in the genomes of filamentous fungi. In Arora, D.K. and Berka, R. Ed. *Applied Mycology and Biotechnology*, Vol. 6, Genes, Genomics and Bioinformatics, Elsevier Science, Amsterdam. Pp. 169-188.

Pringle, A. and Taylor, J.W. (2002). The fitness of filamentous fungi. *Trends in Microbiol. 10*:474-481.

Rosin, I.V. and Moore, D. (1985). Differentiation of the hymenium in *Coprinus cinereus*. *Trans. Br. Mycol. Soc. 84*:621-628.

Trinci, A.P.J., Wiebe, M.G. and Robson, G.D. (1994). The mycelium as an integrated entity. In Wessels, J.G.H. and Meinhardt, F. Ed, *The Mycota* vol. I, Springer-Verlag, Berlin and Heidelberg. Pp. 175-193.

Twigger, S., Lu, J., Shimoyama, M., Chen, D., Pasko, D., Long, H., Ginster, J., Chen, C.F., Nigam, R., Kwitek, A., Eppig, J., Maltais, L., Maglott, D., Schuler, G., Jacob, H. and Tonellato, P.J. (2002). Rat Genome Database (RGD): mapping disease onto the genome. *Nucleic Acids Res. 30*:125-128.

Umar, M.H. and Van Griensven, L.J.L.D. (1997a). Morphological studies on the life span, developmental stages, senescence and death of *Agaricus bisporus*. *Mycol. Res. 101*: 1409-1422.

Umar, M.H., and Van Griensven L.J.L.D. (1997b). Morphogenetic cell death in developing primordia of *Agaricus bisporus*. *Mycologia 89*:274-277.

Umar, M.H., and Van Griensven L.J.L.D. (1998). The role of morphogenetic cell death in the histogenesis of the mycelial cord of *Agaricus bisporus* and in the development of macrofungi. *Mycol. Res. 102*:719-735.

Venter, J.C., Adams, M.D., Martin-Gallardo, A., McCombie, W.R. and Fields, C. (1992). Genome sequence analysis: scientific objectives and practical strategies. *Trends Biotechnol. 10*: 8-11.

Watling, R. and Moore, D. (1994). Moulding moulds into mushrooms: shape and form in the higher fungi. In Ingram, D.S. and Hudson, A. Ed. *Shape and Form in Plants and Fungi*, Academic Press, London. Pp. 270-290.

Wheeler, D.L., Barrett, T., Benson, D.A., Bryant, S.H., Canese, K., Church, D.M., DiCuccio, M., Edgar, R., Federhen, S., Helmberg, W., Kenton, D.L., Khovayko, O., Lipman, D. J., Madden, T.L., Maglott, D.R., Ostell, J., Pontius, J.U., Pruitt, K.D., Schuler, G.D., Schriml, L.M., Sequeira, E., Sherry, S.T., Sirotkin, K., Starchenko, G., Suzek, T.O., Tatusov, R., Tatusova, T.A., Wagner, L. and Yaschenko, E. (2005). Database resources of the National Center for Biotechnology Information. *Nucleic Acids Res. 33* Database Issue: D39-D45.

Wheeler, D.L., Church, D.M., Federhen, S., Lash, A.E., Madden, T.L., Pontius, J.U., Schuler, G.D., Schriml, L.M., Sequeira, E., Tatusova, T.A. and Wagner, L. (2003). Database resources of the National Center for Biotechnology. *Nucleic Acids Res. 31*:28-33.

In: Current Advances in Molecular Mycology
Eds: Y. Gherbawy, R.L. Mach and M.K. Rai

ISBN 978-1-60456-909-4

Chapter II

JOURNEY OF PHYTOPATHOGENIC FUNGI FROM GENETICS TO GENOMICS

A. Pain[1], A. K. Dhar[2] and C. Chattopadhyay[3]

[1]The Pathogen Sequencing Unit, The Wellcome Trust Sanger Institute, Genome Campus, Hinxton, Cambridge CB 10 1SA, UK;

[2]San Diego State University, Biology Department, 5500 Campanile Drive, San Diego, California 92182, USA; Senior Scientist, Advanced Bionutrition Corporation, 7155 Gateway Drive, Columbia, Maryland 21046, USA;

[3]National Research Centre on Rapeseed-Mustard (ICAR), Sewar, Bharatpur 321303, India.

ABSTRACT

Only a small group of fungal species is successful as plant pathogens through employment of a wide array of molecular mercenary to cause disease. Knowledge of molecular mechanisms governing developmental and physiological adaptations of pathogens that allow plant pathogenic fungi to invade plants, colonize tissues and control host defence reactions is limited. This review provides an overview of the genetics of host-pathogen interactions in plants from gene-for-gene theory to genomics approaches for exploring fungal pathogenesis. The utility of functional genomics approaches such as Expressed Sequence Tags (ESTs), microarray analysis, Serial Analysis of Gene Expression (SAGE), high-throughput transposon-mediated insertional mutagenesis, Green Fluorescent Protein (GFP)-based reporter systems, RNA interference (RNAi) and proteomics for elucidating the molecular mechanisms involved in fungal pathogenesis is discussed.

Key words: Fungi, plant, diseases, resistance, genetics, genomics

INTRODUCTION

The ability to cause plant disease is a complex trait that, thankfully, occurs in only a small group of fungal and bacterial species. This small group of fungal and bacterial pathogens are the ones successful in overcoming the host defence and deploy a wide array of molecular techniques to cause disease. Our understanding of the molecular mechanisms governing the developmental and physiological adaptations of these pathogens that allow them to invade plants, colonize tissues and inhibit host defence reactions is limited. However, this situation is set to change as the genome sequencing of a large number of fungal and bacterial plant pathogens is at different stages of completion [8]. Since the genome of *Haemophilus influenzae* was sequenced in 1995 [35], genomes of many prokaryotes have been completely sequenced (www.genomesonline.org). The comparative analysis of genomes of bacterial strains and species provided valuable insights into the evolution of virulence and host adaptations. It appears that there are several forms of co-evolution between hosts and pathogens, some involving reciprocal adaptation and others a combination of adaptations and speciation [118]. Two of these forms of co-evolution have had a major effect on the evolution of interactions between pathogens and host plants. One is the evolution of defence and counter-defence in interactions between pathogens and host plant, which may have led to the radiation of species through a process of escape-and-radiation co-evolution [119]. The other is the gene-for-gene co-evolution [36, 37, 38, 39].

GENE-FOR-GENE INTERACTIONS IN THE EVOLUTION OF FUNGAL PATHOGENS

Much of the work on breeding for resistance to fungal pathogens in agricultural crops has assumed that virulence in fungal pathogens and resistance in plants are governed by gene-for-gene interactions [36]. As different aspects of ecology of species of fungi and plants were incorporated into recent empirical and mathematical studies of gene-for-gene interactions, the original hypothesis has been expanded to accommodate a family of views.

The gene-for-gene hypothesis states that for each gene that conditions a reaction in the host there is a corresponding gene that conditions pathogenicity in the pathogen (i.e., counters the host defence gene). The linked genes of the host and pathogen are known as corresponding genes. Presently, the gene-for-gene hypothesis is interpreted to mean that for each gene determining resistance in the host there is a corresponding gene for avirulence in the pathogen with which it specifically interacts. According to this hypothesis, the occurrence of a resistant or incompatible reaction depends on both the presence of a gene for resistance (R) in the host and the corresponding dominant gene for avirulence (V) in the pathogen. If a host lacks a specific resistance gene, the corresponding avirulence gene cannot be detected in the pathogen and *vice-versa* (Table 1). In diploid host-pathogen associations, the interactions occurring at any particular corresponding gene pair or locus may be further complicated by the occurrence of the resistance or avirulence locus in a heterozygous state. This generates a further five genetic combinations (VvRR, VvRr, Vvrr, VVRr and vvRr). Since good health is the rule and disease the exception (resistance or R dominant to susceptibility or r, avirulence or V dominant to virulence or v), these combinations are generally phenotypically indistinguishable from other compatible or incompatible interactions. The specificity of these interactions, changing from compatible to incompatible as the same host is challenged by different pathogenic genotypes and *vice-versa* (Table 1), is the basic of gene-for-gene interaction. Often a gene-for-gene interaction is inferred from the specificity of a host-pathogen interaction that is coupled with a genetic basis of resistance. Only a few genetic analyses of specific host-pathogen interactions have been done involving only a handful of resistance and avirulence genes.

Table 1. Expected compatibility between homozygous genotypes in a single-locus gene-for-gene interaction

Pathogen genotype	Host genotype	
	RR	rr
VV	Incompatible	Compatible
vv	Compatible	Compatible

R is a dominant host gene conferring resistance to the pathogen and r is a recessive host gene conferring susceptibility. V is a dominant pathogen gene conferring avirulence and v is a recessive pathogen gene conferring virulence.

Table 2. Gene-for-gene interaction between plants and pathogens

Fungal pathogens	Fungal pathogens
Basidiomycetes Uredinales (rusts)	Ascomycetes Erysiphales (powdery mildews)
Avena-Puccinia graminis	*Hordeum-Erysiphe graminis*
Avena-Puccinia coronata	*Secale- Erysiphe graminis*
Coffea-Hemileia vastatrix	*Senecio- Erysiphe fischeri*
Glycine-Phakospora pachyrhizi	*Triticum-Erysiphe graminis*
Helianthus-Puccinia helianthi	
Linum-Melampsora lini	Oomycetes Peronosporales
Triticum-Puccinia graminis	*Lactuca-Bremia lactucae*
Triticum-Puccinia recondita	*Medicago-Peronospora trifoliorum*
Triticum-Puccinia striiformis	*Sorghum-Peronosclerospora sorghi*
	Glycine-Phytophthora megasperma
Basidiomycetes Ustilaginales (bunts and smuts)	*Solanum-Phytophthora infestans*
Avena-Ustilago avenae	Other fungi
Hordeum-Ustilago hordei	*Brassica-Plasmodiophora brassicae*
Triticum-Ustilago tritici	*Malus-Venturia inaequalis*
Triticum-Tilletia caries	*Callistephus-Fusarium oxysporum*
Triticum-Tilletia controversa	*Hordeum-Rhynchosporium sacalis*

[3, 14, 23, 43, 54, 79, 94, 100, 110, 119]

One of the practical implications of studying gene-for-gene interactions is development of strategies to counter virulence in some fungal pathogens with single genes for resistance in plants. This theory has served as a guiding principle in breeding crops for resistance to fungal pathogens. Since the recognition of the dominant, single-gene nature of resistance against some fungal pathogens, breeding in many crops has involved sequential release of new varieties carrying different resistance genes as a form of evolutionary arms race. This has resulted in microevolution as pathogens have responded to these selective forces by acquiring the corresponding genes for virulence. The best examples of these reciprocal responses are found among cereal rusts (*Puccinia*) and potato late blight (*Phytophthora infestans*). The sequential deployment of single-gene resistance has led to changes in particular multilocus virulence phenotypes (races) in the pathogen and/ or appearance of totally new races. Gene-for-gene interactions involving analysis of a few genes for resistance and avirulence have been determined in some plant-pathogen associations (Table 2). Tight linkage groups occur among both resistance and virulence genes [19, 73].

Although breeding programmes and models of gene-for-gene interactions in agriculture often focus on how natural selection will favour increased virulence in local pathogen, studies suggest that the dynamics and evolution of these interactions is sometimes governed by gene flow over long distances. Use of genetic fingerprinting confirms that some epidemics in crops are caused by spread of one or two clones over large areas. Brown *et al.* [12] found that an epidemic of barley powdery mildew (*Blumeria graminis*) in 1980s in UK was initiated largely by two clones able to overcome barley cultivars carrying resistance allele *Mla13*. These results argue for a broader ecological and geographical view of the evolution of gene-for-gene interactions in some crops rather than a view based on minimising the rate of natural selection for virulence in local pathogen populations.

There are two approaches to genetic basis of host-pathogen interaction in natural populations: (i) comparisons of host family lines grown in natural situations, and (ii) detailed study of individual host x pathogen isolates and formal genetic analysis of resistance. The former method compares the total resistance phenotype of individuals or family lines against a genetically variable pathogen population. Thus, the action of major genes for resistance in the host and avirulence in its corresponding pathogen is likely to be masked by interplay between gross morphology and resistance factors. Several virulence and avirulence alleles were identified in a large-scale Europe-wide survey of *Leptosphaeria maculans* isolates [113]. The alternative approach of studying individual host x pathogen isolates has come from increasing interest in the dynamics of individual types of resistance. Many elements of gene-for-gene interactions, such as the resistance conferred by single genes with major phenotypic effects tightly coupled between host lines and specific pathogen genotypes, are found in a wide array of natural host-pathogen associations [14, 15, 54, 59, 98]. Studies that concentrate on interactions in single populations tend to give the impression of tightly co-evolved gene-for-gene associations between host and pathogen. Simultaneous study of variation in hosts and pathogens in a number of populations of *Linum marginale* and *Melampsora lini* in the same geographic region provided a new perspective to our understanding of how plant and pathogens continue to co-evolve [17, 18, 60]. This association is an endemic Australian version of the interaction used by Flor [36] to first elucidate the gene-for-gene concept. It provides further evidence that gene-for-gene interactions do occur in natural populations. The study indicates the spatial scale over which co-evolving associations are likely to develop [16]. In the *Linum-Melampsora* association, migration occurs at a high rate, which ensures interchange between individual populations and rapid re-establishment of those genes lost through extinction. Changes in the plant system are slower than in the pathogen. However, within individual populations all host plants are vulnerable to attack regardless of the resistance genes and the virulence structure of associated pathodeme. Changes in the relative frequency of individual resistant lines appear to show little correlation with the frequency of pathogen races in the associated pathodeme. Thus, at the spatial scale of individual host and pathogen demes, there is no clear pattern of reciprocal response in genetic structure of interacting populations. Alternatively, individual populations could be strongly influenced by genetic drift and gene flow that occurs among demes within the same epidemiological region. The sum of these processes across many plants and pathodemes constitute the metapopulation at which gene-for-gene co-evolution takes place and genetic variation in resistance and virulence genes is maintained.

Several models of gene-for-gene interactions predict the rate at which a pathogen population will overcome a set of major resistance genes in the host plant [72,103]. The metapopulation structure of interacting populations may potentially magnify or mitigate the 'boom-and-bust' cycles expected in some genetic models of plant-pathogen interactions. Combining results from agriculture, natural populations and mathematical models, it appears that the gene-for-gene view of co-evolution is developing into a family of hypotheses that differ slightly in their assumptions and predictions. Thompson and Burdon [119] suggested a hierarchy in this family based on the complexity of population structure and the evolutionary processes driving genetic changes in plants and their pathogens, *viz.*, (a) local agricultural gene-for-gene coevolution, (b) geographic agricultural gene-for-gene coevolution, (c) panmictic gene-for-gene coevolution and (d) metapopulation gene-for-gene coevolution. Continued progress in breeding and management for durable resistance in plant populations [6] and in the understanding of evolutionary dynamics of the gene-for-gene coevolution in natural populations will require more detailed studies.

Functional Genomics Approaches to Explore Fungal Pathogenesis in Plants

Since completion of the *Saccharomyces cerevisiae* genome sequence [44], the genome sequencing of a number of parasitic and symbiotic fungi has been initiated and are completed or nearing completion (Figure 1, http://www-genome.wi.mit.edu/annotation/fungi/). These include the model ascomycetes *Neurospora crassa*, *Ashbya gossypii* and *Aspergillus nidulans*. Analysis of the *N. crassa* genome (~40 megabases) provided insights into aspects of *Neurospora* biology including the identification of genes potentially associated with red light photobiology, genes implicated in secondary metabolism, and possession of the widest array of genome defence mechanisms known for any eukaryotic organism. These defence genes include a process unique to fungi called repeat-induced point (RIP) mutation, that had a profound impact on genome evolution [41]. The genome of *Ashbya gossypii* with 9.2 megabases, encoding 4718 protein-coding genes represents the smallest genome of a free-living eukaryote yet characterized. More than 90% of the *A. gossypii* genes show a pattern of synteny with *S. cerevisiae* that contained 300 inversions and translocations that have occurred since divergence of these two species [27].

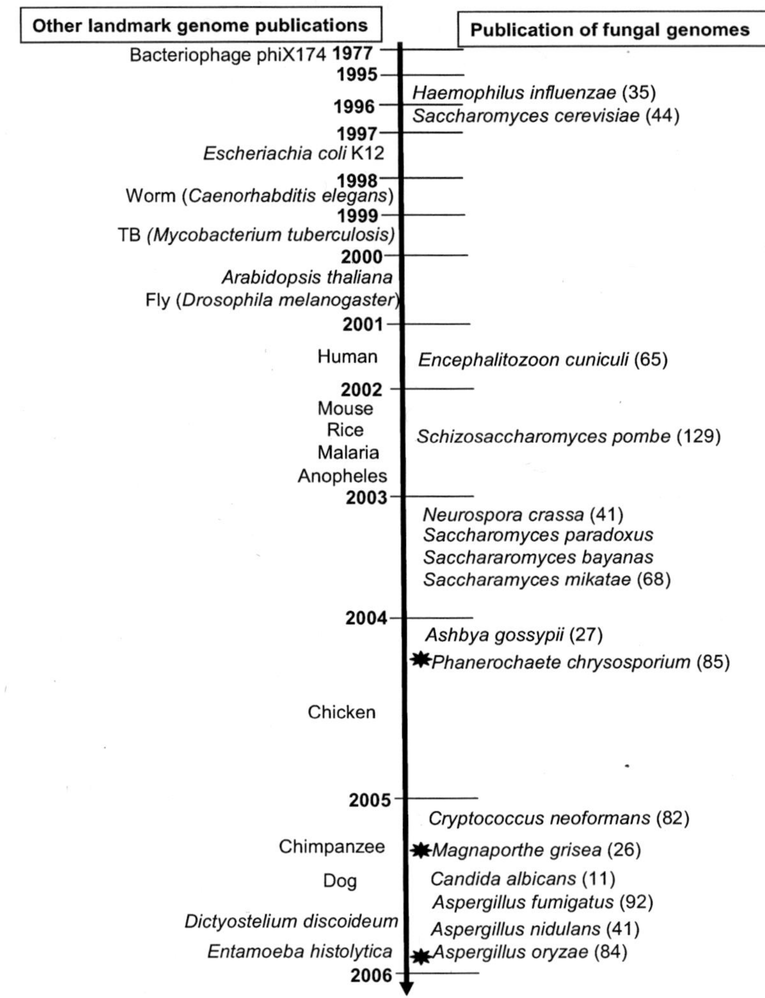

Figure 1. An overview of the evolving chronology of genome publications for fungal pathogens and other major organisms. The plant pathogenic fungal genomes have been indicated with an asterisk (*). A detailed list of ongoing and nearly completed fungal genome projects can be found in the websites: http://www.broad.mit.edu/annotation/fgi and http://www.genomesonline.org/.

Unlike the bacterial pathogens, our current understanding of the genomic adaptations in pathogenic fungi is rather poor. However, this situation is changing rapidly with the availability of the genomes of a few phytopathogenic fungi [96]. Genome analysis of the opportunistic human pathogen *Cryptococcus neoformans* [82], the rice blast pathogen, *Magnaporthe grisea* [26], and the lignocellulose-degrading white rot fungus *Phanerochaete chrysosporium* [85] provided an insight into adaptations required by pathogenic fungi to cause disease. Additionally these data partially answers the query as to whether human and / or plant pathogenic fungi possess a unique set of genes with specific functions dedicated to pathogenic activities. The genome of *M. grisea* encodes a large and diverse set of secreted proteins, including number of proteins containing unusual carbohydrate-binding domains. The rice blast pathogen also possesses an expanded family of G-protein-coupled receptors, several new virulence-associated genes, and large numbers of enzymes involved in secondary metabolism. Consistent with a role in fungal pathogenesis, the expression of several of these genes is up-regulated during early stages of infection [26]. More recently, comprehensive analyses of genomes of three *Aspergillus* species (*A. fumigatus*, *A. nidulans* and *A. oryzae*) have been described [41, 84, 92, 97]. Genome comparison showed that *A. oryzae* has an additional 7-9 Mb in the genome compared to the other two aspergilli. There are several *A. oryzae*-specific gene blocks (i.e. lacking synteny with *A. fumigatus* and *A. nidulans*) scattered in a mosaic pattern throughout the genome. Analyses of the *A. oryzae* specific gene blocks revealed expansion of genes involved in secondary metabolism, amino acid metabolism and amino acid / sugar uptake transporters [84]. In addition, sequence comparison identified over 5,000 non-coding sequences being conserved across all three species, despite about 200 million years of evolution [41].

The availability of genomic sequences for these fungi will provide valuable insight into genome plasticity and the evolution of fungal pathogens. The fungal genome sequence will also serve as a platform to identify gene function and regulatory networks. In addition, genome-wide comparisons coupled with experimental approaches to elucidate the individual gene function will enable us to understand the molecular basis of fungal virulence. Expression profiling efforts identified a number of genes associated with defence signal transduction or antimicrobial action [128]. Large-scale investigation of gene expression using functional genomics approaches such as large-scale ESTs, SAGE, microarray, high-throughput transposon-mediated insertional mutagenesis, RNAi, and proteomic analysis, will unravel the inherent complexity of host-pathogen interactions. Such studies will allow further insight on the interplay of a large number of host and pathogen gene products and the consequence of this interplay on pathogen recognition and the evolution of disease resistance [4, 21, 121, 130].

Expressed Sequence Tags (ESTs)

ESTs are short, single-pass, and partial sequences of randomly selected clones from a cDNA library of a given tissue or cells. EST sequences are derived from either the 5' and/ or 3' ends of the transcript. The 5'-ESTs generally include a portion of the encoded protein, and often provide information about the potential function of the expressed gene. On the other hand, the 3'-ESTs generally include the non-coding region of the mRNA and are more variable. However, 3'-ESTs, due to their inherent variability, are more unique than whole genes and thus can be used to distinguish transcripts within a gene family. The sequencing and analysis of a large number of ESTs provides a snapshot of the total gene activity in a cell under a given condition [2]. The frequency of occurrence of an EST in a cDNA library is largely determined by the abundance of that transcript, if they are obtained from non-normalised cDNA libraries. The EST sequences, after filtering for contaminants, vector sequence, and quality of the reads, are assembled (clustered) into contigs (containing two or more EST sequence reads) and singletons, then searched against appropriate sequence databases to try to identify gene function based on sequence similarity. Due to low cost and the ease of producing and analysing sequence information, EST mining has greatly facilitated new gene discovery and has proven valuable in genomic studies in phytopathogenic as well as symbiotic fungi [102, 106]. Using ESTs, genes involved in the early stages of infection in rice blast fungal pathogen, *M. grisea* were identified [58]. As part of the Consortium for the Genomics of the Microbial Eukaryotes (COGEME) project, a relational database containing EST sequences from several plant pathogenic fungi was developed [111, 112] and is publicly available at http://cogeme.exeter.ac.uk. The current version (version 1.5) of the COGEME Phytopathogenic Fungi and Oomycete EST database contains a total of 59,765 ESTs. In a recent study, comparative clustering of expression patterns of a large set of ESTs from phytopathogenic and saprophytic fungi was used to identify candidate pathogenicity factors [112]. In the absence of the fully sequenced genome of a given fungal plant pathogen, generation of large-scale ESTs and their subsequent analysis often give us a fairly good idea about the genetic diversity and gene structure in that organism.

Microarray Analysis

Microarray analysis has revolutionized our understanding of the dynamics of gene expression during the last decade [108, 109] and is constantly evolving. Gene expression profiling using microarray allows measuring the expression of thousands of genes in parallel in a single experiment and thereby, allows predicting the function of genes, identifying novel transcripts, link genes to biochemical pathways, and developing hypotheses about transcriptional regulation and gene regulatory networks [33]. For example, using an oligonucleotide-based microarray, genes that consistently showed altered mRNA expressions in maize during *Cochliobolus carbonum* pathogenesis were identified [7, 90]. High-density cDNA microarrays were used to analyse transcript profiles of the plant pathogenic fungus *Blumeria graminis* f. sp. *hordei*, the causal agent of barley powdery mildew [10]. The expression profile revealed a group of RNAs whose abundance were correlated with the expression of a gene (*cap20*) known to be required for virulence in *Colletotrichum gloeosporioides*, an important anthracnose causing fungal pathogen of several horticultural crops (avocado, strawberry, papaya, yam, mango, cocoa, olive, lupin, etc.). These candidate genes appeared to be critical for pathogenicity in *B. graminis* [10]. Microarray analysis has the potential to address many important issues in plant-fungal pathogen interactions, such as comparing the spectra of genes affected by different pathogen species, pathotypes, or by same pathogen in different plant genotypes, monitoring the dynamics of gene expression changes over time in a plant from initiation through establishment of fungal pathogenesis, and comparing gene expression at infection site versus sites away from the point of infection. Such studies have begun to reveal new features of plant-pathogen interactions [26, 66, 67, 80, 115] and identify candidate targets for developing therapeutics for economically important fungal diseases in plants. An array based on the *Candida albicans* genome sequence was used to identify components of transcriptional regulators that control the yeast from vegetative to hyphal stage morphogenesis and the complete sexual cycle [89, 123] and to reveal candidate targets for itraconazole treatment in *C. albicans* [24]. A similar microarray-based approach was used to study mutualistic (symbiotic) associations in fungi and to identify novel symbiosis-regulated genes in the *Eucalyptus – Pisolithus* ectomycorrhizal association [126]. Availability of whole genome sequences of *M. grisea* allowed researchers to design oligo-based arrays representing all of the 13,666 predicted genes in the fungus. This array is now commercially available from Agilent Technologies, Inc. (Palo Alto, California). The simultaneous availability of whole genome arrays in rice has been used extensively to identify components of the host's response to microbial colonisation by comparing transcription profiling of rice – *M. grisea*, rice-*Fusarium moniliforme,* and rice-mycorrhizal associations [48]. More recently, temperature-sensitive expression of a distinct set of genes was identified using an array representing the whole genome sequence of *A. fumigatus* [92]. The same array was used for Comparative Genomic Hybridization (CGH) with a closely related sexual species, *Neosartorya fischeri,* which revealed ~700 genes of *A. fumigatus* are either absent or highly diverged in *Neosartorya* [92].

Serial Analysis of Gene Expression (SAGE)

SAGE is a high-throughput strategy for monitoring global gene expression using gene-specific tags. SAGE is not as comprehensive as a method as microarray technique but it does not require the complete genome sequence to be effective. SAGE has proven itself an excellent method for identifying small coding sequences that have been missed during genome annotation by standard *in silico* methods. Briefly, cDNA derived from cellular mRNA is cleaved into precise 15-bp segments (tags) that are ligated together and then sequenced. The transcript abundance is determined by the frequency of occurrence of gene-specific tags, and thus, allows the differentiation of gene expression patterns among the members of a gene family. SAGE, therefore, allows a qualitative and quantitative estimation of the transcriptome [86, 91, 131]. Unlike microarray analysis, the genes that are studied by SAGE are not pre-selected, and therefore, SAGE provides a better representation of gene activity in a cell. SAGE has been utilised to investigate the gene expression patterns during the growth of yeast cells [57, 63] and to explore the relationship between transcript abundance and protein accumulation in yeast [49]. Using SAGE, mRNA expression during pathogenesis-related development and host colonization has been examined in *B. graminis* [117] and in *M. grisea* [56].

High-Throughput Transposon-Based Insertional Mutagenesis

One of the most powerful techniques for functional genomic analysis is the ability to create a large library of single mutants of the organism in question. With the availability of whole genome sequences, this technique could be extended to genome scale to allow the study of the function of thousands of genes. For model organisms such as *S. cerevisiae*, *Caenorhabditis elegans*, *Drosophila melanogaster*, *Arabidopsis thaliana,* and mice, large mutant collections are already available. In the case of plant pathogenic bacteria, methods such as transposon mutagenesis and signature-tagged mutagenesis have allowed high-throughput forward genetics to be carried out to generate libraries of non-pathogenic mutants for subsequent characterisation [55, 120].

The large-scale gene disruption strategies developed for *S. cerevisiae* are yet to be applicable to most of filamentous fungi. However, such efforts are already underway in a few phytopathogenic fungi, including *M. grisea* (e.g., *Magnaporthe grisea - Oryza sativa* Interaction Database at http://www.mgosdb.org). Filamentous phytopathogenic fungal species have proven more recalcitrant to molecular genetic manipulation, largely due to the inefficiency of introducing DNA in to these organisms for transformation. However, there has been major progress in developing high-throughout mutagenesis technologies that are applicable to filamentous fungi. These include transposon-arrayed gene knockout (TAG-KO) system [52], the use of the *Tc1-mariner* transposable element that was originally derived from *Fusarium oxysporum* [77] and subsequently used to generate insertional mutants in *M. grisea* [125]; the *impala* based transposition system (a copy of *impala* inserted in the promoter of *niaD* gene from *A. nidulans*) in generation of mutants in *A. fumigatus* [34]; the *Agrobacterium tumefaciens*-mediated transformation (ATMT) in *F. oxysporum*, *M. grisea* and *Cryphonectria parasitica*, using conidia, and thus avoiding the need to generate protoplasts [87, 88, 99, 107] and a PCR-based tool (called Double-Joint PCR) for rapid gene manipulation in filamentous fungi [132]. New high-throughput insertional mutagenesis approaches have now been taken to generate large library of mutants to identify gene function in many phytopathogenic fungi including *M. grisea* and *Mycosphaerella graminicola* [1, 52, 81].

GFP-Based Reporter Systems

In recent years, Auto Fluorescent Protein (AFP)-based technologies, such as use of Green Fluorescent Protein (GFP) have become important tools for microbiologists and microbial ecologists for studying processes such as microbe-plant interactions, biosensors, biofilm formation, and horizontal gene transfer [78]. The ability to use different AFP markers (e.g., GFP, YFP, CFP and DsRed) with differing fluorescent spectra within a single cell has allowed simultaneous, real-time monitoring of several aspects of microbial physiology and gene expression *in situ*. In practice, a DNA fragment encompassing the promoter region of a candidate gene is fused to a promoterless reporter AFP gene, such as the GFP. This fusion construct is then introduced into the wild-type strain, either on a plasmid or on the chromosome, then the expression of GFP is assayed either in a microplate reader or under fluorescent microscope. This provides a tremendous insight into different aspects of biotic interactions. Furthermore, the integration and use of AFP-based reporter constructs with other markers and technologies is facilitating a systems biology approach to study microbial ecology [78]. GFP-based reporter systems have also been used to analyse promoter activities and to localize candidate fungal genes in plants. Technologies are improving to adapt this system to generate GFP-tagged promoter or gene fusions at the genome scale. There is a considerable amount of literature available on use of these techniques in studying plant pathogen interplay including fungus-plant [76, 104] or fungus-fungus interactions [83].

RNA Interference or RNAi

RNA interference (RNAi) is a phenomenon in which small double-stranded RNA (dsRNA) molecules induce sequence-specific degradation of homologous single-stranded RNA (ssRNA) [28]. Initially, the dsRNA is digested into 21-23 nucleotide fragments called small interfering RNA (siRNA). The siRNAs bind to a nuclease complex to form the RNA-induced silencing complex (RISC) generating an immobilized 21-23 nucleotide ssRNA. The active RISC complex bound RNA then targets the complimentary mRNA and cleaves the target mRNA ~12 nucleotides from the 3'-terminus of the siRNA [51]. The silencing of endogenous gene expression through what is now known as RNAi was first observed in plants [61], when it was called RNA silencing. Later a similar phenomenon was reported in plant resistance to viruses [29, 50, 105]. These phenomena are termed co-suppression, post-transcriptional gene silencing (PTGS), or viral-induced gene silencing (VIGS). Fire *et al.* [32] showed that injection of dsRNA into *C. elegans* was ten-fold more potent in silencing target gene expression than either sense or anti-sense RNA alone and coined the term, RNA interference (RNAi). A genome-wide RNAi project, based on transient expression, was undertaken in *C. elegans* that resulted in description of the phenotype of 10% of the 16,757 genes [5, 40, 64]. There is great potential to apply RNAi-mediated gene silencing technology in the study of plant pathogenic fungi, since RNA-mediated gene silencing has been observed in many fungi, including the filamentous fungus *N. crassa* [20, 114]. Antisense RNA-based methods have been successfully used to control gene expression in many filamentous fungi including *N. crassa, A. oryzae, C. albicans, Cryptococcus neoformans* and *M. grisea* [25, 45, 46, 62, 71, 133]. De Backer *et al.* [25] used a similar antisense-based approach to identify genes critical for growth in the pathogenic yeast *C. albicans*. Recently, using RNAi approach, a cluster of NIF transcriptional activators involved in spore formation in *Phythophthora infestans* has been identified [116].

RNAi-based systems are particularly useful when traditional gene disruption techniques do not work, mainly due to functional redundancy as in cases of paralogous gene families or gene duplications. Since RNAi-mediated gene silencing is a homology-based system, it can be used at the level of gene families or could be used for characterisation at the individual gene level by generating mutants with different gene expression levels as a result of intrinsic variation in the efficacy of interference. In addition, it is possible to target many unrelated genes, allowing epistasis analysis [22]. The ability to create targeted loss-of-function phenotypes by RNAi will greatly expedite the identification of gene function in plant–fungus interactions in the years ahead.

Proteomics

The proteome is defined as the entire protein complement of a genome of a given organism [47]. Analysis of the proteome offers the possibility of identifying protein abundance, localisation, interactions and post-translational modifications. Proteomics includes various approaches that integrate protein separation and characterization. An example is the use of 2D-gel electropheresis of protein with mass spectrometry to generate a catalogue of expressed proteins [69] combined with other analytical techniques, such as mapping protein-protein interactions using yeast two-hybrid (Y2H) screening [127].

The Y2H method aims at identification of physical interactions between proteins by co-expression of gene constructs within a single cell of *S. cerevisiae.* The method was first developed by Fields and Song [31] and used the modular nature of the GAL4 transcriptional activator protein. Two chimeric proteins were produced within the same cell, one was the DNA binding domain of GAL4 fused with Protein A and the other GAL4-activation domain fused with Protein B. The transcription of a given reporter gene is activated only if the Protein A and Protein B physically interact. In a comprehensive study of this kind in *S. cerevisiae*, Uetz *et al.* [124] screened all the 6000 predicted genes in the genome of *S. cerevisiae* and identified 692 interacting pairs. The Y2H method was also applied to *M. grisea* to study signalling pathways regulating appressorium development during infection of rice [74]. Fang *et al.* [30] used a high-throughput Y2H-based system to identify protein-protein interaction maps of the signal transduction pathways in defence reaction of rice. Because of the availability of both the host and pathogen in *M. grisea*-rice system, large-scale Y2H screening studies could be undertaken to identify components of the 'interactome' between rice and the rice blast fungus. One key consideration has to be kept in mind while interpreting Y2H results, the number of false positives can be remarkably high. Thus, protein-protein interactions mapped by Y2H screening methods need to be verified by other techniques, such as protein cross-linking, co-immuno-precipitation, or affinity-column purification. In addition, this method is not suitable for dealing with membrane proteins.

Two-dimensional polyacrylamide gel electrophoresis (2D-PAGE) coupled with tandem mass spectrometry is, by far, the most widely used technique in proteomics research. One of the drawbacks of 2D-PAGE is that integral membrane proteins, low abundance proteins and small proteins cannot be separated by this method. Also, proteins with high molecular weight and extreme isoelectric points are not resolved well by 2D-PAGE. However, common post-translational modifications, such as phosphorylation and glycosylation, can be reliably detected.

Considerable advances in proteomic approaches were made possible by integration of two-dimensional gel electrophoresis with matrix-assisted laser desorption/-ionisation-time of flight-mass spectrometry (MALDI-TOF-MS). This technique enables high-throughput and accurate characterization of proteins [69, 95]. Briefly, a protein in a spot from 2D-PAGE is subjected to trypsin digestion and, then digested peptide fragments are analysed by MALDI-TOF-MS. The observed peptide masses are used to search database of predicted masses derived from translation of genome sequence data of a given organism. Proteomic analysis enables the identification of proteins that are differentially produced during plant-fungus interactions and disease generation. Such proteins could serve as potential markers for disease and at the same time could be used as candidates for developing therapeutics in plant diseases.

There are several published studies on 2D-PAGE-based proteomic approaches in pathogenic fungi *viz.* to study morphogenesis of yeast to filamentous form and to determine multidrug resistance in *C. albicans* [75, 93], to identify GPI-anchored proteins in *A. fumigatus* [13], to study conidial germination, germ-tube growth and appressoria formation in *M. grisea* [70], and to determine mutualistic (symbiotic) association in arbuscular mycorrhizal fungi [53]. These studies have shown the potential of applying proteomics approaches in understanding pathogenesis in pathogenic and non-pathogenic fungi.

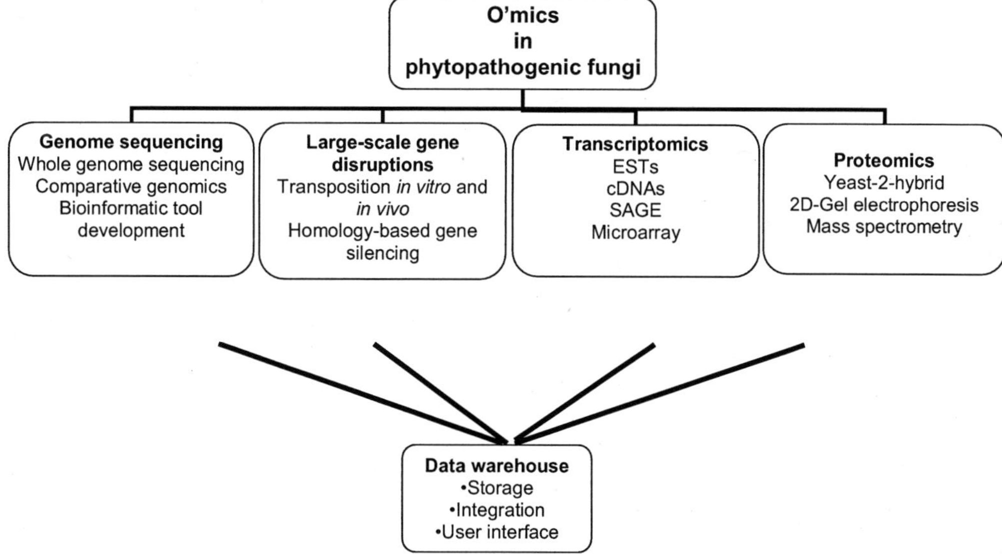

Figure 2. Schematics of the O'mics approaches to the study of phytopathogenic fungi.

PERSPECTIVE

Studies on host-pathogen interactions started with landmark papers of Flor in middle of 20th Century. Application of forward genetic approaches, which once led to the gene-for-gene hypothesis, has now globally evolved through the use of genomic tools. As large amounts of DNA sequence data accumulate in public databases and increasingly powerful computer software is developed for sequence analysis, the biological functions of many of the pathogen and host encoded genes are beginning to emerge. In addition, the availability of the genome sequences of many hosts and their pathogens allow examination of the gene regulatory networks that exist in the hosts and the pathogens and how interplay between them determine the outcome of an interaction. Models are already in place for *Arabidopsis* - fungal systems and in the foreseeable future these models could be extended to test other plant-pathogen systems such as potato and *Brassica* and their associated pathogens as the genome sequencing of these hosts and the pathogens are underway (www.genomesonline.org). Rice and *M. grisea* genomes are now available for studying host-pathogen interaction at the genome level using the systems biology approaches. Large-scale insertional mutagenesis, transcriptional profiling, and proteomic analyses are in progress using the rice - *M. grisea* system to study fungal infection vis-à-vis host reaction. Comparative analysis of the genome sequence of fungal pathogens from diverse taxa will help identify genes unique to each host-pathogen interactions as well as genes common to pathogens across taxa. For example, comparative analysis of *F. oxysporum* (Deuteromycotina; perfect stage *Gibberella*: Ascomycota; Sordariomycetes) with *M. grisea* (Ascomycota; Sordariomycetes) or with the corn pathogen *C. carbonum* (Ascomycota; Dothideomycetes) by reverse genetics and insertional mutagenesis is offering fresh insights into attributes required for this soil fungus to be a successful pathogen. Availability of genome sequences from plant pathogenic and other fungi will therefore help to identify new antifungal drug targets. This will facilitate the design of new ways to control fungal pathogens of plants.

The wealth of genomic information that has accumulated since the sequencing of *H. influenzae* in middle 1990s [35] is enormous and growing at an unprecedented rate. Discussing the findings of each of these papers is far beyond the scope of this review. We, therefore, present a snap shot of the potential of genomics for elucidating the molecular basis of plant-fungus interactions with schematics of the o'mics approaches to the study of phytopathogenic fungi (Figure 2). Using the growing arsenal of available sequence data, combined with bioinformatics and functional genomics, should lead to a greater understanding of the genetic networks activated during a pathogenic fungus's interaction with its plant host, in pathogen virulence, and host resistance. This revolution in genomics is allowing dissection of the classical gene-for-gene hypothesis at the molecular level. Merging of forward genetics approaches with the genomics information for the study of host-pathogen interactions will pave the way towards a global understanding of plant and pathogen biology. This will make major strides in further increasing agricultural production by improving plant health in the years ahead.

REFERENCES

[1] Adachi K, Nelson, GH, Peoples KA, Frank SA, Montenegro-Chamorro MV, DeZwaan TM, Ramamurthy L, Shushter JR, Hamer L, Tanzer MM. 2002. Efficient gene identification and targeted gene disruption in the wheat blotch fungus *Mycosphaerella graminicola* using TAGKO. *Current Genetics 42*: 123-127

[2] Adams MD, Kelly JM, Gocayne JD, Dubnick M, Polymeropoulos MH, Xiao H, Merril CR, Wu A, Olde B, Moreno RF, Kerlavage AR, McCombie WR, Venter JC. 1991. Complementary DNA sequencing: "expressed sequence tags" and the human genome project. *Science 252*: 1651-1656

[3] Allard RW. 1991. The Genetics of host-pathogen coevolution: implications for genetic resource conservation. *Journal Heredity 81*: 1-6

[4] Armstrong MR, Whisson SC, Pritchard L, Bos JIB, Venter E, Avrova AO, Rehmany AP, Böhme U, Brooks K, Cherevach I, Hamlin N, White B, Fraser A, Lord A, Quail MA, Churcher C, Hall N, Berriman M, Huang S, Kamoun S, Beynon JL, Birch PRJ.

2005. An ancestral oomycete locus contains late blight avirulence gene *Avr3a*, encoding a protein that is recognised in the host cytoplasm. *Proc. Natl. Acad. Sci. USA 102*: 7766-7771

[5] Ashrafi K, Chang FY, Watts JL, Fraser AG, Kamath RS, Ahringer J, Ruvkun G. 2003. Genome-wide RNAi analysis of *Caenorhabditis elegans* fat regulatory genes. *Nature 421*: 268-272

[6] Aubertot JN, West JS, Bousset-Vaslin L, Salam MU, Barbetti MJ, Digle AJ. 2006. Improved resistance management for durable disease control: A case study of Phoma stem canker of oilseed rape (*Brassica napus*). *European Journal of Plant Pathology 114*: 91-106

[7] Baldwin D, Crane V, Rice D. 1999. A comparison of gel-based, nylon filter and microarray techniques to detect differential RNA expression in plants. *Current Opinion Plant Biology 2*: 96-103

[8] Birren B, Fink G, Lander E. 2002. *Fungal Genome Initiative*. White paper, Whitehead Institute Centre for Genome Research, Cambridge, MA 02141, USA

[9] Blattner FR, Plunkett G, Bloch CA, Perna NT, Burland V, Riley M, Collado-Vides J, Glasner JD, Rode CK, Mayhew GF, Gregor J, Davis NW, Kirkpatrick HA, Goeden MA, Rose DJ, Mau B, Shao Y. 1997. The complete genome sequence of *Escherichia coli* K-12. *Science 277*: 1453-1474

[10] Both M, Eckert SE, Csukai M, Muller E, Dimopoulos G, Spanu PD. 2005. Transcript profiles of *Blumeria graminis* development during infection reveal a cluster of genes that are potential virulence determinants. *Molecular Plant Microbe Interactions 18*: 125-133

[11] Braun BR, van Het Hoog M, d'Enfert C, Martchenko M, Dungan J, Kuo A, Inglis DO, Uhl MA, Hogues H, Berriman M, Lorenz M, Levitin A, Oberholzer U, Bachewich C, Harcus D, Marcil A, Dignard D, Iouk T, Zito R, Frangeul L, Tekaia F, Rutherford K, Wang E, Munro CA, Bates S, Gow NA, Hoyer LL, KÃhler G, MorschhÃuser J, Newport G, Znaidi S, Raymond M, Turcotte B, Sherlock G, Costanzo M, Ihmels J, Berman J, Sanglard D, Agabian N, Mitchell AP, Johnson AD, Whiteway M, Nantel A. 2005. A human-curated annotation of the *Candida albicans* genome. *PLoS Genet. 1*: 36-57

[12] Brown JKM, Jessop AC, Rezanoor HN. 1991. Genetic uniformity in barley and its powdery mildew pathogen. *Proc. Royal Soc. B246*: 83-90

[13] Bruneau JM, Magnin T, Tagat E, Legrand R, Bernard M, Diaquin M, Fudali C, Latge JP. 2001. Proteome analysis of *Aspergillus fumigatus* identifies glycosylphosphatidylinositol-anchored proteins associated to the cell wall biosynthesis. *Electrophoresis 22*: 2812-2823

[14] Burdon JJ. 1987. *Diseases and Plant Population Biology*. Cambridge University Press, Cambridge, UK

[15] Burdon JJ. 1987. Phenotypic and genetic patterns of resistance to the pathogen *Phakopsora pachyrhizi* in populations of *Glycine canescens*. *Oecologia 73*: 257-267

[16] Burdon JJ, Brown AHD, Jarosz AM. 1990. In *Pests, Pathogens and Plant Communities*. eds. JJ Burdon, SR Leather, Blackwell Publ., Oxford, UK, pp. 233-247

[17] Burdon JJ, Jarosz AM. 1991. Host-pathogen interactions in natural populations of *Linum marginale* and *Melampsora lini*. I. Patterns of resistance and racial variation in a large host population. *Evolution 45*: 205-217

[18] Burdon JJ, Jarosz AM. 1992. Temporal variation in the racial structure of flax rust (*Melampsora lini)* populations growing on natural stands of wild flax (*Linum marginale)*: Local versus metapopulation dynamics. *Plant Pathology 41*: 165-179

[19] Christ BJ, Person CO, Pope DD. 1987. In *Populations of Plant Pathogens: Their Dynamics and Genetics*. eds. MS Wolfe, CE Carten, Blackwell Publ., Oxford, UK, pp. 7-19

[20] Cogoni C. 2001. Homology-dependent gene silencing mechanisms in fungi. *Annual Review of Microbiology 55*: 381-406

[21] Colebatch G, Trevaskis B, Udvardi M. 2002. Functional genomics: tools of the trade. *New Phytologist 153*: 27-36

[22] Cottrell TR, Doering TL. 2003. Silence of the strands: RNA interference in eukaryotic pathogens. *Trends in Microbiology 11*: 37-43

[23] Crute IR, Phelps K, Barnes A, Buczacki ST, Crisp P. 1983. The relationship between genotypes of three *Brassica* species and collections of *Plasmodiophora brassicae*. *Plant Pathology 32*: 405-420

[24] De Backer MD, Ilyina T, Ma XJ, Vandoninck S, Luyten WHM, van den Bosche H. 2001. Genomic profiling of the response of *Candida albicans* to itraconazole treatment using a DNA microarray. *Antimicrob. Agents Chemotherapy 45*:1660-1670

[25] De Backer MD, Nelissen B, Logghe M, Viaene J, Loonen I, Vandoninck S, de Hoogt R, Dewaele S, Simons FA, Verhasselt P, Vanhoof G, Contreras R, Luyten WHM. 2001. An antisense-based functional genomics approach for identification of genes critical for growth of *Candida albicans. Nature Biotechnology 19*: 235-241

[26] Dean RA, Talbot NJ, Ebbole DJ, Farman ML, Mitchell TK, Orbach MJ, Thon M, Kulkarni R, Xu JR, Pan H, Read ND, Lee YH, Carbone I, Brown D, Oh YY, Donofrio N, Jeong JS, Soanes DM, Djonovic S, Kolomiets E, Rehmeyer C, Li W, Harding M, Kim S, Lebrun MH, Bohnert H, Coughlan S, Butler J, Calvo S, Ma LJ, Nicol R, Purcell S, Nusbaum C, Galagan JE, Birren BW. 2005. The genome sequence of the rice blast pathogen *Magnaporthe grisea. Nature 434*: 980-986

[27] Dietrich FS, Voegeli S, Brachat S, Lerch A, Gates K, Steiner S, Mohr C, Pöhlmann R, Luedi P, Choi S, Wing RA, Flavier A, Gaffney TD, Philippsen P. 2004. The *Ashbya gossypii* genome as a tool for mapping the ancient *Saccharomyces cerevisiae* genome. *Science 304*: 304-307

[28] Dykxhoorn DM, Novina CD, Sharp PA. 2003. Killing the messenger: short RNAs that silence gene expression. *Nature Review Molecular Cell Biology 4*: 457-467

[29] English JJ, Mueller E, Baulcombe DC. 1996. Suppression of virus accumulation in transgenic plants exhibiting gene silencing of nuclear genes. *Plant Cell 8*: 179-188

[30] Fang Y, Macool DJ, Xue Z, Heppard EP, Hainey CF, Tingey SV, Miao GH. 2002. Development of a high-throughput yeast two-hybrid screening system to study protein-protein interactions in plants. *Molecular Genetics Genomics 267*: 142-153

[31] Fields S, Song O. 1989. A novel genetic system to detect protein-protein interactions. *Nature 340*: 245-246

[32] Fire A, Xu SQ, Montogomery MK, Kostas SA, Driver SE, Mello CC. 1998. Potent and specific genetic interference by double-stranded RNA in *Caenorhabditis elegans*. *Nature 391*: 806-811

[33] Firnhaber C, Puhler A, Kuster H. 2005. EST sequencing and time course microarray hybridizations identify more than 700 *Medicago truncatula* genes with developmental expression regulation in flowers and pods. *Planta 222*: 269-283

[34] Firon A, Villalba F, Beffa R, d'Enfert C. 2003. Identification of essential genes in the human fungal pathogen *Aspergillus fumigatus* by transposon mutagenesis. *Eukaryotic Cell 2*: 247-255

[35] Fleischmann RD, Adams MD, White O, Clayton RA, Kirkness EF, Kerlavage AR, Bult CJ, Tomb JF, Dougherty BA, Merrick JM, McKenney K, Sutton G, FitzHugh W, Fields C, Gocayne JD, Scott J, Shirley R, Liu LI, Glodek A, Kelley JM, Weidman JF, Phillips CA, Spriggs T, Hedblom E, Cotton MD, Utterback TR, Hanna MC, Nguyen DT, Saudek DM, Brandon RC, Fine LD, Fritchman JL, Fuhrmann JL, Geoghagen NSM, Gnehm CL, McDonald LA, Small KV, Fraser CM, Smith HO, Venter JC. 1995. Whole-genome random sequencing and assembly of *Haemophilus influenzae* Rd. *Science 269*: 496-512

[36] Flor HH. 1942. Inheritance of pathogenicity in *Melampsora lini*. *Phytopathology 32*: 653-669

[37] Flor HH. 1953. Epidemiology of flax rust in the North Central States. *Phytopathology 43*: 624-628

[38] Flor HH. 1956. The complementary genetic systems in flax and flax rust. *Adv. Genet. 8*: 29-54

[39] Flor HH. 1971. Current status of the gene-for-gene concept. *Ann. Rev. Phytopathol. 9*: 275-296

[40] Fraser AG, Kamath RS, Zipperlen P, Martinez-Campos M, Sohrmann M, Ahringer J. 2000. Functional genomic analysis of *Caenorhabditis elegans* chromosome I by systematic RNA interference. *Nature 408*: 325-330

[41] Galagan JE, Calvo SE, Borkovich KA, Selker EU, Read ND, Jaffe D, FitzHugh W, Ma LJ, Smirnov S, Purcell S, Rehman B, Elkins T, Engels R, Wang S, Nielsen CB, Butler J, Endrizzi M, Qui D, Ianakiev P, Bell-Pedersen D, Nelson MA, Werner-Washburne M, Selitrennikoff CP, Kinsey JA, Braun EL, Zelter A, Schulte U, Kothe GO, Jedd G, Mewes W, Staben C, Marcotte E, Greenberg D, Roy A, Foley K, Naylor J, Stange-Thomann N, Barrett R, Gnerre S, Kamal M, Kamvysselis M, Mauceli E, Bielke C, Rudd S, Frishman D, Krystofova S, Rasmussen C, Metzenberg RL, Perkins DD, Kroken S, Cogoni C, Macino G, Catcheside D, Li W, Pratt RJ, Osmani SA, DeSouza CP, Glass L, Orbach MJ, Berglund JA, Voelker R, Yarden O, Plamann M, Seiler S, Dunlap J, Radford A, Aramayo R, Natvig DO, Alex LA, Mannhaupt G, Ebbole DJ, Freitag M, Paulsen I, Sachs MS, Lander ES, Nusbaum C, Birren B. 2003. The genome sequence of the filamentous fungus *Neurospora crassa*. *Nature 422*: 859-868.

[42] Galagan JE, Calvo SE, Cuomo C, Ma LJ, Wortman JR, Batzoglou S, Lee SI, Basturkmen M, Spevak CC, Clutterbuck J, Kapitonov V, Jurka J, Scazzocchio C, Farman M, Butler J, Purcell S, Harris S, Braus GH, Draht O, Busch S, D'Enfert C, Bouchier C, Goldman GH, Bell-Pedersen D, Griffiths-Jones S, Doonan JH, Yu J,

Vienken K, Pain A, Freitag M, Selker EU, Archer DB, Penalva MA, Oakley BR, Momany M, Tanaka T, Kumagai T, Asai K, Machida M, Nierman WC, Denning DW, Caddick M, Hynes M, Paoletti M, Fischer R, Miller B, Dyer P, Sachs MS, Osmani SA, Birren BW. 2005. Sequencing of *Aspergillus nidulans* and comparative analysis with *A. fumigatus* and *A. oryzae*. *Nature 438*: 1105-15

[43] Geiger HH, Schumacher AE, Billenkamp N. 1988. Frequencies of vertical resistances and virulences in the rye-powdery mildew pathosystem. *Plant Breeding 100*: 97-103

[44] Goffeau A, Barrell BG, Bussey H, Davis RW, Dujon B, Feldmann H, Galibert F, Hoheisel JD, Jacq C, Johnston M, Luis EJ, Mewes HW, Murakami Y, Philippsen P, Tettelin H, Oliver SG. 1996. Life with 6000 genes. *Science 274*: 546-567

[45] Goldoni S, Owens RT, McQuillan DJ, Shriver Z, Sasisekharan R, Birk DE, Campbell S, Renato V, Iozzo RV. 2004. Biologically active decorin is a monomer in solution. *J. Biol. Chem. 279*: 6606-6612

[46] Gorlach JM, McDade HC, Perfect JR, Cox GM. 2002. Antisense repression in *Cryptococcus neoformans* as a laboratory tool and potential antifungal strategy. *Microbiology 148*: 213-219

[47] Graves PR, Haystead TAJ. 2002. Molecular biologist's guide to proteomics. *Microbiol. Mol. Biol. Rev. 66*: 39-63

[48] Guimil S, Chang HS, Zhu T, Sesma A, Osbourn A, Roux C, Ioannidis V, Oakeley EJ, Docquier M, Descombes P, Briggs SP, Paszkowski U. 2005. Comparative transcriptomics of rice reveals an ancient pattern of response to microbial colonization. *Proc. Natl. Acad. Sci. USA 102*: 8066 – 8070

[49] Gygi SP, Rochon Y, Franza BR, Aebersold R. 1999. Correlation between protein and mRNA abundance in yeast. *Molecular Cell Biology 19*: 1720-1730

[50] Hamilton AJ, Baulcombe DC. 1999. A species of small antisense RNA in posttranscriptional gene silencing. *Science 286*: 950-952

[51] Hammond SM, Caudy AA, Hannon GJ. 2001. Post-transcriptional gene silencing by double-stranded RNA. *Nature Review Genetics 2*: 110-119

[52] Hamer L, Adachi K, Montenegro-Chamorro MV, Tanzer MM, Mahanty SK, Lo C, Tarpey RW, Skalchunes AR, Heiniger RW, Frank SA, Darveaux BA, Lampe, DJ, Slater TM, Ramamurthy L, DeZwaan TM, Nelson GH, Shuster JR, Woessner J, Hamer JE 2001. Gene discovery and gene function assignment in filamentous fungi. *Proc. Natl. Acad. Sci. USA 98*: 5110 – 5115

[53] Harrier L.A. 2001. The arbuscular mycorrhizal symbiosis: a molecular review of the fungal dimension. *J. Exp. Bot. 52*: 469-478

[54] Harry I, Clarke D. 1987. The genetics of race-specific resistance in groundsel (*Senecio vulgaris* L.) to the powdery mildew fungus, *Erysiphe fischeri* Blumer. *New Phytologist 107*: 715-723

[55] Hensel M, Shea JE, Gleeson C, Jones MD, Dalton E, Holden DW. 1995. Simultaneous identification of bacterial virulence genes by negative selection. *Science 269*: 400-403

[56] Irie T, Matsumura H, Terauchi R, Saitoh H. 2003. Serial analysis of gene expression (SAGE) of *Magnaporthe grisea*: genes involved in appressorium formation. *Mol. Gen. Genomics 270*: 181-189

[57] Jansen R, Gerstein M. 2000. Analysis of the yeast transcriptome with structural and functional categories: characterizing highly expressed proteins. *Nucleic Acid Research 28*: 1481-1488

[58] Jantasuriyarat C, Gowda M, Haller K, Hatfield K, Lu G, Stahlberg E, Zhou B, Li H, Kim H, Yu Y, Dean RA, Wing RA, Soderland C, Wang GL. 2005. Large-scale identification of expressed sequence tags involved in rice and rice blast fungus interaction. *Plant Physiology 138*: 105-115

[59] Jarosz AM, Burdon JJ. 1990. Predominance of a single major gene for resistance to *Phakospora pachyrhizi* in a population of *Glycine argyrea. Heredity 64*: 347-353

[60] Jarosz AM, Burdon JJ. 1991. Host-pathogen interactions in natural populations of *Linum marginale* and *Melampsora lini*: II. Local and regional variation in patterns of resistance and racial structure. *Evolution 45*: 1618-1627

[61] Jorgensen R. 1990. Altered gene expression in plants due to trans interactions between homologous genes. *Trend Biotechnology 8*: 340-344

[62] Kadotani N, Nakayashiki H, Tosa Y, Mayama S. 2003. RNA silencing in the phytopathogenic fungus *Magnaporthe oryzae. Mol Plant Microb Interact 16*: 769-776

[63] Kal AJ, van Zonneveld AJ, Benes V, van den Berg M, Koerkamp M, Albermann K, Strack N, Ruijter JM, Richter A, Dujon B, Ansorge W, Tabak HF. 1999. Dynamics of gene expression revealed by comparison of serial analysis of gene expression transcript profiles from yeast grown on two different carbon sources. *Molecular Cell Biology 10*: 1859-1872

[64] Kamath RS, Fraser AG, Dong Y, Poulin G, Durbin R, Gotta M, Kanapin A, Bot NL, Moreno S, Sohrmann M, Welchman DP, Zipperlen P, Ahringer J. 2003. Systematic functional analysis of the *Caenorhabditis elegans* genome using RNAi. *Nature 421*: 231-237

[65] Katinka MD, Duprat S, Cornillot E, Metenier G, Thomarat F, Prensier G, Barbe V, Peyretaillade E, Brottier P, Wincker P, Delbac F, El Alaoui H, Peyret P, Saurin W, Gouy M, Weissenbach J, Vivares CP. 2001. Genome sequence and gene compaction of the eukaryote parasite *Encephalitozoon cuniculi. Nature 414*: 401-402

[66] Kato-Maeda M, Gao Q, Small PM. 2001. Microarray analysis of pathogens and their interaction with hosts. *Cell. Microbiol. 3*: 713-719

[67] Kazan K, Schenk PM, Wilson I, Manners JM. 2001. DNA microarrays: new tools in the analysis of plant defence response. *Molecular Plant Pathology 2*: 177-185

[68] Kellis M, Patterson N, Endrizzi M, Birren B, Lander ES. 2003. Sequencing and comparison of yeast species to identify genes and regulatory elements. *Nature 423*: 233-234

[69] Kersten B, Bürkle L, Kuhn EJ, Giavalisco P, Konthur Z, Leuking A, Walter G, Eickoff H, Schneider U. 2002. Large-scale plant proteomics. *Plant Molecular Biology 48*: 133-141

[70] Kim ST, Yu S, Kim SG, Kim HJ, Kang SY, Hwang DH, Jang YS, Kang KY. 2004. Proteome analysis of rice blast fungus (*Magnaporthe grisea*) proteome during appressorium formation. *Proteomics 4*: 3579-3587

[71] Kitamoto N, Yoshino S, Ohmiya K, Tsukagoshi N. 1999. Sequence analysis, overexpression, and antisense inhibition of a beta-xylosidase gene, xylA, from *Aspergillus oryzae* KBN616. *Appl. Environ. Microbiol. 65*: 20-24

[72] Kiyosawa SA. 1982. Genetics and epidemiological modeling of breakdown of plant disease resistance. *Ann. Rev. Phytopathol. 20*: 93-117

[73] Kuhn ML, Gout L, Howlett BJ, Melayah D, Meyer M, Bolesdent MH, Rouxel T. 2006. Genetic linkage maps and genomic organization in *Leptosphaeria maculans. European Journal of Plant Pathology 114*: 17-31

[74] Kulkarni RD, Dean RA. 2004. Identification of proteins that interact with two regulators of appressorium development, adenylate cyclase and cAMP-dependent protein kinase A, in the rice blast fungus *Magnaporthe grisea. Mol. Genet. Genomics 270*: 497-508

[75] Kusch H, Biswas K, Schwanfelder S, Engelmann S, Rogers PD, Hecker M, Morschhäuser J. 2004. A proteomic approach to understanding the development of multidrug-resistant *Candida albicans* strains. *Mol. Genet. Genomics 271*: 554-565

[76] Lagopodi AL, Ram AFJ, Lamers GE, Punt PJ, Van den Hondel CAMJJ, Lugtenberg BJJ, Bloemberg GV. 2002. Novel aspects of tomato root colonization and infection by *Fusarium oxysporum* f. sp. *radicis-lycopersici* revealed by confocal laser scanning microscopic analysis using the green fluorescent protein as a marker. *Mol. Plant-Microbe Interact. 15*:172-179

[77] Langin T, Capy P, Daboussi MJ. 1995. The transposable element *impala,* a fungal member of the *Tc1-mariner* superfamily. *Mol. Gen. Genet 246*: 19-28

[78] Larrainzar E, O'Gara F, Morrissey JP. 2005. Applications of autofluorescent proteins for *in situ* studies in microbial ecology. *Ann. Rev. Microbiol. 59*: 257-277

[79] Layton AC, Kuhn DN. 1988. The virulence of interracial heterokaryons of *Phytophthora megasperma* f. sp. *glycinea*. *Phytopathology 78*: 961-966

[80] Leung YF, Cavalieri D. 2003. Fundamentals of cDNA microarray analysis. *Trends Genet. 19*: 649-659

[81] Lo C, Adachi K, Shuster JR, Hamer JE, Hamer L. 2003. The bacterial transposon Tn7 causes premature polyadenylation of mRNA in eukaryotic organisms: TAGKO mutagenesis in filamentous fungi. *Nucleic Acids Res. 31*: 4822-4827

[82] Loftus B, Anderson I, Davies R, Alsmark UC, Samuelson J, Amedeo P, Roncaglia P, Berriman M, Hirt RP, Mann BJ, Nozaki T, Suh B, Pop M, Duchene M, Ackers J, Tannich E, Leippe M, Hofer M, Bruchhaus I, Willhoeft U, Bhattacharya A, Chillingworth T, Churcher C, Hance Z, Harris B, Harris D, Jagels K, Moule S, Mungall K, Ormond D, Squares R, Whitehead S, Quail MA, Rabbinowitsch E, Norbertczak H, Price C, Wang Z, Guillen N, Gilchrist C, Stroup SE, Bhattacharya S, Lohia A, Foster PG, Sicheritz-Ponten T, Weber C, Singh U, Mukherjee C, El-Sayed NM, Petri WA Jr, Clark CG, Embley TM, Barrell B, Fraser CM, Hall N. 2005. The genome of the protist parasite *Entamoeba histolytica. Nature 433*: 865-868

[83] Lu Z, Tombolini R, Woo S, Zeilinger S, Lorito M, Jansson JK. 2004. *In vivo* study of *Trichoderma*-pathogen-plant interactions, using constitutive and inducible green fluorescent protein reporter systems. *Appl. Environ. Microbiol. 70*: 3073-3081

[84] Machida M, Asai K, Sano M, Tanaka T, Kumagai T, Terai G, Kusumoto K, Arima T, Akita O, Kashiwagi Y, Abe K, Gomi K, Horiuchi H, Kitamoto K, Kobayashi T, Takeuchi M, Denning DW, Galagan JE, Nierman WC, Yu J, Archer DB, Bennett JW, Bhatnagar D, Cleveland TE, Fedorova ND, Gotoh O, Horikawa H, Hosoyama A, Ichinomiya M, Igarashi R, Iwashita K, Juvvadi PR, Kato M, Kato Y, Kin T, Kokubun A, Maeda H, Maeyama N, Maruyama J, Nagasaki H, Nakajima T, Oda K, Okada K, Paulsen I, Sakamoto K, Sawano T, Takahashi M, Takase K, Terabayashi Y, Wortman JR, Yamada O, Yamagata Y, Anazawa H, Hata Y, Koide Y, Komori T, Koyama Y, Minetoki T, Suharnan S, Tanaka A, Isono K, Kuhara S, Ogasawara N, Kikuchi H. 2005. Genome sequencing and analysis of *Aspergillus oryzae. Nature 438*: 1157-1161

[85] Martinez D, Larrondo LF, Putnam N, Gelpke MDS, Huang K, Chapman J, Helfenbein KG, Ramaiya P, Detter JC, Larimer F, Coutinho PM, Henrissat B, Berka R, Cullen D, Rokhsar D. 2004. Genome sequence of the lignocellulose degrading fungus *Phanerochaete chrysosporium* strain RP78. *Nature Biotechnology 22*: 695-699

[86] Matsumura H, Reich S, Ito A, Saitoh H, Kamoun S Winter P, Kahl G, Reuter M, Krüger DH, Terauchi R. 2003. Gene expression analysis of plant host-pathogen interactions by SuperSAGE. *Proc. Natl Acad. Sci. USA 100*: 15718-15723

[87] Mullins ED, Kang S. 2001. Transformation: a tool for studying fungal pathogens of plants. *Cell. Mol. Life Sci. 58*: 2043-2052

[88] Mullins ED, Chen X, Romaine P, Raina R, Geiser DM, Kang S. 2001. *Agrobacterium*-mediated transformation of *Fusarium oxysporum*: an efficient tool for insertional mutagenesis and gene transfer. *Phytopathology 91*: 173-180

[89] Murad AMA, d'Enfert C, Claude Gaillardin C, Tournu H, Tekaia F, Talibi D, Marechal D, Marchais V, Cottin J, Brown AJP. 2001. Transcript profiling in *Candida albicans* reveals new cellular functions for the transcriptional repressors CaTup1, CaMig1 and CaNrg1. *Mol. Microbiol. 42*: 981-993

[90] Nadimpalli R, Yalpani N, Johal GS, Simmons CR. 2000. Prohibitins, stomatins and plant disease response genes compose a protein superfamily that controls cell proliferation, ion channel regulation and death. *Journal Biological Chemistry 275*: 29579-29586

[91] Nehls U. 2003. Ectomycorrhizal development and function – transcriptome analysis. *New Phytologist 159*: 5-7

[92] Nierman WC, Pain A, Anderson MJ, Wortman JR, Kim HS, Arroyo J, Berriman M, Abe K, Archer DB, Bermejo C, Bennett J, Bowyer P, Chen D, Collins M, Coulsen R, Davies R, Dyer PS, Farman M, Fedorova N, Fedorova N, Feldblyum TV, Fischer R, Fosker N, Fraser A, Garcia JL, Garcia MJ, Goble A, Goldman GH, Gomi K, Griffith-Jones S, Gwilliam R, Haas B, Haas H, Harris D, Horiuchi H, Huang J, Humphray S, Jimenez J, Keller N, Khouri H, Kitamoto K, Kobayashi T, Konzack S, Kulkarni R, Kumagai T, Lafton A, Latge JP, Li W, Lord A, Lu C, Majoros WH, May GS, Miller BL, Mohamoud Y, Molina M, Monod M, Mouyna I, Mulligan S, Murphy L, O'Neil S, Paulsen I, Penalva MA, Pertea M, Price C, Pritchard BL, Quail MA, Rabbinowitsch E, Rawlins N, Rajandream MA, Reichard U, Renauld H, Robson GD, Rodriguez de Cordoba S, Rodriguez-Pena JM, Ronning CM, Rutter S, Salzberg SL, Sanchez M, Sanchez-Ferrero JC, Saunders D, Seeger K, Squares R, Squares S, Takeuchi M, Tekaia

F, Turner G, Vazquez de Aldana CR, Weidman J, White O, Woodward J, Yu JH, Fraser C, Galagan JE, Asai K, Machida M, Hall N, Barrell B, Denning DW. 2005. Genomic sequence of the pathogenic and allergenic filamentous fungus *Aspergillus fumigatus*. *Nature 438*: 1151-1156

[93] Niimi M, Cannon R, Monk B. 1999. *Candida albicans* pathogenicity: a proteomic perspective. *Electrophoresis 20*: 2299-2308

[94] Nof E, Dinoor A. 1981. The manifestation of gene-for-gene relationships in oats and crown rust. *Phytoparasitica 9*: 240

[95] Nordoff E, Egelhofer V, Giavalisco P, Eickhoff H, Horn M, Przewieslik T, Theiss D, Schneider U, Lehrach H, Gobom J. 2001. Large-gel 2-DE-MALDI-TOF-MS – an analytical challenge for studying complex protein mixtures. *Electrophoresis 22*: 2844-2855

[96] Pain A. 2005. Fungi behaving badly. *Nature Reviews Microbiology 3*: 832-833

[97] Pain A, Bohme U, Berriman M. 2006. Hot and sexy moulds! *Nature Reviews Microbiology 4*: 244-246

[98] Parker MA. 1985. Local population differentiation for compatibility in an annual legume and its host-specific fungal pathogen. *Evolution 39*: 713-723

[99] Park SM, Kim DH. 2004. Transformation of a filamentous fungus *Cryphonectria parasitica* using *Agrobacterium tumefaciens*. *Biotechnology and Bioprocess Engineering 9*: 217-222

[100] Pawar MN, Frederiksen RA, Mughogho LK, Bonde MR. 1985. Survey of the virulence of *Peronosclerospora sorghi* isolates from Ethiopia, Nigeria, Texas (USA), Honduras, Brazil and Argentina. *Phytopathology 75*: 1374

[101] Persiel F, Lein H. 1989. Investigations on the resistance of China asters, *Callistephus chinensis*, to *Fusarium oxysporum* f. sp. *callistephi*. *Journal Plant Disease Protection 96*: 47-59

[102] Peter M, Courty PE, Kohler A, Delaruelle C, Martin D, Tagu D, Frey-Klett P, Duplessis S, Chalot M, Podila G, Matin F. 2003. Analysis of expressed sequence tags from the ectomycorrhizal basidiomycetes *Laccaria bicolor* and *Pisolithus microcarpus*. *New Phytologist 159*: 117-129

[103] Pietravalle S, Lemarie S, van den Bosch F. 2006. Durability of resistance and cost of virulence. *European Journal of Plant Pathology 114*: 107-116.

[104] Ramos HJO, Roncato-Maccari LDB, Souza EM, Soares-Ramos JRL, Hungria M, Pedrosa FO. 2002. Monitoring *Azospirillum*-wheat interactions using the gfp and gusA genes constitutively expressed from a new broad-host range vector. *J. Biotechnol. 97*: 243-252

[105] Ratcliff F, Harrison BD, Baulcombe DC. 1997. A similarity between viral defense and gene silencing in plants. *Science 276*: 1558-1560

[106] Rauyaree P, Choi W, Fang E, Blackmon B, Dan RA. 2001. Genes expressed during early stages of rice infection with the rice blast fungus *Magnaporthe grisea*. *Molecular Plant Pathology 2*: 347-354

[107] Rho HS, Kang S, Lee YH. 2001. *Agrobacterium tumefaciens*-mediated transformation of the plant pathogenic fungus, *Magnaporthe grisea*. *Mol. Cell. 12*: 407-411

[108] Schena M, Shalon D, Davis RW, Brown PO. 1995. Quantitative monitoring of gene expression patterns with a complementary DNA microarray. *Science 270*: 467-470

[109] Schena M, Heller RA, Theriault TP, Konrad K, Davis RW. 1998. Microarrays: biotechnology's discovery platform for functional genomics. *Trends Biotechnology 16*: 301-306

[110] Skinner DZ, Stuteville DL. 1985. Genetics of host-parasite interactions between alfalfa and *Peronospora trifoliorum*. *Phytopathology 75*: 119-121

[111] Soanes DM, Skinner, Keon J, Hargreaves J, Talbot NJ. 2002. Genomics of phytopathogenic fungi and development of bioinformatic resources. *Molecular Plant Microbe Interactions 15:* 421-427

[112] Soanes DM, Talbot NJ. 2006. Comparative genomic analysis of phytopathogenic fungi using expressed sequence tag (EST) collections. *Mol. Plant Pathology 7*: 61-70

[113] Stachowiak A, Olechnowicz J, Malgorzata J, Rouxel T, Balesdent MH, Happstadius I, Gladders P, Latunde-Dada A, Evans N. 2006. Frequency of avirulence alleles in field populations of *Leptosphaeria maculans* in Europe. *European Journal of Plant Pathology 114*: 67-75

[114] Sweigard JA, Ebbole DJ. 2001. Functional analysis of pathogenicity genes in a genomics world. *Curr Opin Microbiol 4*: 387-392

[115] Takano Y, Choi W, Mitchell TK, Okuno T, Dean RA. 2003. Large scale parallel analysis of gene expression during infection-related morphogenesis of *Magnaporthe grisea*. *Mol. Plant Pathol. 4*: 337-347.

[116] Tani S, Kim KS, Judelson HS. 2005. A cluster of NIF transcriptional regulators with divergent patterns of spore-specific expression in *Phytophthora infestans*. *Fungal Genetics Biology 42*: 42-50

[117] Thomas SW, Glaring MA, Rasmussen SW, Kinane JT, Oliver RP. 2002. Transcript profiling in the barley mildew pathogen *Blumeria graminis* by serial analysis of gene expression (SAGE). *Molecular Plant Microbe Interactions 15*: 847-856

[118] Thompson JN. 1989. Concepts of coevolution. *Trends Ecology Evolution 4*: 179-183

[119] Thompson JN, Burdon JJ. 1992. Gene-for-gene co-evolution between plants and parasites. *Nature 360*: 121-125

[120] Titarenko E, Lopez-Solanilla E, Garcia-Olmedo F, Rodriguez-Palenzuela P. 1997. Mutants of *Ralstonia (Pseudomonas) solanacearum* sensitive to antimicrobial peptides are altered in their lipopolysaccharide structure and are avirulent in tobacco. *J. Bacteriology 179*: 6699-6704

[121] Tunlid A. 2003. Exploring plant-microbe interactions using DNA microarrays. *New Phytologist 158*: 235-238

[122] Tunlid A, Talbot N. 2002. Genomics of parasitic and symbiotic fungi. *Curr. Opin. Microbiol. 5*: 513-519

[123] Tzung KW, Williams RM, Scherer S, Federspiel N, Jones T, Hansen N, Bivolarevic V, Huizar L, Komp C, Surzycki R, Tamse R, Davis RW, Agabian N. 2001. Genomic evidence for a complete sexual cycle in *Candida albicans*. *Proc. Natl. Acad. Sci. USA 98*: 3249-3253

[124] Uetz P, Giot L, Cagney G, Mansfield TA, Judson RS, Knight JR, Lockshon D, Narayan V, Srinivasan M, Pochart P, Qureshi-Emili A, Li Y, Godwin B, Conover D, Kalbfleisch

T, Vijayadamodar G, Yang M, Johnston M, Fields S, Rothberg JM. 2000. A comprehensive analysis of protein-protein interactions in *Saccharomyces cerevisiae. Nature 403*: 623-627

[125] Villalba F, Lebrun MH, Hua-Van A, Daboussi MJ, Grosjean-Cournoyer MC. 2001. Transposon *impala*, a novel tool for gene tagging in the rice blast fungus *Magnaporthe grisea. Molecular Plant Microbe Interactions 14*: 308-315

[126] Voiblet C, Duplessis S, Nathalie Encelot N, Martin F. 2001. Identification of symbiosis-regulated genes in *Eucalyptus globulus–Pisolithus tinctorius* ectomycorrhiza by differential hybridization of arrayed cDNAs. *The Plant J. 25*: 181-191

[127] Walhout AJM, Vidal M. 2001. High-throughput yeast two-hybrid assays for large-scale protein interaction mapping. *Methods 24*: 297-306

[128] Wan J, Dunning FM, Bent AF. 2002. Probing plant-pathogen interactions and downstream defense signalling using DNA microarrays. *Functional Integration Genomics 2*: 259-273

[129] Wood V, Gwilliam R, Rajandream MA, Lyne M, Lyne R, Stewart A, Sgouros J, Peat N, Hayles J, Baker S, Basham D, Bowman S, Brooks K, Brown D, Brown S, Chillingworth T, Churcher C, Collins M, Connor R, Cronin A, Davis P, Feltwell T, Fraser A, Gentles S, Goble A, Hamlin N, Harris D, Hidalgo J, Hodgson G, Holroyd S, Hornsby T, Howarth S, Huckle EJ, Hunt S, Jagels K, James K, Jones L, Jones M, Leather S, McDonald S, McLean J, Mooney P, Moule S, Mungall K, Murphy L, Niblett D, Odell C, Oliver K, O'Neil S, Pearson D, Quail MA, Rabbinowitsch E, Rutherford K, Rutter S, Saunders D, Seeger K, Sharp S, Skelton J, Simmonds M, Squares R, Squares S, Stevens K, Taylor K, Taylor RG, Tivey A, Walsh S, Warren T, Whitehead S, Woodward J, Volckaert G, Aert R, Robben J, Grymonprez B, Weltjens I, Vanstreels E, Rieger M, Schafer M, Muller-Auer S, Gabel C, Fuchs M, Dusterhoft A, Fritzc C, Holzer E, Moestl D, Hilbert H, Borzym K, Langer I, Beck A, Lehrach H, Reinhardt R, Pohl TM, Eger P, Zimmermann W, Wedler H, Wambutt R, Purnelle B, Goffeau A, Cadieu E, Dreano S, Gloux S, Lelaure V, Mottier S, Galibert F, Aves SJ, Xiang Z, Hunt C, Moore K, Hurst SM, Lucas M, Rochet M, Gaillardin C, Tallada VA, Garzon A, Thode G, Daga RR, Cruzado L, Jimenez J, Sanchez M, del Rey F, Benito J, Dominguez A, Revuelta JL, Moreno S, Armstrong J, Forsburg SL, Cerutti L, Lowe T, McCombie WR, Paulsen I, Potashkin J, Shpakovski GV, Ussery D, Barrell BG, Nurse P. 2002. The genome sequence of *Schizosaccharomyces pombe. Nature 415*: 871-80

[130] Wren BW. 2000. Microbial genome analysis: insights into virulence, host adaptation and evolution. *Nature Genetics 1*: 30-39

[131] Yamamoto M, Wakatsuki T, Hada A, Ryo A. 2001. Use of serial analysis of gene expression (SAGE) technology. *Journal Immunological Methods 250*: 45-66

[132] Yu JH, Hamari Z, Han KH, Seo JA, Reyes-Domínguez Y, Scazzochio C. 2004. Double-joint PCR: a PCR-based molecular tool for gene manipulation in filamentous fungi. *Fungal Gen. Biol. 41*: 973-981

[133] Zheng XF, Kobayashi Y, Takeuchi M. 1998. Construction of a low-serine-type-carboxypeptidase-producing mutant of *Aspergillus oryzae* by the expression of antisense RNA and its use as a host for heterologous protein secretion. *Appl. Microbiol. Biotechnol. 49*: 39-44.

In: Current Advances in Molecular Mycology
Eds: Y. Gherbawy, R.L. Mach and M.K. Rai
ISBN 978-1-60456-909-4

Chapter III

IN SILICO TOOLS TO STUDY PHYTOFUNGI

K. A. Abd-Elsalam[1,2,3,*]***, J-R. Guo***[2] ***and J.-A. Verreet***[2]
[1]Agricultural Research Centre, Plant Pathology Research Institute, Cotton Diseases Section, Giza, Egypt;
[2]King Saud University, Faculty of Science, Botany and Microbiology Department, P.O. Box: 2455, Riyadh 1145, Saudi;
[3]Institute of Phytopathology, University of Kiel, D-24118, Kiel, Germany.

ABSTRACT

Bioinformatics is indispensable for data management in plant pathology and the quick development of bioinformatic tools allows us to identify new isolates and strains of fungi as well as to study their genetic variability by using public databases. The archived databases like a digital reference library will be very supportive for the management of plant disease as well as for education of plant medicine doctor and their clients. On-line repository of Polymerase Chain Reaction primer sets will enhance the system for the diagnostic of plant pathogenic fungi. Increasing numbers of web sites are accessible to provide up-to-date information and illustrations from nearly every aspect of mycology. These internet resources will allow to search for plant pathogen information, to download marker sequence, to visualize geographical localization together with climate data, to analyze sequences from fungal plant pathogens, to draw phylogenetic trees in addition to to analyze relations between different species and isolates.

Correct and timely identification and control of plant diseases can have a great impact on human health, the environment and agricultural production. A specific repository dedicated to preserve and distribute plant pathogenic organisms would be a valuable tool to achieve the goals of plant protection in the world. Numerous plant pathologists, particularly in developing countries, do not have admittance to books or journals in the field of mycology and therefore they desire to have information's about public database or Internet resources. In this article, we will explore in silico tools to

* Correspondence concerning this article should be addressed to: K. A. Abd-Elsalam, Agricultural Research Centre, Plant Pathology Research Institute, Cotton Diseases Section, 9- Gamaa St., Giza, Egypt E-mail: abdelsalamka@gmail.com.

study phytopathogenic and antagonistic fungi and give a survey over 21 of the most important bioinformatic resources related to plant mycology.

Key words: Computational tools, data mining, genomic sequence, online database, phytopathogens.

INTRODUCTION

The idiom "*in silico*" has been set up into life sciences as a pendant to "*in vivo*" (in the living system) and "*in vitro*" (in the test tube). But, *in silico* is an expression used to mean achieved on computer or via computer simulation (Sieburg, 1990). In silico tools will guide and assist molecular biologists and plant pathology researchers to capitalize on the advantages brought by computational biology.

Plant-associated fungi are critical to agricultural and food security and are key components in preserving the stability of our ecosystems. Some of these diverse microorganisms, which include fungi, cause plant diseases, whereas others prevent diseases or improve plant growth. Fungal pathogens compete for ecological niches and are often present in plants as complexes. Therefore, it is important to be able to detect more than one pathogen at a time.

Pathogenic organisms vary greatly in the extent to which the attacked plant is colonized. For example, the rice blast pathogen *Magnaporthe grisea* and the potato late blight pathogen *Phytophthora infestans* can infect practically every plant organ with devastating consequences (Sesma and Osbourn 2004). Phytopathogenic fungi can infect all main crop plants (Strange and Scott 2005) and lead to food contamination through the production of mycotoxins. As one of the tools suppressing soilborne diseases, plant pathologists have been interested in the effect of microorganisms. A variety of soilborne fungi have demonstrated potential activity for controlling various soilborne plant pathogens (Aberra et al. 1998; Jackson et al. 1994; Larkin and Fravel 1998).

Molecular diagnostics testing is one of the fastest growing field in biotechnology. Although pest management information is available on the Internet and from other sources, plant pest identification and diagnosis are in many cases difficult and often entail consultation with a expert. When there is a disease problem, a general approach is to collect biological samples and post them to a diagnostic laboratory for identification. The delivery process using ordinary mail can take days, leading to setbacks in disease control recommendations. Sometimes, because of sample decomposition in the mail process, samples cannot be used for diagnosis and identification. An accurate and fast diagnosis can avoid costly errors by applying timely and appropriate management practices.

The most widespread approach to identify the origin of the sequences is by using search programs such as BLAST (Altschul et al., 1990) to determine the highest match with sequences in genetic databases such as GenBank at the National Center for Biotechnology Information (NCBI) of the National Library of Medicine, Washington, DC, USA. The DNA DataBank of Japan (DDBJ) and the European Molecular Biology Laboratory (EMBL) also house genetic databases, and the three organizations swap data on a daily basis.

During the past few years, over 40 complete fungal genomes have been sequenced and publicly released, and a similar number of fungi are presently being sequenced (Galagan et al 2005).

Despite their importance, we know little about them on a genomic level (Leach et al 2003). Fungi have small sizes of genome in comparison to plants and animals, and hence represent the highest number of complete or almost complete genomes sequenced (Hsiang and Baillie 2006). Comparative genomics can facilitate different research purposes such as phylogenetics, targeted drugs, gene detection, and gene function (Hsiang and Baillie, 2004, 2005 and 2006). Up to date reviews on fungal genomics have concentrated on food industry applications (Hofmann et al. 2003), pathogenicity (Lorenz 2002; Mitchell et al. 2003; Tunlid and Talbot 2002; Bos et al. 2003), antifungal drug discovery (Jiang et al. 2002; Parkinson 2002).

In 2002, there are over seventy large fungal culture collections worldwide and many smaller collections, often in individual laboratories. Over 385,000 living strains of filamentous fungi and yeasts were already collected. The ability to access an electronic database over the internet makes the smaller collections more relevant (McCluskey 2002). A similar procedure to record fungal nomenclatural novelties (MycoBank), was recently proposed by Crous et al. (2004).

Nearly all biologists do not consider themselves bioinformatics-enabled, but new computer programs should lessen the complexity of bioinformatic tools (Buckingham 2003). These tools are being directed toward the exponentially increasing amounts of genetic data, as well as toward categorizing the ever growing number of publications related to analysis and elucidation of such data (Buckingham 2003). The establishment of an internet-based database that cross-links the digitized genotypic and phenotypic information of plant pathogenic fungi at both the species and population levels could allow us to capably address these problems by coordinating the generation of data and its subsequent archiving (Kang et al 2002). We persuade plant pathologists to discover the use of the novel tools of bioinformatics. These tools are generally freely available and can be downloaded from many websites on the Internet. This review describes the main bioinformatic tools and discusses how they are being used to interpret mycology and phytopathology data and to further applied for disease identification and management. All of the websites described below will be available for the phytofungi-related research.

1. Some Important Websites and Databases

1.1. Fungal Culture Collections on Line

Culture collections are the living DNA banks of the future. Culture collections have long served as foci for biological science research (Samson et al, 2004). Culture collections of plant-associated fungi supply a needed resource of genotypic and phenotypic difference necessary for valuable research in mycology and plant pathology. Collections provide a genetic linkage between the past and present and can offer insight on change that have occurred since previous epidemics.

According to the most up to date data released by the World Data Centre for Microorganisms (WDCM), there are virtually 470 culture collections in sixty two countries at present, which do not include many collections in independent researcher's laboratories. There are over 2,300 people working for culture collections worldwide and these people preserve over one million microbial cultures (McCluskey, 2003).

Resources of culture collections are available electronically and are valuable to plant pathologists looking for sources of cultures for their study. Such as, databases developed at the US National Fungus Collections (http://nt.ars-grin.gov/) provide admittance to information about fungi, mainly those associated with plants or otherwise of agricultural importance. The ability to access an electronic database over the internet makes the smaller collections more pertinent. There have been a number of efforts to publish the holdings of smaller collections online, and these have taken different approaches and met with differing levels of success. Additional databases facilitate genome and taxonomic research. As more and more information becomes available online, the ability to use the correct strain of interest should make research more prolific and reproducible.

The number of collections with online databases is growing (Table 1). The accessibility of taxonomic databases online can serve to decrease potential mis-naming and mis-identification of specimens and strains. The capability to identify specimens in herbaria will allow the comparison with novel specimens with long established materials and should foster collaboration. Visualizing the geographic origins and distributions of documented strains genetically related to a recently isolated pathogen in the form of a map with zoom function may provide insights into the likely origin of this strain. Archiving such data in a format that can be simply accessed and searched is essential for quick assessment of potential risk and can help track the change and movement of pathogens (Kang et al 2006). Fungal testers having phenotypic and genotypic markers can assist detection and identification of emerging pathogens, provide data necessary for forensic study, and be useful in identifying control approaches.

Table 1. Internet sites listing fungal culture collections

Site	Sponsor	URL
WFCC-MIRCEN World Data Centre for Microorganisms	WFCC	wdcm.nig.ac.jp/hpcc.html
Databases by Culture Collections and Cell Banks	WFCC	wdcm.nig.ac.jp/DOC/menu3.xml
Culture Collections of Prokaryotes	Society for Systematic and Veterinary Bacteriology	www.bacterio.cict.fr/collections.html
Microbial Strain Data Network	United Nations	panizzi.shef.ac.uk/msdn
Common Access to Biological Resources and Information	Commission of the European Union	www.cabri.org

1.2. MycoBank

http://www.MycoBank.org

http://www.cbs.knaw.nl/fungi.htm

Nomenclatures of fungi are the key to information management, and access to correct and complete information has a main economic impact on, for instance, agriculture (phytopathology, and epidemiology). New taxonomy of fungi can be published in a wide range of botanical, microbiological, phytopathological or other scientific journals, books or proceedings. This makes it extremely difficult to be aware of new taxa as they are named or as existing species names are recombined into other genera. A probable solution to this predicament is the formal registration of names (Greuter et al. 2000).

An on-line database, MycoBank (http://www.MycoBank.org), is commenced with a remit to document mycological nomenclatural novelties and their associated descriptions and illustrations. The nomenclatural novelties will each be allocated a unique accession number that can be quoted in the publication where the nomenclatural novelty is introduced (Crous et al. 2004a and 2004b).

MycoBank will be a freely obtainable database, but its success will wholly depend on the partnership of mycologist and of the editors of journals insisting on the use of MycoBank accession numbers as part of their publication policies and quality control, with editors and mycologist sharing a vision of an eventual species bank that links all related ecological, molecular, metabolite, publication and additional data with the species, its distribution, biological associations (e.g. as parasites, mycorrhizal partners) or substratum. MycoBank will thus form an essential and fundamental digital link to and for information on fungi (Crous et al 2004b). Authors intending to publish nomenclatural novelties are encouraged to contribute to this new initiative.

1.3. Fungal Plant Pathogen Database

http://fppd.cbio.psu.edu/index.html

The main goal of the Plant Pathogen Database (Figure 1) is to index the emergence of new pathogens in a format simply available and searchable by first responders.

New fungal pathogens can appear through hybridization, universal relocation, unintended release by expanding agricultural activities, or even though deliberate release. The identity of a particular isolate at the species level is often not adequate to predict its traits (virulence, host, chemical resistance, toxin production, etc.) because these traits are often changeable within pathogen species.

The better understanding of the phenotypic and genetic diversity of high-risk pathogens is critical for designing successful long-term disease management strategies. Given the limited resolving power of phenotypic characters for strain and/or species identification, the use of genomic markers for strain identification is critical for the speedy implementation of suitable eradication and regulatory measures.

Developing a searchable database that cross-links digitized genotype (DNA sequence-based genetic fingerprints) and phenotypes (i.e. Morphology, virulence, host range, origin, toxin profile, etc.) of selected high-risk plant pathogens at both the species and population levels will be very valuable. It will allow users to perform phylogenetic analyses in order to identify the closest related species and/or isolates, give genomic data, when available, provide tools for comparative genomics, as well as allow for map-based visualization of the geographic origin of individual pathogen isolates.

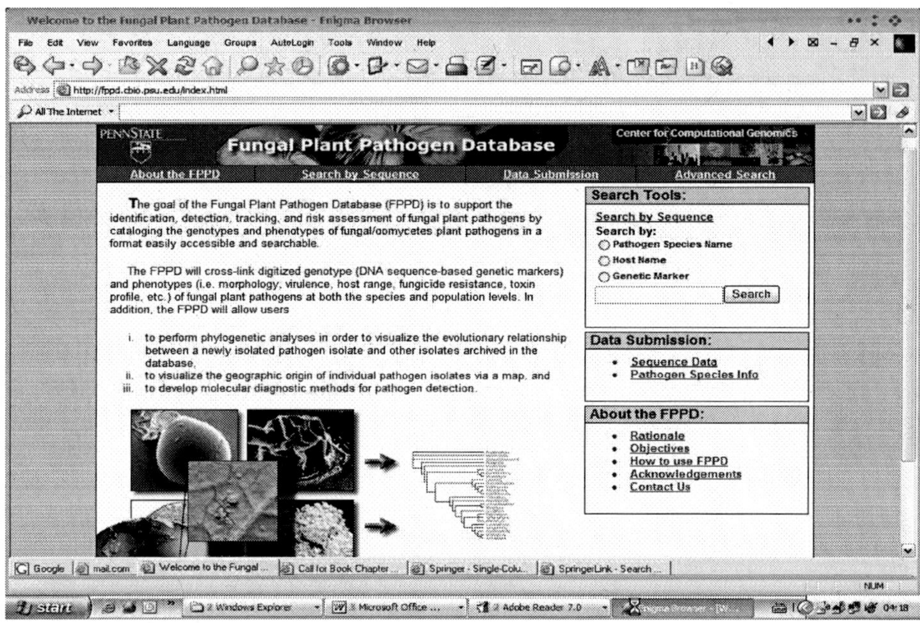

Figure 1. Part of the contents page of the Fungal Plant Pathogen Database.

1.4. Fungus-Host Distribution Database

http://nt.ars-grin.gov/fungaldatabases/fungushost/FungusHost.cfm

This database contains information of fungi on vascular plants and plant products according to their distribution by state. At present the database contains 76,000 fungal taxa on 53,000 vascular plant hosts representing 300,000 unique host-fungus combinations. Over 300 countries and territories are included. Records are continuously added as new publications are received. This database was used to produce the book Fungi on Plants and Plant Products in the United States (Farr et al unpublished data).

1.5. PathoPlant: A Database on Plant-Pathogen Interactions

http://www.pathoplant.de

Pathogen detection and signal transduction through plant pathogenesis is necessary for the activation of plant defense mechanisms. To facilitate easy access to published data and to permit comparative studies of diverse pathogen retort pathways, a database is essential to give a wide indication of the components and reactions so far known. PathoPlant has been developed as a relational database to present relevant components and reactions involved in signal transduction related to plant-pathogen interactions. On the organism level, the tables 'plant', 'pathogen' and 'interaction' are used to describe incompatible interactions between plants and pathogens or diseases (Bülowa et al 2004).

1.6. PHI-Base: Pathogen Host Interaction

http://www.phi-base.org/

PHI-base contains information molecular and biological about pathogenicity, virulence, and effecter genes from fungal and Oomycete pathogens. Each entry is created by domain specialists and supported by experimental evidence plus literature cited to the descriptions of the trials. Information for each gene includes its nucleotide and deduced amino acid sequence and a description of the predicted protein's function during the host infection progression (Baldwin et al 2006; Winnenburg et al 2006).

2. Morphological and Molecular Identification

Molecular phylogeny has shown that pathogenic fungi are found in many taxonomic groups, which proposed that these life cycle have evolved repeatedly within the kingdom fungi (Tunlid and Talbot, 2002). Alignment of sequence data from phytopathogenic fungi with those of closely related non-pathogenic species should therefore supply information about genetic parameters that may be critical for pathogenesis, or as a minimum consistently conserved in species with the capability to cause fungal disease.

2.1. Morphological and Molecular Identification of Trichoderma and Hypocrea

http://www.ısth.info/index.php

This web site will lead you to a morphological and molecular identification and taxonomy of Trichoderma and Hypocrea. It includes three different tools: (1) TrichoBlast, (2) Trichodema morphological key (Figure 2) and DNA barcoded molecular key (*Trich*OKEY v2.0, Figure 3) TrichoBlast is designed for the initial diagnosis of the query sequence. It will detect one or several bands of the user's sequence which correspond to phylogenetic markers stored in the local database. As well as provides a possibility to apply the similarity search to each fragment individually. The Trichoderma morphological key (Figure 2) contains descriptions and over 500 images for the 32 species of *Trichoderma* that represent a majority of the taxa found in moderate regions (Samuels et al 2007). The DNA-barcoded molecular key is designed for the identification of 88 taxa of *Trichoderma* and *Hypocrea* on the basis of existence of hallmark-sequences in ITS1 and ITS2, based on the sequence analysis of 979 strains of *Trichoderma*. The barcode identifies approximately all species but the species pairs

T. tomentosum/*T. cerinum*, *T. longipile*/, *T. crassum* and the species trio *T. koningii*/*T. ovalisporum*/*H. muroinana* from the *H. rufa* species clade (Druzhinina et al 2005).

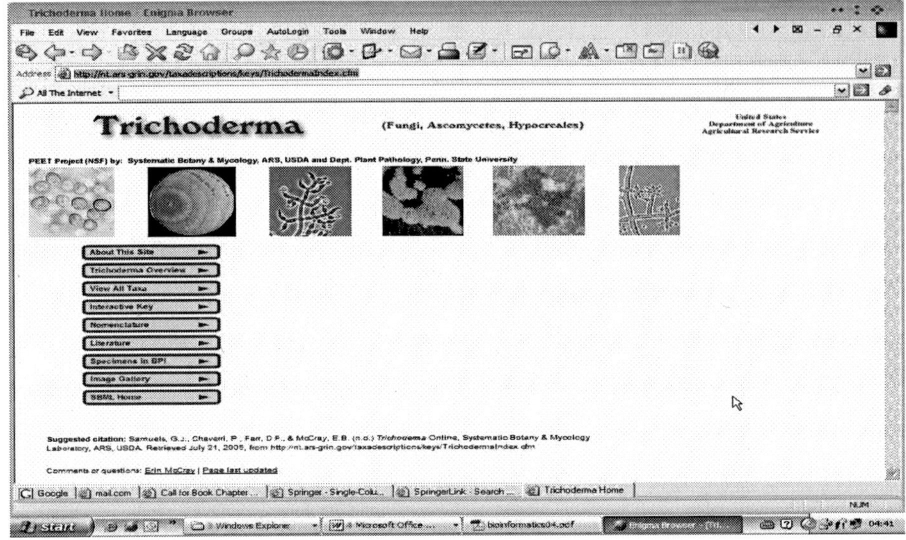

Figure 2. Screen shots from the web-based front of the morphological identification of *Trichoderma* spp.

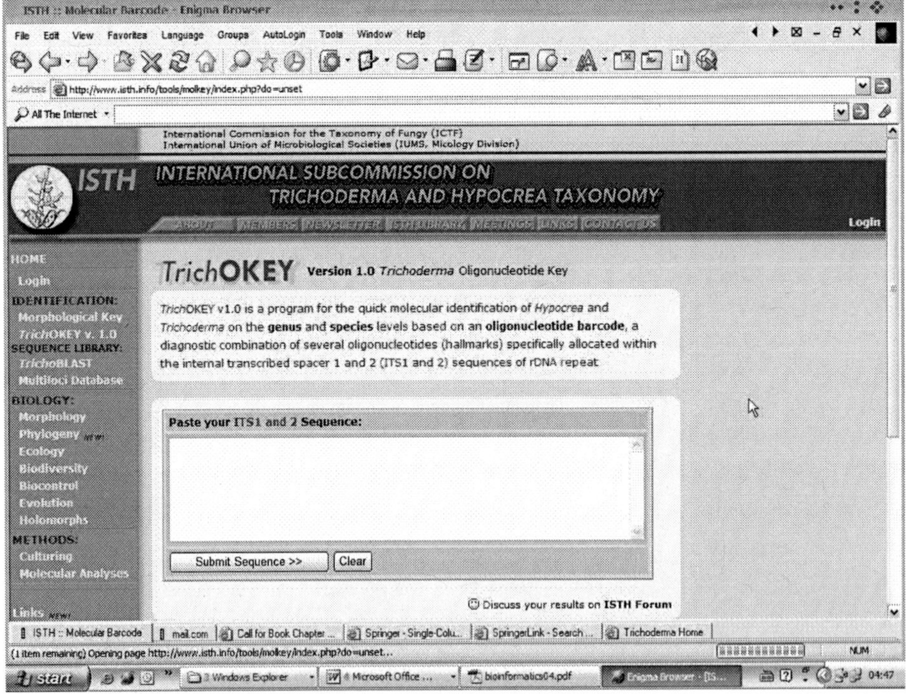

Figure 3. Part of the home page of the TrichOKEY.

2.2. The Database of PCR Primers for Phytopathogenic Fungi

http://www.sppadbase.com/index.php

Ghignone and Migheli (2005) presented the first on-line searchable database of primer sets valuable for the detection and diagonastic of plant pathogenic fungi (Figure 4). This link resource is implemented entirely with open-source software (PHP, MySQL). Specific primer set details can be recovered by fungal name, primer name, nucleotide sequence comparison, target DNA, PCR-based method, author name, journal and year of publication. All records is linked directly to other reference databases to allow easy contact to the accurate nomenclature, taxonomical position and anamorph/teleomorph connections of the fungal pathogen, GenBank-deposited source sequences of the primer sets and reference contents. The database currently counts: 84 genera, 160 species, 411 primer sets, and 200 references. The database is open to user contributions and can be consulted at http://www.sppadbase.com.

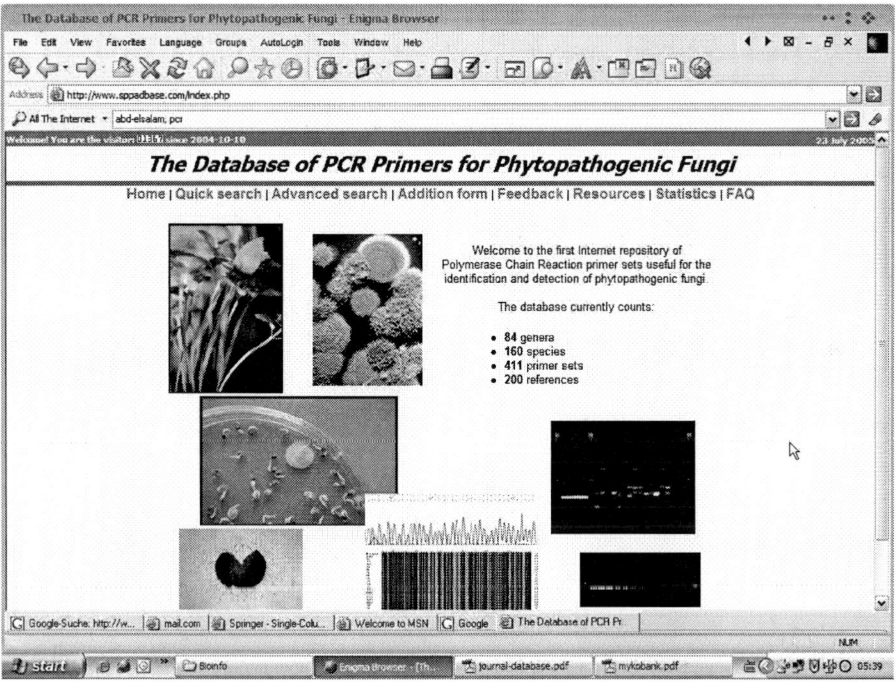

Figure 4. Screen shots from the web-based front of the database of PCR primers for phytopathogenic fungi.

2.3. FUSARIUM-ID v. 1.0: A DNA Sequence Database for Identifying Fusarium

http://fusarium.cbio.psu.edu.

One of the greatest obstructions to the study of Fusarium has been the erroneous and confused application of species names to toxigenic and pathogenic isolates, owing in large part to native limitations of morphological species recognition and its application. To address this problem, we have created FUSARIUM-ID v. 1.0, a publicly available database of partial translation elongation factor 1-alpha (TEF) DNA sequences, currently representing a selected sample of the diversity of the genus diversity, with excellent representation of Type-B trichothecene toxin producers, and the *Gibberella fujikuroi*, *Fusarium oxysporum* and *F. solani* species complexes. Researchers can produce sequences using primers that are conserved across the genus, and use the sequence as a query to BLAST the database, which can be accessed at http://fusari-um.cbio.psu.edu, or in a phylogenetic analysis (Geiser et al 2004).

Table 2. Phytopathogenic fungi genome that have been sequenced

Fungal species	Taxon	Year	Size	WWW	Source and publication
Magnaporthe grisea	Sordariomycetes	June 2002	40 Mb	http://www.broad.mit.edu/cgi bin/annotation/fungi/magnaporthe/ download_license.cgi/magnaporthe_grisea_2.fasta.gz	Broad Institute
Fusarium graminearum	Sordariomycetes	March 2003	40 Mb	http://www.broad.mit.edu/annotation/genome/fusarium_graminearum /Home.html	Broad Institute
Ustilago maydis	Ustilaginomycota	July 2003	20 Mb	http://www.broad.mit.edu/annotation/genome/ustilago_maydis/Home.html	Broad Institute, Bayer CropScience AG, & Exelixis
Stagonospora nodorum	Dothideomycetes	April 2005	37 Mb	http://www.broad.mit.edu/annotation/genome/stagonospora_nodorum/Info.html	Broad Institute & International *Stagonospora nodorum* Genomics Consortium
Botrytis cinerea	Leotiomycetes	Oct. 2005	39 Mb	http://www.broad.mit.edu/annotation/genome/botrytis_cinerea/Home.html	Broad Institute & Syngenta AG; Genoscope
Sclerotinia sclerotiorum	Leotiomycetes	April 2005	38 Mb	http://www.broad.mit.edu/annotation/genome/sclerotinia_sclerotiorum/Home.html	Broad Institute
Nectria haematococca	Sordariomycetes	June 2005	52 Mb	http://genome.jgi-psf.org/Necha2/Necha2.home.html	Joint Genome Institute
Phytophthora ramorum	Oomycete	Oct. 2003	1.4 Gb	http://genome.jgi-psf.org/Phyra1_1/Phyra1_1.info.html	Joint Genome Institute
Phytophthora sojae	Oomycete	Oct. 2003	1.4 Gb	http://genome.jgi-psf.org/Physo1_1/Physo1_1.info.html	Joint Genome Institute

3. THE CURRENT STATUS OF FUNGAL GENOMICS

Genomic information offers the first phase in recognizing the instructions for the synthesis of all of life's molecular machines and the systems needed to control and operate them. The interaction among biological, physical and computing sciences has provided the pedestals for the establishment of a great number of bioinformatic tools (Setubal et al 1997).

Table 3. Phytopathogenic and antagonistic fungi genome sequence in progress

Fungal species	Taxon	Year	Size	WWW	Source and publication
Alternaria brassicicola	Dothideomycetes	Feb 2006	30 Mb	http://genome.wustl.edu/genome_index.cgi	Washington University Genome Sequencing Center
Aspergillus flavus	Eurotiomycetes	Oct. 2005	36 Mb	http://www.aspergillusflavus.org/genomics/	North Carolina State University and USDA National Research Initiative, TIGR
Mycosphaerella fijiensis	Dothideomycetes	ND	ND	http://www.jgi.doe.gov/sequencing/cspseqplans2006.html	Joint Genome Institute
Mycosphaerella graminicola	Dothideomycetes	Nov. 2005	ND	http://genome.jgi-psf.org/Mycgr1/Mycgr1.info.html	Joint Genome Institute
Fusarium verticillioides	Sordariomycetes	June 2003	36 Mb	http://www.broad.mit.edu/annotation/fgi/	Broad Institute
Fusarium oxysporum	Sordariomycetes			http://www.broad.mit.edu/annotation/fungi/fgi/	Broad Institute
Phakopsora pachyrhizi	Urediniomycetes	Dec. 2003	50 Mb	http://genome.jgi-psf.org/	Joint Genome Institute
Phakopsora meibomiae	Urediniomycetes	Dec. 2003	30 MB	http://genome.jgi-psf.org/	Joint Genome Institute
Puccinia graminis f. sp. tritici.	Urediniomycetes	Dec. 2006	81 Mb	http://www.broad.mit.edu/annotation/genome/puccinia_graminis	Broad Institute
Melampsora larici-populina	Urediniomycetes	ND	ND	http://www.jgi.doe.gov/sequencing/cspseqplans2006.html	Joint Genome Institute
Phytopthora infestans	Oomycete	Dec. 2003	237 Mb	http://www.ncbi.nlm.nih.gov/Traces	Broad Institute
Peronospora parasitica	Oomycete			http://genome.wustl.edu/home.cgi	Washington University Genome Sequencing Center
Pyrenophora tritici-repentis	Dothideomycetes	Feb. 2007	38 Mb	http://www.broad.mit.edu/annotation/genome/pyrenophora_tritici_repentis	Broad Institute
Trichoderma virens	Sordariomycetes	ND	ND	http://www.jgi.doe.gov/sequencing/cspseqplans2006.html	Joint Genome Institute

ND= not detected.

In the last ten years, fungal genomics have been speedy developed. Of all Eukaryotic phyla with genome projects, fungi are the most commonly used, with nearly 200 diverse species (Figure 4, Liolios et al 2006). Recently, full genome sequencing projects have been completed for nine phytopathogenic fungi (Xu et al 2006). These genome sequences are now publicly available (Table 2). Most of them are economically important phytopathogenic fungal pathogens, for instance, the first basidiomycete genome sequence of the white rot fungus *Phanerochaete chrysosporium* (Martinez *et al.*, 2004), the rice-blast fungus *Magnaporthe grisea* (Dean *et al* ., 2005), *Gibberella zeae* and *Ustilago maydis.* Furthermore, the thirty-six Mb genome of *A. flavus* was sequenced at TIGR; *S. nodorum* was sequenced at the Broad Institute and released to the public in 2005. In addition, about 14 phytopathogenic fungal genomes are underway (Table 3).

There are also some confidentially-held complete or almost complete fungal genomic data, including *Cochliobolus heterostrophus* and *Gibberella fujikuroi* by Syngenta Biotechnology at the Research Triangle Park, NC (Turgeon et al. 2002). Syngenta also published an approximately 4X genome sequence of *F. verticillioides* strain 7600, a causal agent of kernel and ear rot of maize and a maker of fumonisin mycotoxins.

At this exciting time, the completion of large-scale genomic and post-genomic methods presents unmatched opportunities to additional understand functional genomics and clarify the pathogenic traits associated with fungal infections in addition to provide a framework for the development of novel diagnostic tools for these important pathogens. Conditionally, therefore, there is requiring to employ current genomic data in a quantitative manner and to develop bioinformatics tools that will guide further researches. Admittance to these fungal genomic data is accessible through a growing number of online resources. Several examples of online resources were presented below.

3.1. Fungal Genome Initiative (Broad Institute)

http://www.broad.harvard.edu/annotation/fgi/

The Fungal Genome Initiative (FGI) produces and analyzes sequence data from fungal organisms that are important to medicine, agriculture and industry. Over 25 fungi have been sequenced or are being sequenced, including human and plant pathogens as well as fungi that serve as basic models for molecular and cellular biology. Annotated genomes are available for *Aspergillus, Cryptococcus, Fusarium, Magnaporthe, Neurospora, Rhizopus, Sclerotinia*, and numerous other fungi.

3.2 The Genomes OnLine Database (GOLD)

http://www.genomesonline.org

GOLD is in silico resource for comprehensive access to information concerning complete and in progress genome was sequencing projects worldwide. The database currently incorporates information on over 1500 sequencing projects, of which 294 have been completed and the data deposited in the public databases (Liolios et al 2006). There are currently 2481 ongoing and completed sequencing projects. Of those, 200 fungal, 697 are bacterial, 38 archaeal and 526 are eukaryotic projects (Figure 5).

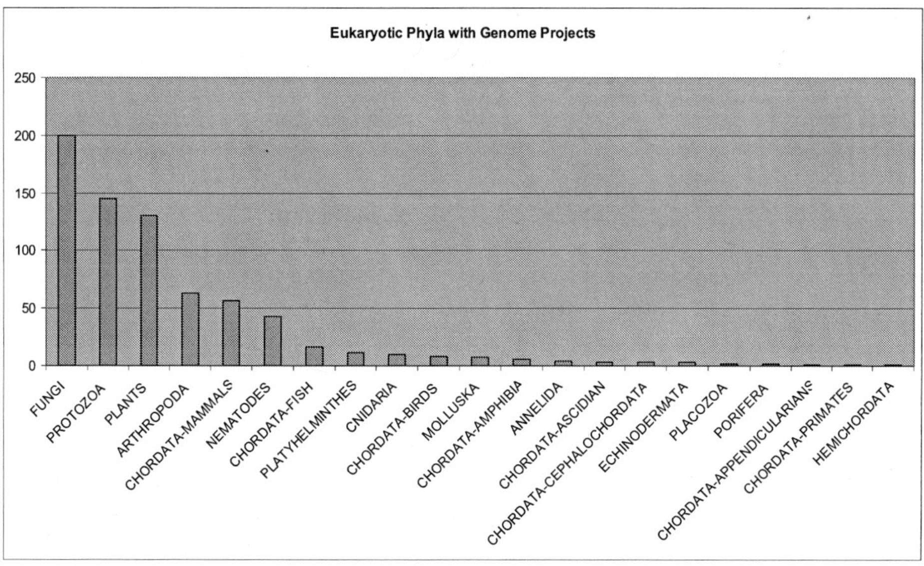

Figure 5. Eukaryotic phyla with genome projects (Liolios et al 2006).

3.3. Phytopathogenic Fungi and Oomycete EST Database, Version 1.5

http://cbr-rbc.nrc-cnrc.gc.ca/services/cogeme/

http://cogeme.ex.ac.uk/

This resource, which is freely obtainable on the web, has been incorporated into the Consortium for Functional Genomics of Microbial Eukaryotes (COGEME) EST data storehouse (http:// cogeme.ex.ac.uk/), an online comparative and functional genomics resource developed for the phytopathogenic fungi research society (Soanes et al., 2002). Expressed sequence tags (ESTs) have been achieved from thirteen plant pathogenic fungi, three species of saprophytic fungi and two species of phytopathogenic oomycete. Hierarchical clustering software was used to categorize together ESTs representing the identical gene and create a single contig, or consensus sequence. The unisequence set for each pathogen therefore represents a set of unique gene sequences, each one consisting of either a single EST or a contig sequence made from a group of ESTs. Putative functions were assigned to each unisequence based on top hits against the NCBI non-redundant protein database using blastx.

3.4. Munich Information Centre for Protein Sequences (MIPS)

http://mips.gsf.de/projects/fungi

This link contains the *Fusarium graminearum* Genome Database (FGDB) and *Ustilago maydis* Database (MUMDB). FGDB is a comprehensive genome database on one of the most destructive fungal plant pathogens of wheat and barley. It provides information on two gene sets separately derived by automated annotation of the *F. graminearum* genome sequence (Güldener et al 2006). Complete sets of *F. graminearum* sequences and annotation can be downloaded from ftp://ftpmips.gsf.de/fusarium/. These include lists of genetic elements and the contig sequences of the automatically predicted gene sets as well as the current valid gene set. The functional classification can be found on ftp://ftpmips.gsf.de/fusarium/catalogues/. The MUMDB intends to present information on the molecular structure and functional network of the entirely sequenced, filamentous fungus *Ustilago maydis.*

3.5. Two Genomic Online Resources for *Magnaporthe grisea*

http://www.fungalgenomics.ncsu.edu/Projects/mgdatabase/int.htm

www.mgosdb.org

M. grisea causes the most destructive disease of rice and has emerged as a central model organism for the study of fungal plant diseases. Rice blast, the disease caused by *M. grisea* is estimated to devastate enough rice annually to feed 60 million inhabitants (Zeigler et al. 1994). Here we introduce two essential databases of this pathogen. The Magnaporthe DB (http://www.fungalgenomics.ncsu.edu/Projects/mgdatabase/int.htm) was constructed for genome studies of *Magnaporthe grisea* by integrating end sequence data from BAC clones, genetic marker data and BAC contig gathering data. A library of 9216 BAC clones provided that >25-fold coverage of the entire genome was end sequenced and fingerprinted by *Hind*III digestion. The Image/FPC software package was then used to create an assembly of 188 contigs covering >95% of the genome. The database includes the results of this assembly integrated with hybridization data of genetic markers to the BAC library. AceDB was used for the core database engine and a MySQL relational database, populated with numerical representations of BAC clones within FPC contigs, was used to create appropriately sized images. The database is being used to assist sequencing efforts. The database also permits researchers mapping known genes or other sequences of notice, quick and simple access to the fundamental organization of the *M. grisea* genome (Stanton et al 2002).

The MGOS database (www.mgosdb.org) was developed to store all investigational data and supply a schematic approach to discover the results. To augment these results, the database also contains genome searchers for both the rice and *M. grisea* genomic sequences (Soderlund et al 2006).

3.6 Databases for Phytophthora

3.6.1. Phytophthora Functional Genomics Database (PFGD)

http://www.pfgd.org/

PFGD is a freely accessible source that includes functional assay and expression data along with transcript analysis and annotation. Sequence data come from a number of resources. For *P. sojae*, transcript data is obtainable for various different tissues, developmental phases, and growth conditions. Related transcript data is available for *P. infestans*, along with genomic data (Gajendran et al 2006).

3.6.2. PhESTDB v 1.0 (Phytophthora Soybean EST Database)

http://phytophthora.vbi.vt.edu/EST/

The database contains ESTs from *P. sojae, P. infestans*, and soybean with sequence annotation and additional data (Tripathy and Tyler 2007).

3.6.3. PhytophthoraDB

www.phytophthoradb.org

The *Phytophthora* DataBase project aims to improve our ability of quick identification and diagnosis of *Phytophthora* spp. by recording known genotypic and phenotypic diversity in a highly incorporative database (Kang et al 2007).

3.7. Resources for Fungal Comparative Genomics

http://fungal.genome.duke.edu/

This website is aimed to present genome annotation data for the available fungal genomes. Extra genome annotations, similarity searching, and best mutual orthologous genes have been created for the on hand species and will be made downloadable this fall. Analyses of orthologous gene relationships, gene trees, and gene structures can be carried out. A tree representing the phylogenetic rapport of the sequenced species is available.

3.8. Comprehensive Plant Pathogen Genomic Warehouse

http://cpgr.tigr.org/cgi-bin/warehouse/cpgr_warehouse.cgi

The Comprehensive Plant Pathogen Genomic storehouse is a database of completed, rough copy and in progress genome sequencing projects and EST projects for plant pathogenic organisms.

3.9. CPGR Plant Pathogen Ribosomal DNA (rDNA) Database

http://cpgr.tigr.org/cgi-bin/cpgr_rdna/cpgr_rdna_db.pl

The CPGR Plant Pathogen Ribosomal DNA (rDNA) Database contains all the ribosomal DNA in GenBank for Bacterial, Stramenopile, Nematode and Fungal plant pathogens. The rDNA sequences for other species in the genus of each pathogen are also stored in the database.

4. ELECTRONIC JOURNALS AND BIBLIOGRAPHIC DATABASES

Recently, electronic publishing in science has become the spotlight of an increasing number of workshops and conferences, typically including representatives from qualified societies and other scholarly publishing concerns, and members of the library community; but only a small or diminishing participation from real researchers.

A accurately configured completely on-line scholarly journal can be activated at a fraction of the cost of a conventional print journal, and could for example be wholly sustained by author subsidy (page charges or related mechanism, as already paid to some journals), ideally allowing for free network distribution and maximal profit both to authors and readers.

Some links to journals in which articles concerning mycology and plant pathology are listed (Table 4). The sites will almost always have a Table of Contents that usually contain summary of articles or the full text online. A few sites are free, while others necessitate registration or paid subscriptions. Several journals have free "trial" periods, or bundle online with print subscriptions, so the researchers can verify the sites to test accessibility.

Bibliographic databases usually include references and abstracts, indexed by subject-specific keywords so users can recover searches for a particular theme or species of interest. For example, the CAB ABSTRACTS database, published by CAB International, provides abstracts of internationally published scientific research literature in agriculture and the biosciences. The Review of Plant Pathology is assemble from this database and is available as a CD-ROM, online and as a printed journal (see www.cabi.org or http://pest.cabweb.org/ for further information). AGRICOLA (AGRICultural OnLine Access) is a machine-readable database of bibliographic records created by the National Agricultural Library of the US Department of Agriculture and its co-operators (see www.nalusda.gov/ for further information). Some bibliographic databases include the full text of the original paper, allowing users greater access to the world's literature. Software packages are also available to help scientists handle databases of references they need for writing report. Plant pathologist can readily find their journals on the World Wide Web, those with a general interest in the discipline of plant pathology need a service that consolidates what is available across the world's publishers and language.

Table 4. List of plant pathology and mycology journals and their URL and impact factors (2007)

Full Journal Title	URL	Impact Factor
Annual Review of Phytopathology	http://arjournals.annualreviews.org/loi/phyto	9.000
Archives of Phytopathology and Plant Protection	http://www.tandf.co.uk/journals/titles/03235408.html	ND
Australasian Plant Pathology	http://www.publish.csiro.au/nid/39.htm	0.766
Canadian Journal of Plant Pathology	http://www.cps-scp.ca/journals.htm	ND
European Journal of Plant Pathology	http://www.blackwellpublishing.com/journal.asp?ref=0032-0862&site=1	1.217
Fungal Diversity	http://www.fungaldiversity.org/fdp/FD/index.htm	2.297
Fungal Genetics and Biology	http://www.elsevier.com/wps/find/journaldescription.cws_home/622835/description#description	3.121
Journal of General Plant Pathology JGPP	http://sciserver.lanl.gov/cgi-bin/sciserv.pl?collection=journals&journal=13452630	ND
Journal of Phytopathology	http://www.blackwellpublishing.com/journal.asp?ref=0931-1785	0.817
Journal of Plant diseases and Protection	www.jpdp-online.com/	0.176
*Journal of Plant Pathology (JPP)	http://www.agr.unipi.it/sipav/jpp/	0.783
Journal of Plant Protection Research	http://www.plantprotection.pl/	ND
Molecular Plant Pathology	http://www.blackwellsynergy.com/servlet/useragent?func=showIssues&code=mpp&code=MPP&goto=journal	2.963
Molecular Plant-Microbe Interactions(MPMI)	http://www.apsnet.org/mpmi/top.asp	3.936
Mycobiology	http://mycology.or.kr/button3_3.asp	ND
Mycologia	http://www.mycologia.org/	1.574
Mycological Progress	http://www.springeronline.com/journal/11557/about	ND
Mycological Research	http://journals.cambridge.org/action/displayJournal?jid=MYC	1.860
Mycopathologia	http://www.springer.com/west/home/biomed/medical+microbiology?SGWID=4-129-70-35759034-0	0.915
Mycorrhiza	http://www.springerlink.com/content/1432-1890/	1.813
Mycoscience	http://springerlink.metapress.com/content/1618-2545/	ND
*New Disease Reports	http://www.bspp.org.uk/ndr/index.htm	ND
Physiological and Molecular Plant Pathology	http://www.elsevier.com/wps/find/journaldescription.cws_home/622932/description#description	1.288
Phytoparasitica	http://www.phytoparasitica.org/	0.632
Phytopathologia Mediterranean	http://www.unifi.it/istituzioni/mpu/phymed.htm	ND
Phytopathology	http://www.apsnet.org/phyto/top.asp	2.195
Plant Disease	http://www.apsnet.org/pd/current/top.asp	1.795
Plant Health Progress	http://www.plantmanagementnetwork.org/php/	ND
Plant Pathology	http://www.blackwellpublishing.com/journal.asp?ref=0032-0862&site=1	2.198
*Plant Pathology Journal	http://www.ansijournals.com/3/c4p.php?id=1&theme=3&jid=ppj	ND
Review of Plant Pathology (RoPP)	http://www.cabi.org/AbstractDatabases.asp?PID=52	ND
Studies in Mycology	http://www.studiesinmycology.org/	2.957
*Sydowia	http://www.sydowia.at/index.htm	0.375

Table 4. (Continued)

Full Journal Title	URL	Impact Factor
The Plant Pathology Journal	http://www.kspp.org/ejournal/journal.php?a=intro&id=1	ND

*Journals accessible without a fee; ND= not detected.

5. Other Applications of Bioinformatics

Aside from analysis of genome sequence data, computational biology is now being used for a huge array of other important errands, including analysis of gene difference and expression, scrutiny and prediction of gene and protein structure and function, prediction and discovery of gene regulation networks, simulation environments for whole cell modelling, intricate modeling of gene regulatory dynamics and networks, and presentation and analysis of molecular pathways to facilitate comprehend gene-disease interactions. While on a minor level, simpler bioinformatics tasks precious to the molecular plant pathologist researcher can vary from designing primers to predicting the function of gene products.

A major aim of agricultural biotechnology is the detection of genes or genetic loci which are associated with characteristics advantageous to crop production. This acquaintance of genetic loci possibly will be applied to improve crop breeding. Agriculturally Important genes may also benefit crop production through transgenic technologies. Recent years have seen an application of high throughput technologies to agricultural biotechnology leading to the production of large amounts of genomic data.

6. Viewpoint

Plant pathology has made significant progress over the years, a process that involved overcoming a variety of conceptual and technological hurdles. The growth of bioinformatics has been a global venture, creating computer networks that have allowed easy access to biological data and enabled the development of software programs for effortless analysis. Multiple international projects aimed at providing gene and protein databases are available freely to the whole scientific community via the internet.

The major benefit coming from the development of such a kind of bioinformatics tool is that it significantly reduces the times to perform bio-sequences retrieval and analyses through the user-friendly web query and processing system that helps and guides end-users to find the data and the analyses that best fit to their need with the implementation of a work flow logic. This has a great importance in the time-consuming operations in which the researcher has to perform data retrievals and analysis manually, in many steps.

Developing countries are not only economically poor but also informatically poor. Bioinformatics resources are often seriously limited in these countries. Technological innovation is an essential approach for their development. It would also help them to improve their own research methodologies and therapies for serious phytofungal diseases. Researchers in the developing countries generally have dilemmas to know the information of databases. Hence, better information resources are urgently needed particularly for the developing countries to enhance crop protection. The free use of bioinformatics information and tools via internets will greatly help them to study phytofungi.

REFERENCES

Aberra MB, Seah S, Sivasithamparam K (1998) Suppression of the take-all fungus (*Gaemannomyces graminis* var. *tritici*) by a sterile red fungus through induced resistance in wheat (*Triticum aestivum*) seedling roots. *Soil Biol Biochem 30*:1457–1461.

Altschul, S.F., Gish, W., Miller, W., Myers, E.W., and Lipman, D.J. 1990. Basic local alignment search tool. *J. Mol. Biol. 215*:403–410.

Baldwin, T.K., Winnenburg, R., Urban, M., Rawlings, C., Köhler, J. and Hammond-Kosack, K.E. 2006. The Pathogen-Host Interactions database (PHI-base) provides insights into generic and novel themes of pathogenicity. *Mol Plant Microbe Interact. 19*:1451-62

Bos JIB, Armstrong M, Whisson SC, Torto TA, Ochwo M, Birch PRJ, and Kamoun S (2003) Intraspecific comparative genomics to identify avirulence genes from *Phytophthora. New Phytol 159*:63-72.

Buckingham, S. 2003. Programmed for success. *Nature (London), 425*: 209–215.

Bülowa L, Schindlera M, Choi C and Hehla R. (2004) PathoPlant: A database on plant-pathogen interactions. *In Silico Biology 4*: 529–536.

Crous PW, Gams W, Stalpers JA, Cannon PF, Kirk PM, David JC, Triebel D (2004 a). An online database of names and descriptions as an alternative to registration. *Mycological Research 108*: 1236–1238.

Crous P. W. , Gams W., Stalpers A. J, Robert V. and Stegehuis G. 2004 b. MycoBank: an online initiative to launch mycology into the 21st century. *Studies in Mycology 50*: 19–22.

Dean, R.A., Talbot, N.J., Ebbole, D.J., Farman, M.L., Mitchell, T.K., Orbach, M.J., Thon, M., Kulkarni, R., Xu, J.R., Pan, H., Read, N.D., Lee, Y.H., Carbone, I., Brown, D., Oh, Y.Y., Donofrio, N., Jeong, J.S., Soanes, D.M., Djonovic, S., Kolomiets, E., Rehmeyer, C., Li, W., Harding, M., Kim, S., Lebrun, M.H., Bohnert, H., Coughlan, S., Butler, J., Calvo, S., Ma, L.J., Nicol, R., Purcell, S., Nusbaum, C., Galagan, J.E. and Birren, B.W. (2005) The genome sequence of the rice blast fungus Magnaporthe grisea. *Nature, 434*, 980–986.

Druzhinina IS, Kopchinskiy A. G, Komoj M, Bissett J, Szakacs G, Kubicek C.P. 2005 An oligonucleotide barcode for species identification in *Trichoderma* and *Hypocrea. Fungal Genetics and Biology 42*: 813–828.

Farr, D.F., Rossman, A.Y., Palm, M.E., & McCray, E.B. (2007) Fungal Databases, Systematic Botany & Mycology Laboratory, ARS, USDA. Retrieved March 5, 2007, from http://nt.ars-grin.gov/fungaldatabases/.

Gajendran K, Gonzales M. D., Farmer A., Archuleta E., Win J., Waugh M. E., and Kamoun S. 2006 Phytophthora functional genomics database (PFGD): functional genomics of *phytophthora*–plant interactions. *Nucleic Acids Res. 34(Database issue)*: D465–D470.

Galagan JE, Henn MR, Ma LJ, Cuomo CA, Birren B. 2005. Genomics of the fungal Kingdom: insights into eukaryotic biology. *Genome Res. 15*:1620–31

Ghignone S and Q Migheli (2005) The database of PCR primers for phytopathogenic fungi. *European Journal of Plant Pathology 113*:107–109

Greuter W, McNeill J, Hawksworth DL, Barrie FR (2000). *Report on botanical nomenclature* – Saint Louis 1999. Englera 20: 1–253.

Güldener U, Mannhaupt G, Münsterkoütter M, Haase D, Oesterheld M, Stümpflen V, Mewes H-W and Adam G. 2006 FGDB: a comprehensive fungal genome resource on the plant pathogen Fusarium graminearum *Nucleic Acids Research, 34*:456-459-8.

Hofmann, G., McIntyre, M., and Nielsen, J. 2003. Fungal genomics beyond *Saccharomyces cerevisiae. Curr. Opin. Biotechnol. 14*:226–231.

Hsiang T and Baillie DL (2004) Recent progress, developments and issues in comparative fungal genomics. *Can J Plant Pathol 26*:19-30.

Hsiang T and Baillie DL (2005) Comparison of the yeast proteome to other fungal genomes to find core fungal genes. *J Mol Evol 60*:475-483.

Hsiang T and Baillie DL (2006) Issues in comparative fungal genomics. *Applied Mycology and Biotechnology*, an international series volume 6. Bioinformatics Pages 1-26.Elsevier Science.

Hyman RW (2001) Sequence data: posted vs. published. *Science* 291:827.

Jackson AJ, Walters DR, Marshall G (1994) Evaluation of *Penicillium chrysogenum* and its antifungal extracts as potential biological control agents against *Botrytis fabae* on faba beans. *Mycol Res 98*:1117– 1126

Jiang B, Bussey H and Roemer T (2002) Novel strategies in antifungal lead discovery. *Curr Opin Microbiol 5*:466-471.

Kang, S., Ayers, J. E., DeWolf, E. D., Geiser, D. M., Kuldau, G., Moorman, G. W., Mullins, E., Uddin, W., Correll, J. C., Deckert, G., Lee, Y.-H., Lee, Y.-W., Martin, F. N., and Subbarao, K. 2002. The internet based fungal pathogen database: A proposed model. *Phytopathology 92*:232-236.

Kang, S., Blair, J. E., Geiser, D. M., Khang, C.-H., Park, S.-Y., Gahegan, M., O'Donnell, K., Luster, D. G., Kim, S. H., Ivors, K. L., Lee, Y.-H., Lee, Y.-W., Grünwald, N. J., Martin, F. M., Coffey, M. D., Veeraraghavan, N., and Makalowska, I. 2006. Plant pathogen culture collections: It takes a village to preserve these resources vital to the advancement of agricultural security and plant pathology. *Phytopathology 96*:920-925.

Kang S., J.Blair, D.Geiser, I.Makalowska,S-Y Park, B.Park, M.Coffey, K.Ivors, Y.ee4, J-S Park, K.O'Donnell and F.Martin (2007) *Phytophthora* Database: An integrated resource for detecting, monitoring, and managing *Phytophthora* diseases. Oomycete Molecular Genetics Network Workshop Asilomar Conference Center, Pacific Grove, CA, USA March 18-20, 2007.p11.

Larkin RP, Fravel DR (1998) Efficacy of various fungal and bacterial biocontrol organisms for control of *Fusarium* wilt of tomato. *Plant Dis 82*:1022–1028

Leach, J. E., Gold, S., Tolin, S., and Eversole, K. 2003. A plant associated microbe genome initiative. *Phytopathology 93*:524-527.

Lorenz MC (2002) Genomic approaches to fungal pathogenicity. *Curr Opin Microbiol 5*:372-378.

Martinez, D., Larrondo, L.F., Putnam, N., Gelpke, M.D., Huang, K., Chapman, J., (2004) Genome sequence of the lignocellulose degrading fungus *Phanerochaete chrysosporium* strain RP78. *Nat. Biotechnol. 22*, 695–700.

Geiser DM, Jime´nez-Gasco M Kang S Makalowska I, Veeraraghavan N, Ward T. J., Zhang N, Kuldau G.A. and O'Donnel K. 2004.FUSARIUM-ID v. 1.0: A DNA sequence database for identifying Fusarium. *European Journal of Plant Pathology 110*: 473-479.

McCluskey, K. 2003. Fungal germplasm and databases. Pages 295-310 in: *Applied Mycology & Biotechnology: Fungal Genomics*. G. G. Khachatourians, D. K. Arora, eds. San Diego Technical Books, Inc. San Diego, CA.

Mitchell TK, Thon MR, Jeong JS, Brown D, Deng J and Dean RA (2003) The rice blast pathosystem as a case study for the development of new tools and raw materials for genome analysis of fungal plant pathogens. *New Phytol 159*:53-61.

Parkinson T (2002) The impact of genomics on anti-infectives drug discovery and development. *Trends Microbiol 10*:22-26.

Samson R. A., van der Aa H. A. and de Hoog G. S. (2004) Hundred years microbial resource centre. *Studies in Mycology 50*: 1–8.

Samuels, G.J., Chaverri, P., Farr, D.F., & McCray, E.B. (2007) *Trichoderma* Online, Systematic Botany & Mycology Laboratory, ARS, USDA. Retrieved April 25, from http://nt.ars-grin.gov/taxadescriptions/keys/TrichodermaIndex.cfm.

Sesma, A. & Osbourn, A. E. (2004).The rice blast pathogen undergoes developmental processes typical of root-infecting fungi. Nature 431, 582–586

Setubal, João Carlos, Medanis, João, 1997 "Introduction to computational molecular biology", *PWS Pub.*, Boston,

Sieburg, H. B. (1990). Physiological Studies *in silico. Studies in the Sciences of Complexity 12*, 321-342.

Soanes DM, Skinner W, Keon J, Hargreaves J, Talbot NJ. 2002 Genomics of phytopathogenic fungi and the development of bioinformatic resources. *Mol Plant Microbe Interact. 15*:421-7.

Soderlund Carol, Karl Haller, Vishal Pampanwar, Daniel Ebbole, Mark Farman, Marc J. Orbach, Guo-liang Wang,5 Rod Wing, Jin-Rong Xu, Doug Brown, Thomas Mitchell, and Ralph Dean (2006) MGOS: A Resource for Studying *Magnaporthe grisea* and *Oryza sativa* Interactions. *Molecular Plant-Microbe Interactions 19*, 1055-1061.

Liolios K, Nektarios Tavernarakis, Philip Hugenholtz and Nikos C. Kyrpides (2006) The Genomes On Line Database (GOLD) v.2: a monitor of genome projects worldwide. *Nucleic Acids Research 34*: (Database issue):D332-334.

Stanton ML., Blackmon, Barbara P., Rajagopalan, Ravi, Houfek, Thomas D., Sceeles, Robert G., Denn, Sheila O., Mitchell, Thomas K., Brown, Douglas E., Wing, Rod A., Dean, Ralph A. (2002) *Magnaporthe* DB: a federated solution for integrating physical and genetic map data with BAC end derived sequences for the rice blast fungus *Magnaporthe grisea. Nucl. Acids Res. 30*: 121-124

Strange, R.N. and Scott, P.R. 2005. PLANT DISEASE: A threat to global food security. *Annu. Rev. Phytopathol. 43*: 83–116.

Tripathy S. and Tyler B. M. 2007 DB: An Integrated resource for Phytophthora and Soybean EST sequences. Oomycete Molecular Genetics Network Workshop Asilomar Conference Center, Pacific Grove, CA, USA March 18-20, 2007. p 26.

Tunlid A, and Talbot NJ (2002) Genomics of parasitic and symbiotic fungi. *Curr Opin Microbiol 5*:513-519.

Turgeon BG, Kroken S, Lee BN, Bsaker SE, Amedeo P, Catlett N, Gunawardena U, Wagner E, Robbertse B, Wu J, Yoder OC, Glass NL and Taylor JW (2002) Comparative genomic analysis of fungal plant pathogens: secondary metabolites and mechanisms of pathogenesis. APS Symposium on Functional Genomics of Plant Pathogen Interactions, Milwaukee, Wisconsin, July 27-31, 2002.

Winnenburg, R., Baldwin, T.K., Urban, M., Rawlings, C., Köhler, J. and Hammond-Kosack, K.E. 2006. PHI-base: a new database for pathogen host interactions. *Nucleic Acids Research. 34* (Database issue):D459-D464

Xu Jin-Rong, You-Liang Peng, Martin B. Dickman, and Amir Sharonv (2006) The dawn of fungal pathogen genomics. *Annu. Rev. Phytopathol. 44*:337–66

Zeigler, R.S., Leong, S.A., and Teeng, P.S. 1994. *Rice Blast Disease.* CAB International, Wallingford, UK.

In: Current Advances in Molecular Mycology
Eds: Y. Gherbawy, R.L. Mach and M.K. Rai
ISBN 978-1-60456-909-4

Chapter IV

FUSARIA: MYCOTOXIN PRODUCTION AND SPECIES DIFFERENTIATION BY MOLECULAR MARKERS

***A. P. Ingle*[1]*, A. S. Karwa*[1]*, M. K. Rai*[1] *and Y. Gherbawy*[2]**

[1]Department of Biotechnology, SGB Amravati University, Amravati 444 602, Maharashtra, India;

[2]Biological Sciences Department, Faculty of Science, Taif University, Saudi Arabia.

ABSTRACT

The genus *Fusarium* is a common plant pathogen occurring worldwide, mainly associated with cereal crops, pulses, and other agronomic crops such as cotton, coffee, etc. *Fusarium* species are responsible for wilts, blights, root rots, and cankers in these plants. The importance of *Fusarium* species in current context is that infection may sometimes occur in developing seeds, especially in cereals, and also in maturing fruits, stored foodgrains and vegetables.

Fusarium species can produce over one hundred secondary metabolites, some of which can unfavourably affect human and animal health. The most important *Fusarium* mycotoxin, which can frequently occur at biologically significant concentrations in cereals are fusanisins, zearalenone and trichothecenes (deoxynivalenol, nivalenol and T-2 toxin). The compounds have been implicated as the causative agents in a variety of animal disease, such as pulmonary oedema, infertility, diarrhea, vomiting, anorexia, leucopenia, immunosuppression, skin and gastrointestinal irritation etc. and have been associated with some human diseases. Practical strategies are required to eliminate these mycotoxins and to overcome the problems of plant infections so as to save our economy and health. There is a pressing need to develop tools for rapid identification of Fusaria.

This can be achieved by producing the methods for early detection of *Fusarium* by using different techniques involving molecular markers; hence methods based on polymerase chain reaction (PCR) offer many new efficient tools for rapid detection and determination of relationships among *Fusarium* species, so the loss of crop can be minimized up to safe level.

Key words: Fungi, plant, diseases, resistance, genetics, genomics.

INTRODUCTION

Fusarium is one of the most important genera of plant pathogenic fungi on the earth. Not only plant but *Fusarium* species also cause infections in animals and in immunocompromised patients as a secondary infection (Spiewak, 1998). *Fusarium* mostly causes devastating infections in many kinds of economically important crop plants. *Fusarium* species are responsible for wilts, blights, root rots, and cankers in coffee, pine trees, wheat, corn, rice, cotton, raddish, tomato, onion, carnations and grasses. The importance of *Fusarium* species in current context is that infection may sometimes occur in developing seeds, especially in cereals, and also in maturing fruits and vegetables (Hocking and Andrews, 1987 and Chimbekujwo, 1999). The taxonomy and identification of the genus is complicated by the extreme variability of *Fusarium* species in their habitat (Snyder and Hansen, 1940; Nelson, 1992; Nelson *et al.*, 1994 and Summerell, 2003).

Fusarium also causes infections in stored foodgrains like maize, wheat, rice, barley, oat, sorghum, pulses, oil seeds and also in the dry fruits leading to change in the quality and nutrition's and ultimately great economic loss (Girish and Goyal, 1986).

India is predominantly an agrarian country with nearly three fourth of the population dependent on agriculture. The most outstanding achievement of Indian agriculture since independence is the phenomenal growth of foodgrains. Among them wheat, corn, sorghum, rice are the important foodgrains in India. But the fungal attack has been a major problem associated with these plants and more predominant is attack of *Fusarium*, which causes tremendous loss in Indian economy. To avoid these problems there is a need to control the fungal attack in order to increase the production of foodgrains (Girish and Goyal, 1986).

It was claimed that our knowledge of *Fusarium* began with the discovery of "diseases of cereals" on which a paper was published by Bennett in 1928, but Link was the first man who described the genus *Fusarium* and he described *Fusarium roseum* as the first species in 1809 (Moss and Smith, 1984). *Fusarium* is a filamentous fungus widely distributed on plants and in the soil. It is found in normal mycoflora of commodities, such as rice, bean, soybean, and other crops (Pitt and Hocking, 1985).

More than 40 to 50 species of *Fusarium* have been discovered of which more than 24 species are reported to cause devastating infections in plants, animals and in human beings (Keith, 1996).

Fusarium is one of the most drug-resistant fungi. The species *Fusarium solani* in general tends to be most resistant of all. *Fusarium* strains yield quite high MICs for the antibiotics like flucytosine, ketoconazole, miconazole, fluconazole, itraconazole, and posaconazole.

Among the *Fusarium* species *F. solani* is the most pathogenic species and other than this *F. moniliforme, F. graminearum, F. oxysporum, F. affine, F. dimerum, F. radicicola, F. roseum, F. vasinfectum, F. poae,* etc. are the most predominant pathogenic species infecting economically important crop plants. And sub species or strains of certain above *Fusaria* are also found to cause great affect on their respective host crop plants. The list of some pathogenic *Fusaria* and their respective host are shown in table 1 and 2.

Table 1. Plant pathogenic species of *Fusarium* and their respective host plants (Raabe *et al.*, 1981)

***Fusarium* species**	**Host plant**	**Disease type**
F. oxysporum	Garden bean (*Phaseolus vulgaris*), Potato (*Solanum tuberosum*)	Wilt
F. oxysporum f.sp. *lycopersici*	Tomato (*Lycopersicon esculentum*)	Wilt
F. oxysporum f.sp. *melonis*	Muskmelon (*Cucumis melo*)	Wilt
F. oxysporum f.sp.*tracheiphilum*	Soybean (*Glycine max*)	Wilt
F. oxysporum f.sp.*zingiberi*	Ginger (*Zingiber officinalis*)	Fusarium yellow
F. oxysporum f.sp. *niveum*	Watermelon (*Citrullus vulgaris*)	Wilt
F. oxysporum f.sp. *dianthi*	Carnation (*Dianthus caryophyllus*)	Wilt
F. oxysporum f.sp. *cubense*	Banana (*Musa acuminate*)	Panama disease
F. moniliforme	Pineapple (*Ananas comosus*) Banana (*Musa acuminate*) Sorghum (*Sorghum bicolor*) Sugarcane (*Saccharum officinarum*) Maize (*Zea mays*)	Fruitlet core rot Heart rot Blight Root rot Seed rot
F. graminearum	Cultivated oats (*Avena sativa*)	Head Blight
F. solani	Potato (*Solanum tuberosum*) Papaya (*Carica papaya*) Sweet Potato (*Ipomoea batatas*)	Dry Rot Root Rot Root Rot
F. roseum	Carnation (*Dianthus caryophyllus*)	Root Rot
F. radicicola	*Hibiscus* sp.	Root Rot
F. culmorum	Potato (*Solanum tuberosum*)	Dry Rot
F. affine	Pineapple (*Ananas comosus*)	Root Rot

Table 2. Some other *Fusarium* species and their host plants

Fusarium	**Host**	**Disease**	**Reference**
F. graminearum	Wheat	Head scab	Saharan *et al.*, 2007
F. oxysporum	Raddish	Wilt	De Boer *et al.*, 2003
F. oxysporum f. sp. *gladioli*	*Gladiolus*	Yellowing and corm rot	Dallavallel *et al.*, 2002
F. oxysporum f. sp. *pisi*	*Pisum sativum*	Wilt	Okubara *et al.*, 2002
F. oxysporum f. sp. *basilici*	Basil	Wilt	Chiocchetti *et al.*, 2001
F. oxysporum f. sp. *canariensis*	*Phoenix canariensis (Date Palms)*	Wilt	Smith *et al.*, 2003

MOLECULAR DETECTION OF *FUSARIUM*

For food born toxigenic *Fusaria* it is very necessary to develop PCR based molecular methods for early and accurate detection so that it helps to minimize the problems associated with it. Now a day's different methods are available for the molecular detection of *Fusarium*, which are also helpful to differentiate a particular strain of any species from other strains of the same species (Shuxian *et al.*, 2002).

The polymerase chain reaction (PCR) for amplification of specific nucleic acid sequences was introduced by Saiki *et al.* in 1985, and has subsequently proved to be one of the most important scientific innovations.

PCR uses a thermostable polymerase to produce multiple copies of specific nucleic acid regions quickly and exponentially, including non-coding regions of DNA as well as particular genes. For example, starting with a single copy of a 1 KB DNA sequence, 10^{11} copies (or 100 ng) of thc same sequence can be produced within a few hours. Once the reaction has occurred, number of methods for identification and characterization of the amplification products according to their size following migration on agarose gels were used.

The PCR based method requires a DNA template containing the region to be amplified and two oligonucleotide primers flanking this target region. The amplification is based on the use of a thermostable DNA polymerase isolated from *Thermus aquaticus*, called *taq* polymerase (Saiki *et al.*, 1988). All PCR reaction components are mixed and the procedure consists of a succession of three steps, which are determined by temperaure conditions: template denaturation, annealing of primer and extension. In the first step, the incubation of the reaction mixture at a high temperature (90-95°C) allows the denaturation of the double-stranded DNA template. By cooling the mixture to an annealing temperature, which is typically around 55°C, the target-specific oligonucleotide primers anneal to the 5' end of the two single-stranded templates. For the extension step, the temperature is raised to 72°C and the primer-target strands. The time of incubation for each step is usually 1-2 min. This sequence newly synthesized DNA strands which are separated from the original strands by denaturation and each strand serves again as template in the annealing and extension steps. Theoretically, *n* cycles of PCR allow a 2^n-fold amplification of the target DNA sequence. Now a days different PCR based methods like RAPD, RFLP, AFLP etc. can be effectively used for the molecular identification of *Fusaria.*

1. Random Amplified Polymorphic DNA (RAPD)

The random amplified polymorphic DNA (RAPD) or arbitrarily primed polymerase chain reaction (AP-PCR) fingerprinting assay detects small inverted nucleotide sequence repeats through genomic DNA (Welsh and McClelland, 1990 and Williams *et al.*, 1990).

Amplification with RAPD primers is extremely sensitive to single-base changes in the primer-target site. This feature suggests that RAPD-PCRs should be highly useful for phylogenetic analysis among closely related individuals, but less useful for analysis of genetically diverse individuals. RAPD primers are useful for distinguishing species, for discriminating between different isolates of the same species, and to measure similarity among individuals within natural or artificial (through breeding) populations within a species. The identification of shared characters in this analysis is limited to the resolution of the agarose gel. In fact, many more bands are present among the amplification products that are detected by ethidium bromide staining. Caetano-Annoles *et al.,* (1991) were able to detect over 100 bands amplified with a single random primer, by resolving the reaction products on a polyacrylamide gel and staining with silver.

Random amplified polymorphic DNA (RAPD) offers several advantages that may be useful in studying formae specialis and races of *F. oxysporum*, to identify RAPD markers for formae specialis and race identification (Grajal-Martin *et al.*, 1993 and Gherbawy, 1999). RAPD reduces the time needed for race identification in diseased plants, and provides genetic information on isolates studied, allowing for fingerprinting of isolates (Welsh and McClelland, 1990; Weising *et al.*, 1991; Gherbawy, 1999; and Gherbawy and Abdelzaher, 2002). Random PCR approaches are being increasingly used to generate molecular markers, which are useful for taxonomy and for characterizing fungal populations. The main advantage of these approaches is that previous knowledge of DNA sequences is not required, so that any random primers can be tested to amplify any fungal DNA. RAPD primers are chosen empirically and tested experimentally to find RAPD banding patterns, which are polymorphic between the taxa studied. These advantages also include quickness, and small amount of template DNA.

RAPD-PCR assays have been used extensively to define fungal populations at species, intraspecific, race and strain levels. In general, most studies have concentrated on intraspecific grouping, although others have been directed at the species level. Some examples of RAPD-PCR at species level include the production of species-specific probes and primers from RAPD data for *F. oxysporum* f. sp. dianthi, *Phytophthora cinnamomi, Tuber magnatum, Glomus mosseae, F. sambucinum,* members of *Fusarium* section *Fusarium*, *F. oxysporum* (Dobrowolski and O'Brein, 1993; Lanfranco *et al.,* 1993 and 1995; Manulis *et al.*, 1994; Hering and Nirenberg, 1995; Yoder and Christianson, 1998; O'Donnell *et al.*, 1999; Gherbawy and Yaser 2003).

2. Restriction Fragment Length Polymorphism (RFLP)

Restriction Fragment Length Polymorphism (RFLP) is a technique in which organisms may be differentiated by analysis of patterns derived from cleavage of their DNA. If two organisms differ in the distance between sites of cleavage of a particular restriction endonuclease, the length of the fragments produced will differ when the DNA is digested with a restriction enzyme. The similarity of the patterns generated can be used to differentiate species and even strains from one another.

Restriction endonucleases are enzymes that cleave DNA molecules at specific nucleotide sequences depending on the particular enzyme used. Enzyme recognition sites are usually 4 to 6 base pairs in length. Generally, the shorter the recognition sequence, the greater the number of fragments generated. If molecules differ in nucleotide sequence, fragments of different sizes may be generated. The fragments can be separated by gel electrophoresis. Restriction enzymes are isolated from a wide variety of bacterial genera and are thought to be part of the bacterial defense against invading viruses. These enzymes are named by using the first letter of the genus, the first two letters of the species, and the order of discovery.

For fungal mitochondrial genomes, which are usually small (19-121 kb) and exist in relatively high copy number (Grossman and Hudspeth, 1985), RFLPs are usually easy to detect. Total DNA of the fungus is extracted and the mitochondrial DNA separated from nuclear DNA by centrifugation in CsCl gradients containing bis-benzimide (Garber and Yoder, 1985). The relatively small size of mitochondrial genomes results in a simple banding pattern when cut with restriction endonucleases. Mitochondrial RFLP markers have been used in a number of fungi although the extent of variation differs considerably. Variation in mitochondrial (mt) DNA has been used to assess inter- and intra- species relationships in various fungi (Taylor *et al.*, 1986).

RFLP of amplified fragments (PCR-RFLP) analysis has been used to address the research problems of fungal population biology such as the differentiation of species; species forms and isolates (Marcon and Powell, 1987 and Carter *et al.*, 2000). This method has often utilized the analysis of nuclear ribosomal DNA (nrDNA) sequences that are found in all eukaryotic cells and contain both regions that showed substantial resolution at different taxonomic units in a majority of fungi including *Fusarium* species (Nicholson *et al.*, 1993; Appel and Gordon, 1995; Donaldson *et al.*, 1995; Edel, *et al.*, 1996a and 1996b, Waalwijk, *et al.*, 1996; Blanz and Unseld, 1987; Laguerre *et al.*, 1994; Brayford, 1996; Hyun and Clark, 1998 and Mishra *et al.,* 2003). PCR-RFLP is a simple and inexpensive method compared to traditional RFLP or sequence analyses as it avoids the need for blotting, probing and/or sequencing.

To minimize the problems associated with toxigenic *Fusaria* developing early detection methods were necessary. In this regard a new method was described for identification of *F. graminearum* in cereal samples by DNA Detection Test Strips. DNA Detection Test Strips^{TM} were used for PCR-product detection and the method was compared to agarose gel electrophoresis. A minimum of 0·26 ng of purified target DNA was detectable with the Test Strip^{TM} Detection limit in less contaminated samples was slightly lower when gel electrophoresis was used for amplicon detection. In highly contaminated samples, detection limits of both methods were similar (Knoll *et al.*, 2002). PCR detection of *F. oxysporum* f. sp. *basilici* can be easily done so that it helps in early detection of infection in Basil (Chiocchetti *et al.,* 2001).

3. rRNA (rDNA) Sequence Comparisons:

Sequence comparisons of ribosomal RNA (rRNA) and its template ribosomal DNA (rDNA) have been used extensively to assess both close and distant relationships among many kinds of organisms. The interest in rRNA/rDNA comes from two important properties; First, ribosomes are present in all cellular organisms and appear to share a common evolutionary origin, thus providing a molecular history shared by all organism; Second, some rRNA / rDNA sequences are sufficiently conserved so that they are homologous for all organisms and serve as reference point that enable alignment of the less conserved areas used to measure evolutionary relationships (Kurtzman, 1994).

Sequences of rDNA are often used for taxonomic and phylogenetic studies because they are found universally in living cells in which they have an important function; thus, their evolution might reflect the evolution of the whole genome. These sequences also contain both variable and conserved regions, allowing the comparison and discrimination of organisms at different taxonomic levels. The nuclear rDNA in fungi is organized as rDNA unit, which is tandemly repeated. One unit includes three rRNA genes: the small nuclear (18S-like) rRNA, the 5.8S rRNA, and the large (28S-like) rRNA genes. In one unit, the genes are separated by two internal transcribed spacers (ITS1 and ITS2). The 18S rRNA evolves relatively slowly and is useful for comparing distantly related organisms whereas the non-coding regions (ITS) evolve faster and are useful for comparing fungal species within a genus or strains within a species. Some regions of the 28S rDNA are also variable between species.

Ribosomal RNA occurs in several size classes in eukaryotes. The genes coding for large-subunit (25S to 28S), small-subunit (16S to 18S), and 5.8 rRNAs occur as tandem repeats with as many as 100 to 200 copies. The separately transcribed 5S r RNA gene may also be included in the repeats (Garber *et al*, 1988). Each of the RNA size classes have been examined for extent of phylogenetic information present.

The first rRNA to be sequenced was the 5S rRNA. Because of the conserved nature and small size (ca. 120 nucleotides) of 5S rRNAs, their sequences were easily determined and have been widely used for estimating broad phylogenetic relationships (Hori and Osawa, 1979).

The 5.8S rRNA sequence analysis was not often used in the studies of yeast and fungal phylogeny. Its sequence includes only about 160 nucleotides and do not offer much more information than the 5S rRNA. Besides, that the 5.8S rRNA has modified nucleotides and is therefore less prone to sequencing.

There are different types of conserved sequences present on rRNA (rDNA) as described below:

ITS Regions

The internal transcribed spacers (ITS) are noncoding regions of DNA sequence that separate genes coding for the 28S, 5.8S, and 18S ribosomal RNAs. These ribosomal RNA (rRNA) genes are highly conserved across taxa while the spacers between them may be species-specific. The conservation of the rRNA genes allow for easy access to the ITS regions with "versatile" primers for polymerase chain reaction (PCR) amplification. The variation in the spacers has proven useful for distinguishing among a wide diversity of difficult-to-identify taxa. The ITS region is subdivided into the ITS1 region, which separates the 18S and 5.8S rRNA genes, and the ITS2 region, which is found between the 5.8S and 28S rRNA genes (James *et al*., 1996).

Two taxon-selective primers have been developed for quick identification of the *Fusarium* genus. These primers, ITS-Fu-f and ITS-Fu-r were designed by comparing the aligned sequences of internal transcribed spacer regions (ITS) of a range of *Fusarium* species. The primers showed good specificity for the genus *Fusarium*, and approximately 389-bp product was amplified exclusively. PCR sensitivity ranged from 100 fg to 10 ng for DNA extracted from *F. oxysporum* mycelium (Kamel, 2003).

IGS Regions

The research team of Bhaba Atomic Research Institute, Mumbai also developed a simple and rapid molecular method for distinguishing between races of *F. oxysporum* f.sp. *ciceris* from India. It was confirmed by PCR-amplification of the IGS region followed by digestion with *EcoRI* and a set of other enzymes. It is suggested that amplification of the IGS region and digestion with restriction enzymes could be used to study polymorphism in *F. oxysporum* f.sp. *ciceris* (FOC), and to rapidly identify the races existing in India (Chakrabarti *et al.*, 2001).

A method was developed to obtain genus-specific DNA probes. It consists of specific amplification of the intergenic spacer between the 18S and 5.8S ribosomal RNA genes, using primers deduced from conservative ribosomal DNA sequences. The utility of the method is demonstrated on isolation of the 209 b.p. spacer fragment from the genomic DNA of a plant pathogenic fungus *F. oxysporum* (Irisbaev, 1991).

Method for molecular characterization was developed for *F. oxysporum* f. sp. *ciceri* causing wilt of chickpea. Thirty isolates of *F. oxysporum* f. sp. *ciceri* were isolated from rhizosphere soil of chickpea from different locations in Northern India. The amount of genetic variation was evaluated by polymerase chain reaction (PCR) amplification with a set of 40 RAPD primers and 2 IGS primers. Less than 10% of the amplified fragments in each case were polymorphic. Genetic similarity between each of the isolates was calculated and results indicate that there was little genetic variability among the isolates collected from the different locations. At the 0.75 similarity index the isolates divides into three groups. Isolates Foc-A18, Foc-A19, Foc-A20 forming a similar group and far different from other isolates. (Singh *et al.*, 1998 and 2006).

MYCOTOXINS

The toxins produced by Fungi are called as mycotoxins ("mykes" means"fungus"). Mycotoxins are secondary metabolites of fungal origin. Diseases that results from ingestion of mycotoxin-contaminated feed or food by animals or humans is called mycotoxicoses.

Mycotoxicoses (poisoning due to mycotoxins) have several common symptoms that are shared from species to species and toxin-to-toxin. These include:

1. Drugs and antibiotics are not effective in treatment.
2. The symptoms can be traced (associated) to food or feedstuffs.
3. Testing of food/feedstuffs reveals fungal activity.
4. The symptoms are not transmissible to control subjects.
5. The degree of toxicity in subjects is influenced by age, sex, and the nutritional status of the host.
6. Outbreak of symptoms is seasonal.

The mycotoxin producing fungi are aerobic, microscopic and colonise many kinds of food from the field to the table. Mycotoxins can appear in food and animal feed as a result of fungal infection of the crop, for example *Fusarium* ear diseases in cereals, or the infection of stored products. Not all fungi can produce mycotoxins, it is estimated that among the thousands of species of fungi, only about 100 are known to produce mycotoxins under favorable environmental conditions such as at specific levels of moisture, stress and the correct temperature. Even those with the ability to produce mycotoxins may not produce them all the time. The absence of mycotoxins doesn't ensure the absence of fungal spores, so it is possible for fungi to 'appear' when the temperature and humidity are favourable. In addition, the mycotoxins are very resistant to temperature treatments and to conventional food processes such as cooking, freezing, etc.

Mycotoxins in Plants

There are five different mycotoxins generally produced in plants as follows:

1) Deoxynivalenol / Nivalenol
2) Zearalenone
3) Ochratoxin
4) Fumonisins
5) Aflatoxin

Out of these five mycotoxins Deoxynivalenol / Nivalenol, Zearalenone and Fumonisins are the three mycotoxins predominantly produced by *Fusarium* species. Deoxynivalenol / Nivalenol are produced by *F. graminearum, F. crookwellense, F. culmorum* in wheat, maize, barley, etc. Zearalenone is also produced by same *Fusarium* species and its presence is mainly reported in wheat and maize. The third Fumonisins i.e. fumonisin B1, fumonisin B2 are reported in maize and produced by *F. moniliforme.* While rest of two mycotoxins are produced by *Apsergillus* and *Penicillium* Species. Other than this T-2 toxin is also found in a variety of grains but its occurrence, to date is less frequent than the preceding five mycotoxins.

Cotton (*Gossypium*) is a major crop of global significance which is grown primarily for fiber and seeds all over the world. *Fusarium* species such as *F. solani* and *F. equiseti* are highly pathogenic to cotton, it was observed in survey of cotton fungi in Nigeria in 1992 and 1993 by counting the number of isolates in each 100 infected plants per plot. Approximately 90% of the isolated fungi were *F. solani* and *F. equiseti,* both pathogenic; *F. solani* isolates were more virulent and frequent than *F. equiseti*. The high frequency and virulence of both fungi make them important pathogens of cotton in the area (Chimbekujwo, 1999).

Fusarium head blight (scab) is an important disease affecting wheat and barley caused by *F. graminearum.* Small grain producers in the United States have been struggling to manage Fusarium head blight since early 1900s. Recent epidemics of the disease have caused devastating yield losses in many states where wheat and barley are produced. Fusarium head blight affects the developing heads of small grains directly, and yield losses that exceed 45 percent are common during years when disease is severe. Fusarium head blight also negatively affects grain quality, often resulting in lower test weights and mycotoxin contamination (Wolf, 2003).

F. oxysporum f.sp. *cepae* causes the basal rot in onion in which it affects the root of onion and leads to infection. Fusarium basal rot can cause crop failure and economic losses.

F. moniliforme is associated with disease at all stages of maize plant development, infecting the roots, stalk, and kernels (Cole *et al.*, 1973). This fungus is not only the most common pathogen of maize; it also is among the most common fungi found colonizing symptomless maize plants. *F. moniliforme* is an almost constant companion of maize plants and seed. In many cases, its presence is ignored because it is not causing visible damage. Symptomless infection can exist throughout the plant, and seed-transmitted strains of the fungus can develop systemically to infect the kernels (Munkvold *et al.*, 1997; Torres, 2001 and Miller, 2002).

F. verticilliodes can also cause the ear rot in maize, which also leads to crop failure and loss (Bush *et al.*, 2003). *F. oxysporum* f. sp. *ciceri* cause wilt in the chickpea which is the world's third most important pulse crop after bean and pea, which is greatly affected by this infection and cause great economic loss in developing countries like India in which this crop is of most importance (Chakrabarti *et al.*, 2001 and Singh *et al.,* 2006).

Many *Fusarium* species like *F. oxysporum* and their strains, *F. solani*, *F. subglutinans* causes wilt and crown rot of tomatoes, wilting in cucumber, cotton, watermelon, tobacco, crucifers, soybean, sweet potato and root and lower stem rot in many other vegetable and crop plants. Wilting is main infection caused by *Fusarium* because *Fusarium* plugs the water conducting tissue (xylem) in root or stem due to which plant doesn't get sufficient water uptake and it leads in to wilt (Kucharek *et al.*, 2000).

Mycotoxins in Animals and Humans

There are different types of mycotoxins reported in animals and human beings. These are either produced by *Fusarium* after getting entry in the body of respective host or these can be transferred through the intake of contaminated food like infected grain, vegetables where they are present during their infection (King *et al.*, 1986). *Fusarium* also produces many toxins in animals and human being, which are responsible for many severe infections and disorders in both animals and human beings. In animals mainly trichothecenes, T-2 toxin, fusarenon X, deoxynivalenol and nivalenol are the toxins secreted by *Fusarium* found in association with infections in poultry animals, in pigs, dogs, horses, cattle, sheep and fish etc. Mycotoxins normally occur in feed stuff which leads to animal infection (Mirocha *et al.*, 1976).

Only four species were judged to be most important from the viewpoint of human health, *F. sporotrichioides, F. equiseti, F. graminearum and F. moniliforme.* Infection of *F. solani* have been also found in immunocompromised patients. These species also produces trichothecenes, T-2 toxin, fusarenon X, deoxynivalenol, nivalenol, diacetoxyscirpenol, butenolide, fumonisin B etc. (Maracas *et al.*, 1984 and Hocking and Andrews, 1987). These mycotoxins produce severe infection in human beings.

Human *Fusarium* infection or Fusariosis usually occurs in immunocompromised individuals, such as those affected by other diseases like AIDS (HIV) or even a severe case of the common cold. Extreme exhaustion can also produce an immunocompromised state. The first reference of *Fusarium* infection in human's dates back in 1916 in an article published in French by Dr. N.V. Greco in an Argentine Medical Journal in which he described a fungal infection of the nose, which he believed to be caused by a *Fusaria. Fusarium* also causes other infections like skin infection, onychomycosis, keratitis, endophthalmitis, arthritis and alimentary toxic aleukia (ATA) which is caused by most notorious T-2 toxin (Joffe, 1978; Larone, 1995 and Spiewak, 1998). *F. solani* was found to cause breast abscess (Anandi *et al.*, 2005). *F. dimerum* is reported to cause human eye infection (Vismer, 2002).

Mycotoxins and Health Hazards

Trichothecenes:

Pathogen: The trichothecenes are a chemically related family of compounds that are produced by fungi such as *Fusarium, Trichoderma, Myrothecium,* and *Stachybotrys.* The trichothecene mycotoxins have been isolated and found in Canada, England, Japan, South Africa, and the United States. The most common mycotoxins in the trichothecene family found in grain are DON and T-2, with ZEN and Fumonisin also commonly found (Krogh, 1987). *F. gramminearum*, the parent fungi that produces DON, causes both Gibberella Ear Rot in corn, and head scab in wheat. Over a ten-year period the Mycotoxin Laboratory at North Carolina State University found *Fusarium* species of fungi in almost every lot of corn tested. DON was detected in over 60 percent of poultry and dairy feed tested, and ZEN was found in 15 to 20 percent of feeds tested.

The processing of grain with toxins in the trichothecene group generally does little to remove the toxin. Milling, baking or boiling has only a slight effect in removing the toxins. Tests conducted on finished products contaminated with these mycotoxins have shown that 50 to 60 percent of the toxins are transmitted to the finished product. In some cases, such as the tempering of grain to reach a desired moisture level, the toxins have actually increased due to the proper environmental conditions needed for toxin production. The toxins can be transmitted to final products such as flour, bread, crackers, and cereal (Krogh, 1987).

Health Hazards. The trichothecene family of mycotoxins affects each species of animals in different ways. Testing done so far indicates that poultry has the highest resistance to these toxins. Cattle, sheep, and goats have some level of resistance due to their multiple digestive processes, while animals of the monogastric digestive process seem to have the least resistance to the toxins. Swine seem to be the most sensitive, partly due to their increased sense of smell, which leads to feed refusa (Miller and Trenholm, 1994). The greatest problems associated with these toxins are from prolonged feed intake at low contamination levels. The effects depend on the specific toxin, the duration of exposure, and the type of animal involved. All animal species suffering from chronic toxicoses show very good to excellent signs of improvement when the contaminated feed is removed. Few long-term side effects remain with most of this group of toxins if diagnosis is made quickly before the general health of the affected animals is compromised (Krogh, 1987).

Structure

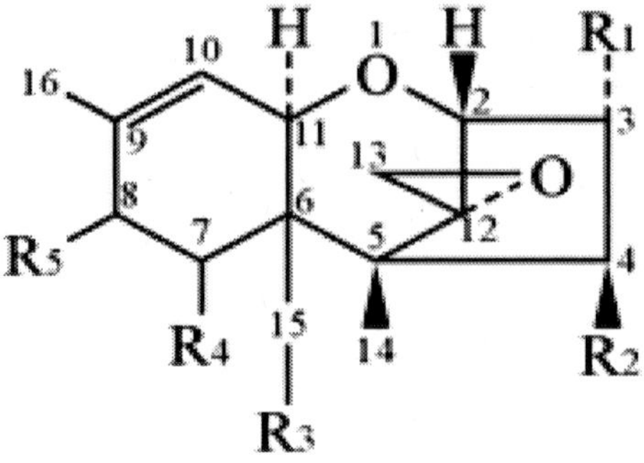

Deoxynivalenol (DON):

Pathogen: F. graminearum is the parent fungi of deoxynivalenol (DON) or vomitoxin. Wheat and barley are the most commonly effected grain crops but the same fungus does infect corn. In the field, it shows up as a brown discoloration at the base of barley glumes, a pink to reddish mold on the glumes and kernels of the wheat heads and the tips of the ears of corn. Spores from the mold stage of the fungi can stay dormant on infected residues left on or in the soil. Contamination is most severe in fields where corn follows corn, or where corn follows wheat, especially if the previous crop was infected (Woloshuk, 1994).

Ecology: The optimal temperature range for the DON mold is 70 to 85° F with moisture levels preferred to be greater than 20 percent. There are exceptions to be noted. The mold can survive temperatures as low as 0 ° F for short periods of time. This particular fungi has two distinct growth cycles, with the mold growing during the warm temperatures of daytime, while the toxins are produced during the cooler temperatures of the night (Cheeke and Lee, 1985).

Health Effects: The symptoms associated with DON poisoning are many and varied which sometimes leads to its misdiagnosis as a problem. At low levels of toxicoses the symptoms may include behavioral and skin irritations, feed refusal, lack of appetite, and vomiting. In later stages, symptoms may include hemorrhage and necrosis of the digestive tract, neural problems, suppression of the immune system, lack of blood production in the bone marrow and spleen, and possible reproductive problems including birth defects and abortion (Miller and Trenholm, 1994) (Table. 3).

Structure and Chemical name:

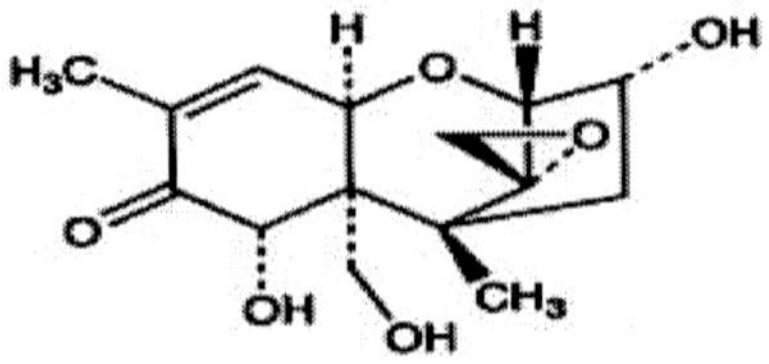

Deoxynivalenol (Vomitoxin) = 3",7",15-trihydroxy-12,13-epoxytricho thec-9-en-8-one.

Table 3. FDA advisory levels for DON

Class of Animal	Portion of Diet	Maximum DON level
Humans	Finished wheat products (flour, bran and germ)	1 ppm
Beef and feedlot cattle older than 4 months	Grains and grain byproducts not to exceed 50% of diet	10 ppm
Chickens	Grains and grain byproducts not to exceed 50% of diet	10ppm
Swine	Grains and grain byproducts not to exceed 20% of diet	5 ppm
All other animals	Grains and grain byproducts not to exceed 40% of diet	5 ppm

Zearalenone (ZEN):

Pathogen: Zearalenone is very similar to deoxynivalenol (DON) in most aspects with a few exceptions.

Ecology: The growing conditions of ZEN are very comparable to DON, with the optimal temperature range of 65° to 85° F. A drop in temperature during growth also stimulates the production of toxins (Cheeke and Lee, 1985). The moisture content required by ZEN is also similar to DON at 20 percent or greater. But if the moisture content during growth drops below 15 percent the production of toxins is halted. This is one of the reasons that corn for storage must be dried to moisture levels less than 15 percent (Woloshuk, 1994).

Health Effects: The greatest difference between ZEN and DON is the way the toxin acts in animals. ZEN mimics the hormone estrogen in the way it affects animal tissue. Swine are the most sensitive to its effects with levels of 1 ppm causing feed refusal. Continued consumption of contaminated grain will cause estrogenism (health problems related to the reproductive system). These effects include swelling of the reproductive organs including the genital and mammary glands, interruption of the reproductive cycles, birth defects, and atrophy of the ovaries and testes. In male animals, feminization occurs with enlargement of the mammary glands and loss of sex drive (Cheeke and Lee, 1985).

Structure and Chemical name:

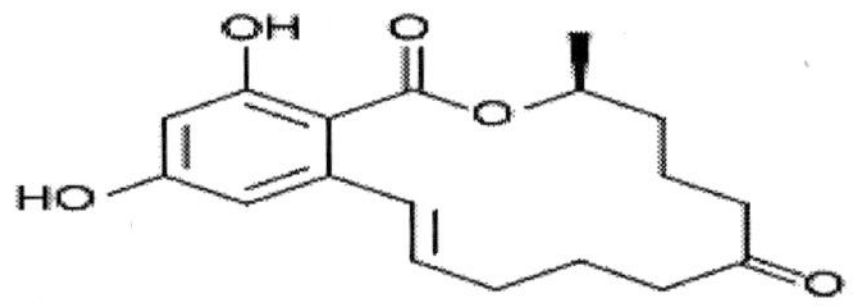

Zearalenone = 6-(10-hydroxy-6-oxo-trans-1-undecenyl)-β-resorcylic acid lactone

Fumonisin:

There are two types of fumonisins viz. Fumonisin B1 and Fumonisin B2. Fumonisin B2 is a mycotoxin produced by the fungus *F. moniliforme*. It is a structural analog of Fumonisin B1. Fumonisin B2 is more cytotoxic than Fumonisin B1. Fumonisin B2 inhibits sphingosine acyl-transferase. Fumonisine B2 and other umonisines frequently contaminate maize and other crops.

Pathogen: F. moniliforme is the parent fungi species of fumonisin. This fungus causes Fusarium Ear Rot in corn, which is the most common disease of corn in the United States Midwest region. Testing of corn-fields has shown that over 90 percent of fields are affected by these fungi in one of its various strains (Woloshuk, 1994). The mold appears on the corn ears as cottony white to light grey filaments between the corn kernels. As the mold progresses the kernels will turn grey to light brown. The fumonisin toxin can grow in the kernels even with no apparent outward signs of mold. Testing of the grain is the only positive means of verifying whether fumonisin is present or not (Woloshuk, 1994).

Ecology: Growing conditions vary widely. The temperature and moisture ranges are so wide spread as to include most of the Northern and Southern Hemispheres. The one common factor associated with fumonisin is that higher incidence of infections seem to occur after periods of drought which stress the plants immune system (Woloshuk, 1994). Fumonisin causes the corn kernels to become brittle and crack more frequently than is normal. The more the grain is handled the more cracking and breaking occurs, giving the fungi more host material to grow on. For this reason corn screenings should be very suspect when used as feed, especially in horses. Testing has shown that screenings contain a higher level of fumonisin toxin (and mycotoxins in general) than the whole grain product.

Health Effects: Fumonisin is one of the mycotoxins that has only recently been discovered and has been little studied. The related health effects have shown few effects in humans and most animals other than swine and horses. While in depth research is lacking, fumonisin has shown a high degree of toxicity in preliminary studies conducted in horses. Swine have shown little or no effects with the only preliminary symptoms to be possible respiratory problems and possible links to the liver and kidneys. Toxin levels of as low as 5 ppm have shown direct links in horses with symptoms which include: disorientation, walking/agitation, derangement, colic, head pressing, blindness, and death. The toxin seems to attack the liver and kidneys, which is similar to other mycotoxins except in the severity. Fumonisin has also been linked to equine leukoencephalomalicia, also known as “Blind Staggers” (a complete breakdown of the neural system in the brain) that has a high mortality rate.

Structure and Chemical name:

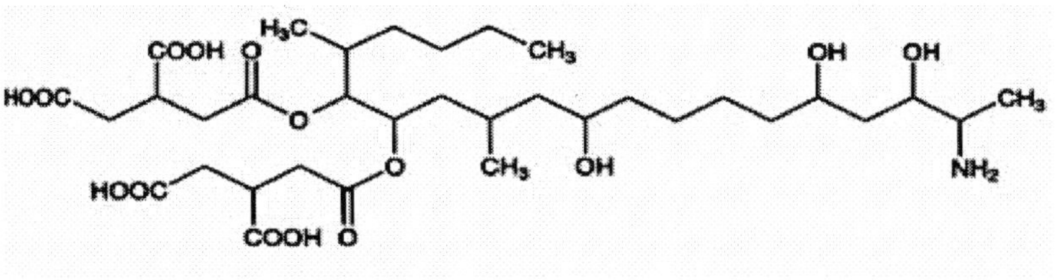

FumonisinB1= 1,2,3-propanetricarboxylic acid 1,-l-[1-(12-amino-4,9,11-trihydroxy-2-methyltridecyl)-2-(1-methylpentyl) -1,2-ethanediyl] ester; macrofusine $^{+}$.

Nivalenol (NIV):

Pathogen: Nivalenol is produced by the *F. nivale* fungi and has also only recently been isolated. Little is known of its growth cycle or habitat range. Studies have shown it to be much rare in occurrence and have only been found in a few samples of barley, wheat, wheat flour, and rice (Krogh, 1987).

Health Effects: Though little actual test data has been produced, the results so far cause scientists to be extremely cautious. Preliminary testing shows that NIV is thought to be 10 times more potent than DON. If the advisory level for DON is used as a guide for toxicity, NIV would have an advisory level of only 0.2 ppm (Krogh, 1987).

Structure and Chemical name:

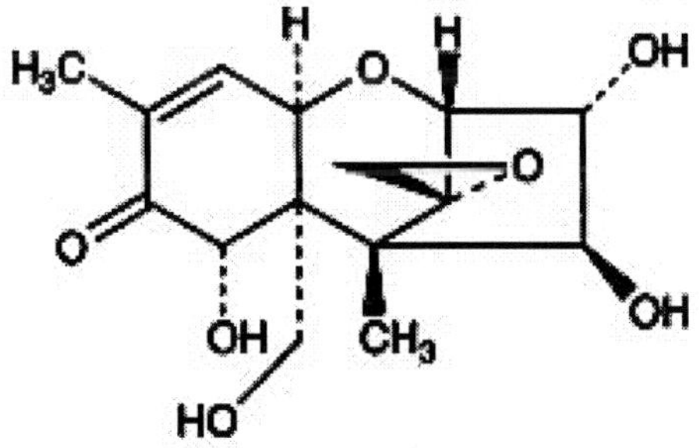

Nivalenol = 3", 4$, 7", 15-tetrahydroxy-12, 13-epoxytrichothec-9-en-8-one.

T-2 toxin:

Pathogen: F. tricinctum and some strains of *F. roseum* produce T-2. T-2 has been found in corn in the field, silage, and prepared feeds made with corn.

Health Effects: During WWII, a very severe human disease occurred in the former Soviet Union. Alimentary toxic aleukia (ATA) is believed to have been caused by T-2 and HT-2 in grain left to over winter in the field. When this grain was consumed, severe mycotoxicoses occurred. ATA results in a burning sensation in the mouth, tongue, esophagus, and stomach. Eventually the blood making capacity of the bone marrow is destroyed and anemia develops. In the final stages hemorrhaging of the nose, gums, stomach, and intestines develops and the mortality rate is high. In poultry, T-2 may produce lesions at the edges of the beaks, abnormal feathering, reduced egg production, eggs with thin shells, reduced body weight, and mortality.

Structure and Chemical name:

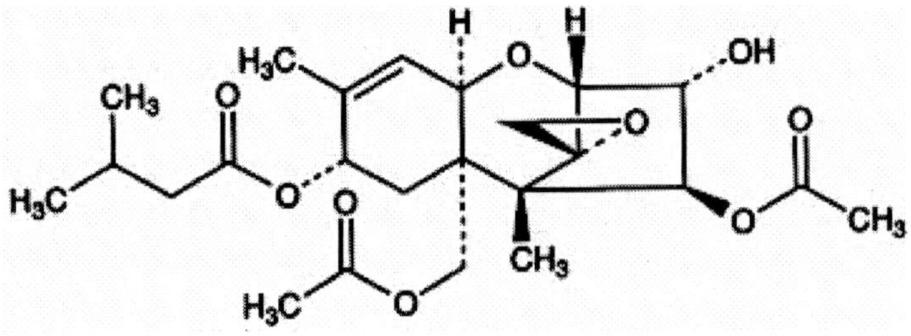

T-2 toxin = 4$, 15-diacetoxy-3"-hydroxy-8"-(3-methylbutyrlyloxy)-12,13- epoxytrichothec-9-ene.

Mycotoxin Production

Much research work has been done on *Fusarium* and new techniques for its production, isolation, detection and characterization of their mycotoxin. The temperature, pH, moister and growth period are the important physical parameters for production of mycotoxins in *Fusarium* (Paulo, 2002). Many methods developed are for the detection of mycotoxins produced by *Fusarium* (Richardson *et al.*, 1984).

Radiolabeled T-2 toxin can be synthesized biologically by *F. tricinctum.* By incubating the *F. tricinctum* NRRL 3299 on a solid rice medium in the presence of [1-14C] sodium acetate, [2-3H] mevalonic acid, [2-14C] mevalonic acid, or [5-3H] mevalonic acid yielded preparations of radiolabeled T-2 toxin with specific activities (Hagler, 1981).

Daniel *et al.* (1982) discovered that strains of *Fusarium* produced high levels of T-2 toxin when cultured on certain media absorbed into vermiculite. Modified Gregory medium was nutritionally complex (2% soya meal, 0.5% corn steep liquor, 10% glucose) and, when inoculated with the appropriate fungal strain, yielded maximum T-2 toxin within 24 days of incubation at 19°C.

Production of fumonisins B_1 (FB_1) and B_2 (FB_2) by two Brasilian strains (LAMIC 2999/96 and 113F) and one American strain (NRRL 13616) of *F. moniliforme* were evaluated in laboratory cultures subjected to different temperatures (20, 25, and 30°C), and moisture contents (25, 34, and 42%) on corn substrate. The cultures were grown during 10,

20, 30, 45, and 60 days, totalizing 135 treatments with two repetitions for each one (Maria *et al.*, 1991).

A method for the preparative-scale isolation of the fumonisin B (FB) mycotoxins, from corn cultures of *F. moniliforme*, is described and quantitatively evaluated. Eighty percent of FBI and 60% of FBz were recovered after extraction with CH_3OH/H_2O (3:l). The fumonisins, including the newly discovered FB3 and FB4, were purified using Amberlite XAD-2, silica gel, and reverse-phase CIS chromatography.

Trichothecene mycotoxins can be produced by *Fusarium* species in shake culture and twelve T-2 toxin-producing isolates and four fusarenon-X-producing isolates of *Fusarium* species were examined for their ability to produce trichothecene mycotoxins in shake culture and jar fermentation. T-2 toxin producers such as *F. solani*, *F. sporotrichioides*, and *F. tricinctum* produced T-2 toxin and neosolaniol in semi synthetic medium, and he was found that *F. solani* M-1-1 produced the largest amount of the mycotoxins in a nutrient medium consisting of 5% glucose (or sucrose), 0.1% peptone, and 0.1% yeast extract in either shake culture or jar fermentation at 24 to 27°C for 5 days. And its presence can be detected by using chemical and biological methods (Ueno *et al.*, 1973 and 1975). Trichothecenes and other secondary metabolites can be easily produced *by F. culmorum* and *F. equiseti* on common laboratory media and a soil organic matter agar (Hestbjerg *et al.*, 2002).

Conclusion

Ours being agriculture based economy and broad host range of *Fusarium* sp. leads to great economic losses. There is need to control the infection caused by *Fusarium* sp. by either a chemical agent or by a biological agent. The biological agent will score over the chemical agent by being environmental friendly, biodegradable and ease of preparation on large scale. The taxonomy and identification of the *Fusarium* genus is complicated due to the extreme variability of *Fusarium* species in their habitat and morphological appearance.

Hence methods based on polymerase chain reaction (PCR) offer many new tools that are directly applicable to fungal systematics at the species level. These tools can be used to differentiate and to determine relationships among species, either by direct comparison or through phylogenetic analysis. PCR-based methods have given a greater approach into molecular variability within fungi and have highlighted the need to consider carefully sampling strategies and sample size, prior to making taxonomic decisions.

References

Anandi, V., Viswanathan, P., Sasikala, S., Rangarajan, M., Subramaniyan, C.S and Chidambaram, N (2005). *Fusarium solani* breast abscess. *Indian Journal of Medical Microbiology*. *23*: 198-199.

Appel, D.J. and Gordon, T.R. (1995). Intraspecific variation within populations of *Fusarium oxysporum* based on RFLP analysis of the intergenic spacer region of the rDNA. *Exp Mycol. 19*: 120-128.

Blanz, P.A. and Unseld, M. (1987). Ribosomal RNA as a taxonomic tool in mycology. *In*: *The Expanding Realm of Yeast Like Fungi* (De Hoog, G.S., Smith, M.T. and Weijman, A.C.M., Eds.). pp. 247 - 258.CBS/Elsevier, Amsterdam.

Brayford, D. (1996). *Fusarium*-molecules maketh the mould. *Sydowia. 48*: 163 - 183.

Bush, B, J., Carson, M. L., Cubeta, M.A., Hagler, W. M., Payne, G. A (2003). Infection and fumonisin production by *Fusarium verticilliodes* in developing maize kernels. *Phytopathology.94*: 88-93.

Caetano-Anolles, G., Bassam, B.J. and Gresshoff, P.M. (1991). DNA amplification fingerprinting using very short arbitrary oligonucleotide primers. *Biol Technology. 9*: 553-557.

Carter, J.P., Rezanoor, H.N., Desjardins, A.E. and Nicholson, P. (2000). Variation in *Fusarium graminearum* isolates from Nepal associated with their host of origin. *Plant Pathol. 49*: 452- 460.

Chakrabarti, A., Mukherjee, P., Sherkhane, D., Bhagwat, A and Murthy, B (2001). A simple and rapid molecular method for distinguishing between races of *Fusarium oxysporum* f.sp. *ciceris* from India. *Current Science.80*: 571-575.

Cheeke, P. R and Lee R. S. (1985). *Natural Toxicants in Feeds and Poisonous Plants*.

Chimbekujwo, I.B. (1999). Frequency and pathogenicity of *fusarium* wilts (*Fusarium solani, Fusarium equiseti*) of cotton (*Gossypium hirsutum)* in Nigeria. Department of Biological sciences, Federal University of Technology, Yola, Nigeria.

Chiocchetti, A., Sciaudone, L., Durando, F., Garibaldi, A and Migheli, Q. (2001). PCR detection of *Fusarium oxysporum* f. sp. *basilici* on Basil. *Plant Disease.85*: 607-611.

Cole, R.J., Kirdsey, J.W., Cutler, H.G., Doupnik, B.L. and Peckham, J.G. (1973). Toxin from *Fusarium moniliforme*: effect on plants andanimals. *Science. 179*: 1324-1326.

Dallavallel, E., D'auleriol, A. Z., Verardil, E and Bertaccini, A. (2002). Detection of rapid polymorphisms in *Gladiolus* cultivars with differing sensitivities to *Fusarium oxysporum* f. sp. *gladioli. Plant Molecular Biology Reporter. 20*: 305–305.

Daniel, C., Eugene, B., Smalley, and Caldwell, R.W. (1982). New process for T-2 toxin production. *Applied and Environmental Microbiology. 44*: 371-375.

De Boer, M., Bom, P., Kindt, F., Keurentjes, J. J.B., Sluis, I. V., Loon, L.C and Bakker, P.A. H.M. (2003). Control of *Fusarium* wilt of Radish by combining *Pseudomonas putida* strains that have different disease-suppressive mechanisms. *Phytopatholog. 93*: 626-632.

Dobrowolski, M.P. and O'Brein, P.A. (1993). Use of RAPD-PCR to isolate a species-specific DNA probe for *Phytophthora cinnamomi. FEMS Microbiology Letters. 113*: 43-48.

Donaldson, G.C., Ball, L.A., Axelrood, P.E. and Glass, N.L. (1995). Primers sets to amplify conserved genes from filamentous ascomycetes are useful in differentiating *Fusarium* species associated with conifers. *Applied and Environmental Microbiology. 61*: 1331-1340.

Edel, V., Steinberg, C., Gautheron, N. and Avelange, I. (1996a). Differentiation of *Fusarium* by restriction fragment length polymorphism analysis of PCR- amplified ribosomal DNA. First International *Fusarium* Biocontrol Workshop. Beltsville Agricultural Research Center, USA. pp. 21.

Edel, V., Steinberg, C., Gautheron, N. and Alabouvette, C. (1996b). Evaluation of restriction analysis of polymerase chain reaction (PCR) amplified ribosomal DNA for the identification of *Fusarium* species. *Mycological Res. 101*: 179-187.

Garber, R and Yoder, O.C. (1985). *Curr. Genet. 8*:621-628. *In*: Kinhorn, J.R. and Turner, G. (eds). Applied molecular genetics of filamentous fungi Blackie Academic and Professional, UK; p259.

Garber, R.C., Turgeon, B.G., Selker, E.U. and Yoder, O.C. (1988). Organization of ribosomal RNA genes in the fungus Cochliobolus heterstrophus. *Curr Genet. 14*:573-582.

Gherbawy, YAMH. (1999). RAPD profile analysis of isolates belonging to different formae speciales of *Fusarium oxysporum. Cytologia. 64*:269-276.

Gherbawy, YAMH and Abdelzaher, H.M. (2002). Using of RAPD-PCR for Separation of *Pythium spinosum* Sawada into two varieties: var. *spinosum* and var. *sporangiferum. Cytologia. 67*: 83-94.

Gherbawy, YAMH and Yaser M. (2003). Fungicides and some biological controller agents effects on the growth of *Fusarium oxysporum* causing wilt disease of paprika. *Archives of Phytopathology and Plant Protection. 36*: 235-245.

Girish, G.K. and Goyal, R.K (1986). Prevention and control of mycotoxins in food grains in India. *Bull.Grain Tech. 24*: 157-77.

Grajal-Martin, M.J., Simon, C.J. and Muchlbauer, F.J. (1993). Use of random amplified polymorphic DNA (RAPD) to characterize race 2 of *Fusarium oxysporum* f. sp. *pisi. Phytopathology. 83*: 612-614.

Grossman, L and Hudspeth, M. (1985). *In*: *Gene Manipulations in fungi* (Bennett, J.W. and Lasure, L. eds), Academic Press, Orlando, FL; pp. 65-103.

Hagler, W M (1981). Biosynthesis of radiolabeled T-2 toxin by *Fusarium tricinctum. Appl Environ Microbiol*; *41*: 1049–1051.

Hering, O. and Nirenberg, H.I. (1995). Differentiation of *Fusarium sambucinum* Fuckel sensu lato and related species by RAPD PCR. *Mycopathologia. 129*: 159-164.

Hestbjerg, H. (2002). Production of trichothecenes and other secondary metabolites *by Fusarium culmorum* and *Fusarium equiseti* on common laboratory media and a soil organic matter agar: an ecological interpretation. *J Agric Food Chem. 50*: 7593-7599.

Hocking, A.D. and Andrews, S. (1987). Dichloran chloramphenicol peptone agar as an identification medium for *Fusarium* species and some dematiaceous Hyphomycetes. *Trans. Br. Mycol. Soc. 89*: 239-244.

Hori, H. and Osawa, S. (1979). Evolutionary changes in 5S RNA secondary structure and a phylogenetic tree of 54 5S RNA species. *Proc. Natl. Acad. Sci. USA; 76*:381-385.

Hyun, J.W. and Clark, C. (1998). Analysis of *Fusarium lateritium* using RAPD and rDNA RFLP techniques. *Mycol Res. 102*: 1259 -1264.

Irisbaev, B.K. (1991). Cloning a species-specific DNA probe from *Fusarium oxysporum* fungus. . *Mol.Bio (Mosk). 25*: 1667-1669.

James, S.A., Collins, M.D. and Roberts, L.N. (1996). Use of an rRNA internal transcribed spacer region to distinguished phylogenetically closely related species of the genera *Zygosaccharomyces* and *Torulaspora. Int J Syst Bacteriol. 46*:189-194.

Joffe, A.Z. (1978). *Fusarium poae* and *F. sporotrichioides* as principal causal agents of alimentary toxic aleukia. In *"Mycotoxic Fungi, Mycotoxins, Mycotoxicoses: an*

Encyclopaedic Handbook. Vol. 3", eds. T.D. Wyllie and L.G. Morehouse. New York: *Marcel Dekker*. 21-86

Kamel, A. (2003). PCR identification of *Fusarium* genus based on nuclear ribosomal-DNA sequence data. *African Journal of Biotechnology. 2*: 82-85.

Keith, S. (1996). *Fusarium* interactive key. *Agriculture and Agri food:* 1-65

King, A., Pitt, J., Beuchat, L and Corry, J (1986*). Methods for the Mycological Examination of Food.* New York: Plenum Press. pp. 256-262.

Knoll, S., Vogel, R.F and Niessen, L (2002). Identification of *Fusarium graminearum* in cereal samples by DNA detection test strip. *Applied Microbiology. 34*: 144.

Krogh, P. (1987). *Mycotoxins in Food.* San Diego: Academic Press Limited.

Kucharek, T., Jones, J.P., Hopkins, D., Stranberg, J. (2000). Some diseases of vegetables and agronomic crops caused by *Fusarium* in Florida. Institute of Food and Agricultural Science, University of Florida. pp. 656-677.

Kurtzman, C.P. (1994). Molecular taxonomy of the yeasts. *Yeast. 10*: 1727-1740.

Laguerre, G., Allard, M.R., Revoy, F. and Marger, N. (1994). Rapid identification of rhizobia by restriction fragment length polymorphism analysis of PCR-amplified 16S rRNA genes. *Appl Environ Microbiol. 60*: 56 -63.

Lanfranco, L., Wyss, P., Marzachi, C. and Bonfante, P. (1993). DNA probes for the identification of the ectomycorrhizal fungus *Tuber magnatum* Pico. *FEMS Microbiology Letters. 114*:245-251.

Lanfranco, L., Wyss, P., Marzachi, C. and Bonfante, P. (1995). Generation of RAPD-PCR primers for the identification of isolates of *Glomus mosseae*, an arbuscular mycorrhizal fungus. *Molecular Ecology. 4*:61-68.

Larone, D. H (1995). *Medically Important Fungi - A Guide to Identification*, 3rd ed. ASM Press, Washington, D.C.

Manulis, S., Kogan, N., Reuven, M. and Ben-Yephet, Y. (1994). Use of RAPD technique for identification of *Fusarium oxysporum* f.sp. *dianthi* from carnation. *Phytopathology. 84*: 98-101.

Maracas, W.F.O., Nelson, P.E and Tousson, T.A. (1984). *"Toxigenic Fusarium Species: Identity and Mycotoxicology".* University Park, Pennsylvania: Pennsylvania State University Press.

Marcon, M.J. and Powell, D.A (1987). Epidemiology, diagnosis, and management of *Malassezia furfur* systemic infection. *Diagn Microbiol Infect Dis. 7*: 161 - 175.

Maria, E.C., Wentzel, C.A., Vleggaar, R., Behrend, Y., Thiel, P.G and Marasas, F.O. (1991). Isolation of the Fumonisin Mycotoxins: A Quantitative Approach. *J. Agrlc, Food Chem. 39*:1958-1962.

Miller, J.D and Trenholm, H.L. (1994). *Mycotoxins in grain: Compounds Other Than Aflatoxin* St. Paul: Eagan Press 1994.

Miller, J.D (2002). Aspects of the ecology of *Fusarium* toxins in cereals. *Adv Exp Med Biol. 504*:19-27.

Mirocha, C.J., Pathre, S.V., Schauerhamer, B and Christensen, M. (1976). Natural Occurrence of *Fusarium* Toxins in Feedstuff. *Applied and Environmental Microbiology. 32*: 553-556.

Mishra, P.K., Fox, R.T.V. and Culham, A. (2003). Restriction analysis of PCR amplified nrDNA regions revealed intraspecific variation within populations of *Fusarium culmorum*. *FEMS Microbiology Letters*. *215*: 291-296.

Moss, M.O and Smith, J.E. (1984). The Applied Mycology of Fusarium: Symposium of the British Mycological Society held at Queen Mary College, London, September 1982. Cambridge University Press: Cambridge. pp. 264.

Munkvold, G.P. (1997). Fumonisins in Maize. *Plant Disease*. *81*: 556-564.

Nelson, P.E (1992). Taxonomy and biology of *Fusarium moniliforrme*. *Mycopathologia 117* (1-2): 29-36.

Nelson, P.E., Dignani, M.C and Anaissie, E.J. (1994). Taxonomy, biology and clinical aspects of *Fusarium* species. *Clin. Micrabiol Rev.7*: 479-504.

Nicholson, P., Jenkinson, P., Rezanoor, H.N. and Parry, D.W. (1993). Restriction fragment polymorphism analysis of variation in *Fusarium* species causing ear blight of cereals. *Plant Pathol*. *42*: 905-914.

O'Donnell, K., Gherbawy, Y., Schweigkofler, W., Adler, A. and Prillinger, H. (1999). Phylogenetic analyses of DNA sequence RAPD data compared in *Fusarium oxysporum* and related species from maize. *Phytopathology*. *147*: 445-452.

Okubara, P.A., Inglis, D.A., Muehlbauer, F.J. and Coyne, C.J. (2002). A novel RAPD marker linked to the *Fusarium* wilt race 5-resistance gene (*Fwf*) in *Pisum sativum*. *Pisum Genetics*. *34*: 1-3.

Paulo, D. (2002). Production of fumonisins by strains of *Fusarium moniliforme* according to temperature, moisture and growth period. *Braz. J. Microbiol*. *33*: 27-36

Pitt, J and Hocking, A.D. (1985). "*Fungi and Food Spoilage*". Sydney, N.S.W.: Academic Press.

Raabe, R. D., Conners, I.L and Martinez, A.P. (1981). Checklist of Plant Diseases in Hawaii. Hawaii Institute of Agriculture and Human Resources, College of Tropical Agriculture and Human Resources, University of Hawaii (Information Text Series 022).

Richardson, K. E., Hagler, W.M and Hamilton P.B. (1984). Methods for detecting production of Zearalenone, Zearalenol, T-2 toxin and Deoxynivalenol by *Fusarium* isolates. *Applied and Environmental Microbiology*. *14*: 643-646.

Saharan, M. S., Naef, A, J. Kumar1, J and Tiwari, R. (2007). Characterization of variability among isolates of *Fusarium graminearum* associated with head scab of wheat using DNA markers. *Current Science*. *92*: 230- 234

Saiki, R., Scharf, S., Faloona, F., Mullis, K.D., Horn, G.T., Erlich, H.A. and Arnheim, N. (1985). Enzymatic amplification of B-globin genomic sequences and restriction site analysis for diagnosis of sickle cell anemia. *Science*. *230*:1350-1354.

Saiki, R., Gelfand, D.H., Stoffel, S., Scharf, S.J., Higuchi, R., Horn, G.T., Mullis, K.B. and Erlich, H.A. (1988). Primer-directed enzymatic amplification of DNA with a thermostable DNA polymerase. *Science*. *239*: 487-491.

Shuxian Li, Tam, Yan Kit and Hartman, G.L (2002). Molecular Differentiation of *Fusarium solani* f. sp. glycines from Other *Fusarium solani* Based on Mitochondrial Small Subunit rDNA Sequences.*Phytopathology*. *90*: 491-497.

Singh, B.P., Saikia, R., Yadav, M., Singh, R., Chauhan, V.S and Arora, D.K (2006). Molecular characterization *of Fusarium oxysporum* f.sp. *ciceri* causing wilt of chickpea. *African Journal of Biotechnology. 5*: 497-502

Singh, P. P., Shin, Y. C., Park, C. S., and Chung, Y. R. (1998). Biological control of *Fusarium* wilt of cucumber by chitinolytic bacteria. *Phytopathology. 89*: 92-95

Snyder, W.C and Hansen, H.N. (1940). The species concept *in Fusarium. Am. J. Bot. 27*: 64-67.

Smith, D, I., Smith, I. W andClements, P. (2003). Visible symptons and sampling techniques for identification of Fusarium wilt of Canari island Date Palm. State of Victoria, Department of Sustainability and Environment. 1- 3.and

Spiewak, R. (1998). Zoophilic and geophilic fungi as a cause of skin disease in farmers. *Ann Agric Environ Med. 5*: 97-102.

Summerell, B.A., Salleh, B and Leslie, J.F (2003). A Utilitarian Approach to *Fusarium* Identification. *Plant Disease. 87*: 117-128.

Taylor, J.W., Smolich, B.D. and May, G. (1986). Evolution of mitochondrial DNA in *Neurospora crassa. Evolution. 49*: 716-739.

Torres, A. M. (2001). *Fusarium* species (section Liseola) and its mycotoxins in maize harvested in northern Argentina. *Taylor & Francis. 18*: 836 – 843.

Ueno, Y., Sato, N., Ishii, K.S., Tsunoda, H and Enomoto, M. (1973). Biological and Chemical Detection of Trichothecene Mycotoxins of *Fusarium* Species. *Applied Microbiology. 25*: 699-704.

Ueno, Y., Sawano, M and Ishii, K. (1975). Production of Trichothecene Mycotoxins by *Fusarium* Species in Shake Culture. *Applied Microbiomogy. 30*: 4-9.

Vismer, H. F. (2002). *Fusarium dimerum* as a cause of human eye infections. *Med Mycol. 40*: 399-406.

Waalwijk, C., de Koning, J.R.A., Baayen, R.P. and Gams, W. (1996). Discordant groupings of *Fusarium* spp. From sections *Elegans Liseola* and *Dlaminia* based on ribosomal ITS1 and ITS2 sequences. *Mycologia. 88*: 361-368.

Weising, K., Kaemmer, D., Epplen, J., Weigand, F., Saxena, M. and Kahl, G. (1991). DNA fingerprinting of *Ascochyta rabiei* with synthetic oligodeoxynucleotides. *Curr Genet. 19*: 483-489.

Welsh, J. and McClelland, M. (1990). Fingerprinting genomes using PCR with arbitrary primers. *Nucleic Acid Res. 18*:7213-1218.

Williams, J.G.K., Kubiek, A.R., Livak, K.J., Rafalski, J.A. and Tingey, S.V. (1990) DNA polymorphisms amplified by arbitrary primers are useful as genetic markers. *Nucleic Acids Res. 20*:303-306.

Wolf, E.de (2003). Fusarium head blight. Agriculture research and cooperative extension Division, College of Agricultural Science, Penn state University, U.S.A.

Woloshuk, C. P. (1994). Dept. of Botany & Plant Pathology, Purdue University "Mycotoxins and Mycotoxin Test Kits", Cooperative Extension Service, Corn Diseases BP-47: pp.12.

Yoder, W.T. and Christianson, L.M. (1998). Species-specific primers resolve members of *Fusarium* section *Fusarium. Fungal Genetics and Biology. 23*: 68-80.

In: Current Advances in Molecular Mycology
Eds: Y. Gherbawy, R.L. Mach and M.K. Rai

ISBN 978-1-60456-909-4

Chapter V

FUSARIUM DNA LEVELS IN FINNISH CEREAL GRAINS

T. Yli-Mattila[1], P. Parikka[2], T. Lahtinen[1], S. Rämö[3], M. Kokkonen[4], A. Rizzo[4], M. Jestoi[4] and V. Hietaniemi[3]

[1]Lab. of Plant Physiology and Molecular Biology, University of Turku, Turku, Finland;
[2]Plant Production/Plant Protection, MTT Agrifood Research Finland, Jokioinen, Finland;
[3]MTT Agrifood Research Finland, Laboratories, Finland, Jokioinen, Finland;
[4]Finnish Food Safety Authority (Evira), Research Department, Chemistry and Toxicology Unit, Helsinki , Finland .

ABSTRACT

A highly significant correlation was found between *Fusarium graminearum* DNA and deoxynivalenol (DON) levels in Finnish oats, barley and wheat. In barley the correlation was improved, when *F. culmorum* DNA was added to *F. graminearum* DNA, especially when the DNA was extracted from whole ground grains. It thus seems that *F. culmorum* plays a more important role in DON production in barley than in spring wheat or oats. A significant correlation was also found between *F. avenaceum* DNA (TMAV) and enniatins in spring wheat, barley and oats and between *F. avenaceum* DNA and moniliformin (MON) in barley and spring wheat. The correlation between *F. langsethiae/F. sporotrichioides* DNA (TMLAN) and HT-2+T-2 levels was highly significant in oats, in which the highest HT-2+T-2 levels were found. The correlation between *F. poae* DNA and NIV was significant in both barley and oats.

The correlation results of the present study between *Fusarium* DNA and mycotoxin levels are in accordance with the idea that each *Fusarium* species has its own mycotoxin profile. The correlation between mycotoxins and DNA extracted from ground grains was generally in agreement with the correlation between mycotoxins and DNA extracted from grain surfaces. The relative amounts of TMTRI (trichothecene-producing *Fusarium* fungi) and TMAV DNA per total DNA obtained from grain surfaces and ground grains were usually at the same level, although the TMTRI and TMAV concentrations were higher from DNA samples extracted from grain surfaces.

The coefficient of determination (R^2) between *Fusarium graminearum* DNA and DON was about 0.50 (0.49-0.61), when DNA was extracted from the grain surfaces. Higher R^2 values (0.78-0.99) between *Fusarium graminearum* DNA and DON in oats, barley and spring wheat were obtained when DNA was extracted from ground grains. The R^2 values between *F. langsethiae/F. sporotrichioides* DNA (TMLAN) and HT-2+T-2 levels and between *F. avenaceum* DNA (TMAV) and enniatins were also clearly higher when DNA was extracted from ground grains.

Fusarium DNA levels started to increase after flowering. In most cases this increase continued till harvesting, but especially in 2005 *F. langsethiae* levels decreased before harvesting; this was probably caused by other *Fusarium* fungi which overgrew *Fusarium langsethiae.*

High levels of *F. langsethie/ F. sporotrichioides* DNA were detected already two weeks after flowering. *F. poae* DNA levels were higher in plots with tillage as compared to those without tillage in 2005 and 2006 in both oats and barley; the difference was greater a few weeks before harvesting than during it. TMLAN (*F. langsethie + F. sporotrichioides*) DNA levels were higher in plots without tillage in 2004 in oats two weeks after flowering and in 2006 during harvesting in both oats and barley. Fungicide treatment one week before flowering decreased the amount of TMLAN DNA in 2005 in both oats and barley, while *F. poae* DNA levels were increased after fungicide treatment in oats in 2004 and 2005. The highest *F. avenaceum*-DNA levels were found during harvesting.

Key words: barley, fungicide, mycotoxins, oat, TaqMan real-time qPCR, tillage, wheat.

INTRODUCTION

Fusarium is probably economically the most important phytopathogenic and mycotoxigenic genus of filamentous fungi in cereals worldwide (O'Donnell, 1996; McMullen et al., 1997; Langseth et al., 1999; Bottalico and Perrone, 2002) and in Finland (Ylimäki et al., 1979; Eskola et al., 2001; Yli-Mattila et al., 2004b). Several *Fusarium* species are involved in *Fusarium* head blight (FHB), which reduces both yield and quality in cereal crops. FHB was first described in England in 1884. Since then FHB has increased worldwide; recent oubreaks have been reported in Asia, Canada, USA, South America and Europe (McMullen et al., 1997; Goswami and Kistler, 2004). In addition, *Fusarium* species are able to produce mycotoxins harmful to humans and animals. The mycotoxins are stable and may still be present in plants after the mycelium has died. The detection of *Fusarium* species in plants is currently based on morphological, microbiological and antibody-based techniques. The resolution of these techniques is not always sufficient. Many of them detect only living mycelia and spores or are semi-quantitative. DNA is relatively stable even after the mycelium has died, and molecular methods can be used for quantitative detection of fungi grown in plants.

Molecular DNA Analyses

The basis for molecular DNA analyses of plant pathogenic *Fusarium* fungi is DNA extraction and purification. In fungi and plants the cell walls should be broken and different PCR inhibitors should be removed. DNA can be extracted e.g. with the chloroform-octanol method (Yli-Mattila et al., 1998) or with CTAB methods (Weising et al., 1995). Younger plants and fungi usually contain fewer inhibitory compounds than older ones. RNA can be degraded by RNAase. Different commercial kits are also now available for DNA extraction and should be optimised for the fungal DNA from plant material and soil. The quality and quantity of DNA can be estimated by gel electrophoresis and spectrophotometry (Yli-Mattila et al., 2007a, 2008).

Species-specific primers can be designed by comparing known *Fusarium* DNA sequences from different species. Alternatively, primers can be designed based for species-specific RAPD-PCR (random amplified polymorphic DNA), UP-PCR (universally primed PCR), RAMS (random amplified microsatellites) etc. products (Yli-Mattila and Paavanen-Huhtala, 2007b). The use of species-specific primers (e.g. Doohan et al., 1998, Waalwijk et al., 2003; Yli-Mattila et al. 2004c), microarrays and SNP (single nucleotide polymorphisms) analysis (e.g. Kristensen, 2006) has made it possible to identify isolates of *Fusarium* species based on molecular data. Molecular data have also been used for phylogenetic studies (e.g. Paavanen-Huhtala 2000; O'Donnell et al., 2000, 2004; Yli-Mattila et al., 2004a; Kristensen, 2006). It is important that enough well-identified isolates and/or DNA sequences from GenBank, from the species for which the primers or probes are designed and from closely related species, are available for testing species-specific primers and probes before they are used for identification; the isolates should be single-spore isolates, identified by both morphological and molecular characters.

For AFLPs (amplified fragment length polymorphism) the overall similarity of the bands is usually more important than individual bands. Within the *Liseola* section, strains that share 60 % usually belong to the same species, those that share <40 % of the bands belong to different species and those that share 40-60 % of their bands need additional evaluation by different molecular and morphological methods (Leslie et al., 2007). In the most reliable species definitions morphological, biological and phylogenetic species concepts are consistent, but the concept of biological species cannot be used for anamorphic species, while the subjective nature of the concept of morphological species may also cause problems, especially with closely related species.

TaqMan real-time PCR can be used for quantifying DNA of *Fusarium* species in plants. The advantages of TaqMan qPCR analyses have been discussed e.g. in Waalwijk et al. (2004), Sarlin et al. (2006) and Yli-Mattila et al. (2007a, 2008). There are still only few studies dealing with fluorogenic TaqMan qPCR detection assays for the quantification of *Fusarium* species in grain samples (Reischler et al., 2004; Waalwijk et al., 2004; Bluhm et al., 2004; Yli-Mattila et al., 2004c, 2006, 2008; Sarlin et al., 2006; Halstensen et al., 2006; Leisova et al., 2006; Hogg et al., 2007; Burlakoti et al., 2007). Most of these studies have concentrated on DON-producing *F. graminearum* and *F. culmorum*. Thus a considerable amount of work still remains to be done with other *Fusarium* species.

The advantage of primers based on ribosomal DNA is that there are numerous ribosomal DNA copies in the genome, making the reaction more sensitive. Internal transcribed spacer (ITS) and intergenic spacer (IGS) region sequences of ribosomal DNA are also highly variable. On the other hand, there may be different ITS sequences present in the same species, which in certain cases makes ITS sequences unreliable in species identification (O'Donnell et al., 1997; Leslie et al., 2007). According to Jurado et al. (2005), species-specific primers based on IGS sequences can be used for separating *F. culmorum* isolates from *F. graminearum* isolates, while according to O'Donnell et al. (2004) *F. culmorum* is nonmonophyletic based on IGS sequences. Thus further studies with more isolates are required to determine how well these species can be separated on the basis of IGS sequences. In the present work TMpoae primers and probes were based on IGS sequences, while TMLAN primers and probes were based on ITS sequences (Yli-Mattila et al., 2004a, 2008).

Fusarium Species and their Mycotoxins in Northern Europe

According to morphological and molecular studies (Ylimäki et al., 1979; Eskola et al., 2001; Bottalico and Perrone, 2002; Yli-Mattila et al., 2004b,c, 2008; Elen et al., 2007), the most common *Fusarium* species in cereal grains in Northern Europe, including Finland, are the *F. aveanaceum/F. arthrosporioides* species complex (Yli-Mattila et al., 2002, 2004c), *F. tricinctum, F. poae, F. culmorum, F. graminearum, F. sporotrichioides* and the new species *F. langsethiae* (Torp and Nirenberg, 2004; Yli-Mattila et al., 2004a). It is interesting that *F. graminearum* was found in Finland already in the 1970s (Ylimäki et al., 1979), while in northwestern Russia it was found only after the year 2000 (Gagkaeva et al., 2006). Recently *F. graminearum* has been spreading northward in Europe (e.g. Waalwijk et al., 2003; Nicholson et al., 2003, Elen et al., 2007, personal communication) and has been replacing the closely related *F. culmorum*, which is less effective in producing mycotoxins (Langseth et al., 1999; Jestoi et al., 2007).

Before exploring the correlation between *Fusarium* DNA and mycotoxin levels, it is useful to have some knowledge about mycotoxin profiles in pure cultures. Mycotoxin profiles have been found to be species-specific both in worldwide investigations (e.g. Langseth et al., 1999; Thrane et al., 2004) and in Finland (Jestoi et al., 2004, 2007), although some quantitative and qualitative variation has been found within *Fusarium* species. Mycotoxin levels are also affected by environmental factors.

The most important *Fusarium* mycotoxins in Northern Europe are trichothecenes and zearalenone (ZEN), (Langseth et al., 1999; Bottalico and Perrone 2002; Thrane et al., 2004, Yli-Mattila et al., 2004b). Trichothecenes can be divided into two types: type A (e.g. T-2 toxin, HT-2 toxin, diacetoxyscirpenol (DAS), monoacetoxyscirpenol (MAS) and type B (e.g. deoxynivalenol (DON) and nivalenol (NIV) and their mono- and di-acetylated derivatives) (Bottalico, 1998). Some trichothecene toxins are connected with pathogen aggressiveness, disease severity, and colonization of a plant tissue (Proctor et al., 2002; Jansen et al., 2005; Maier et al., 2006). Maximum limit values in food have already been established in the EU for DON and ZEN; in the near future they will be enacted for HT-2 and T-2 as well.

Fusarium head blight caused by *Fusarium graminearum* (sexual state *Gibberella zeae*) is an important disease of cereals (Goswami and Kistler, 2004; Gagkaeva and Yli-Mattila, 2004). O'Donnell et al. (2004) have divided *F. graminearum* into nine lineages, of which the northern lineage 7 is the dominant one in Europe and North America (O'Donnell et al., 2000). *F. graminearum* is often found in grains together with *F. culmorum*, which can also cause root and foot rot in cereals. *F. graminearum* and *F. culmorum* produce only type B trichothecenes.

Trichothecene biosynthetic pathways have been studied using molecular and chemical analyses (Desjardins, 2006), which together with genetic data have made it possible to distinguish different chemotypes (3AcDON, 15AcDON and NIV) and lineages of *F. graminearum* (O'Donnell et al., 2004; Ward et al., 2008). The chemotypes have also been called chemotype IA (producing 3AcDON), IB (producing 15AcDON) and II (producing NIV) (Miller et al., 1991). In North America isolates from 3AcDON populations have been found to grow more quickly and to produce more conidia and trichothecenes than those from the 15AcDON population (Ward et al., 2008). NIV is the end product of trichothecene biosynthesis (Bottalico and Perrone, 2002; Jennings et al., 2004, Desjardins, 2006). According to Ueno (1983), Ryu et al. (1987) and Perkowski et al. (1997) NIV is acutely more toxic than DON.

Fusarium poae and *F. sporotrichioides* are among the most common plant-pathogenic *Fusarium* species in the cereal grains which are involved in *Fusarium* head blight in northern Europe (Bottalico and Perrone, 2002). They are closely related to the new species *F. langsethiae*, which is found in various European countries (Knutsen & Holst-Jensen 2004; Schmidt et al., 2004; Yli-Mattila et al., 2004a). *F. langsethiae* is morphologically closer to *F. poae* than to *F. sporotrichioides*, but according to the mycotoxin profile and molecular sequence data it is more closely related to *F. sporotrichioides*. No teleomorphs have been found for these species (Torp and Nirenberg, 2004).

The mycotoxins produced by *F. poae* include both type A (e.g. DAS) and type-B trichothecenes (e.g. NIV and fusarenon X (FX)), beauvericin (BEA) and enniatins (ENNs) (Pettersson, 1991; Liu et al., 1998; Torp and Langseth 1999; Thrane et al., 2004; Jestoi et al, 2004). *F. langsethiae* and *F. sporotrichioides* produce type A trichothecenes, including DAS and HT-2 and T-2 toxins (Torp and Langseth, 1999; Torp and Nirenberg, 2004; Thrane et al., 2004; Jestoi et al., 2004, 2007). According to Thrane et al. (2004) a few *F. poae* isolates can produce small amounts of T-2 and HT-2, but only *F. sporotrichioides* and *F. langsethiae* isolates can produce large amounts of these mycotoxins. Most *F. sporotrichioides* isolates also produce BEA (Jestoi et al., 2004, 2007; Thrane et al., 2004).

Information regarding the effects of T-2 mycotoxin on humans has been collected from many incidents caused by moldy wheat or corn. One such incident took place in the Ural region in the former Soviet Union during World War II, when wheat could not be harvested before the winter. The moldy wheat, which was harvested after the winter, caused the clinical syndrome alimentary toxic aleukia (ATA), with a mortality rate of 10-60 % (Sarkisov, 1954; Joffe A.Z., 1974; T. Gagkaeva, personal communication).

Fusarium avenaceum, *F. arthrosporioides* and *F. tricinctum* are very closely related species, which together with other *Fusarium* species are associated with FHB and seedling (stem and root rot) diseases of all cereals in northern Europe (Bottalico and Perrone 2002).

Using a combination of the three primer pairs designed for the main groups of *F. avenaceum/F.arthrosporioides/F. tricinctum* species complex, it is possible to distinguish most of the strains of the three species, including degenerated strains (Turner et al. 1998, Yli-Mattila et al. 2002, Yli-Mattila et al. 2004c).

F. avenaceum (teleomorph *Gibberella avenacea*), *F. arthrosporioides* and *F. tricinctum* (teleomorph *Gibberella tricincta*) are considered trichothecene nonproducers (Langseth et al. 1999, Edwards et al. 2001, Thrane et al. 2001, Jestoi et al. 2004, 2007), but they are able to produce other mycotoxins, such as moniliformin (Golinski et al. 1996, Jestoi et al. 2004b; Uhlig 2005) and enniatins (Langseth et al. 1999, Jestoi et al. 2004b, Uhlig 2005, Uhlig et al. 2006, 2007). According to Logrieco et al. (2002) they also produce beauvericin, but this has not been confirmed in subsequent investigations. According to preliminary metabolite analyses based on HPLC chromatograms, there are also differences between the metabolite profiles of *F. avenaceum, F. arthrosporioides* and *F. tricinctum* strains (Yli-Mattila et al., 2006); this is in accordance with the morphological and molecular differences between these species (Yli-Mattila et a., 2002), although their mycotoxin profiles are almost identical (Uhlig et al., 2005; Jestoi et al., 2007).

Investigations of Fusarium Species and Toxins in Finland

Figure 1. Rye field during harvesting after the rainy and cold growing season of 1998. The fields were beaten down by rain already before flowering in June. New green shoots have grown in the field.

In Finland there have been heavy *Fusarium* contaminations in barley, oats and spring wheat in the 1930s, in 1972 (Ylimäki et al., 1979), during 1982-1984 (Rizzo, 1993) and in 1998 (especially in rye, Eskola et al., 2001, Yli-Mattila et al., 2002). In 1998, however, the mycotoxin (trichothecenes and ZEN) levels were low, except for the high BEA and ENNs levels found in rye (Figure 1) (Logrieco et al., 2002; in the original paper the crop was erroneously reported as wheat). The 1998 *Fusarium* contamination in rye was probably connected to the cool growing season with heavy rainfall, which started already before flowering in June. In 2002 DON levels were high, which was probably due to high *F. graminearum* and *F. culmorum* levels (Yli-Mattila et al., 2004b). Likewise in 2004 high mycotoxin levels (DON and HT-2/T-2) were found in oats. The highest DON levels in Finland have usually been found in oats (Yli-Mattila 2004b, www.agronet.fi/cerveg). High NIV levels have been found in barley, while high HT-2 and T-2 levels have been found in oats (Yli-Mattila et al., 2004b, 2008, www.agronet.fi/cerveg). There is considerable variation between years and between different cereals and cultivars. The lowest mycotoxin levels have usually been found in winter wheat and winter rye, which are harvested early.

In most investigations in Finland mycotoxins (e.g. www.agronet.fi/cerveg) and *Fusarium* species have been studied separately from different grain samples. During the most recent years collaboration between chemists, mycologists and molecular biologists has increased (Eskola et al., 2001; Jestoi et al., 2004, 2008, Yli-Mattila et al., 2004b, 2008). The design and use of species- and strain-specific primers in the identification of *Fusarium* species started at the end of the 1990s (Paavanen-Huhtala, 2000; Konstantinova and Yli-Mattila 2004), and was soon followed by phylogenetic studies (e.g. Yli-Mattila et al., 1997, 2002, Paavanen-Huhtala et al., 1999). Species-specific TaqMan qPCR primers and probes (Yli-Mattila et al. 2004c, 2006, 2008) were first developed and used for detecting and quantifying the amount of *Fusarium* fungi in grain samples collected in 2002-2004 for a project funded by Tekes (National Technology Agency of Finland) and published by Sarlin et al. (2006), Jestoi et al. (2004, 2008) and Yli-Mattila et al., (2004a, 2006, 2008).

Aim of the Paper

Our purpose in the present paper is to give up-to-date information on molecular *Fusarium* research on Finnish cereal grain samples after the year 2002 and to compare it to mycotoxin and mycological data from the same grain samples. The results are also compared to those obtained in other countries with the same species.

Our work is part of two larger projects, monitoring both mycotoxin and *Fusarium* levels in Finland during 2005-2006 (Finmyco) and the effect of tillage and fungicide (prochloraz) on *Fusarium* fungi and mycotoxins in Finland during 2004-2006 (Development of safety indicators for Finnish cereal grain). Some unpublished results from the Tekes project are also included. In addition, we have collaborated within the Nordic network project (http://fou02.planteforsk.no/NordforskNetworkMycotox/) in developing molecular methods for *Fusarium* species. According to the previous work of Yli-Mattila et al. (2008) there is a correlation between the DNA levels of most *Fusarium* species and their mycotoxins in Finnish grain samples, which is in agreement with qPCR results from other countries.

The main aim of the present investigation was to determine whether it is possible to detect high levels of *Fusarium* DNA in grain samples during harvesting, and how they correlate with high mycotoxin levels in field. In addition, *Fusarium* DNA levels were studied during and before harvesting in grain samples, to explore the effect of tillage and fungicide treatment on *Fusarium* DNA levels. Reischler et al. (2004) have previously used qPCR to detect *F. graminearum* in wheat before harvesting; the present study, however, is the first one in which the effects of different tillage and fungicide treatments are compared to the DNA levels of *Fusarium* fungi before and during harvesting.

MATERIALS AND METHODS

Grain Samples and Fungal Strains

During the years 2005-2006, 173 grain samples (33 wheat, 67 barley, 73 oats) collected from different parts of Finland were analysed for *Fusarium* DNA levels. Most of them were randomly chosen from among 342 dried grain samples which were provided to the Finmyco project by ProAgria, MTT and farmers.

The effect of artificial inoculation was studied in barley and spring wheat field plots at Marttila (Figure 2). The artificially inoculated field samples were from the years 2002, 2003 and 2004 (Yli-Mattila et al., 2004b, 2008, Jestoi et al., 2008).

Figure 2. Plant pathogenic *Fusarium* fungi can produce mycotoxins in plants; most of the mycotoxins are also present in the food produced from these plants. Left: wheat artificially infected with *F. culmorum*. Right: barley artificially infected with *F. graminearum*.

The effect of tillage and fungicide was studied in field plots at Jokioinen. In 2004 32 fresh samples for qPCR were taken during and two weeks after flowering, while in 2005 and 2006 (Figure 3) the samples were taken four weeks after flowering (32 fresh samples) and during harvesting (32 dried samples). For mycotoxin analyses the number of samples was threefold (96 samples taken during harvesting).

The fungal strains used to generate standard curves for quantitative PCR were collected from grain samples in Finland during 2001-2003 as part of the Tekes project. The morphological identification and purity of each standard strain for quantitative PCR was confirmed by species-specific primers as described by Yli-Mattila et al. (2004b).

Figure 3. Field where the oats and barley plots were situated at Jokioinen, August 19 2005.

DNA Extraction

DNA was extracted from the surfaces of grain samples (10 g) according to the modified method of Taylor et al. (2001) as described by Yli-Mattila et al. (2008). The filtered seed soak (15 ml) was centrifuged (10000 rpm at 4°C) and DNA was extracted with the GenElute™ Plant Genomic DNA Kit of Sigma. Ground grain samples (50-100 mg) were also extracted directly with the the GenElute™ Plant Genomic DNA Kit of Sigma. For the isolation of genomic DNA from pure cultures, fungi were grown for four to six days at 24 °C on PDA. DNA was extracted by the chloroform/octanol method as described by Yli-Mattila et al. (1998).

TaqMan Primers and, Probes and qPCR:

The TMpoae primers and probe have been designed for the IGS region (Yli-Mattila et al., 2004c) of *F. poae*. The TMFg12 primers and probe have been designed for the *F. graminearum* specific RAPD-PCR product (Doohan et al., 1998, Yli-Mattila et al., 2008). The TMLAN primers and probe for *F. langsethiae/F. sporotrichioides*, the TMAV primers and probe for *F. avenaceum* and the TMTRI primers and probe for trichothecene-producing *Fusarium* fungi have been designed by another research group (Halstensen et al., 2006), and the culmorumMGB primers and probe for *F. culmorum* by a third research group (Waalwijk et al., 2004). A GeneAmp 5700 cycler was used for running qPCR samples as described by Yli-Mattila et al. (2006, 2008).

Mycological Identification

The mycological identification of *Fusarium* isolates was performed at MTT as described by Yli-Mattila et al. (2004b) and Parikka et al. (2005).

Analysis of Mycotoxins and Statistical Analyses

NIV, DON, 3AcDON, 15AcDON, DAS, HT-2, T-2, FX, ZEN, MON, ENNs and BEA were extracted and determined as described by Eskola et al. (2001), Hietaniemi et al. (2004), and Jestoi *et al.* (2003, 2005). R^2 (= coefficient of determination), regression slope and P (= significance of regression slope) were calculated using the program SigmaPlot 2001 version 7.1 (SPSS Inc.). The original DNA and toxin concentrations were transformed to logarithmic values in order to obtain a more normal distribution for the values of toxin and DNA concentrations.

Table 1. Mycotoxins (ppb) in artificially contaminated Scarlett barley and Mahti wheat field plots as compared to contamination % results from the same field plots in 2002 (Yli-Mattila et al., 2004b, 2008, unpublished results). FX and DAS were not detected in any of the samples. n.d. = not detected

		Mycotoxins							
cereal	**sample**	**BEA**	**ΣENNs**	**DON**	**3AcDON**	**NIV**	**HT-2 +T-2**	**ZEN**	**MON**
barley	control	<10	3700	940	100	1260	<20	17	150
barley	*F.tricinctum*	23	15200	<10	n.d.	2890	n.d.	n.d.	450
barley	*F.arthrosporioides*	<23	10700	94	28	4790	n.d.	n.d.	470
barley	*F.avenaceum*	52	11320	500	72	5510	n.d.	<6	550
barley	*F.poae*	40	10200	370	47	5460	n.d.	<6.	460
barley	*F.sporotrichioides*	44	14300	510	62	6200	<20	16	560
barley	*F.culmorum*	16	2800	3060	230	1320	n.d.	95	150
wheat	control	<10	210	1060	39	60	n.d.	n.d.	70
wheat	*F.tricinctum*	<10	1480	720	33	220	n.d.	n.d.	150
wheat	*F.arthrosporioides*	<10	1270	650	32	110	n.d.	n.d.	100
wheat	*F.avenaceum*	18	850	590	37	550	20	n.d.	210
wheat	*F.poae*	<10	1400	200	n.d.	330	<20	n.d.	40
wheat	*F.sporotrichioides*	22	710	33	n.d.	100	530	n.d.	n.d.
wheat	*F.culmorum*	<10	400	22000	450	190	<20	45	62

		Contamination %					
cereal	**sample**	***Fusarium %***	***F.av /F.arthr /F.tric %***	***F.sport.%***	***F. poae %***	***F.culm. %***	***F.gram.%***
barley	control	15	7/0/2	0.5	0	7	2
barley	*F.tricinctum*	33	3/2/17	0.5	3	0.5	2
barley	*F.arthrosporioides*	52	7/1/28	0	1	4	3
barley	*F.avenaceum*	37	5/3/18	0	8	4	0
barley	*F.poae*	41	7/2/20	0	5	5	0
barley	*F.sporotrichioides*	31	7/0/9	4	3	11	0
barley	*F.culmorum*	33	5/0/5	2	2	32	0.5
wheat	control	17	1/1/1	0.5	0	13	0.5

Table 1. (Continued)

cereal	sample	*Fusarium %*	*F.av /F.arthr /F.tric %*	*F.sport.%*	*F. poae %*	*F.culm. %*	*F.gram.%*
wheat	*F.tricinctum*	10	2/0.5/2	0	0.5	3	0
wheat	*F.arthrosporioides*	18	5/3/3	1	0	8	0
wheat	*F.avenaceum*	25	9/3/7	0.5	2	2	3
wheat	*F.poae*	19	3/0/5	0	2	4	2
wheat	*F.sporotrichioides*	25	3/0/3	18	0	1	1
wheat	*F.culmorum*	54	1/0.5/2	0	0	47	0

Table 2. Mycotoxins (ppb) in artificially contaminated Scarlett barley and Mahti wheat field plots as compared to molecularly determined *Fusarium* DNA levels (10^{-6} ng ng^{-1} total DNA) from the same field plots in 2003. TMAV = *F. avenaceum/F. arthrosporioides* DNA, Tmpoae = *F. poae* DNA, TMTRI = DNA from trichothecene-producing *Fusarium*-species, TMFG12 = *F. graminearum* DNA, TMLAN = *F. sporotrichioides/F. langsethiae* DNA. (Jestoi et al., 2008, unpublished results). FX and DAS were not detected in any of the samples

		Mycotoxins							
cereal	**sample**	**BEA**	**ΣENNs**	**DON**	**3AcDON**	**NIV**	**HT-2 +T-2**	**ZEN**	**MON**
barley	control	<10	150	23	n.d.	<30	<20	n.d.	n.d.
barley	*F.tricinctum*	<10	440	29	n.d.	<30	n.d.	n.d.	n.d.
barley	*F.arthrosporioides*	<10	130	31	n.d.	n.d.	n.d.	n.d.	n.d.
barley	*F.avenaceum*	<10	1400	16	n.d.	67	n.d.	n.d.	70
barley	*F.poae*	<10	130	13	n.d.	82	n.d.	n.d.	n.d.
barley	*F.sporotrichioides*	210	66	1010	130	210	1430	n.d.	n.d.
barley	*F.culmorum*	<10	150	9800	650	160	<20	45	n.d.
barley	*F.graminearum*	<10	55	320	28	<30	n.d.	n.d.	n.d.
barley	*F.langsethiae*	15	75	46	n.d.	130	40	n.d.	n.d.
wheat	control	<10	25	40	n.d.	<30	25	n.d.	n.d.
wheat	*F.tricinctum*	<10	310	39	n.d.	<30	n.d.	n.d.	n.d.
wheat	*F.arthrosporioides*	<10	130	66	n.d.	60	n.d.	n.d.	n.d.
wheat	*F.avenaceum*	<10	550	18	n.d.	65	20	n.d.	140
wheat	*F.poae*	<10	40	19	n.d.	65	<20	n.d.	n.d.
wheat	*F.sporotrichioides*	59	32	<10	n.d.	<30	490	n.d.	n.d.
wheat	*F.culmorum*	<10	35	28000	380	77	<20	n.d.	n.d.
wheat	*F.graminearum*	<10	47	140	n.d.	54	<20	n.d.	n.d.
wheat	*F.langsethiae*	<10	43	13	n.d.	75	<20	n.d.	n.d.

		***Fusarium* DNA**				
cereal	**sample**	**TMAV**	**TMpoae**	**TMTRI**	**TMFg12**	**TMLAN**
barley	control	20	30	10	0	0
barley	*F.tricinctum*	<10	0	0	0	0
barley	*F.arthrosporioides*	200	0	0	0	0
barley	*F.avenaceum*	30	<10	<10	0	0
barley	*F.poae*	50	<10	<10	0	0
barley	*F.sporotrichioides*	400	0	<10	0	<10
barley	*F.culmorum*	400	<10	70	<10	0
barley	*F.graminearum*	100	0	<10	0	0
barley	*F.langsethiae*	100	<10	<10	<10	0

Table 2. (Continued)

		***Fusarium* DNA**				
cereal	**sample**	**TMAV**	**TMpoae**	**TMTRI**	**TMFg12**	**TMLAN**
wheat	control	10	<10	<10	0	<10
wheat	*F.tricinctum*	10	<10	<10	0	0
wheat	*F.arthrosporioides*	70	<10	<10	0	0
wheat	*F.avenaceum*	2000	<10	20	0	10
wheat	*F.poae*	100	0	<10	0	0
wheat	*F.sporotrichioides*	600	200	60	0	60
wheat	*F.culmorum*	1000	300	5000	<10	0
wheat	*F.graminearum*	0	<10	<1	0	0/0
wheat	*F.langsethiae*	<10	<10	<10	0	0

Table 3. DNA (ng/μl), mycotoxin (ppb), contamination % and DNA levels (ng/ng total DNA) in Scarlett barley and Mahti wheat in 2004 (Yli-Mattila, Burkin & Gagkaeva, unpublished results). Mycotoxins were analysed with ELISA analyses. Fungal isolates were grown on potato-sucrose medium (PSA) and/or SNA (Nirenberg 1981) for 14 days before mycological identification

Cereal	**Treatment**	**Mycotoxins**				**Contamination %**				
		DNA	**ZEN**	**DON**	**T-2/ HT-2**	***F.aven/F. arthrosp***	***F. tric.***	***F. poae.***	***F. spor. /F.lan.***	***F. gram. /F. culm.***
barley	control	5	n.d.	n.d.	0	2/0	1	4	1/0	0/0
barley	*F. avenaceum*	15				12/1	23	0	3/0	7/0
barley	*F. graminearum*	8	355	794		0/0	0	7	0/0	23/1
barley	*F. langsethiae*	8			0	0/0	1	0	0/0	0/0
spring wheat	control	1	n.d.	n.d.	27	2/2	2	6	0/0	0/0
spring wheat	*F. avenaceum*	1				21/8	14	3	4/0	1/0
spring wheat	*F. graminearum*	3	66	832		2/0	0	1	0/0	15/1
spring wheat	*F. langsethiae*	1			0	2/0	1	8	1/0	0/0

Cereal	**Treatment**	***Fusarium* DNA**			
		TMAV	**TMpoae**	**TMLAN**	**TMFg12**
barley	control	$3x10^{-4}$	$2x10^{-5}$	n.d.	$5x10^{-4}$
barley	*F. avenaceum*	$1.3x10^{-3}$	$6x10^{-5}$	n.d.	$2x10^{-2}$
barley	*F. graminearum*	$6.7x10^{-5}$	$<10^{-5}$	n.d.	$4x10^{-2}$
barley	*F. langsethiae*	$8.4x10^{-5}$	$<10^{-5}$	n.d.	n.d.
spring wheat	control	$3.9x10^{-4}$	$3x10^{-4}$	n.d.	n.d.
spring wheat	*F. avenaceum*	$5.2.x10^{-3}$	10^{-5},	n.d.	$2x10^{-5}$
spring wheat	*F. graminearum*	$8.0x10^{-5}$	$3x10^{-5}$	n.d.	10^{-2}
spring wheat	*F. langsethiae*	0	$2x10^{-5}$	n.d.	$2x10^{-4}$

DNA LEVELS IN ARTIFICIALLY INOCULATED FIELD PLOTS

It was possible to increase the amount of several *Fusarium* species and corresponding mycotoxins by artificial inoculation in barley and wheat during flowering (Yli-Mattila et al., 2004b, Jestoi et al., 2007, Tables 1-3), but DAS and FX were not produced after artificial inoculation by any species in field plots (Tables 1, 2). At the DNA level, the effect of artificial inoculation was clear with *F. culmorum* (TMTRI), *F. sporotrichioides* (TMLAN)

Table 4. Mean *Fusarium* DNA (10^{-6} ng ng^{-1} total DNA) levels in grain samples in Finland in 2003, 2005 and 2006 as compared to mean mycotoxin (ppb) levels in the same samples. *F. avenaceum* contamination % (year 2003) includes *F. arthrosporioides* isolates. Mean *Fusarium* contamination % levels of the samples are shown in 2003. Results for 2003 are from the papers of Yli-Mattila et al., 2006, 2008; results for 2005-2005 are from Finmyco project. In 2005 and 2006 samples with highest TMTRI levels were chosen for TMpoae, TMLAN, TMFg12 and culmorum MGB analyses. n.d. = not detected. n.a. = not analysed. LOQ (limit of quantification) was 25 ppb for each trichothecene, 20 ppb for MON and 0.6 ppb for ENN A, 4.0 pppb for ENN A1, 3.8 ppb for ENN B and 10.8 ppb for ENN B1

Cereal	**Mycotoxins**					**Contamination %**
	DON	**NIV**	**T-2+HT-2**	**MON**	**ΣENNs**	***F. aven./tric./ poae./spor./ langs./gram./culm.***
Oats						
2003 (n=15)	420	97	503	n.a.	n.a.	18/2/16/7/2/4/13
2005 (n=34)	1430	48	112	<20	130 (n=27)	
2006 (n=36)	51	59 (n=12)	286 (n=12)	n.d. (n=30)	<40 (n=29)	
Barley						
2003 (n=19)	25	<30	<30	<20 (n=9)	470 (n=9)	
						23/8/7/16/0/0.5/9
2005 (n=29)	510	<30	<30	71	1900 (n=24)	
2006 (n=33)	<30	<30	64	<20	230	
Spring Wheat						
2003 (n=17)	70	<30	<30	50 (n=9)	480 (n=6)	27/2/8/6/1/1/5
2005 (n=13)	790	n.d.	n.d.	186 (n=13)	1006 (n=13)	
2006 (n=14)	<30	<30 (n=13)	<30 (n=13)	n.d.	<40	
Rye						
2005-2006 (n=7)	<30	<30	<30	<20	<40	
Cereal	***Fusarium* DNA**					
	TMAV	**TMTRI**	**TMpoae**	**TMLAN**	**TMFg12**	**culmorum MGB**
Oats						
2003 (n=15)	142	51	123	86	557	n.a.
2005 (n=34)	396	51	98 (n=16)	183 (n=9)	658 (n=13)	30 (n=13)
2006 (n=36)	83 (n=35)	129	62000 (n=12)	51000 (n=12)	25 (n=12)	<2 (n=12)
Barley						
2003 (n=19)	915	1264	250	640	407	n.a.
2005 (n=29)	1475	15	2 (n=16)	42 (n=5)	230 (n=5)	<2 (n=5)
2006 (n=33)	2138	1270	439000 (n=16)	403000 (n=16)	21 (n=12)	8 (n=16)
Spring Wheat						
2003 (n=17)	313	220	33	21	2	n.a.
2005 (n=13)	284	19	<2 (n=5)	13200 (n=5)	1025 (n=5)	<2 (n=2)
2006 (n=14)	779	63	50386 (n=7)	7810 (n=7)	n.d. (n=7)	14 (n=7)
Rye						
2005-2006 (n=7)	274	2	n.a.	n.a	n.a	n.a

and *F. avenaceum* (TMAV) in 2003 and with *F. graminearum* and *F. avenaceum* in 2004. With *F. poae, F. langsethiae, F, tricinctum* and *F. arthrosporioides* no clear effect at the DNA level (Table 2) and only a slight effect on mycotoxin (Tables 1, 2) and morphologically determined contamination % levels was observed (Jestoi et al., 2007, Table 1).

Natural infection also took place in artificially infected field plots, and in 2002 it resulted in high background levels of BEA, ENNs, NIV and MON, especially in barley (Table 1). The
increase in BEA, ENNs and NIV levels was connected to relatively high levels of natural *F. poae* contamination, while the increase in ENNs and MON was connected to unusually high *F. tricinctum* contamination levels.

DNA Levels in Cereal Grains in Field Samples During 2003-2006

In oat samples the main *Fusarium* species as determined by qPCR analyses were *F. graminearum, F. sporotrichioides/F.langsethiae* and *F. avenaceum*. In 2006 high levels of *F. poae* DNA were also detected (Table 4). In barley and spring wheat the main *Fusarium* species were *F. avenaceum* and *F. graminearum.* High levels of *F. sporotrichioides/F.langsethiae* and *F. poae* DNA were also detected in 2006 (Table 5). In rye (Table 5) the main *Fusarium* species was *F. avenaceum*.

Table 5. Combined results of the correlation analyses between *Fusarium* DNA and mycotoxins of the years 2005-2006 (Finmyco project). n=number of grain samples

Cereal	TMTRI/DON R^2, p (n)	*F.gram.*/DON R^2, p (n)	*F.gram.+F.culm.*/DON R^2, p (n)	*F. poae*/NIV R^2, p (n)
Wheat	.007, 0.68 (25)	0.61, 0.002** (12)	0.56, 0.006** (11)	No correlation (11)
Surface	0.66, 0.06 (6)	0.88, 0.06 (4)	No correlation (3)	No correlation (4)
Ground				
Barley	0.078, 0.03* (64)	0.53, 0.0009*** (17)	0.60, 0.0003*** (17)	0.07, 0.21 (21)
Surface	0.28, 0.18 (8)	0.54,0.06 (6)	0.99, 0.001** (4)	0.69, 0.041* (6)
Ground				
Oats	0.0005, 0.86 (69)	0.49, 0.0002*** (25)	0.46, 0.0004*** (25)	0,20, 0.03* (24)
Surface	0.16, 0.06 (23)	0.78, 0.0001*** (11)	0.57, 0.012* (11)	0.42, 0.03* (11)
Ground				

Cereal	*F. langs.+ F.sporot.*/ HT-2+T-2 R^2, p (n)	*F. avenaceum*/MON R^2, p (n)	*F. avenaceum*/ENNs R^2, p (n)
Wheat	No correlation (11)	0.30, 0.003** (27)	0.54, <0.0001*** (27)
Surface	No correlation (4)	0.60, 0.12 (5)	0.73, 0.06 (5)
Ground			
Barley	0.19, 0.075 (20)	0.29, <0.0001*** (55)	0.45, <0.0001*** (55)
Surface	0.51, 0.11 (6)	No correlation (5)	0.60, 0.20 (4)
Ground			
Oats	0.41, 0.0008*** (24)	0.10, 0.10 (60)	0.25, <0.0001*** (60)
Surface	0.84, <0.0001*** (11)	Not analysed	Not analysed
Ground			

The main mycotoxins in the oat samples which were analysed for DNA levels were DON (2003 and 2005) and T-2/HT-2 (2003 and 2006). In barley the main mycotoxins were enniatins and in some cases also DON. In spring wheat the main mycotoxins were ENNs, DON and MON. NIV and HT-2/T2 levels were always low in spring wheat (Table 4). In rye and winter wheat (results not shown, www.agronet.fi/cerveg), mycotoxin levels were always low. This is probably due to the early harvesting time. 3AcDON was detected in samples with high DON levels, but 15AcDON, DAS and FX were not detected in any field samples.

Correlation between *Fusarium* DNA from Grain Surfaces and Mycotoxin Levels in Grains

Oats

A highly significant correlation was found in oats between *F. graminearum* DNA and and DON (R^2=0.49, p=0.0002***), *F. langsethiae+F.sporotrichioides* DNA and HT-2+T-2 (R^2=0.41, p=0.0008***) and *F. avenaceum* DNA and ENNs (R^2=0.25, p=<0.0001) during 2005-2006 (Figure 4, Table 5). A few false positive samples were found between *F. langsethiae+F.sporotrichioides* DNA and HT-2+T-2. The correlation was slightly lower (R^2=0.56, p=0.006**) between *F. graminearum+ F. culmorum* DNA and DON than between *F. graminearum* DNA and DON. Only a slight correlation was found between *F. poae* DNA and NIV levels. No clear correlation was found between TMTRI DNA and DON, TMTRI DNA and HT-2+T-2 or *F. avenaceum* DNA and MON.

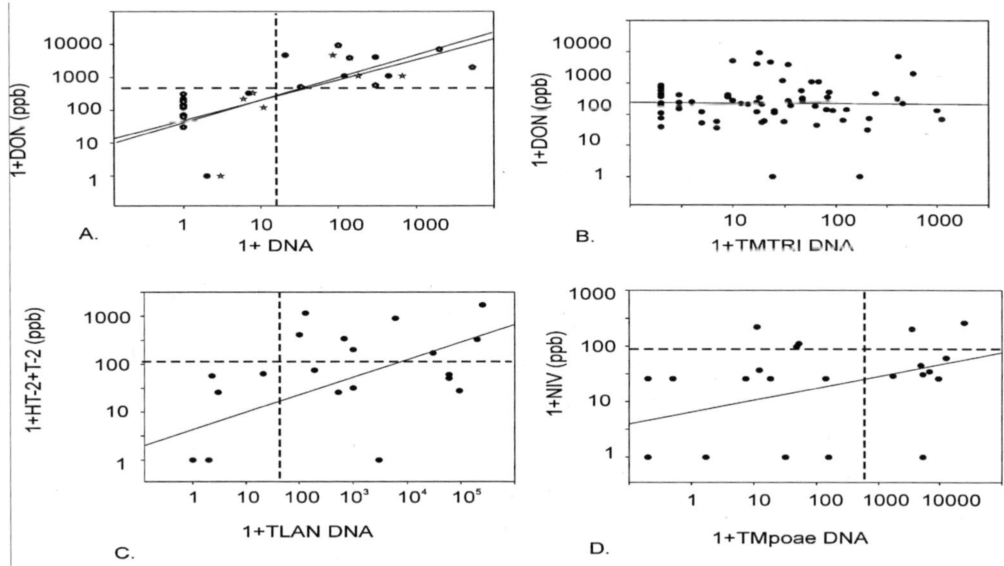

Figure 4. Correlation between log *Fusarium* DNA (10^{-6} ng ng^{-1}) on grain surfaces and mycotoxin (ppb) levels in oats in 2005-2006. A. *F. graminearum* DNA/DON (black circles) and *F. graminearum+F. culmorum* DNA/DON (white stars). B.TMTRI/DON (black circles). C. *F. langsethiae/F. sporotrichioides* DNA/HT-2+T-2. D. *F. poae* DNA/NIV. Regression slopes are shown.

Barley

A highly significant correlation (R^2=0.53, p=0.0009***) was also found between *F. graminearum* DNA and DON in barley samples from 2005-2006 (Figure 5, Table 5). The correlation was better, when the total amount of *F. graminearum+ F. culmorum* DNA was compared to DON (R^2=0.60, p=0.003***). Only a slight correlation was found between TMTRI and DON levels, and the regression slope was negative. The correlation between *F. avenaceum* DNA and MON and ENNs was highly significant, although a few false positive samples were found when TMAV and MON levels were compared.

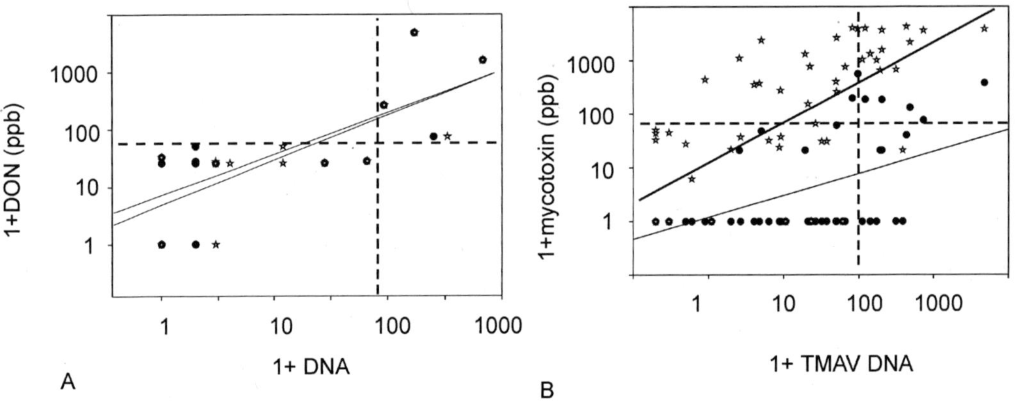

Figure 5. Correlation between log *Fusarium* DNA (10^{-6} ng ng^{-1}) on grain surfaces and mycotoxin (ppb) levels in ground grains of barley in 2005-2006. A. *F. graminearum* DNA/DON (black circles) and *F. graminearum+F. culmorum* DNA/DON (white stars). B. *F. avenaceum* DNA/MON (black circles) and *F. avenaceum* DNA/ENNs (white stars). Regression slopes are shown.

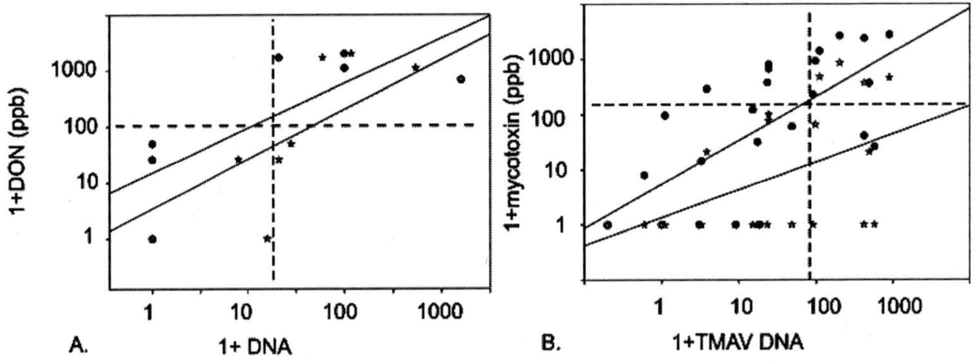

Figure 6. Correlation between log *Fusarium* DNA (10^{-6} ng ng^{-1}) on grain surfaces and mycotoxin (ppb) levels in ground grains of spring wheat in 2005-2006. A. *F. graminearum* DNA/DON (black circles) and *F. graminearum+F. culmorum* DNA/DON (white stars). B. *F. avenaceum* DNA/MON (black circles) and *F. avenaceum* DNA/ENNs (white stars). Regression slopes are shown.

Spring Wheat

A significant correlation (R^2=0.61, p=0.002**) was found between *F. graminearum* DNA and DON in spring wheat samples from 2005-2006 (Figure 6, Table 5). The correlation was slightly lower (R^2=0.56, p=0.006**) between *F. graminearum+ F. culmorum* DNA and DON. No correlation was found between TMTRI and DON levels. In addition a less significant correlation was found between *F. avenaceum* DNA and MON and ENNs. A few false positive samples were found when TMAV and MON levels were compared. R^2 and p were not calculated with NIV and HT-2+T-2, as these mycotoxins were detected in only one spring wheat sample.

FUSARIUM DNA LEVELS ON GRAIN SURFACES AS COMPARED TO GROUND GRAINS

It was possible to obtain higher concentrations of *Fusarium* DNA by extracting DNA from grain surfaces, while the total concentration of DNA was higher from ground grains (Table 6). However the relative amounts of TMTRI and TMAV DNA per total DNA obtained from grain surfaces and ground grains were usually similar, although the sensitivity (DNA levels of *Fusarium* species) was higher with DNA samples extracted from grain surfaces.

Table 6. Examples of *Fusarium* DNA levels (10^{-6} ng ng^{-1} total DNA) from grain surfaces as compared to ground samples (Finmyco project)

Cereal	ground	DNA	TMTRI	TMAV	surface	DNA	TMTRI	TMAV
Rye	267-1	>10	<1	<1	267-1	<1	<1	6
W.wheat	267-3	>10	<1	<1	267-3	<1	3	24
Barley	291-2	>10	<1	<1	291-2	<1	197	105
w.wheat	298-1	5	<1	<1	298-1	<1	24	178
Oats	300-2	>10	32	<1	300-2	<1	4	>1
Rye	301-1	>10	<1	<1	301-1	<1	<1	24
Barley	299-1	>10	<1	<1	299-1	<1	45	<1
Barley	307-1	>10	12	<1	307-1	<1	1080	500
s.wheat	308-1	>10	<1	2	308-1	<1	44	5720
Barley	309-1	5	2	75	309-1	<1	2090	47500
S.wheat	314-1	>10	3.3	3	314-1	<1	99	480

Correlation between *Fusarium* DNA and Mycotoxin Levels in Ground Grains

Oats

A significant correlation was found between *F. graminearum* DNA and DON in ground grains from 2005-2006 (Figure 7, Table 5). The correlation was somewhat slighter between *F. graminearum*+ *F. culmorum* DNA and DON. The correlation between *F. langsethiae*+*F.sporotrichioides* DNA and HT-2+T-2 was highly significant and no false positive samples were found. A significant correlation was also found between *F. poae* DNA and NIV and between TMTRI DNA and DON. There was no correlation between TMTRI DNA and HT-2+T-2. *F. avenaceum* DNA levels were measured from only one sample of ground grains.

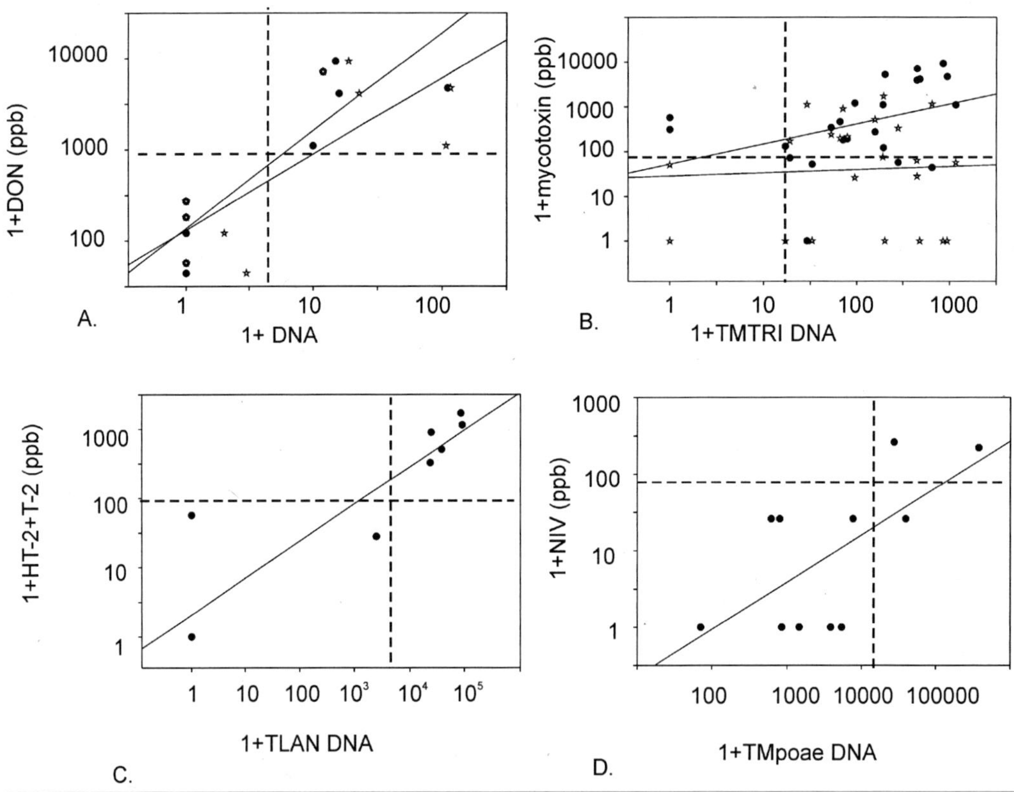

Figure 7. Correlation between log *Fusarium* DNA (10^{-6} ng ng^{-1}) and mycotoxin (ppb) levels in ground grains of oats in 2005-2006. A. *F. graminearum* DNA/DON (black circles) and *F. graminearum*+*F. culmorum* DNA/DON (white stars). B. TMTRI DNA/DON (black circles), TMTRI DNA/HT-2+T-2 (white stars). C. *F. langsethiae*/*F. sporotrichioides* DNA/HT-2+T-2. D. *F. poae* DNA/NIV. Regression slopes are shown.

Barley

The correlation was significant between *F. graminearum*+ *F. culmorum* DNA and DON and between *F. poae* DNA and NIV (Figure 8, Table 5). High R^2 values were also obtained between *F. graminearum* DNA and DON and between *F. avenaceum* DNA and ENNs. The correlation between TMAV and MON could not be calculated, since none of the five samples contained MON.

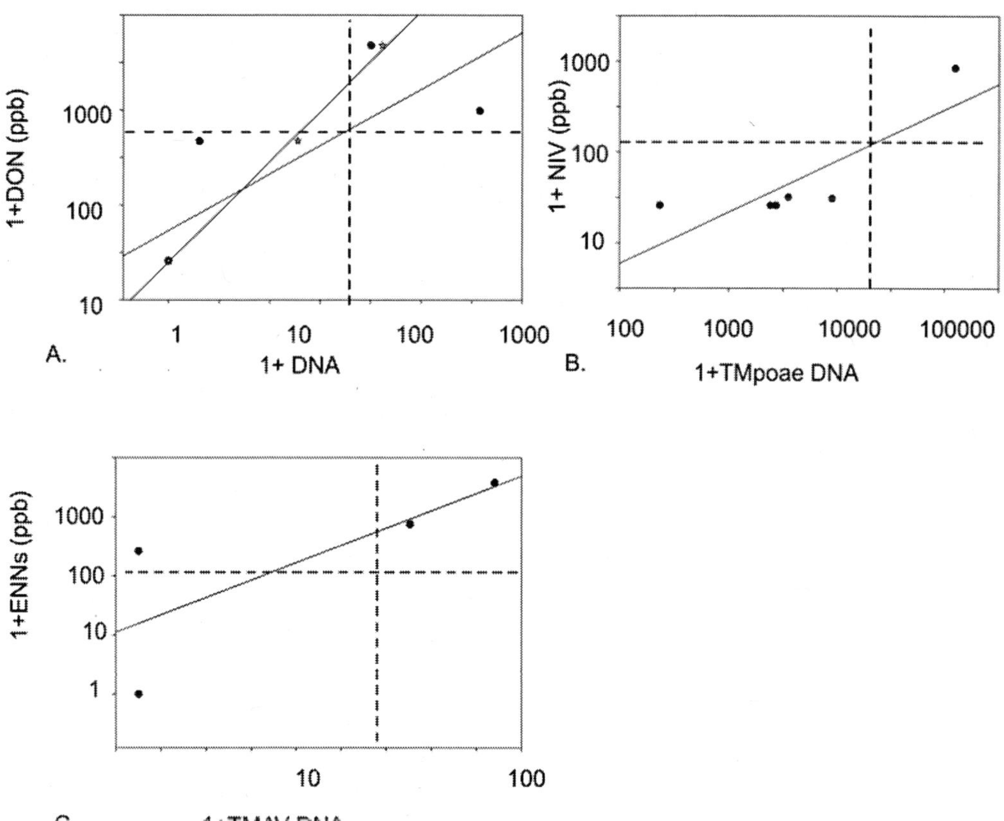

Figure 8. Correlation between log *Fusarium* DNA (10^{-6} ng ng^{-1}) and mycotoxin (ppb) levels in ground grains of barley in 2005-2006. A. *F. graminearum* DNA/DON (black circles) and. *F. graminearum*+*F. culmorum* DNA/DON (white stars). B. *F. poae* DNA/NIV. C. *F. avenaceum* DNA/ENNs. Regression slopes are shown.

Spring Wheat

High R^2 values were obtained between *F. graminearum* DNA and DON and between TMTRI DNA and DON (Figure 9, Table 5). High R^2 values were also obtained between *F. avenaceum* DNA and MON and ENNs Due to the small number of ground samples and the low levels of other mycotoxins in these samples, it was not possible to study the correlation between other *Fusarium* DNA levels and mycotoxins.

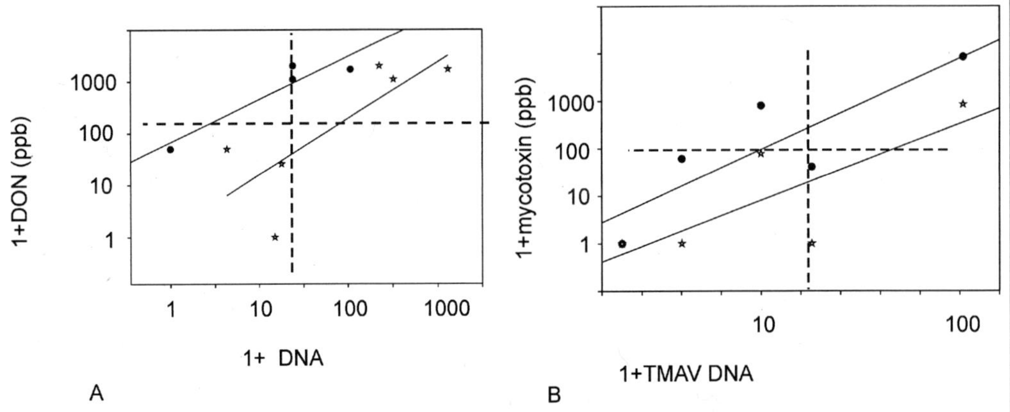

Figure 9. Correlation between log *Fusarium* DNA (10^{-6} ng ng^{-1}) and mycotoxin (ppb) levels in ground grains of spring wheat in 2005-2006. A. *F. graminearum* DNA/DON (black circles) and TMTRI DNA/DON (white stars). B. *F. avenaceum* DNA/ENNs (black circles) and *F. avenaceum* DNA/MON (white stars). Regression slopes are shown.

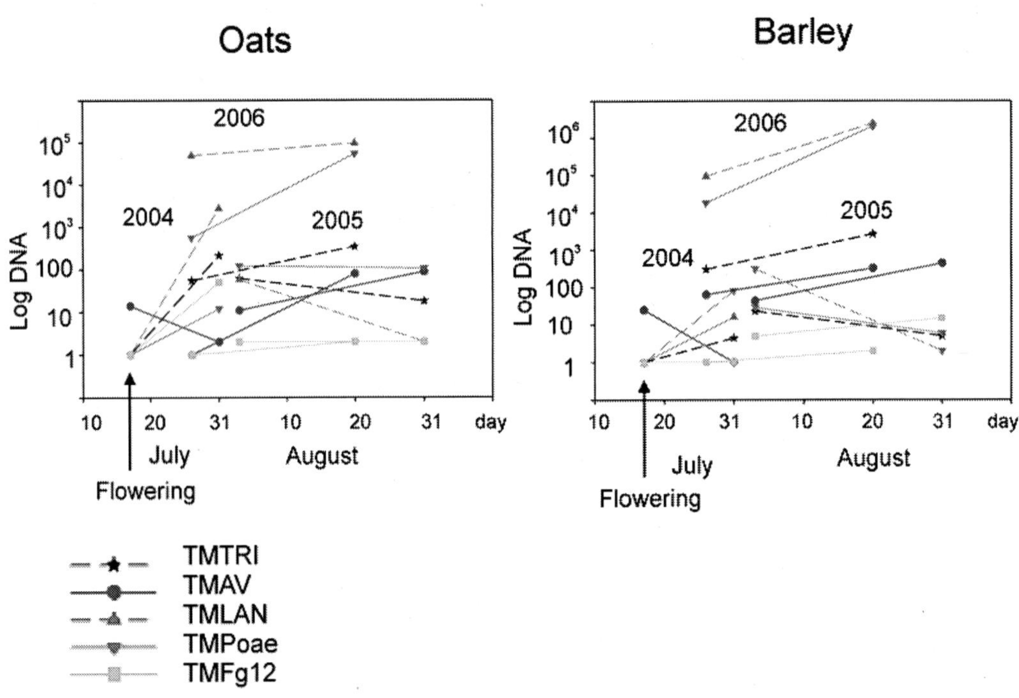

Figure 10. Means of *Fusarium* DNA levels (10^{-6} ng ng^{-1} total DNA) in oats and barley plots during the growing seasons of 2004, 2005 and 2006.

Table 7. TMTRI, TMAV, TMpoae, TMFg12, TMLAN and MGBculm DNA levels (10^{-6} ng ng^{-1} total DNA) in oats and barley in 2004 two weeks after flowering (above) and in 2005 three-four weeks before harvesting and during harvesting (below). Results from Safety Indicators Project

Cereal	2004 2 weeks after flowering	TMTRI (n)	TMAV (n)	TMpoae (n)	TMFg12 (n)	TMLAN (n)
Oats	Tillage	67.0±39.4 (8)	<10 (8)	7.2±5.8 (5)	59±45 (5)	1120±470 (5)
	Direct drilling	360±160 (8)	<10 (8)	16±9 (7)	39±21 (7)	4400±2180 (7)
	Fungicide	120±90 (8)	<10 (8)	22±10 (6)	24±24 (6)	3040±1940 (6)
	Without fungicide	300±150 (8)	<10 (8)	2.0±1.0 (6)	71±35(6)	3030±2020 (6)
Barley	Tillage	0 (8)	<10 (8)			
	Direct drilling	6.8±5.9 (8)	<10 (8)	15±9 (3)	0 (3)	81±76 (3)
	Fungicide	0.8±0.6 (n=8)	<10 (8)	6.0 (2)	0 (1)	5 (2)
	Without fungicide	6.0 ±6.0	<10 (8)	32.0 (1)	0 (2)	232 (1)

Cereal	2005 3-4 weeks before harvesting	TMTRI (n)	TMAV (n)	TMpoae (n)	TMFg12 (n)	TMLAN (n)	MGBculm (n)
Oats	Tillage	115±80 (8)	<10 (8)	240±140 (8)	<1 (4)	9±8 (4)	<1 (4)
	Direct drilling	8.2±4.1 (8)	<10 (8)	7.4±2.4 (8)	<1 (1)	250 (1)	
	Fungicide	110±80 (8)	<10 (8)	200±150 (8)	<1 (3)	12±11 (3)	<.1 (3)
	Without fungicide	16±5 (8)	<10 (8)	42±18.(8)	<1 (2)	126 (2)	<.1 (1)
Barley	Tillage	31±15 (8)	<10 (8)	41±23 (8)	7 (2)	240 (2)	<1 (2)
	Direct drilling	12±5 (8)	76±52 (8)	15±9 (8)	<1 (2)	380 (2)	
	Fungicide	18±8 (8)	<10 (8)	14±9 (8)	2.9 (2)	2.5 (2)	<1 (1)
	Without fungicide	26±14 (8)	76±52 (8)	17±8 (8)	<1 (2)	460 (2)	<1 (1)
	2005 during harvesting						
Oats	Tillage	30±11 (8)	110±80 (8)	200±94 (8)	<1 (5)	<1 (5)	<1 (3)
	Direct drilling	4.6±1.3 (8)	66±12 (8)	13±6(8)	<1 (1)	<1 (1)	<1 (1)
	Fungicide	17±7 (8)	130±80(8)	90±84	<1 (3)	<1 (3)	<1 (3)
	Without fungicide	21±11 (8)	33±11 (8)	11±7	<1 (2)	<1 (2)	<1 (2)
Barley	Tillage	4.9±3.0 (8)	600±460 (8)	7.6±1.8 (8)	<1 (1)	<1	<1 (1)
	Direct drilling	2.9±1.5 (7)	270±53 (8)	3.1±1.5 (8)	26 (1)	<1	<1 (1)
	Fungicide	1.2±0.2	200±50 (8)	6.3±2.6	<1 (1)	<1 (3)	<1 (1)
	Without fungicide	<1	670±450 (8)	4.3±1.8 (8)	26 (1)	<1 (4)	<1 (1)

DNA Levels in Oats and Barley between Flowering and Harvesting

DNA levels (from grain surfaces) between flowering and harvesting were investigated in field plots at Jokioinen during 2004-2006. In 2004 samples for qPCR were taken during and two weeks after flowering, while in 2005 and 2006 they were taken ca. three to four weeks before harvesting and during it (Tables 7, 8). The growing season in 2004 was more rainy than in 2005, especially before and during flowering and several times afterwards. No DNA of trichothecene-producing *Fusarium* species and only low levels of *F. avenaceum* DNA were detected during flowering. The DNA levels of *F. avenaceum, F. graminearum, F. poae* and *F. sporotrichioides/F. langsethiae* in 2004 increased clearly within two weeks after flowering (Figure 10).

Table 8. TMTRI, TMAV, TMpoae, TMFg12, TMLAN and MGBculm DNA levels (10^{-6} ng ng^{-1} total DNA) in oats and barley in 2006 three-four weeks before harvesting (above) and during harvesting (below). Results from Safety Indicators Project

Cereal	3-4 weeks before harvesting	TMTRI (n)	TMAV (n)	TMpoae (n)	TMFg12 (n)	TMLAN (n)	MGBculm (n)
Oats	Tillage	60±31 (8)	<10 (8)	1270±1120 (4)	<1 (4)	35430±34860(5)	0 (5)
	Direct Drilling	52±40 (8)	<10 (8)	4±1 (5)	<1 (5)	59800±57600(4)	<1 (5)
	Fungicide	56±40 (8)	<10 (8)	99±95 (5)	<1 (5)	58400±57900 (5)	<.2 (5)
	Without fungicide	53±31 (8)	<10 (8)	1150±1150 (4)	<1 (4)	37160±34290 (4)	<.1 (4)
Barley	Tillage	310±94 (8)	<10 (8)	45400±32900(3)	<1 (3)	2820±2600 (3)	<1 (3)
	Direct Drilling	290±130 (8)	125±125 (8)	1220±1140 (6)	<1 (6)	138700±86730(6)	<1 (6)
	Fungicide	470±110 (8)	<10 (8)	26860±21300(5)	<1 (5)	117400±81900(5)	<1 (5)
	Without fungicide	130±70 (8)	125±125 (8)	2340±1610 (4)	<1 (4)	63360±12230 (4)	<1 (4)
	During harvesting						
Oats	Tillage	240±110(8)	150±120 (8)	74300±55500 (4)	1±1 (4)	1610±1100(4)	<10 (4)
	Direct Drilling	17460±10280 (8)	12±8 (8)	61170±26480 (6)	1±0.6(6)	167000±78600(6)	<10 (6)
	Fungicide	410±140 (8)	130±80 (8)	87320±41380(5)	<1 (5)	115680±96690 (5)	<10 (5)
	Without fungicide	15300±13800 (8)	33±11 (8)	44040±31940(5)	2±1 (5)	113010±53460 (5)	<10 (5)
Barley	Tillage	4050±1410 (8)	20±16 (8)	3500000±1002000 (3)	2.0±1.5 (3)	56940±56530 (3)	3.3±3.3(3)
	Direct Drilling	9330±2930 (8)	600±350 (8)	1729150±1500220 (6)	100±100 (6)	3615000±1994170 (6)	6±6 (6)
	Fungicide	3830±1240 (8)	390±360 (8)	1626000±1068000 (5)	<1 (3)	632170±267530 (5)	<10 (3)
	Without fungicide	99540±250 (8)	210±120 (8)	3186000±2073000 (4)	<1 (3)	4675000±2984500 (4)	<10 (3)

It was interesting that in 2005 the DNA levels of *F. sporotrichioides/F. langsethiae* and the total DNA amount of trichothecene-producing *Fusarium* fungi (TMTRI) decreased clearly before harvesting, while at the same time the TMAV level continued to increase both in oats and barley. The year 2006 was exceptionally dry after flowering, and *F. sporotrichioides/F. langsethiae* and *F. poae* DNA levels continued to increase together with *F. avenaceum* DNA levels down to harvesting. In 2006 (Table 8) the DNA levels of

Fusarium species were higher than in 2005 (Table 7), and unlike 2005 were still increasing during the last weeks before harvesting (Figure 10). *F. graminearum* and *F. culmorum* levels were low.

The highest *F. graminearum* DNA levels were found in barley during harvesting in 2005 and in oats two weeks after flowering in 2004. Only low levels of *F. culmorum* DNA were found in 2005-2006. TMTRI DNA levels were affected by *F. langsethiae* and *F. poae* DNA levels. In 2004 the main trichothecene-producing fungi were *F. langsethiae/F. sporotrichioides* and *F. graminearum* in oats, while in 2005 the main species were *F. poae* and *F. langsethiae/F. sporotrichioides* (Figure 10).

EFFECT OF TILLAGE AND FUNGICIDE TREATMENT ON *FUSARIUM* DNA LEVELS

In 2004 tillage decreased the amount of TMTRI DNA (mainly TMLAN) in oats two weeks after flowering as compared to direct drilling. In 2005 tillage increased the amount of TMTRI DNA in both oats (more clearly) and barley; this was due to TMpoae DNA (Table 7). Fungicide treatment one week before flowering decreased the amount of TMLAN DNA and increased the amount of TMpoae DNA present 3-4 weeks before harvesting in oats in 2005. Also in 2004 the amount of TMpoae DNA was higher 2 weeks after flowering in oats plots with the fungicide treatment. The TMAV level was lower 3-4 weeks before and during harvesting in barley after the fungicide treatment.

In 2006 tillage with ploughing increased the amount of *F. poae* DNA in both oats and barley 3-4 weeks before harvesting as compared to direct drilling, while in barley TMLAN levels were clearly higher in plots with direct drilling (Table 8). During harvesting, TMLAN levels were higher in plots with direct drilling in both oats and barley. Based on *F. sporotrichioides* and *F. langsethiae* contamination % levels (Parikka et al., 2008), the changes in TMLAN levels were mainly due to changes in *F. langsethiae* levels. TMTRI levels were also higher during harvesting in plots with direct drilling in oats. TMAV levels were higher in oats with tillage during harvesting, while in barley they were higher with direct drilling. No clear effect on *Fusarium* DNA levels was found after the fungicide treatment in 2006, except for a transient increase of TMpoae level in barley 3-4 weeks before harvesting.

High levels of DON (> 1000 ppb) were not found in the plots in 2004-2006, which is in agreement with the low TMFg12 levels found. Changes in NIV and HT-2/T-2 levels were in accordance with *F. poae* and TMLAN DNA levels respectively. The higher *F. poae* DNA levels in plots with tillage before harvesting were connected to higher NIV levels in both oats and barley during harvesting in 2005-2006. The difference in NIV levels in 2005/2006 was ca. 7-fold/16-fold in oats and ca. 3-fold/3-fold in barley. The higher TMLAN levels in plots with direct drilling during harvesting were connected to higher HT-2+T-2 levels with direct drilling in oats and barley during harvesting in 2006. The difference in HT-2+T-2 levels was ca. 5-fold in oats and ca. 9-fold in barley.

DISCUSSION

Based on the R^2 results between qPCR and chemical determinations, *F. graminearum* seems to be the main DON producer in Finnish oats, barley and spring wheat, especially in samples with high DON levels. This is in accordance with the results obtained with Norwegian grain samples (Elen et al., 2007, unpublished results) and with the previous investigation in Finland (Yli-Mattila et al. 2008), but it differs from the results of Sarlin et al. (2006) with Finnish barley. In barley the R^2 value was higher when *F. culmorum* DNA was added to *F. graminearum* DNA. The DON produced by *F. culmorum* in barley might explain the poor correlation between *F. graminearum* DNA and DON and the better correlation between TMTRI DNA and DON levels in the study of Sarlin et al. (2006). It should also be taken into account that all DON levels in this study were relatively low (<1000 ppb); this may also explain the poor correlation between *F. graminearum* DNA and DON, as *F. graminearum* is a indicator species for high levels of DON.

According to the mycotoxin results the 3AcDON chemotypes of *F. graminearum* and *F. culmorum* seem to be prevalent in Finland, since no 15AcDON was found in grain samples; this is in agreement with the earlier results of Hietaniemi et al. (1991), Jestoi et al. (2007) and Yli-Mattila et al. (2007c, unpublished results). The 3AcDON chemotype has been prevalent in Europe and China, while the 15AcDON chemotype is prevalent in North America (Mirocha et al., 1989; Miller et al., 1991) and England (Jennings et al., 2004). In Austria the 15AcDON chemotype has become more common (Adler et al. 2002), while in North America the 3AcDON chemotype has become more common (Ward et al., 2008).

F. poae, F. sporotrichioides and *F. langsethiae* are the main producers of NIV, HT-2 and T-2 toxins (Pettersson, 1991; Torp and Langseth, 1999; Salas et al., 1999; Jestoi et al., 2004; Bottalico and Perrone, 2002; Thrane et al., 2004). The present study confirmed the correlation between *F. poae* DNA and NIV levels in barley and oats found in the previous study, while the correlation between TMLAN DNA and HT-2+2T-2 levels was clearly higher in oats than in the previous study (Yli-Mattila et al., 2008).

The results of the present work are in agreement with previous research (Wilson et al. 2004; Parikka et al., 2005; Yli-Mattila et al, 2008, Jestoi et al, 2007), according to which *F. langsethiae* is one of the pioneer *Fusarium* fungi on grains during flowering, especially in oats. It grows, however, more slowly than *F. poae* and *F. sporotrichioides* (Torp and Nirenberg 2004; Thrane et al., 2004), and may be overgrown by other *Fusarium* species later in the growing season, which may explain the decrease of TMLAN DNA levels in field plots in 2005.

Natural infection took place in artificially infected field plots, and in 2002 it resulted in high background levels of BEA, ENNs, NIV and MON, especially in barley (Table 1). The increase in BEA and NIV levels was connected to relatively high levels of natural *F. poae* contamination, while the increase of ENNs and MON was also connected to unusually high *F. tricinctum* contamination levels. The connection between BEA and *F. poae* contamination levels and the highly significant correlation between *F. avenaceum* DNA and ENNs/MON levels in the present work are in accordance with the results obtained previously in Norway (Uhlig 2005) and Finland (Yli-Mattila et al., 2006).

The idea that the high BEA, NIV and ENNs levels in artificially infected barley plots in 2002 were due to the natural contamination by *F. poae* is supported by the high *F. poae* DNA levels in two of these samples (Yli-Mattila et al., 2008). In 2003 there was less natural contamination and high BEA (and HT-2+T-2) levels were only produced after artificial inoculation with *F. sporotrichioides*, while high ENNs and MON levels were only produced after artificial inoculation with *F. avenaceum*. The effect of artificial inoculation on *F. culmorum* could be followed by *F. culmorum* contamination %, TMTRI, DON and 3AcDON levels. *F. graminearum*, *F. poae* and *F. langsethiae* inoculation was less effective in 2003. In 2004 the amount of *F. graminearum* DNA increased clearly in *F. graminearum*-inoculated barley and wheat field plots, when higher amounts of spores were used for inoculation (Yli-Mattila et al., 2008). This also resulted in high ZEN and DON levels in artificially contaminated field plots as measured in the present work. Higher amounts of spores from three *F. langsethiae* isolates (Yli-Mattila et al., 2008), however, did not give better artificial inoculation with *F. langsethiae* isolates in 2004.

According to the results of the present work, fluorogenic PCR detection can also be used for the detection and quantification of *Fusarium* fungi in cereals before harvesting in order to predict the risk of mycotoxins in mature grain, which is in accordance with the results of Reischler et al. (2004). The changes in DNA levels in oats and barley between flowering and harvesting were in agreement with the changes in NIV and HT-2+T-2 levels (S. Rämö and V. Hietaniemi, manuscript in preparation). NIV levels were clearly higher during harvesting in oats and barley plots in 2005 than in 2004 or 2006. In oats HT-2+T-2 levels were highest in 2006, while in barley they were highest in 2005.

TaqMan real-time quantitative PCR assays gave reproducible results with both DNA extraction methods used. The advantage of the surface extraction method is that the DNA is extracted from the surfaces of approximately 200 grains, while in the other method only 0.1 g of ground grains is used for extraction. With the surface method it is also possible to avoid possible PCR-inhibitory compounds from the inner parts of the grains, and the proportion of plant DNA is smaller in the DNA extract, while with the ground seed method fungal DNA from the inner parts of the grains can also be included in the analysis. According to the results of the present work, PCR inhibition was not a problem with the GenElute^{TM} Plant Genomic DNA Kit of Sigma, while fungal DNA from the inner parts of the grains improved the correlation. This is in agreement with the previous work of Sarlin et al. (2006).

Future Perspectives

In future the current molecular methods will be improved, validated and automated, making it possible to use them in the routine identification, detection and quantification of *Fusarium* species. Molecular analyses require expensive equipment, but they allow quicker and more reliable identification with fewer people than traditional morphological identification. Internal standards should also be used in future to discriminate between uninfected samples and possible PCR inhibition (Waalwijk et al., 2004). The computer analysis of molecular data is also expected to improve and the total amount of sequence data in the databases will increase. The whole genome of *F. graminearum*, for instance, has already been sequenced (Guldener et al. 2006, Cuomo et al., 2007, http://www.broad.mit.edu/annotation/genome/fusarium_graminearum/Home.html, http://mips.gsf.de/genre/proj/fusarium). The most promising techniques for the large-scale identification of different fungal species and chemotypes are microarray-based methods (Wilson et al., 2002; Kristensen, 2006) and other hybridisation methods (e.g. Luminex, Ward et al., 2008, Yli-Mattila et al., 2007b,c) detecting single nucleotide polymorphisms.

Work dealing with molecular studies of biosynthetic genes of different *Fusarium* toxins will continue and will result in finding new genes, which will be used for designing new primers and probes specific e.g. for zearalenone, enniatin and moniliformin production. The infection process will also be studied by different molecular methods and using *Fusarium* isolates labelled by fluorescent labels.

ACKNOWLEDGEMENTS

This study was supported by grants from the National Technology Agency of Finland (No. 40168/03) and from The Academy of Finland (No. 52104), and by two grants from the Ministry of Agriculture (Dnro 5159/501/2003 and 4759/501/2004). The authors want to thank Taina Lahtinen and Maiju Saarinen for their assistance with DNA extraction and qPCR analyses, Kristiina Kuitunen, Kati Lehti and Sanna Yli-Karjula for their assistance in mycotoxin analyses, Dr. Sonja Klemsdal for designing TMTRI, TMAV and TMLAN primers and probes, Dr. Tatiana Gagkaeva for mycological analyses and for critically reading through the manuscript, Dr. Alexey Burkin for ELISA analyses and Dr. Jouni Ahlholm for help in the statistical analyses. The English was checked by Ellen Valle of the Department of English, University of Turku.

REFERENCES

Adler, A., Lew, H, Brunner, S., Oberforster, J., Hinterholzer, J., Kullnig-Gradinger CM, Mach RL and Kubicek CP (2002). *Fusaria* in Austrian cereals – change in species and toxins spectrum. *Journal of Applied Genetics 43 A*: 11-16.

Bluhm, B.H., Cousin, M.A. and Woloshuk, C.P. (2004) Multiplex real-time PCR detection of fumonisin-producing and trichothecene-producing groups of *Fusarium* species. *Journal of Food Protection 67*: 536-543.

Bottalico, A. and Perrone, G. (2002) Toxigenic *Fusarium* species and mycotoxins associated with head blight in small-cereals in Europe. *European Journal of Plant Pathology 108*: 611-624.

Burlakoti, R.R., Estrada, R., Jr., Rivera, V.V., Boddela, A., Secor, G.A. and Adhikari, T.B. (2007) Real-time PCR quantification and mycotoxin production of *Fusarium graminearum* in wheat inoculated with isolates collected from potato, sugar beet, and wheat. *Phytopathology 97*: 835-841.

Cuomo, C.A., Guldener, U., Xu, J-R, Turgeon, B.G., Pietro, A.D., Walton, J.D., Ma, L-J. Baker, S.E., Rep, M., Adam, G., Antoniw, J., Baldwin, T., Calvo, S., Chang, Y-L.,

DeCaprio, D., Gale, L.R., Gnerre, , S., Goswami, R.s., Hammond-Kosack, K, Harri., L.J., hilburn, K., Kennell, J.C;, Kroken, S., Magnuson, K.k., Mannhaupt, G., Munsterkotter, M., Nelson, D., O'Donnell, K., Quellet, T., Qi, W., Quesneville, H., roncero, I.G., Seong, KY., Tetko, I.V., Urban, M., Waalwijk, C., Ward, T.J., Yao, J., Birren, B.W. and Kistler, H.C. (2007) The *Fusarium graminearum* genome reveals a link between localized polymorphism and pathogen specialization. *Science 317*: 1400-1402.

Desjardins, A.E. 2006. *Fusarium mycotoxins. Chemistry, Genetics and Biology.* – The American Phytopathological Society. 260 pp.

Doohan, F.M, Parry, D.W., Jenkinson, P. and Nicholson P. (1998) The use of species-specific PCR-based assays to analyse *Fusarium* ear blight of wheat. *Plant Pathology 47*: 197-205.

Edwards, S.G., Pirgozliev, S.R., Hare, M.C., Jenkinson, P. (2001) Quantification of trichothecene-producing *Fusarium* species in harvested grain by competitive PCR to determine efficacies of fungicides against *Fusarium* head blight of winter wheat. *Applied and Environmental Microbiology 67*: 1575-1580.

Elen, O., Klemsdal, S.S., Clasen, P.-E. and Razzaghiani, M.J. (2007) *Fusarium* spp. and mycotoxins in Norwegian wheat grain (2001-2004). XV Congress of European Mycologists, Saint Petersburg, Russia, September 16-21: 275-276 (abstract).

Eskola, M., Parikka, P. and Rizzo, A. (2001) Trichothecenes, ochratoxin A and zearalenone contamination and *Fusarium* infection in Finnish cereal samples in 1998. *Food Additives and Contaminants 18*: 707-718

Gagkaeva, T. and Yli-Mattila, T. (2004) Genetic diversity of *Fusarium graminearum* in Europe and Asia. *European Journal of Plant Pathology 110*: 551-562.

Gagkaeva, T., Kononenko G., Burkin, A. and Yli-Mattila, T. (2006) *Comparative analysis of tri7 and tri13 genes in Fusarium graminearum isolates.* 9th European *Fusarium* seminar 19-22 September 2006, Wageningen, The Neteherlands (book of abstracts): 179.

Golinski, P., Kostecki, M., Lasocka, I., Wisniewska, H., Chelkowski, J. and Kaczmarek, Z. (1996) Moniliformin accumulation and other effects of *Fusarium avenaceum* (Fr.) Sacc. on kernels of winter wheat cultivars. *Journal of Phytopathology 144*: 495-499.

Goswami, R.S. and Kistler, C., (2004). Heading for disaster: *Fusarium graminearum* on cereal crops. *Molecular Plant Pathology 5*: 515-525.

Guldener, U., Mannhaupt, G., Munsterkotter, M., Haase, D., Oesterheld, M. Stumpflen, V., Mewes, H.W. and Adam, G. (2006) FGB: a comprehensive fungal genome resource on the plant pathogen. *Nucleic Acids Research 34*: 456-458.

Hietaniemi, V., Kontturi, M., Rämö, S., Eurola, M., Kangas, A., Niskanen, M. and Saastamoinen, M. (2004) Contents of trichothecenes in oats during official variety, organic cultivation and nitrogen fertilization trials in Finland. *Agricultural and Food Science 13*: 54-67.

Halstensen, A.S., Nordby K.C., Eduard, W. and Klemsdal S.S. (2006b) Real-time PCR detection of toxigenic *Fusarium* in airborne and settled grain dust and associations with trichothecene mycotoxins. *Journal of Environmental Monitoring 8*: 1235-1241.

Hietaniemi, V. and Kumpulainen, J. (1991) Contents of *Fusarium* toxins in Finnish and imported grains and feeds. *Food Additives and Contaminants 8*: 171-182.

Hogg, A.C., Johnston, R.H. and Dyer, A.T. (2007) Applying real-time quantitative PCR to *Fusarium* crown rot of wheat. *Plant Disease 91*: 1021-1028.

Jansen, C., von Wittstein, D., Schäfer, W., Kogel, K-H., Felk, A. and Maier F.J. (2005) Infection patterns in barley and wheat spikes inoculated with wild-type and trichodiene synthase gene disrupted *Fusarium graminearum. Proceedings of the National Academy of Science 102*: 16892-16897.

Jennings P., Coates M.E., Walsh K., Turner J.A., Nicholson P. (2004) Determination of deoxynivalenol- and nivalenol-producing chemotypes of *Fusarium graminearum* isolates from wheat crops in England and Wales. *Plant Pathology 53*: 643-652.

Jestoi, M., Paavanen-Huhtala, S., Parikka, P.. and Yli-Mattila, T. 2008. *In vitro* and *in vivo* mycotoxin production of *Fusarium* isolated from Finnish grains. *Archives of Phytopathology and Plant Protection (in press).*

Jestoi M, Paavanen-Huhtala S, Uhlig S, Rizzo A and Yli-Mattila T (2004) Mycotoxins and cytotoxicity of Finnish *Fusarium* strains grown on rice cultures. In: Canty SM, Boring T, Wardwell J and Ward RW (Eds.), *Proceedings of the 2nd International Symposium on Fusarium Head Blight; incorporating the 8th European Fusarium seminar*; 2004, 11-15; Orlando, Fl, USA. Michigan State University, East Lansing, MI. pp. 405-409.

Jestoi, M., Rokka, M., Rizzo, A., Peltonen, K., Parikka, P. and Yli-Mattila, T. (2003) Moniliformin in Finnish grains: Analysis with LC-MS/MS. *Aspects of Applied Biology 68*: 211-216

Jestoi, M., Rokka, M., Rizzo, A., Peltonen, K. and Aurasaari, S. (2005) Determination of *Fusarium* mycotoxins beauvericin and enniatins with liquid chromatography –tandem mass spectrophotometry (LC-MS/MS). *Journal of Liquid Chromatography and Related Technologies 28*: 369-381.

Joffe, A.Z. (1974) Toxicity of *Fusarium poae* and *F. sporotrichioides* and its relation to alimentary toxic aleukia. In: Purhcase: I.F.H., ed. *Mycotoxins*, Amsterdam, Elsevier, pp. 229-262.

Jurado, M., Vazques, C., Patino, B. and Gonzales-Jaen, M.T. (2005) PCR detection for the trichothecene-producing species *Fusarium graminearum, Fusarium culmorum, Fusarium poae, Fusarium equiseti* and *Fusarium sporotrichioides. Systematical and Applied Microbiology 28*, 562-568.

Konstantinova, P. and Yli-Mattila, T. (2004) IGS-RFLP analysis and development of molecular markers for identification of *F. poae, F. pulverosum, F. sporotrichioides* and *F. kyushuense. International Journal of Food Microbiology 95*: 321-331.

Kristensen, (2006) Rapid multiplex detection of toxigenic *Fusarium* species. Ph.D. thesis, University of Oslo, Norway.

Langseth, W., Bernhoft, A., Rundberget, T., Kosiak, B. and Gareis, M. (1999) Mycotoxin production and cytotoxicity of *Fusarium* strains isolated from Norwegian cereals. *Mycopathologia 144*: 103-113.

Leisova, L., Kucera, L., Chropova, J., Sykorova, V., Sip, V. and Ovesna, J. (2006) Quantification of *Fusarium culmorum* in wheat and barley tissues using real-time PCR in comparison with DON content. *Journal of Phytopathology 154*: 603-611.

Leslie, J.L., Anderson, L.L., Bowden, R.L., and Lee, Y.-W. 2007. Inter- and intra-specific genetic variation in *Fusarium. International Journal of Microbiology 119*: 25-32.

Liu, W.Z., Sundheim, L. and Langseth, W. (1998) Trichothecene production and the relationship to vegetative compatibility groups in *Fusarium poae*. *Mycopathologia 140*: 105-114.

Logrieco, A., Rizzo, A., Ferracane, R. and Ritieni, A. (2002) Occurrence of beuavericin and enniatins in wheat affected by *Fusarium avenaceum* head blight. *Applied and Environmental Microbiology 68*: 82-85.

Maier F.J., Miedaner, T., Hadeler, B., Felk, A., Salomon, S., Lemmens, M., Kasser, H. and Schaefer, W. (2006) Involvement of trichothecenes in fusarioses of wheat, barley and maize evaluated by gene disruption of the trichodiene synthase (Tri5) gene in three field isolates of different chemotype and virulence. *Molecular Plant Pathology 7*: 449-461.

McMullen M., Jones, R. and Gallenberg, D. 1997. Scab of Wheat and Barley: A Re-emerging Disease of Devastating Impact. *Plant Disease 81*: 1340-1346.

Miller, J.D., Greenhalgh, R., Wang, Y. and Lu, M. (1991) Trichothecene chemotypes of three *Fusarium* species. *Mycologia 83*: 121-130.

Mirocha, C.J.H.K., Abbas, C.E., Windels, E.C. and Xie, W. (1989) Variation in deoxynivalenol, 15-acetyldeoxynivalenol, 3-acetyldeoxynivalenol, and zearalenone production by *Fusarium graminearum* isolates. *Applied and Environmental Microbiology 55*: 1315-1316.

Mishra, P.K., Fox, R.T.V. and Culham, A. (2003) Development of a PCR-based assay for rapid and reliable identification of pathogenic *Fusaria*. *FEMS Microbiology Letters 218*: 329-332.

Nirenberg, H.I. (1981) A simplified method for identifying *Fusarium* spp. occurring on wheat. *Canadian Journal of Botany 59*: 1599-1609.

O'Donnell, K (1996) Progress towards a phylogenetic classification of *Fusarium*. *Sydowia 48*: 57-70.

O'Donnell, K, Kistler, H.C., Tacke, B.K. and Casper, H.H. (2000) Gene genealogies reveal global phylogeographic structure and reproductive isolation among lineages of *Fusarium graminearum*, the fungus causing wheat scab. *Proceedings of the National Academy of Sciences, USA 97*: 7905-7910.

O'Donnell, K. and Cigelnik, E. (1997): Two divergent intragenomic rDNA ITS2 types within a monophyletic lineage of the fungus *Fusarium* are nonorthologous. *Molecular Phylogenetics and Evolution 7*: 103-116

O'Donnell, K., Ward, T.J., Geiser, D.M., Kistler, H.C. and Aoki, T. (2004) Genealogical concordance between the mating type locus and seven other nuclear genes supports formal recognition of nine phylogenetically distinct species within the *Fusarium graminearum* clade. *Fungal Genetics and Biology 41*: 600-623.

Paavanen-Huhtala, S. (2000). Molecular-based Assays for the Determination of Diversity and Identification of *Fusarium* and *Gliocladium* fungi. Ph.D. thesis, University of Turku, Finland.

Paavanen-Huhtala, S., Hyvönen, J., Bulat, S.A. and Yli-Mattila, T. (1999). RAPD-PCR, isozyme,rDNA RFLP and rDNA sequence analyses in identification of Finnish *Fusarium oxysporum* isolates. *Mycological Research 103*: 625-634.

Parikka, P., Hietaniemi, V. and Rämö, S. (2005) The effect of tillage on *Fusarium* infection and mycotoxins on barley and oats. *Proceedings of the BCPC International Congress-*

Crop Science and Technology 2005, pp 423-428. Glasgow, Scotland, 31 Oct-2 Nov 2005.

Parikka, P., Hietaniemi, V., Rämö, S. and Jalli, H. (2008) *Fusarium* infection and mycotoxin contents under different tillage treatment. *Journal of Plant Pathology 90:* S3.75

Perkowski, J., Kiecana, I., Schumacher, U., Muller, H-M., Chelkowski, J. and Golinski, P. (1997) Head infection and accumulation of *Fusarium* toxins in kernels of 12 barley genotypes inoculated with *Fusarium graminearum* isolates pf two chemotypes. *European Journal of Plant Pathology 103*: 85-90.

Pettersson, H. (1991) Nivalenol production by *Fusarium poae*. *Mycotoxin Research 7A*: 26-30.

Proctor, R.H., Desjardins, A.E., McCormick, S.P., Plattner, R.D., Alexander, N.J, Brown, D.W. (2002) Genetic analysis of the role of trichothecene and fumonisin mycotoxins in the virulence of *Fusarium*. *European Journal of Plant Pathology* 108: 691-698.

Reischler, G.H., Lemmens, M., Famleitner, A., Adler, A. and Mach, R.L. (2004) Quantification of *Fusarium graminearum* in infected wheat by species specific real-time PCR applying a TaqMan probe. *Journal of Microbiological Methods 59*: 141-146.

Rizzo, A.F. (1993) Determination of major naturally occurring *Fusarium* toxins in Finnish grains and feeds. The hemolytic activity of DON and T-2 toxin, and the lipid peroxidation induced in experimental animals. Ph.D. thesis, University of Helsinki.

Ryu, J.C., Ohtsubo, K., Izumiyarna, N., Mori, M., Tanaka, T. and Ueno, Y. (1987) Effects of nivalenol on the bone marrow in mice. *Journal of Toxicology 12*: 11-21.

Salas, B, Steffensson, B.J., Casper, H.H. and Tacke, B. (1999) *Fusarium* species pathogenic to barley and their associated mycotoxins. *Plant Disease 83*: 667-674

Sarkisov, A.Ch. 1954. *Mycotoxicoses*. Moscow, Russia, pp. 216 (in Russian).

Sarlin, T., Yli-Mattila, T., Jestoi, M., Rizzo, A., Paavanen-Huhtala, S. and Haikara, A. (2006). Real-time PCR for quantification of toxigenic *Fusarium* species in barley and malt. *European Journal of Plant Pathology 114*: 371-380.

Schmidt, H., Holst-Jensen, A., Klemsdal, S., Kullnig, Gradinger, C.M., Kubicek, C.P., Mach, R., Niessen, L., Nirenberg, H., Niessen, L., Nirenberg, H., Thrane, U., Torp, M., Yli-Mattila, T. and Vogel, R.F. (2004) An integrated taxonomic study of *Fusarium langsethiae, F. poae* and *F. sporotrichioides* based on the use of composite datasets. *International Journal of Food Microbiology 95*: 341-349.

Taylor, E.J.A., Stevens, E.A., Bates, J.A., Morreale, D., Lee, D., Kenyon, D.M. and Thomas, J.E. (2001) Rapid-cycle PCR detection of *Pyrenophora graminea* from barley seed. *Plant Pathology 50*: 347-355.

Thrane, U. (2001) Development in the taxonomy of *Fusarium* species based on secondary metabolites. In: Summerell BA, Leslie JF, Backhouse D, Bryden WL, Burgess LW (eds) *Fusarium*. Paul E. Nelson Memorial Symposium. APS Press, St Paul: 29-49.

Thrane, U., Adler, A., Clasen, P.-E., Galvano, F., Langseth, W., Lew, H., Logrieco, A., Nielsen, K.F. and Ritieni, A. (2004) Diversity in metabolite production by *Fusarium langsethiae*, *Fusarium poae* and *Fusarium sporotrichioides*. *International Journal of Food Microbiology 95*: 257-266

Torp, M. and Langseth, W. (1999) Production of T-2 toxins by a *Fusarium* resembling *Fusarium poae*. *Mycopathologia 147*: 89-96

Torp, M. and Nirenberg, H. (2004) *Fusarium langsethiae* sp. nov. on cereals in Europe. *International Journal of Food Microbiology 95*: 247-256.

Turner, A.S., Lees, A.K., Rezanoor, H.N. and Nicholson, P. (1998) Refinement of PCR-detection of *Fusarium avenaceum* and evidence from DNA marker studies for phenetic relatedness to *Fusarium tricinctum. Plant Pathology 47*: 278-288.

Ueno Y. (1983) In: *Trichothecenes: Chemical, Biological and Toxicological Aspects*. Ed. By Y.Ueno. Amsterdam: 135-146.

Uhlig, S. (2005) *Fusarium avenaceum* Toxic metabolites, their occurrence and biological effects (Ph.D. thesis, The Norwegian School of Veterinary Science), National Verinary Institute, Oslo, Norway .

Uhlig, S., Jestoi, M., Knutsen, A.K. and Heier B.T. 2006. Multiple regression analysis as a tool for the identification of relations between semi-quantitative LC-MS data and cytotoxicity of extracts of the fungus *Fusarium avenaceum* (syn. *F. arthrosporioides*). *Toxicon 48*: 567-597.

Uhlig, S., Jestoi, M. and Parikka, P. 2007. *Fusarium avenaceum* – the North European situation. *International Journal of Food Microbiology 119*: 17-24.

Waalwijk, C., Kastelein, P., de Vries I., Kerenyi, Z., van der Lee, T., Hesselink, T. and Kema, G. (2003) Major changes in *Fusarium* spp. in wheat in the Netherlands. *European Journal of Plant Pathology 109*: 743-754.

Waalwijk, C., van der Heide, R., de Vries, I., van der Lee, T. Schoen, C., Costrel-de Corainville, G., Häuser-Hahn, I., Kastelein, P., Köhl, J., Lonnet, P., Demarquet, T. and Kema, G.H.J. (2004) Quantitative detection of *Fusarium* species in wheat using TaqMan. *European Journal of Plant Pathology 110*: 481-494

Ward, T.J., Clear, R., Rooney, A., O'Donnell, K., Gaba, D., Patrick, S., Starkey, D., Gilbert, J., Geiser, D. and Nowicki T. (2008). An adaptive evolutionary shift in *Fusarium* head blight pathogen populations is driving the rapid spread of more toxigenic *Fusarium graminearum* in North America. *Fungal Genetics and Biology 45:* 473-484.

Weising K., Nybom H., Wolff K. and Meyer W. (1995). *DNA fingerprinting in plants and fungi,* CRC Press, Boca Raton.

Wilson, A., Simpson, D., Chandler, E., Jennings P. and Nicholson P., (2004) Development of PCR assays for the detection and differentiation of *Fusarium sporotrichioides* and *Fusarium langsethiae. FEMS Microbiology letters 233*. 69-76.

Yli-Mattila, T., Mach, R., Alekhina, I.A., Bulat, S.A., Koskinen, S., Kullnig-Gradinger, C.M., Kubicek, C.P., and Klemsdal, S.S. (2004a) Phylogenetic relationship of *Fusarium langsethiae* to *Fusarium poae* and *F. sporotrichioides* as inferred by IGS, ITS, β-tubulin sequence and UP-PCR hybridization analysis. *International Journal of Food Microbiology 95*: 267-285.

Yli-Mattila, T., Mironenko, N.V., Alekhina, I.A., Hannukkala, A. and Bulat, S.A. (1997a) Universally primed polymerase chain reaction analysis of *Fusarium avenaceum* isolated from wheat and barley in Finland. *Agricultural and Food Science in Finland 6*: 25-36.

Yli-Mattila, T., Mironenko, N.V., Alekhina, I., Hannukkala, A., Hyvönen, J. and Bulat, S.A. (1998) Identification of *Fusarium avenaceum* isolates by RAPD-PCR and UP-PCR. In: Manceau C and Spak J (Eds) Advances in the detection of plant pathogens by polymerase chain reaction, Proceedings of COST 823 Workshop Ceske Budejovice, 20-21 June, 1996 (pp. 114-123)

Yli-Mattila, T., O'Donnell, K., Ward, T. and Gagkaeva, T.Yu., (2007b) Trichothecene chemotype composition of *Fusarium graminearum* and related species in Finland and Russia. *XV Congress of European Mycologists, Saint Petersburg, Russia*, September 16-21: 286-287 (abstract).

Yli-Mattila, T. and Paavanen-Huhtala, S. (2007a). Molecular identification and detection of plant pathogenic and toxigenic *Fusarium* fungi. In " *The Biodiverse World of Fungi* (Ed. by B.N. Ganguli and S.K. Deshmukh) Anamaya Publishers, New Delhi- 110030, India and their counter part in UK (pp. 82-111).

Yli-Mattila, T., Paavanen-Huhtala, S., Bulat, S.A., Alekhina, I.A. and Nirenberg, H.I. 2002. Molecular, morphological and phylogenetic analysis of *Fusarium avenaceum/F.arthrosporioides/F. tricinctum* species complex – a polyphasic approach. *Mycological Research 106*: 655-669.

Yli-Mattila, T., Paavanen-Huhtala, S., Parikka, P., Hietaniemi, V., Jestoi, M., Gagkaeva, T., Sarlin, T., Haikara, A, Laaksonen, S. and Rizzo, A. (2008). Real-time PCR detection and quantification of *Fusarium poae, F. graminearum, F. sporotrichioides* and *F. langsethiae* as compared to mycotoxin production in grains in Finland and Russia. – *Archives of Phytopathology and Plant Protection* 41:313-322.

Yli-Mattila, T., Paavanen-Huhtala, S., Parikka, P., Hietaniemi, V., Jestoi, M. and Rizzo, A. (2004b) Toxigenic fungi and mycotoxins in Finnish cereals. In: Logrieco A and Visconti A (eds) *An overview on toxigenic fungi and mycotoxins in Europe*. (pp. 83-100) Kluwer Academic Publishers, The Netherlands.

Yli-Mattila, T., Paavanen-Huhtala, S., Parikka, P., Hietaniemi, V., Jestoi, M. and Rizzo, A. (2004d) Real-time PCR detection and quantification of *Fusarium poae* as compared to mycotoxin production in grains in Finland. In: Canty SM, Boring T, Wardwell J and Ward RW (Eds.), Proceedings of the 2nd International Symposium on *Fusarium* Head Blight; incorporating the 8th European *Fusarium* seminar; 2004, 11-15; Orlando, Fl, USA (pp. 422-425). Michigan State University, East Lansing, MI.

Yli-Mattila, T., Paavanen-Huhtala, S., Parikka, P., A., Jestoi, M., Klemsdal, S. and Rizzo, A., (2006) Genetic variation, real-time PCR, metabolites and mycotoxins of *Fusarium avenaceum* and related species. - *Mycotoxin Research 22*: 79-86.

Yli-Mattila, T., Paavanen-Huhtala, S., Parikka, P., Konstantinova, P. and Gagkaeva, T. (2004c) Molecular and morphological diversity of *Fusarium* species in Finland and northwestern Russia. *European Journal of Plant Pathology 110*: 573-585.

Ylimäki, A., Koponen, H., Hintikka, E.-L., Nummi, M., Niku-Paavola, M.-L., Ilus T., Enari, T.M. (1979) Mycoflora and occurrence of *Fusarium* toxins in Finnish grain. Technical Research Centre of Finland, *Materials and Processing Technology publication 21* (pp. 1-28) Valtion Painatuskeskus, Helsinki.

In: Current Advances in Molecular Mycology
Eds: Y. Gherbawy, R.L. Mach and M.K. Rai

ISBN 978-1-60456-909-4

Chapter VI

MOLECULAR MARKERS FOR *FUSARIUM OXYSPORUM* FORMAE SPECIALES CHARACTERIZATION

Y. Gherbawy[1], B. El-Deeb[1], S. Hassan[1], A. Altalhi[1] and M. Rai[2]
[1]Biological Sciences department, Faculty of Science, Taif University, KSA;
[2]Dept. of Biotechnology, SGB Amravati University , Amravati , Maharashtra , India.

ABSTRACT

Fusarium oxysporum is a causative agent of wilt disease in a wide range of economically important crops. *Fusarium oxysporum* Schlechtend.: Fr. is an anamorphic species circumscribed by different morphological criteria: principally the size and shape of the macroconidium, the presence or absence of microconidia and chlamydospores, colony colour, and conidiophore structure. *Fusarium* species can be identified morphologically, but differences between species are very difficult to establish. Based on morphological criteria, it is sometimes difficult to distinguish *F. oxysporum* from several other species belonging to the sections Elegans and liseola. Futhermore, plant pathogenic, saprophytic and biocontrol strains of *F. oxysporum* are morphologically indistinguishable. *F. oxysporum* is among the more commonly isolated fungi from asymptomatic roots of crop plants. *Fusarium oxysporum* formae specifies and races within a species can be determined only by time consuming procedures such as pathogenicity tests in host plants, and vegetative compatibility group (VCG) tests. Several molecular markers were used for accurate identification and characterization of this species and its different formae speciales. In this chapter we focus on the molecular tools that are used by researchers for *Fusarium* oxysporum formae speciales diagnostic, taxonomic and phylogenetic studies.

Key words: RFLP, IGS. RAPD, SSCP, Microsatellites SSRs, PG. AFLP, Transition elongation factor.

FUSARIUM OXYSPORUM

Fusarium oxysporum Schlechtend.; Fr. is an anamorphic species circumscribed by a suite of morphological criteria (Nelson et al., 1983), principally the shape of the macroconidium, the structure of the microconidiophore, and the formation and disposition of chlamydospores, colony colour, and conidiophore structure (Nelson et al., 1981; Windels, 1992). Notwithstanding these unifying criteria, considerable morphological and physiological variation within *F. oxysporum* has long been recognized; many such variants were initially assigned specific status (Wollenweber and Reinking, 1935). Subsequently, Snyder and Hansen (1940) proposed to consolidate all of these species (the entire section elegans) into *F. oxysporum*, a concept that has received wide acceptance. Nevertheless, it is clear that *F. oxysporum* comprises a wide diversity of strains, and the extent to which these differences are sufficient to justify recognition of separate species is by no means a sealed issue. Here, the broad concept *of F. oxysporum* applied as it is consistent with usage in most of the recent literature.

Although *Fusarium oxysporum* has been extensively studied because of its ability to cause diseases of economically important plant hosts, its distribution and ecological activities reflect a much more diverse repertoire. For example, *F. oxysporum* was commonly isolated fungus from asymptomatic roots of crop plants (Taylor, 1965; Hancock, 1985; Gordon et al., 1989). Typically, isolates obtained in this way cannot be demonstrated to be pathogenic on the plant species from which they were recovered (Katan ,1971). Whereas, they may be incapable of causing disease, nonpathogenic strains of *F. oxysporum* are aggressive colonizers of the root cortex (Alabouvette et al., 1979; Schneider, 1984; Gordon et al., 1989). These strains do not cause a wilt disease is presumably due to either, their inability lo enter the vascular tissue or to a rapid response of the host that localizes the infection (Gao et al., 1995). Nonpathogenic strains are also capable of colonizing crop residue (Gordon and Okamoto, 1990) and quickly reoccupying fumigated soils (Marois et al., 1983). Thus, *F. oxysporum* has a well-documented ability to persist without recourse to pathogenesis. Given the widespread occurrence of *F. oxysporum* strains that are apparently nonpathogenic, it is reasonable to suspect that pathogenic forms were derived from originally nonpathogenic antecedents. If pathogens have developed as specialized forms from a larger pool of generalist nonpathogenic (and possibly endophytic) strains, the former might have lost some of the ecological versatility that is the hallmark of the nonpathogens. Garret (1970) observed that, given their ability to grow within the xylem of a living plant, at least initially without causing extensive damage, vascular wilt fungi must be relatively specialized parasites. Thus, given the necessity of survival within a living host and a lessening of the selective pressure to preserve saprophytic competence, diminished abilities to compete with nonpathogenic strains might be expected. Indeed numerous studies confirm that pathogens may, in some situations, be out-competed by nonpathogenic strains (Alabouvette et al., 1979; Schneider , 1984; Larkin et al., 1996). However, these observations may reflect the range of variation among strains of *F. oxysporum* rather than a fundamental difference between pathogens and nonpathogens.

Formae Speciales and Races

The *forma specialis* concept was introduced by Snyder and Hansen (1940). This designation was intended to describe the physiological capabilities of the fungi to attack a specific group of plants and was not a part of the formal taxonomic hierarchy. It has been useful to plant pathologists because it identifies a subset of isolates that are of concern to the production of a crop susceptible to Fusarium wilt. Pathogenic *F. oxysporum* is very host specific attacking only one or a few species of plants, and in many cases only certain cultivars of that plant. In other cases, the same pathogen may be pathogenic on a different plant families. The specificity for a particular host and for cultivars of that host is designated, respectively, as *forma specialis* and as a race of the pathogen. These pathogenic fungi are morphologically indistinguishable from each other as well as from non-pathogens. *Forma specialis* is determined by testing the fungus for pathogenicity on various plant species, while race is determined by testing for pathogenicity on cultivars of a single plant species. Although bioassays are very useful in verifying pathogenicity, they are highly time consuming. More than 150 host-specific formae speciales have been described in the *F. oxysporum* complex (FOC), each of them consisting of one more vegetative compatibility groups (VCGs) and often distinct pathogenic races (Fravel et al., 2003). These formae speciales of *F. oxysporum* are recognized and many are further divided into pathological races based on pathogenicity to a set different host cultivars. Host specialization historically has been considered the most important trait in *F. oxysporum* and, because of its practical application, has been used as the basis of classification. However, some *formae speciales* have broader host ranges. The categorization of strains by host range may or may not lead to a natural subdivision within species (Kistler, 1997). Such pathogenic variation suggests an extreme genetic diversity within the species despite highly conserved morphology. Also, because more than 150 *formae speciales* and races have been described, it would be necessary to inoculate the unidentified strains to an endless number of different plant species and cultivars (Fravel et al., 2003).

Molecular Markers

1. Restriction Fragment Length Polymorphisms (RFLP)

Restriction fragment length polymorphisms (RFLP) of amplified fragments (PCR-RFLP) analysis has been used lo address the research problems of fungal population biology such as the differentiation of species, species forms, and isolates (Marcon and Powell, 1987; Carter et al., 2000). This method has often utilized the analysis of nuclear ribosomal DNA (nrDNA) sequences that are found in all eukaryotic cells and contain both regions showed substantial resolution at different taxonomic units in a majority of fungi including *Fusarium* species (Nicholson et al., 1993; Appel and Gordon, 1995; Hyun and Clark, 1998). PCR-RFLP is a simple and inexpensive method compared with traditional RFLP or sequence analyses as it avoids the need for blotting, probing and /or sequencing. Correll et at. (1992) used RFLP to study genetic diversity in California and Florida populations of the pitch canker fungus

Fusarium subglutinans f.sp. *pini.* Kim et al. (1993a) described that when thirty-nine isolates were examined for genetic similarity by restriction fragment length polymorphism analysis of mitochondrial DNA. *F. oxysporum* f. sp. *cucumerinum* was the most diverse and may be the oldest forma speciales. Fernandez et al. (1994) used rDNA and mtDNA restriction fragment length polymorphism (RFLP) to separate *F. oxysporum* f. sp. *vasinfectum* isolates into four rDNA haplotypes and seven mtDNA haplotypes. Appel and Gordon (1995) used RFLP method to study the intraspecific variation within populations of *F. oxysporum* f. sp. *melonis.* RFLP was used to distinguish 3 races of *F. oxysporum* f. sp. *pisi* (Coddington et al., 1987) and to distinguish formae speciales on curcurbitaceae using mitochondrial DNA (Kim et al., 1993a). Nel et al. (2006) used a PCR-based RFLP analysis of the intergenic spacer region of the ribosomal RNA operon to characterize the nonpathogenic *Fusarium oxysporum* isolates from the rhizosphere of healthy banana plants.

2. Intergenic Spacer (IGS)

Nuclear rDNA provides useful inter- and intra-specific polymorphisms in eukaryotic organisms. There are multiple copies of the ribosomal genes, which are arranged as head-to-tail repeats separated by non-coding spacers. The larger intergenetic spacer (IGS) or non-transcribed spacer (NTS) lies between the large subunit and small subunit coding regions of consecutive cistrons (Reed et al., 2000). The IGS, which separates rDNA repeat units, appears to be the most rapidly evolving spacer region. Closely related species may show considerable diversity in IGS, often reflecting both length and sequence variation (Hills and Dixon, 1991). Variations in rDNA among closely related taxa are found in the intergenic spacer (IGS), which separates the repeated ribosomal units (Avelange, 1994; Fernandez et al., 1994). IGS sequences might be good candidates for the differentiation of strains at the intraspecific level (Hills and Dixon, 1991; Edel et al., 1995) presumably due to relative lack of selective constrains, at least in a large part of its sequence.The IGS region contains greater sequence variation than other genomic regions and, thus may have greater utility in phylogenetic analyses. However, analyses to date have been limited to restriction fragment length polymorphisms or partial sequences, or have focused on human pathogens (Appel and Gordon, 1995; 1996; Fujinaga et al., 2005; McCreight et al., 2005).

Appel and Gordon (1995) studied the intraspecific variation within 56 isolates of *Fusarium oxysporum* including *F. oxysporum* f. sp. *melonis* and nonpathogenic strains, those isolates were chosen from a larger collection to represent diversity in vegetative compatibility groups (VCGs), mitochondrial (mtDNA) haplotype, geographic distribution, and virulence. Using PCR, they amplified 26- kb fragment including the intergenic spacer (IGS) region of the ribosomal DNA from each isolate. The enzymes *EcoRI, Sau3A, CfoI* and *AvaII*, cut this fragment differentially, revealing 5. 6. 6. and 7 patterns, respectively. Among the 56 isolates, a total of 13 unique IGS haplotype were identified. Among most *F. oxysporum* f. sp. *melonis* isolates. IGS haplotype - correlated with VCG and mtDNA haplotype, but did not differentiate among races. Appel and Gordon (1996) studied the evolution of the different races of *F. oxysporum* f. sp. *melonis*. They reported that, bootstrapped parsimony analysis of the partial IGS sequence data identified a phylogenetic tree with highly significant branches.

Nonpathogens that were vegetatively compatible with the pathogen were not closely related to the pathogen based on 1GS sequence data. Thus, nonpathogens and pathogens may share common alleles at vegetative compatibility loci by coincidence rather than because of recent clonal derivation from a common ancestor (Appel and Gordon 1996) .

Namiki et al. (1994) reported that genetic differences between the two groups of formae speciales *melonis* which were divided into two different IGS groups could be due to geographic isolation and to their dispersal throughout the world. Alves-Santos et al. (1999) used intergenic spacer (IGS) region polymorphism of ribosomal DNA, electrophoretic karyotype patterns, and vegetative compatibility and pathogenicity analyses to assess the genetic diversity within *Fusarium oxysporum* isolates recovered from common bean plants growing in fields around El Barco de Avila. Ninety-six vegetative compatibility groups (VCGs) were found among 128 isolates analyzed; most of these VCGs contained only a single isolate. The strains belonging to pathogenic VCGs and the most abundant nonpathogenic VCGs were further examined for polymorphisms in the IGS region and electrophoretic karyotype patterns. Isolates belonging to the same VCG exhibited the same IGS haplotype and very similar electrophoretic karyotype patterns. These findings are consistent with the hypothesis that VCGs represent clonal lineages that rarely, if ever, reproduce sexually. The *F oxysporum* f. sp. *phaseoli* strains had recovered the same IGS haplotype and electrophoretic karyotype patterns, which are different from those found for *F. oxysporum* f. sp. *phaseoli* from the Americas, and were assigned to these new VCGs (VCCs 0166, 0167, and 0168). Based on these results, they do not consider these strains belonging to *F. oxysporum* f. sp. *phaseoli* to be a monophyletic group within *F. oxysporum.* Also, there is no correlation between pathogenicity and VCG. IGS restriction fragment length polymorphism, or electrophoretic karyotype. This diversity of the IGS haplotype within *F. oxysporum* suggests that sexual reproduction is infrequent or absent in this fungus (Alves-Santos et al., 1999).

*Eco*RI restriction patterns of the nuclear ribosomal DNA from four isolates of *F. oxysporum* f. sp. *ciceris* (FOC) representing four races prevalent in India indicated that these races could be grouped into three distinct groups; race 1 and 4 representing one group and race *2* and race 3 representing the others (Chakrabarti et al., 2002). The restriction pattern indicated presence of three *Eco*RI sites on the nuclear rDNA of this species, two of these sites were found in the 5.8S and the 25S regions. The other one, i.e the variable site, was found in the intergenic spacer (IGS) region of the nuclear rDNA (Chakrabarti el al., 2002). These findings were confirmed by PCR-amplification of the IGS region followed by digestion with *Eco*RI and a set of other enzymes. It is suggested that amplification of the IGS region and digestion with restriction enzymes could be used to study polymorphism in FOC and to rapidly identify the races existing in India. They also, proposed that out of the four types of races described from India, races 1 and 4 are the same.

Kim et al. (2001) indicated that, there are different length and restriction site variations among IGS in *F. oxysporum* and its formae specifies. They reported that, amplification of the IGS region resulted in an interspecific size polymorphism, the size of the fragments being 2.5 kb for *F. oxysporum* f. sp. *cucumerinum* Korea and *F oxysporum* f. sp. *niveum*, and 2.6 kb for the other. They identified a total of nine IGS haplotypes among of the 1GS-RFLP data divided into two groups at the similarity level of about 44%. *Fusarium oxysporum* f. sp.

cucumerinum Korea and *niveum* formed a distinct cluster with the lowest similarity. In *cucumerinum,* isolates from ATCC and from Korea showed different haplotype, type I and type VII (Kim et al., 2001). These results are consistent with the finding of Kim et al. (1993b). Kim et al. (2005) showed that there were nine haplotypes among California Isolates of *Fusarium oxysporum* f. sp. *vasinfectum*, based on restriction digests of the IGS region.

Fusarium oxysporum f. sp. *radicis-lycopersici* isolates were divided into five IGS types (Hibar et al., 2007). From the 53 *F. oxysporum* f. sp. *radicis-lycopersici* isolates, 34 isolates have the same IGS types (IGS type 25), and the remaining 19 isolates were distributed into four IGS types. However, the only nine isolates of *F. oxysporum* f. sp. *lycopersici* have six different IGS types (Hibar et al., 2007). This difference of diversity between the two formae speciales suggests that *F. oxysporum* f. sp. *radicis-lycopersici* isolates have a foreign origin and may have been accidentally introduced into Tunisia. Analysis of the IGS region revealed numerous sequence polymorphisms among *F. oxysporum* f. sp. *lactucae* consisting of insertions, deletions, and single nucleotide transitions and substitutions (Mbofung et al., 2007).

3. Random Amplified Polymorphic DNA (RAPD)

The RAPD technique is a variation of the polymerase chain reaction (PCR) that has been widely used as a molecular marker since 1990. Two groups have developed the RAPD assay. One group at Dupont Co, (Wilmington. U) called the new method RAPD (random amplification polymorphic DNA) (Williams et al., 1990) and described its genetic mapping applications. The other group at the California Institute of Biological Research USA (Welsh and McCelland, 1990) focused on genome fingerprinting and collection of their assay arbitrary primed polymerase chain reaction (AP PCR). Both these assays are based on the observation that a single short oligo deoxynucleotide of a randomly chosen sequence when mixed with genomic DNA, dNTPs, buffer and thermostable DNA polymerase and subjected to temperature cycling, amplified several DNA fragment (Innis et al., 1990). RAPD assay is a modification of the basic polymerase chain reaction (PCR) technique (Mullis et al., 1987). However, this assay, unlike the PCR, does not require knowledge of the target DNA sequence, and a single arbitrary primer will support DNA amplification from a genomic template if binding sites on opposite strands of the template exist within a distance that can be traversed by the thermostable equation usually random oligonucleotides (or 10 bases) used as primer to amplify discrete fragments of genomic DNA. The primers are generally of random sequence, contain at least 50% G C and without internal inverted repeats. The products are easily separated by standard electrophoretic technique and visualized under ultraviolet (UV) light. Polymorphism results from changes in either the sequence of the primer binding site (e.g., point mutation) which prevent stable association with the primer or from changes which alter the size or prevent amplified on of target DNA (e,g.. insertions, deletions, inversions). However, the polymorphisms between individuals results from sequence differences in one or both of the primer binding sites, and are visibility as presence or absence of a particular RAPD band such polymorphisms is known as dominant genetic markers. RAPDs offers many advantages; (1) non-radioactive detection (2) no prior DNA sequence information for a genome is required (3) universal primers work in any genome (4) very small amounts of genomic DNA are sufficient (5,25 ng) (5) experimental simplicity (6) no need for expensive equipment beyond a thermocycler and transilluminator. Given all of the advantages, one might think that RAPDs are the perfect marker system. Unfortunately there are a few disadvantages which limit the utility of RAPDs as genetic markers. For instance, because the typical polymorphism observed (presence or absence of a band), RAPD detect dominant loci (not–codominant). Therefore, it is difficult or impossible to determine the heterozygous condition in an individual. This severely limits the amount of genetic information derived from each individual in a self population, however, they are useful in backcross and recombinant inhybride populations. RAPD can only evidently amplify within a certain size range of DNA, and Taq DNA-polymerase will introduce errors. The accumulated mutation occured after 20-30 cycles was reported to be as high as 0.3-0.8 % (El-Badawy 2001). Also, some of the minor fragments are unstable which have been suggested that produced from non-specific amplification when template/primer homology is not perfect (He et al., 1992). Therefore, some modifications have been introduced to improve the RAPD technique and to over come many of its limitations. Genetic similarity has been successfully assessed with RAPD markers in several formae specials of *Fusarium oxysporum* (Grajal-Mariin et al.. 1993; Assigbetse et al., 1994; Kelly et al., 1994; Manulis et al., 1994; Bentley et al., 1994; Bentley et al., 1995; Nelson el al., 1997; Chiocchetti et al., 1999a,b; Jimenez-Gasco et al., 2004).

RAPD analysis has been used for example to distinguish non pathogenic isolates from *F. oxysporum* f. sp. *dianthi* (Manulis et al., 1994; Migheli et al., 1998), to distinguish *F. oxysporum* f. sp. *vasinfectum* (Assigbetse *et al.,* 1994) two pathotype of f. sp. *ciceris* (Kelly *et al.,* 1994) and race 2 of f. sp. *pisi* (Grajal-Martin *et al.,* 1993), the f. sp. *albedinis* (Tantoui et al., 1996), the f. sp. *baslici* (Chiocchetti et al., 1999b). Gherbawy (1999) used RAPD to assess the genetic diversity among 20 isolates of different formae speciales of *Fusarium oxysporum*. He reported that 3 distinct groups were differentiated, the first group included *F. oxysporum* f. sp. *lycopersici* , f. sp. *passiflorae* and *tuberose* and exhibited little differences in RAPD products. The percentage of similarity among those isolates matched to 67.3- 69.1%. The second group included 13 isolates and subdivided into two subgroups. The first subgroup included 9 strains. From these strains, strain 3 (f. sp. *pisi,* from Netherlands) and strain 4 (f. sp. *pisi,* from Germany) showed the highest percentage of similarity (94.9%). The second subgroup contained 4 strains. Strain 7 and 8 (f. sp. *cyclaminis*) showed identical patterns of DNA fragments RAPD with different primers.

Fernandez et al. (1998) developed a method to diagnosis *F. oxysporum* f. sp. *dianthi* infections using RAPD technique. They reported that OPA 17 primer permitted to distinguish between the *F. oxysporum* f. sp. *dianthi* and other special forms of *F. oxysporum.* Their results also indicated that no direct correlation was observed between the RAPD pattern and the race of the samples. They proved that molecular hybridization analysis, using the OPA 17 primer amplified fragments as probes, showed the feasibility of identifying molecular markers that can be used for developing a PCR method for diagnosis of *F. oxysporum* f. sp. *dianthi.*

RAPD technique found to be a useful tool in providing sources of sequences to develop the SCAR technique in order to create a simple PCR test based on specific sequences (de

Haan *et al.,* 2000; Chiocchetti *et at.,* 2001; Alves- Santos *et al.,* 2002). De Haan et al. (2000) tested 160 arbitrary 10-mer oligonucleotide primers on FOC by PCR to find RAPD markers specific for race I, the RAPD primer G 12 amplified two discriminating DNA fragments, AB (609 bp) and EF (1196 bp), in race 1 isolates only. Both fragments were cloned and sequenced. Two pairs of race 1 -specific primers for multiplex PCR were designed. Tests of 112 *F. oxysporum* isolates by PCR showed that, in almost all cases, race 1 isolates of vegetative compatibility group 0340 could be distinguished with these primers.

Jimenez-Gasco et al. (2002) clustered the isolates of *F. oxysporum* f. sp. *ciceris* into three groups by RAPD analysis. The yellowing isolates were separated into two branches while races 1A, 2, 3, 4, 5, and 6, causing wilt constituted the third cluster. Zamani et al. (2004) reported that RAPD analysis revealed considerable genetic variation among *F. oxysporum* obtained from chickpea and these isolates were divided into three different groups by vegetative compatibility and virulence assay. Conversely, they did not find a relationship among the RAPD analysis, vegetative compatibility and virulence assay. Amplification of *F. oxysporum* f. sp. *ciceris* using the RAPD technique produced reproducible and polymorphic bands that allowed the characterization of isolates examined by Bayraktar et al. (2008). Their UPGMA cluster analysis revealed a high degree of genetic diversity among the isolates and separated 74 isolates into three groups at an arbitrary level of 65% similarity.

Pasquali *et al.* (2003) used random amplified polymorphic DNA (RAPD) technique to analyze a total genomic DNA of ten isolates of a new *Fusarium oxysporum* pathogenic on *Argyranthemum frutescens* (Paris daisy), by comparing them to representative of the *formae speciales basilica, chrysanthemi, cyclaminis, dianthi, gladioli, lilii, lycopersici, melonis, pisi, radicis-lycopersici, tracheiphilum,* and a nonpathogenic isolate of *F. oxysporum.* They reported that, all the new isolates from *A. frutescens,* with the exception of the single divergent one, could be identified by their characteristic amplification profile, using selected random primers. All the *F. oxysporum* isolates obtained from diseased *A. frutescens* in Liguria showed very similar RAPD profiles and some of the tested primers (e.g. OPE-17, OPM-06, and OP-T17) generated a strong amplification signal which enabled their clear distinction from representatives of other formae speciales. Based on this evidence, Pasquali et al. (2003) decided to couple the RAPD amplification with primer OPL-I 7 and a fast protocol for template DNA extraction directly from *F. oxysporum* cultured on Komada's selective medium. As previously shown in the case of *F. oxysporum* f. sp. *basilici* (Chiocchetti et al., 1999b), this method proved suitable in the identification of the pathogen, allowing the complete analysis. These data, combined with additional molecular analysis, will certainly contribute in understanding the phylogeny of those pathogen, as already determined for other representatives of the *F. oxysporum* complex (Migheli et al., 1998; Baayen et al., 1998; 2000; Chiocchetti et al., 1999b; Mes et al., 1999; Vakalounakis and Fragkiadakis, 1999).

Cramer et al. (2003) used a set of 34 isolates, including pathogenic *F. oxysporum* f. sp. *betae* (Fob) isolates pathogenic on sugar beet, and non-pathogenic (Fo) isolates, were selected for random-amplified polymorphic DNA (RAPD) analysis. A total of 12 RAPD primers, which generated 105 polymorphic bands, were used to construct an unweighted paired group method with arithmetic averages dendrogram based on Jaccard's coefficient of similarity. All CHP (Central High Plains region of the USA) Fob isolates had identical RAPD banding patterns, suggesting low genetic diversity for Fop in this region. CHP Fob isolates showed a greater degree of diversity, but in general clustered together in a grouping distinct from Fop isolates.

The occurrence of different band profiles in pathogenic and non-pathogenic isolates was reported in other formae speciales, such as albedinis (Tantaoui et al., 1996), dianthi (Manulis et al., 1994) and erythroxyli (Nelson et al., 1997). However, in *F. oxysporum* f. sp. *lycopersici* (Manulis et al., 1994; Suleman et al., 1994) an identical profile was reported. Several authors reported correlations between RAPD amplification pattern and race. Manulis et al. (1994) obtained identical band pattern with 22 oligonucleotides for isolates of *F. oxysporum* f. sp. *dianthi* belonging to the same race in Israel, confirming the existence of just one race in that country. In *F. oxysporum* f. sp. *vasinfectum* from different geographical regions, 11 oligonucleotides amplified genetic patterns, which divided the isolates into three clusters corresponding to the races (Assigbetse et al., 1994). Bentley et al. (1995) observed the same with physiological races of *F. oxysporum* f. sp. *cubense* originated from different countries, where the amplification pattern did not show correlation with the races. Zanotti et al. (2006) used twenty isolates of *Fusarium oxysporum* from Brazil, pathogenic and non-pathogenic to common bean, were analysed using random amplified polymorphic DNA (RAPDs) to study the genetic diversity. RAPD analysis using 23 oligonucleotides resulted in the amplification of 229 polymorphic and 7 monomorphic DNA fragments ranging from 234 to 2590 bp. High genetic variability was observed among the isolates, with the distances varying between 8% and 76% among pathogenic, 2% and 63% among the non-pathogenic and 45% and 76% between pathogenic and non-pathogenic isolates. The analysis of genetic distance data showed that the pathogenic isolates tended to associate in one group and the non-pathogenic in another. Bayraktar et al. (2008) used RAPD in the detection of genetic variation and population Structure among *Fusarium oxysporum* f. sp. *ciceris* isolates on Chickpea in Turkey.

4. Amplification Fragment Length Polymorphism (AFLP)

Amplification fragment length polymorphism (AFLP) is a new powerful marker based on the detection of DNA restriction fragments by PCR amplification (Zabeau and Vos , 1993; Vos et al., 1995). In this technique amplification of restriction fragments is accomplished by the ligation of double- strands (ds) adapter sequences to the ends of the restriction sites which subsequently serve as binding sites for primer annealing in PCR. In this way, restriction fragments of a particular DNA can be amplified with universal AFLP primers corresponding to the restriction site and an adapter sequence. Adding 1-3 selective bases to these oligonucleotide adapters used as primers can restrict the number of DNA fragments, which are amplified. This marker was originally conceived to allow the construction of very high density DNA marker maps for application in genome research and positional cloning of genes. It is equally suitable for application in genetic analysis, which require more modest DNA marker densities. It is apparent that the AFLP approach is now widely used for developing polymorphic markers. High frequency of identifiable AFLP coupled with high reproducibility makes this technology attractive tool for identifying polymorphism and for determining linkages by analyzing individuals from a segregating population.

AFLP is a PCR-based DNA analysis technique that can detect variations in RFLPs on a genome-wide basis (Vos et al., 1995), Like RFLP analysis, AFLPs can detect size differences in restriction fragments caused by DNA insertions, deletions or changes in target restriction site sequence, but with less labour required. AFLPs have been increasingly used in analysis of fungal population structure (Majer et al., 1996; Gonzalez et al., 1998; DeScenzo et al., 1999; Purwantara et al., 2000; Zeller et al., 2000). The complex DNA fingerprinting patterns produced by the AFLP technique are reproducible and subsets of these data appear to show higher correlations to one another compared with the sets of RFLP or RAPD data (Spooner et al., 1996; Gonzalez et al., 1998). The advantage of AFLP over other techniques is that multiple bands are derived from all over the genome. Basically, this prevents over interpretation or misinterpretation due to point mutations or single-focus recombination, which may affect other genotypic characteristics. The main disadvantage of AFLP markers is that alleles are not easily recognized (Majer et al, 1998). PCR has proven to be successful in detecting plant-pathogenic fungi as well as bacteria (Majer et al. 1998; Restrepo et al., 1999). The utility, repeatability, and efficiency of AFLP are leading to broader application of this technique to the analysis of *Fusarium* population (Baayen et al., 2000; Abd-Elsalam et al., 2002 a,b; Kiprop et al., 2002; Sivaramakrishnan et al., 2002; Abdel-Satar et al., 2003; Zeller et al., 2003; Belabid et al., 2004, Leslie et al., 2004).

Bao et al., (2002) reported that pathogenic and nonpathogenic strains of *Fusarium oxysporum* were separated into different clusters based on AFLP data, however, some nonpathogenic strains grouped with pathogenic strains. The population of pathogenic strains was less diverse than that of nonpathogenic strains, suggesting that the pathogenic strains were possibly of monophyletic origin. For both pathogenic and nonpathogenic *F. oxysporum* strains, no relationships was observed between the genetic profiles and geographic origin; this may indicate that pathogens did not originate independently at each locality (Bao et al., 2002).

Differentiation between *F. oxysporum* f.sp. *melonis* and *F. oxysporum* f. sp. *radicis-cucumerinum* can be achieved by pathogenicity test on a set of differential hosts. The results of Vakalounakis et al. (2005) showed a strong correlation between VCG and AFLP haplotypes on the one hand, and formae speciales on the other. Therefore, VCGs as well as AFLP can effectively be used to disntinguish between these two pathogens of melon. Up to now, VCGs have been used extensively to differentiate these special forms of *F. oxysporum* formae speciales (Katan and Primo, 1999). Groenewald et al. (2006) reported that the AFLP technique is a powerful tool to perform detailed analysis of genetic diversity in *Fusarium oxysporum* f. sp. *cubense* (banana pathogen).

5. The Single-strand Conformational Polymorphism (SSCP)

This is a powerful and rapid technique for gene analysis particularly for detection of point mutations and typing of DNA polymorphism (Orita et al., 1989a). SSCP can identify heterozygosity of DNA fragments of the same molecular weight and can even detect changes of a few nucleotide bases as the mobility of the single-stranded DNA changes with change in its GC content due to its conformational change. To overcome problems of reannealing and complex banding patterns, an improved technique called asymmetric-PCR SSCP was developed (Ainsworth et al., 1991), wherein the denaturation step was eliminated and a large-sized sample could be loaded for gel electrophoresis, making it a potential tool for high throughput DNA polymorphism. It was found useful in the detection of heritable human diseases. In plants, however, it is not well developed although its application in discriminating progenies can be exploited, once suitable primers are designed for agronomically important traits (Fukuoka et al., 1994).

Universal primers common to all fungi have been used as a promising approach for clinical microbiological diagnosis (White et al., 19990). Thus, so far two techniques have been reported to separate different fungi detected by universal primers. These include restriction fragment length polymorphism (Hopfer et al., 1993; Maiwald et al., 1994) and hybridization of the amplicon with a specific probe (Kan, 1993). For example, Hopfer et al. (1993) amplified a segment of a ribosomal DNA gene that is highly conserved throughout the fungal kingdom. Using restriction endonucleases digestion of the amplified product, the authors classified medically important fungi into five groups. A similar approach employing the combination of PCR and restriction enzyme analysis was reported by Maiwald et al.(1994).

The single-strand conformational polymorphism (SSCP) technique includes PCR amplification of a conserved region of the 18S rRNA with further separation of genus and species on the basis of exploiting small but phylogenetically important base pair differences among medically important fungi (Maiwald et al., 1994). As reported by Orita et al. (1989a,b) and further demonstrated by others (Mashiyama et al., 1990; Dockhorn-Dworniczak et al., 1991; Hayashi, 1991; Yandell, 1991), minor sequence variations in highly conserved DNA segments will cause subtle changes in the tertiary structure that forms in short-stranded DNA fragments after they are denatured. These conformationally different fragments can then be separated electrophoretically under nondenaturing conditions.

Single-strand length conformational polymorphism analysis is performed by electrophoresis of the amplicon from the PCR on a polyacrylamide gel, thus eliminating the need for restriction digestion. The technique of SSCP analysis displays migration of the amplified DNA fragment as a function of its conformational structure as well as its size. Given that the tertiary structure is extremely sensitive to single-nucleotide base substitutions, this method can distinguish different DNA fragments that may differ by as little as a single base pair. Such substitutions may be indicative of a different species or strain, which otherwise may not be detected by restriction fragment length polymorphism on agarose gel electrophoresis. In addition to improved sensitivity for detection of different fungal pathogens, the method eliminates the step of restriction digestion prior to gel electrophoresis and thus permits the amplicon to be run directly on polyacrylamide gels. The use of SSCP analysis allowed large numbers of samples to be rapidly screened for sequence variation without the need for sequencing. This approach therefore greatly reduced the effort needed to screen large sample size, whilst retaining the high level of sensitivity required for the investigation of genetic diversity in natural populations (Wong and Jeffries, 2006).

Dong et al. (2005) reported that the SSCP patterns of *F. oxysporum* isolates proved to he highly reproducible. Sequencing data confirmed that this SSC'P method could detect one single base change within the 550 bp PCR fragment from the ribosomal internal transcribed spacer region of *F. oxysporum*. Over of 360 fusaria isolates were obtained from fields showing symptoms of asparagus decline, and most were easily differentiated by SSCP into four principal species *F. oxysporum* f. sp. *asparagi*, *F. proliferatum*, *F. redolens* and *F. solani* (Wong and Jeffries, 2006).

6. Microsatellites or Simple Sequence Repeats (SSRs)

Microsatellites, also called simple sequence repeats (SSRs), are tandem repeated arrays of short core sequence. They are present in the vast majority of eukaryotic genomes. The total number of different dinucleotide blocks has been estimated for several species (Wu and Tanksley, 1993; Morgante and Olivieri, 1993 Ma et al., 1996). The number of sites ranged from 103 to 105 depending on the species and repeat motif. Polymorphism produced by a variable number of tandem repeats has been demonstrated in a large number of species. This feature has made microsatellites a very attractive molecular marker for species with a narrow genetic background such as wheat and barley. This methodology is based on the use of primers complementary to SSRs. Multilocus profiles have been generated using different kinds of oligonucleotide containing simple sequence repeats as single primer (Gupta et al., 1994; Nagaoka and Ogihara, 1997) or in combination with arbitrary sequence oligonucleotides (Wu et al., 1994). These studies have shown the reproducibility of the patterns generated the Mendelian inheritance of the polymorphic amplified bands and their usefulness in the investigation of the genetic relationships.

Fusarium oxysporum f. sp. *ciceri,* the causal agent of chickpea wilt, is an important fungal pathogen in India. Thirteen oligonucleotide probes complementary to microsatellite loci, in combination with 11 restriction enzymes, were used to assess the potential of such markers to study genetic variability in four Indian races of *Fusarium oxysporum* f. sp. *cicer* (Barve et al., 2001). Hybridization patterns, which were dependent upon both the restriction enzymes and oligonucleotide probe used, revealed the presence of different repeat motifs in the *Fusarium oxysporum* f. sp. *ciceri* genome. Dependent upon the levels of polymorphism detected, Barve et al. (2001) have identified $(AGT)_5$, $(ATC)_5$ and $(GAT)_4$ as the best fingerprinting probes for the *Fusarium oxysporum* f. sp. *ciceri* races. The distribution of microsatellite repeats in the genome revealed races 1 and 4 to be closely related at a similarity index value of 76.6%, as compared to race 2 at a similarity value of 67.3%; race 3 was very distinct at a similarity value of 26.7%. Their study demonstrates the potential of oligonucleotide probes for fingerprinting and studying variability in the *Fusarium oxysporum* f. sp. *ciceri* races and represents a step towards the identification of potential race diagnostic markers.

7. Inter-Simple Sequence Repeat (ISSR)

Zietkiewics et al. (1994) and Kantety et al. (1995) described a marker system now referred as Inter-Simple Sequence Repeat (ISSR) amplification. This makes use of anchored primers to amplify simple sequence repeats without the requirement for prior sequence information. This technique is more reliable than the RAPD technique and generates larger numbers of polymorphisms per primer (Zietkiewics et al., 1994; Hantula et al., 1996; Charters et al., 1996). Theoretically, polymorphisms should be easier to detect because variable regions in the genome are targeted. The technique is quicker and more straightforward than AFLPs and does not require the high development cost of conventional SSRs. Although the ISSR technique also yields dominant markers, it has been reported that a longer 50-anchor can yield markers which are codominant (Fisher et al., 1996).

ISSR fingerprints show a higher level of polymorphism and reproducibility because of the long primers and high annealing temperatures compared with the RAPD and have been used extensively in other fungal population analysis (Meyer et al., 1993; Hantula et al., 2000). Bayraktar et al. (2008) reported that a cluster analysis of ISSR fingerprints with UPGMA divided the isolates into three main groups as the dendrogram obtained from RAPD data. ISSR markers generally provided a similar discrimination between these isolates but the differences were observed in the relative positions of isolates in the different groups. Likewise the separate dendrograms of RAPD and ISSR data, the UPGMA cluster analysis of RAPD + ISSR data separated 74 isolates into three groups. Bogale et al. (2005) described nine simple sequence repeat (SSR) markers developed for studying *Fusarium oxysporum*. Allelic diversity at the nine loci ranged from 0.003 to 0.895, with a total of 71 alleles among 64 isolates. They reported that, these markers will facilitate studies on relationships amongst isolates of *F. oxysporum*.

8. Sequencing of Specific Genes

8.1. Endopolygalacturonase (PG)

The primary plant cell-wall consists mainly of cellulose micro-fibrils embedded in a matrix of hemicellulose pectic polysaccharides, as well as a number of glycoproteins (Carpita and Gibeaut, 1993). Since this structure represents a formidable barrier to the entry of microbes, enzymatic breakdown of the cell-wall has been traditionally associated with plant pathogenesis. However, whereas in bacterial pathogens the causal link between secretion of cell-wall degrading: enzymes (CWDEs) and virulence has now been firmly established (Hugouvieux- Cotte-Pattat et al., 1996), their role in fungal pathogenesis is still under debate. The main reason for this is the frequently observed that transformation-mediated inactivation of individual CWDE encoding genes has no detectable effect on virulence. The most likely cause is the presence of multiple genes encoding similar and functionally redundant enzyme activities in most plant pathogenic fungi (Walton, 1994). During root penetration and host plant colonization, *Fusarium* secretes an array of CWDEs, such as polygalacturonases (PGs), pectate lyases (PLs), xylanases and proteases, that might contribute to infection (Beckman, 1987). EndoPG is the first enzyme activity detected in *F. oxysporum* cultures on tomato cell-walls (Jones et al., 1972). EndoPGs efficiently macerate plant tissue by depolymerizing homogalacturan, a major component of the plant cell-wall (Collmer and Keen, 1986). A specific endoPG, PG1, was found to be the major endoPG secreted by *F. oxysporum* f.sp. *lycopersici* during culture on pectin and during infection of tomato plants (Di Pietro and Roncero, 1996). Cloning of the encoding gene allowed the identification of a naturally existing PG1-deficient strain from *F. oxysporum* f.sp. *melonis* (Di Pietro et al., 1998). No difference in the virulence pattern toward muskmelon was observed between PG1-overproducing transformants of this isolate and the wild-type strain lacking PG1, suggesting that PG1 is not essential for pathogenicity in this forma specialis (Di Pietro and Roncero, 1998). Subsequently, two additional polygalacuronase genes were identified in *F. oxysporum*, pg5 encoding a second endoPG (Garcia-Maceira et al., 2001), and pgx4 encoding an exopolygalacturonase (Garcia-Maceira et al., 2000). Targeted inactivation of either gene had no effect on virulence (Garcia-Maceira et al., 2000; Garcia-Maceira et al., 2001).

It has been proposed that endopolygalacturonases (endoPGs) (poly-α-1,4-galacturonide glycanohydrolases; EC 3.2.1.15) play a key role in fungal pathogenicity for plants by depolymerizing homogalacturonan, a major component of the plant cell-wall (Cooper , 1984). They may also function as avirulence determinants by releasing oligogalacturonide inducers of plant defense mechanisms (Davis et al., 1984) and interacting with plant proteins that modulate polygalacturonase (PG) activity (Cervone et al. 1989). *Fusarium oxysporum* Schlecht. is an economically important soilborne plant pathogen that has a worldwide distribution and causes vascular wilt disease in a wide variety of crops. This species includes more than 120 described formae speciales that are defined on the basis of specificity for host species (Armstrong and Armstrong, 1981). The mechanisms of pathogenicity and wilt symptom induction by this fungus are poorly understood, although it has been suggested that endoPGs may be involved (Beckman , 1987). Until now, no information concerning the occurrence and distribution of specific pectinolytic isozymes in this species has been available. Recently, PG1, the major endoPG produced by *F. oxysporum* f. sp. *lycopersici* during *in vitro* growth on pectin, was purified and characterized (Di Pietro and Roncero, 1996), and the corresponding gene was cloned (Di Pietro and Roncero, 1998). Di Pietro et al. (1998) compared 12 *F. oxysporum* isolates belonging to seven different formae speciales to determine the occurrence and diversity of PG1 and the corresponding gene in the species and determined if PG1 production by *F. oxysporum* f. sp. *melonis* was correlated with virulence for muskmelon.

PG1, the major endopolygalacturonase of the vascular wilt pathogen *Fusarium oxysporum*, was secreted during growth on pectin by 10 of 12 isolates belonging to seven formae speciales, as determined with isoelectric focusing zymograms and sodium dodecyl sulfate-polyacrylamide gel electrophoresis gels. A Southern analysis of genomic DNA and PCR performed with gene-specific primers revealed that the *pg1* locus was highly conserved structurally in most isolates. Two PG1-deficient isolates were identified; one lacked the encoding gene, and the other carried a *pg1* allele disrupted by a 3.2-kb insertion with sequence homology to *hAT* transposases. The virulence for muskmelon of different *F. oxysporum* f. sp. *melonis* isolates was not correlated with PG1 production *in vitro*. They concluded that PG1 is widely distributed in *F. oxysporum* and that it is not essential for pathogenicity (Di Pietro et al., 1998).

Hirano and Arie (2006) compared the partial nucleotide sequences for endo polyglacturonase (pg1) and exo-polyglacturonase (pgx4) genes from isolates of *F. oxysporum* f. sp. *lycopersici* (FOL) and radicis-lycopersici (FORL) from Japan and designed a specific primer sets based on the nucleotide differences that appeared among the pathogentic types.

Polygalacturonases (PGs) secreted by plant pathogenic fungi have received much attention owing to their potential role in various aspects of fungal pathogenicity (Di Pietro and Roncero 1996, 1998; Arie et al. 1998; Huertas-González et al. 1999; Ruiz-Roldán et al. 1999; Garcia-Maceira et al. 2000; Gomez-Gomez et al. 2002). Peever et al. (2002) used *endo*polygalacturonase (*endo*PG) gene sequences to infer a phylogeny among closely related *Alternaria* spp. Associated with citrus. Similarly, a sequence variation in *pg1*, which encodes the major extracellular *endo*PG (PG1) of *F. oxysporum* f. sp. *lycopersici* (Arie et al. 1998; Di Pietro and Roncero 1998), could provide insight into the phylogenetic relationships among *Fusarium* spp.

8. 2. The Translation Elongation Factor 1-α (TEF) gene

The translation elongation elongation factor 1-α (TEF) gene, which encodes an essential part of the protein translation machinery, has high phylogenetic utilily because it is (i) highly informative at the species level in *Fusarium*; (ii) non-orthologous copies of the gene have been detected in the genus; and (iii) universal primers have been designed that work across the phylogenetic breadth of the genus. This gene was first used as a phylogenetic marker to inter species- and generic-level relationships among Lepidoptera (Cho et al., 1995; Mitchell et al., 1997). Primers were first developed in the fungi to investigate lineages within the *F. oxysporum* complex (O'Donnell et al,. 1998a). The efl and ef2 primers were designed based on sites shared in exons between *Trichoderma reesei* (Hypocreales/ Sordariomycetes/ Pezizomycotina/ Ascomycotina) and *Histoplasma capsulatum* (Eurotiales/ Eurotiomycetes/ Pezizomycotina/ Ascomycotina), and they can be applied to a wide variety of filamentous ascomycetes. These primers amplify an ~ 700 bp region of' TEF, flanking three introns that total over half of the amplicon's length, in all known fusaria. This gene appears to be consistently single-copy in *Fusarium*, and it shows a high level of sequence polymorphism among closely related species, even in comparison to the intron-rich portions of protein-coding genes such as calmodulin, beta-tubulin and histone H3. For these reasons, TEF has become the marker of choice as a single-locus identification tool in *Fusarium*. O'Donnell et al. (1998a) indicated that for the genus *Fusarium*, the ribosomal regions contain less interspecific variation than the translation elongation factor 1 alpha gene (EF-1 alpha). Due to its high discriminating power at the species level, the EF-1 alpha gene has also been used as genetic marker for phylogenetic studies, allowing the accurate discrimination of formae speciales or strains for specific Fusarium species such as *F.oxysporum* and *F. solani* (O'Donnell et al.1998a; O'Donnell, 2000).

Analysis of EF-1-α sequences resulted in moderate resolution, grouping seven formae speciales with the Lactucae iaolates (Mbofung et al., 2007). Their results indicated that the lactucae race 1 isolates grouped together with seven other formae speciales (*rhois, matthiolae, cepae, phaseoli, albedinis, heliotrope* and *fabae*), forming the major clade supported by a bootstrap value of 63%. Also, the nonpathogenic isolates from soil formed a distinct clade with f.sp. *asparagi, callistephi, batatas, vasinfectum* 1 and *lycopersici*). Although the EF-1α gene has been used in most studies for phylogenetic resolution within and between some formae speciales of *F. oxysporum*, it did not contain enough variation to separate the f. sp. *lactucae* race 1 isolates from seven other formae speciales. In previous studies using EF-1α and mtSSU sequences, *F. oxysporum* was shown to comprise three major clades, within which could be found a combination of formae speciales forming distinct lineages (O'Donnell, et al., 1998a; Baayen, et al., 2000).

8.3. The Transposable Elements

Members of the *Fusarium oxysporum* complex provide an interesting example of interspecific variation, and effort has been focused on understanding the molecular mechanisms generating population variability. In the absence of a sexual stage, the origin and maintenance of variability probably require mechanisms ensuring high mutation rates. In this respect, the activity of transposable elements has been postulated to explain part of the genetic variability in this complex. Indeed, 12 families of transposable elements have been identified, including representatives of both retroelements and DNA transposons (Daboussi and Langin 1994; Hua-Van et al. 2000). Daboussi and Langin, (1994) reported that, the genome of the fungal plant pathogen*Fusarium oxysporum* contains at least six different families of transposable elements. Representatives of both DNA transposons and retrotransposons have been identified, either by cloning of dispersed repetitive sequences (*Foret* and*palm*) or by trapping in the nitrate reductase gene (*Fot1, Fot2 Impala* and*Hop*). *Fot1* and*Impala* elements are related to the*Tc1* and*mariner* class of transposons. These transposable elements can affect gene structure and function in several ways: inactivation of the target gene through insertion, diversification of the nucleotide sequence by imprecise excisions, and probably chromosomal rearrangements as suggested by the extensive karyotype variation observed among field isolates. Comparisons of the distribution of these elements in *Fusarium* populations have improved our understanding of population structure and epidemiology and provided support for horizontal gene transfer. Also they could be developed as genetic tools for tagging genes, a cloning strategy that is particularly promising in imperfect fungi (Daboussi and Langin, 1994).

More recent studies have used new tools such as fingerprinting with repeated sequences (usually transposable elements), proven to be valuable for several fungal species, including *Magnaporthe grisea* (Hamer et al., 1989; Dobinson et al., 1993; Farman et al., 1996), *Mycosphaerella graminicola* (McDonald and Martinez, 1990), *Erisyphe graminis* (O'Dell et al. 1989), *Cryphonectria parasitica* (Milgroom et al., 1992), and *F. oxysporum* (Lievens et al., 2008). Transposons are discrete DNA segments that are able to jump or replicate to other locations within a genome. They are ubiquitous in virtually all organisms examined, and are a common cause of spontaneous genetic changes that can affect the biology of the organism (McDonald, 1993).

In *Fusarium* genus, fingerprinting allowed researchers to distinguish *formae speciales* (Namiki et al. 1994), to track the origin of new infestation (Mouyna et al., 1996), to detect a given *forma specialis* (Fernandez et al. 1998), and to identify races within a *forma specialis* (Chiocchetti et al. 1999a,b). Moreover, PCR assays based on several transposable element insertion sites provided a useful diagnostic tool for quickly identifying *formae speciales* and races.

Transposable elements appear especially abundant within *F. oxysporum* that exhibit a high degree of genetic variability. This is illustrated by numerous transposon families (about 17) characterized thus far, representing the major classes of retroelements and DNA transposons (Julien et al., 1992; Daboussi and Langin, 1994 ; Mouyna et al., 1996 ; Okuda et al., 1998 ; Gomez-Gomez et al., 1999 ; Hua-Van et al., 2000 ; Mes et al., 2000). One of these active DNA transposon families, named *impala,* is composed of few (8–10) elements and is typically 1,280 bp long with 37-bp inverted terminal repeats (ITRs). *Impala* contains a single open reading frame encoding a transposase of 340 amino acids (aa). This transposase is related to those found in elements belonging to the widespread *Tc1-mariner* superfamily (Langin et al., 1995; Hua-Van et al. 1998). In strain FOM24 (herein called M24), in which *impala* was first identified, about 10 different *impala* copies have been characterized. Three subfamilies, named E, D, and F, have been detected (Hua-Van et al. 1998). The E and D subfamilies are represented by several copies, which are autonomous, inactive, or truncated (Hua-Van et al. 1998; Hua-Van et al. 2001a,b). Within each subfamily, nucleotide divergence between the full-length elements is relatively low (around 1%), while the truncated elements are more polymorphic. The F subfamily contains only one full-length-but-inactivated element. These three subfamilies differ by as much as 20% at the nucleotide level (Hua-Van et al. 1998). This result is intriguing when compared with the 0.3%–5% polymorphism observed within the *F. oxysporum* complex for the *nia* and EF1 genes and internal transscribed sequence or intergenic sequence of ribosomal DNA (Avelange 1994; Appel and Gordon 1996; O'Donnell et al. 1998a). Two non-mutually-exclusive hypotheses may be proposed to explain the evolutionary origin of *impala* subfamilies. First, the presence of different subfamilies in a genome might be the result of ancestral polymorphism. The subfamilies present in the common ancestor would then be expected to be present in genetically diverse strains associated with diversification of these pathogens. Another possibility is the occurrence of one or more horizontal transfers. Such events have been described for other transposable elements (Kidwell 1992; Robertson and Lampe 1995*a*), notably for *mariner* elements, for which horizontal transfer appears to have played a major role in evolution (Garcia-Fernandez et al. 1995; Lohe et al. 1995 ; Robertson and Lampe 1995*b*). In this case, the foreign *impala* element(s) might be expected to be present in a small number of related strains, all derived from the same ancestor in which the transfer occurred.

Although the genome of *F. oxysporum* is still largely uncharacterized, at least 5% is estimated to be composed of transposons (Roncero et al., 2003). Among these are both class-I and class-II transposons, the first mobilizing via retroposition (via an RNA intermediate) and the second via a DNA cut-and-paste mechanism (Daboussi and Capy, 2003). Class-II transposons are reported to play an important role in the evolution of fungal genomes (Daviere et al., 2001) and have been used for various research purposes, including knockout mutagenesis and variability analysis.46 In some cases, pathogenicity of *F. oxysporum* isolates could be linked to the presence of certain transposons. For example, a specific insertion of the *Fot1* transposon has been exploited to develop specific markers for *F. oxysporum* f. sp. *Albedinis* (Fernandez et al., 1998) as well as for a new VCG of *F. oxysporum* f. sp. *chrysanthemi* highly pathogenic on Paris daisy (VCG 0052) (Pasquali et al., 2004a,b). In addition, this transposon provided the source of target sequences to discriminate certain races of *F.oxyporum dianthi* (races 1, 2 and 8), (Chiocchetti et al., 1999a) whereas a copy of the transposon *Impala* was used to identify race 4 strains of this *forma specialis* (Chiocchetti et al., 1999a). Furthermore, the transposable element *Palm* displayed suitable variability for population analysis of *F. oxysporum* f. sp. *Elaeidis* (Mouyna et al., 1996). Apart from these class-II transposons, class-I transposons such as *Foxy* have also been proposed for the discrimination of *F. oxysporum formae speciales* and its races (Di Pietro et al., 1994; Mes et al., 2000). This abundant retrotransposon (200–300 copies per genome) was used to construct the first mitotic linkage map of *F. oxysporum*. 54 In addition, the genomic regions between the insertions of long terminal repeat retrotransposon copies were used to develop a diagnostic assay for *F. oxysporum* f. sp. *lactucae* race 1 strains based on inter-retrotransposon amplified polymorphisms (Pasquali et al., 2007). Because class-II transposons can move around the genome through complete excision,46 inactivity of the transposable element should be verified in order to be used as a molecular marker for reliable pathogen diagnosis (Fernandez et al., 1998; Chiocchetti et al., 1999a; Lievens et al., 2007). Analysis of a large collection of strains, preferably isolated from different geographic areas at different time points, should reduce the risk of selecting an instable transposon (Chiocchetti et al., 1999a). Apart from this, incomplete copies, either truncated or containing internal deletions, are often inactive (Mills et al., 2007).

CONCLUSION

Isolates within a *forma specialis* are generally more similar genetically than isolates with different host specificities and have been assumed to have a monophyletic origin (Tantaoui et al.1996; Kistler, 1997). However, a few gene genealogy studies have shown that some *formae speciales* of *F. oxysporum* can have multiple independent origins (i.e. polyphyletic), with pathogenicity and virulence evolving more than once (O'Donnell et al. , 1998a,b; Baayen et al. , 2000). Also, clonality in *F. oxysporum* has been associated with vegetative compatibility (Gordon and Martyn, 1997; Kistler, 1997), with isolates belonging to a vegetative compatibility group (VCG) showing high genetic similarity, as determined by mitochondrial (mt) DNA or intergenic spacer region (IGS) haplotyping (Gordon and Martyn, 1997). Similarly, notable exceptions have been found in the correlation between VCG and mtDNA or IGS haplotypes. For example, a *F. oxysporum* f. sp. *melonis* isolate was identified in VCG 0131 that shared mtDNA and IGS haplotypes with pathogenic isolates from VCG 0134, instead of with other isolates representative of VCG 0131 (Appel and Gordon, 1995). Also, nonpathogenic *F. oxysporum* isolates vegetatively compatible with *F. oxysporum* f. sp. *melonis* in VCGs 0131 and 0134 had nucleotide sequences in the IGS region of rDNA that were distinct from the pathogenic isolates (Appel and Gordon, 1996). The latter example demonstrates that in some cases vegetative compatibility may be coincidental, possibly arising by convergence rather than common descent. Although isolates within a forma specialis are related by their pathogenicity to a given host, genetic heterogeneity within certain formae speciales often has revealed a polyphyletic origin (Gordon and Martyn, 1997). This heterogeneity has important implications on the evolution of pathogenic forms within the species, and on the development of resistant host cultivars. Therefore, it is critical to be able to quickly and accurately differentiate genetic diversity. Phylogenetic analyses based on DNA sequences of the mitochondrial small subunit ribosomal RNA gene (mtSSU), rDNA, and translation elongation factor 1a (EF1a) have helped to elucidate the evolutionary relationships within several formae speciales of *F. oxysporum* (O'Donnell et al. 1998a). They showed that the *F. oxysporum* complex is strongly supported as monophyletic, but that many formae speciales in the species were found to be polyphyletic, and suggested that host pathogenicity has evolved convergently.

REFERENCES

Abdel-Satar M.A., Khalil M.S., Mohmed I.N., Abd-Elsalam K.A., Verreet J.A. (2003). Molecular phylogeny of *Fusarium* species by AFLP fingerprint. *African Journal of Biotechnology 2*: 51-55.

Abd-Elsalam K.A., Schnieder F., Verreet J.A. (2002b). Population analysis of *Fusarium* spp. *Phytomedizin 3*:18-19.

Abd-Elsalam K.A., Khalil M.S., Aly A.A., Asran-Amal A. (2002a). Genetic diversity among *Fusarium oxysporum* f. sp. *vasinfectum* isolates revealed by UP-PCR and AFLP markers. *Phytopathol. Mediterr. 41*:1-7.

Ainsworth P.J., Surh L.C., Coulter-Mackie M.B. (1991). Diagnostic single strand conformational polymorphism, (SSCP): a simplified non-radioisotopic method as applied to a Tay-Sachs B1 variant. *Nucleic Acids Res.19*:405-406.

Alabouvette C., Rouxel F., Louvet J. (1979). Characterization of *Fusarium* wilt suppressive soils and prospects for their utilization in biological control. In *Soil-Borne Plant Pathogens*, ed. B. Schippers, W Gams, pp. 165–82. New York: Academic. 686 pp.

Alves-Santos, F.M., B. Ramos, M.A. Garcia-Sanchez, A.P. Eslava, and J.M. Diaz-Minguez. 2002. A DNA based procedure for *in planta* detection of *Fusarium oxysporum* f. sp. *phaseoli*. *Phytopathology 92*:237–244.

Alves-Santos F. M., Benito E. P., Eslava A. P., Diaz-Minguez J. M. (1999). Genetic diversity of *Fusarium oxysporum* strains from common bean fields in Spain *Appl. Environ. Microbiol. 65*:3335-3340.

Appel D., Gordon T.R. (1995). Intraspecific variation within populations of *Fusarium oxysporum* based on RFLP analysis of the intergenic spacer region of the rDNA. *Experimental Mycology 19*, 120–8.

Appel D., Gordon T.R. (1996). Relationships among pathogenic and nonpathogenic isolates of *Fusarium oxysporum* based on the partial sequence of the intergenic spacer region of the ribosomal DNA. *Molecular Plant–Microbe Interactions 9*: 125–38.

Arie T, Gouthu S, Shimazaki S, Kamakura T, Kimura M, Inoue M, Takio K, Ozaki A, Yoneyama Y, Yamaguchi I (1998). Immunological detection of *endo*polygalacturonase secretion by *Fusarium oxysporum* in plant tissue and sequencing of its encoding gene. *Ann. Phytopathol. So.c Jpn. 64*: 7–15.

Armstrong G. M., Armstrong J. K. (1981). Formae speciales and races of *Fusarium oxysporum* causing wilt diseases, p. 391–399. *In* P. E. Nelson, T. A. Toussoun, and R. J. Cook (ed.), *Fusarium: diseases, biology, and taxonomy*. Pennsylvania State University Press, University Park.

Assigbetse, K.B., Fernandez, D., Dubois, M.P., Geiger, J.P. (1994). Differentiation of *Fusarium oxysporum* f.sp. *vasinfectum* races on cotton by random amplified polymorphic DNA (RAPD) analysis. *Phytopathology 84*: 622-626.

Avelange I. (1994). Recherche d'un outil de caractérisation visant à construire une classification phylogénétique englobant l'ensemble des souches de l'espèce *Fusarium oxysporum* Ph.D. dissertation, Université Paris-Sud, Orsay, France.

Baayen, R. P., O'Donnell, K., Bonants, J. M., Cigelnik, E., Kroon, L. P. N., Roebroeck, M E. J. A., Waalwijk, C. (2000). Gene genealogies and AFLP analyses in the *Fusarium oxysporum* complex identify monophyletic and nonmonophyletic formae speciales causing wilt and rot disease. *Phytopathology 90*:891-900.

Baayen P. , Förch M.G. , Waalwijk C., Bonants P.J.M. , Löffler H.J.M. , Roebroeck E.J.A. (1998). Pathogenic, genetic and molecular characterization of *Fusarium oxysporium* f. sp. *lilii*. *European Journal Plant Pathology, v.104*, p.887-894, 1998.

Bao J.R., Fravel D.R., O'Neill N.R., Lazarovits G., Berkum P.V. (2002). Genetic analysis of pathogenic and nonpathogenic *Fusarium oxysporum* from tomato plants. *Canadian Journal of Botany 80*: 271-279.

Barve M.P., Haware M.P., Sainani M.N., Ranjekar P.K., Gupta V.S. (2001). Potential of microsatellites to distinguish four races of *Fusarium oxysporum* f. sp. *Ciceri* prevalent in India. *Theoretical Application of Genetics102*: 138–147.

Bayraktar H., Dolar F.S., Maden S. (2008). Use of RAPD and ISSR Markers in Detection of Genetic Variation and Population Structure among *Fusarium oxysporum* f. sp. *ciceris* Isolates on Chickpea in Turkey. *Journal of Phytopathology 156*:146-154.

Beckman C.H. (1987). The Nature of Wilt Diseases of Plants. St Paul, MN: American Phytopathological Society.

Belabid L., Baum M., Forts Z., Bouznad Z.. Eujayl I. (2004). Pathogenic and genetic characterization of Algerian isolates of *Fusarium oxysporum* f. sp. *lentis* by RAPD and AFLP analyses. *African Journal of Biotechnology 3*: 25-31.

Bentley S., Pegg K.G., Dale J.L. (1994). Optimization of RAPD-fingerprinting 1 to analyze genetic variation within populations of *Fusarium oxysporum* f. sp. *cubense*. *J. Phytopathol. 142*:64–78.

Bentley S., Pegg K.G., Dale J.L. (1995). Genetic-variation among a world-wide collection of isolates of *Fusarium oxysporum* f. sp. *cubense* analyzed by RAPD-PCR fingerprinting. *Mycol. Res. 99*:1378–84.

Bogale M., Wingfield B.D., Wingfield M.J., Steenkamp E.T.(2005). Simple sequence repeat markers for species in the *Fusarium oxysporum* complex. *Mol. Ecol. Notes 5*: 622–624.

Carpita N.C., Gibeaut D.M. (1993). Structural models of primary cell walls in flowering plants: consistency of molecular structure with the physical properties of the walls during growth. *Plant J. 3*, 1–30.

Carter, J.P., Rezanoor H.N., Desjardins A.E., Nicholson. P. (2000). Variation in *Fusarium gramineaum* isolates from Nepal associated with their host of origin. *Plant Pathol. 49*: 452-460.

Cervone F., Hahn M.G., De Lorenzo G., Darvill A., Albersheim P. (1989). Host-pathogen interactions. XXXIII. A plant protein converts a fungal pathogenesis factor into an elicitor of plant defense responses. *Plant Physiol. 90*:542–548.

Chakrabarti A., Mukherjee P.K., Sherkhane P.D., Bhagwat A.S. Murthy N.B.K. (2002). A simple and rapid molecular method for distinguishing between races of *Fusarium oxysporum* f.sp. *ciceris* from India. *BARC Newsletter 149*-152.

Charters Y.M., Robertson A., Wilkinson M.J. (1996). PCR analysis of oilseed rape cultivars (*Brassica napus* L. ssp Oleifera) using 5′- anchored simple sequence repeat (SSR) primers. *Theoretical Applied Genetics 92*: 442-447.

Chiocchetti A., Bernardo I., Daboussi M. J., Garibaldi A., Lodovica Gullino M., Langin T., Migheli Q. (1999a). Detection of *Fusarium oxysporum* f. sp. *dianthi* in carnation tissue by PCR amplification of transposon insertions *Phytopathology 89*:1169-1175.

Chiocchetti A., Ghignone S., Minuto A., Gullino M.L., Garibaldi A. , Migheli Q. (1999b). Identification of *Fusarium oxysporum* f. sp. *basilici* isolated from soil, basil seed, and plants by RAPD analysis. *Plant Dis. 83*:576–581.

Chiocchetti, A., Sciandone, L., Durando, F., Garibaldi, A., Migheli, Q. (2001). PCR detection of *Fusarium oxysporum* f.sp. basilici on basil. *Plant Disease 85*: 607-611.

Cho S., Mitchell A., Regier J.C., Mitter C., Poole R.W., Friedlander T.P., Zhao S. (1995). A highly conserved nuclear gene for low-level phylogenetics: elongation factor-1α recovers morphology-based tree for heliothine moths. *Molecular Biologyand Evolution 12*: 650–656.

Coddington A., Matthews P. M., Cullis C., Smith K. H. (1987). Restriction digest patterns of total DNA from different races of *Fusarium oxysporum* f.sp. *pisi* - An improved method for race classification. *Journal of Phytopathology, 118*: 9-20.

Collmer A. , Keen N.T. (1986). The role of pectic enzymes in plant pathogenesis. *Annu. Rev. Phytopathol. 24*: 383–409.

Cooper R M. (1984). The role of cell wall-degrading enzymes in infection and damage. In: Wood R K S, Jellis G J. , editors; Wood R K S, Jellis G J. , editors. *Plant diseases: infection, damage and loss*. Oxford, United Kingdom: Blackwell Scientific Publications; pp. 13–27.

Correll, J. C. Gordon, T. R., McCain, A. H. (1992). Genetic diversity in California and Florida populations of the pitch canker fungus *Fusarium subglutinans* f. sp. *pini*. *Phytopathology 82*:415-420.

Cramer R.A., Byrne P.F., Brick M.A., Wickliffe E., Schwartz H.F. (2003). Characterization of *Fusarium oxysporum* isolates from common bean and sugar beet using pathogenicity assays and random amplified polymorphic DNA markers. *J. Phytopathol .151*:352–360.

Daboussi M. J., Langin T. (1994). Transposable elements in the fungal plant pathogen *Fusarium oxysporum*. *Genetica 93*:49-59.

Daboussi M.J., Capy P. (2003). Transposable elements in filamentous fungi. *Annu Rev Microbiol 57*:275–299.

Daviere J.M., Langin T., Daboussi M.J. (2001). Potential role of transposable elements in the rapid reorganization of the *Fusarium oxysporum* genome. *Fungal Genet Biol 34*:177–192.

Davis K.R., Lyon G.D., Darvill A.G., Albersheim P. (1984). Host-pathogen interactions. XXV. Endopolygalacturonic acid lyase from *Erwinia carotovora* elicits phytoalexin accumulation by releasing plant cell wall fragments. *Plant Physiol. 1984*;74:52–60.

de Haan, Numansen, Roebroeck, van Doorn (2000). PCR detection of *Fusarium oxysporum* f.sp. *gladioli* race 1, causal agent of *Gladiolus* yellows disease, from infected corms. *Plant Pathology 49*: 89–100.

DeScenzo R.A., Engel S.R., Gomez G., Jackson E.L., Munkvold G.P., Weller J., Irelan N.A. (1999) Genetic analysis of *Eutypa* strains from California supports the presence of two pathogenic species. *Phytopathology 89*: 884–893.

Dong D., Liu H., Peng H., Huang X., Zhang X., Xu Y. (2005). Rapid differentiation of *Fusarium oxysporum* isolates using PCR-SSCP with the combination of pH-variable electrophoretic medium and low temperature. *Electrophoresis 26*:4287 – 4295.

Di Pietro A., Roncero M.I.G. (1996). Endopolygalacturonase from *Fusarium oxysporum* f. sp. *lycopersici*: purification, characterization, and production during infection of tomato plants. *Phytopathology 86*:1324–1330.

Di Pietro A., Roncero M.I.G. (1998). Cloning, expression, and role in pathogenicity of *pg1* encoding the major extracellular endopolygalacturonase of the vascular wilt pathogen *Fusarium oxysporum* f. sp. *lycopersici. Mol Plant-Microbe Interact.* 1998;11:91–98.

Di Pietro A., Anaya N., Roncero M.I.G. (1994). Occurrence of a retrotransposon-like sequence among different *formae speciales* and races of *Fusarium oxysporum. Mycol. Res. 98*:993–996.

Di Pietro A., García-Maceira F.I., Huertas-González M. D., Ruíz-Roldan M. C., Caracuel Z., Barbieri A.S., Roncero M.I.G. (1998). Endopolygalacturonase PG1 in Different Formae Speciales of *Fusarium oxysporum. Appl Environ Microbiol.64*(5): 1967–1971.

Dobinson K. F., Harris R. E., Hamer J. E. (1993). *Grasshopper,* a long terminal repeat (LTR) retroelement in the phytopathogenic fungus *Magnaporthe grisea* Mol. *Plant Microbe Interact. 6*:114-126.

Dockhorn-Dworniczak B., Dworniczak B., Brommelkamp L., Bulles J., Horst J., Bocker W. W. (1991). Non-isotopic detection of single strand conformation polymorphism (PCR-SSCP): a rapid and sensitive technique in diagnosis of phenylketonuria. *Nucleic Acids Res. 19*:2500.

Edel E., Steinberg C., Avelange I., Lagurre G., Alabouvette C. (1995). Comparison of three molecular methods for the characterization of *Fusarium oxysporum* strains. *Phytopathology 85*: 579–585.

El-Badawy M.E.M. (2001). Localization and characterization of quantitative trait loci for fusarium head blight resistance in wheat by means of molecular markers. *Ph.D. Thesis*, Technischen Universität München eingereicht und durch die Fakultät Wissenschaftszentrum Weihenstephan für Ernährung, Landnutzung und Umwelt

Farman M. L., Taura S., Leong S. A. (1996). The *Magnaporthe grisea* DNA fingerprinting probe MGR586 contains the 3' end of an inverted repeat transposon *Mol. Gen. Genet 251*:675-681.

Fernandez D., Assigbetse K., Dubois M., Geiger J. P. (1994). Molecular characterization of races and vegetative compatibility groups in *Fusarium oxysporum* f. sp. *vasinfectum. Appl. Environ. Microbiol.* 60:4039-4046.

Fernandez D., Ouinten M., Tanaoui A., Geiger J.P., Langin T. (1998). *Fot1* insertions in the *Fusarium oxysporum* f. sp. *Albedinis* genome provide diagnostic PCR target for detection of the date palm pathogen. *Appl. Environ. Microbiol. 64*:633–636.

Fisher P.J., Gardner R.C., Richardson T.E. (1996). Single locus microsatellites isolated using 5' anchored PCR. *Nucleic Acids Research 24*: 4369–4371.

Fravel, D., Olivain C., Alabouvette, C. (2003). *Fusarium oxysporum* and its biocontrol. *New Phytologist 157*: 493-502.

Fujinaga M., Ogiso H., Shinohara H., Tsushima S., Nishimura N., Togowa M., Saito H., Nosue M. (2005). Phylogenetic relationships between the lettuce root rot pathogen *Fusarium oxysporum* f. sp. *lactucae* races 1, 2, and 3 based on the sequence of the intergenic spacer region of its ribosomal DNA. *J. Gen. Plant Pathol. 71*:402-407.

Fukuoka S., Inoue T., Miyao A., Monna L., Zhong M. S., Sasaki T., Minobe, Y. (1994). Mapping of sequence-tagged sites in rice by single strand conformation polymorphism. *DNA Res. 1*:271-277.

Gao H., Beckman C.H., Mueller W.C. (1995). The rate of vascular colonization as a measure of the genotypic interaction between various cultivars of tomato and various formae speciales of *Fusarium oxysporum. Physiol. Mol. Plant Pathol. 46*:29–43.

Garcia-Fernandez J., Bayascas R. J., Marfany G., Munoz M. A., Casali A., Baguna J., Salo E. (1995). High copy number of highly similar mariner-like transposons in planarian (Platyhelminthe): evidence for a trans-phyla horizontal transfer *Mol. Biol. Evol 12*:421-431.

Garcia-Maceira F.I., Di Pietro A., Roncero M.I.G. (2000). Cloning and distribution of *pgx4* encoding an in planta expressed exopolygalacturonase from *Fusarium oxysporum. Mol. Plant Microbe Interact. 13*:359–365.

Garcia-Maceira F.I., Di Pietro A., Huertas-Gonzalez M.D., Ruiz-Roldan M.C., Roncero M.I. (2001) Molecular characterization of an endopolygalacturonase from *Fusarium oxysporum* expressed during early stages of infection. *Appl. Environ. Microbiol. 67*: 2191–2196.

Garrett S.D. (1970). *Pathogenic Root-InfectingFungi.* London: Cambridge Univ. Press. 294 pp.

Gherbawy Y.A.M.H. (1999). RAPD profile analysis of isolates belonging to different formae speciales of *Fusarium oxysporum. Cytologia 64*:269-276.

Gomez-Gomez E., Anaya N., Roncero M. I., Hera C. (1999). *Folyt1,* a new member of the *hAT* family, is active in the genome of the plant pathogen *Fusarium oxysporum Fungal Genet. Biol 27*:67-76.

Gomez-Gomez E., Ruíz-Roldán M.C., Di Pietro A., Roncero M.I., Hera C. (2002). Role in pathogenesis of two endo-b-1,4-xylanase genes from the vascular wilt fungus *Fusarium oxysporum. Fungal Genet. Biol. 35*:213–333.

Gonzalez M., Rodriguez M.E.Z., Jacabo J.L., Hernandez F., Acosta J., Martinez O., Simpson J. (1998). Characterization of Mexican isolates of *Colletotrichum lindemuthianum* by using differential cultivars and molecular markers. *Phytopathology 88*: 292–299.

Gordon T.R., Martyn R.D. (1997). The evolutionary biology of *Fusarium oxysporum. Annu. Rev. Phytopathol. 35*:111-128.

Gordon T.R., Okamoto D. (1990). Colonization of crop residue by *Fusarium oxysporum* f. sp. *melonis* and other species of *Fusarium. Phytopathology 80*:381–86.

Gordon T.R., Okamoto D., Jacobson D.J. (1989). Colonization of muskmelon and non host crops by *Fusarium oxysporum* f. sp. *melonis* and other species of *Fusarium. Phytopathology 79*:1095–100.

Grajal-Martin M.J., Simon C.J., Muehlbauer F.J. (1993). Use of random amplified polymorphic DNA (RAPD) to characterize race 2 of *Fusarium oxysporum* f. sp. *pisi. Phytopathology 83*:612–614.

Groenewald S., Van Den Berg N., Marasas W.F.O., Viljoen A. (2006) The application of high-throughput AFLP's in assessing genetic diversity in *Fusarium oxysporum* f. sp. *cubense. Mycological research 110*: 297-305.

Gupta M., Chyi J., Romero-Severson J., Owen J.L. (1994). Amplification of DNA markers from evolutionary diverse genomes using single primers of simple sequence repeats. *Theor. Appl. Genet. 89*:998-1006.

Hamer J. E., Farrall L., Orbach M. J., Valent B., Chumley F. G. (1989). Host species-specific conservation of a family of repeated DNA sequences in the genome of a fungal plant pathogen *Proc. Natl. Acad. Sci. USA 86*:9981-9985.

Hancock J.G. (1985). Fungal infection of feeder rootlets of alfalfa. *Phytopathology 75*:1112–20.

Hantula J., Dusabenygasani M., Hamelin R.C. (1996). Random amplified microsatellites (RAMS) - a novel method for characterizing genetic variation within fungi. *Eur. J. For. Path. 26*:15-166.

Hantula J., Lilja A., Nuorteva H., Parikka P., Werres S. (2000). Pathogenicity, morphology and genetic variation of *Phytophthora cactorum* from strawberry, apple, rhododendron and silver birch. *Mycol. Res. 104*:1062–1068.

Hayashi K.(1991). PCR-SSCP: a simple and sensitive method for detection of mutations in the genomic DNA. *PCR Methods Appl. 1*:34–38.

He S., Ohm H., Mackenzie S. (1992). Detection of DNA sequence polymorphisms among wheat varieties. *Theor. Appl. Genet. 84*:573-578.

Hibar K., Edel-Herman V., Steinberg Ch., Gautheron N., Daami-Remadi M., Alabouvette C., El Mahjoub M. (2007). Genetic Diversity of *Fusarium oxysporum* Populations Isolated from Tomato Plants in Tunisia. *J. Phytopathology 155*: 136–142.

Hills D.M., Dixon M.T. (1991). Ribosomal DNA: molecular evolution and phylogenetic inference. *Q. Rev. Biol. 66*, 411-453.

Hirano Y., Arie T. (2006). PCR-based differentiation of *Fusarium oxysporum* f. sp. *lycopersici* and *radicis-lycopersici* and races of *F. oxysporum* f. sp. *lycopersici. J. Gen. Plant Pathol.72*:273-283.

Hopfer, R. L., Walden P., Setterquist S., Highsmith W. E. (1993). Detection and differentiation of fungi in clinical specimens using polymerase chain reaction (PCR) amplification and restriction enzyme analysis. *J. Med. Vet. Mycol. 31*:65–75.

Hua-Van A., Davière J. M., Kaper F., Langin T., Daboussi M. J.(2000). Genome organization in *Fusarium oxysporum*: clusters of class II transposons *Curr. Genet 37*:339-347.

Hua-Van A., Hericourt F., Capy P., Daboussi M. J., Langin T. (1998). Three highly divergent subfamilies of the *impala* transposable element coexist in the genome of the fungus *Fusarium oxysporum Mol. Gen. Genet 259*:354-362.

Hua-Van A., Langin T., Daboussi M.-J. (2001a). Evolutionary History of the *impala* Transposon in *Fusarium oxysporum. Molecular Biology and Evolution 18*:1959-1969.

Hua-Van A., Pamphile J., Langin T., Daboussi M. J. (2001b). Transposition of autonomous and engineered *impala* transposons in *Fusarium oxysporum* and a related species *Mol. Gen. Genet. 264*:724-731.

Huertas-González M.D., Ruiz-Roldán M.C., Maceria F.I.G., Roncero M.I.G., Di Pietro A. (1999). Cloning and characterization of *pl1* encoding an in planta-secreted pectate lyase of *Fusarium oxysporum. Curr. Genet. 35*:36–40.

Hugouvieux-Cotte-Pattat N., Condemine G., Nasser, W., Reverchon S. (1996). Regulation of pectinolysis in *Erwinia chrysanthemi. Annu. Rev. Microbiol. 50*: 213–257.

Hyun JW, Clark CA (1998). Analysis of *Fusarium lateritium* using RAPD and rDNA RFLP techniques. *Mycolo. Res. 102*: 1259-1264.

Innis M.A., Gelfand D.H., Snisky J.J., White T.J. (1990). *PCR protocol. A guide to methods and application.* San Diego, CA, USA, Academic Pres: 482.

Jimenez-Gasco M. M., Milgroom M. G., Jiménez-Díaz R. M. (2002). Gene genealogies support *Fusarium oxysporum* f. sp. *ciceris* as a monophyletic group. *Plant Pathology 51*: 72–77.

Jimenez-Gasco, M. M., Navas-Cortes, J. A., and Jimenez-Diaz, R. M. 2004. The *Fusarium oxysporum* f. sp. *ciceris / Cicer arietinum* pathosystem: a case study of the evolution of plant-pathogenic fungi into races and pathotypes. *International Microbiology 7*: 95-104.

Jones T.M., Anderson A.J., Albersheim, P. (1972). Host–pathogen interactions. IV. Studies on the polysaccharide-degrading enzymes secreted by *Fusarium oxysporum* f.sp. *lycopersici. Physiol. Plant Pathol.2*, 153–166.

Julien J., Poirier-Hamon S., Brygoo Y. (1992). *Foret1,* a reverse transcriptase-like sequence in the filamentous fungus *Fusarium oxysporum Nucleic Acids Res. 20*:3933-3937.

Kan V.L. (1993). Polymerase chain reaction for the diagnosis of candidemia. *J. Infect. Dis. 168*:779–783.

Kantety R.V., Zeng X., Bebbetzen J.L., Zehr B.E. (1995). Assessment of genetic diversity in dent and popcorn (*Zea mays* L.) inbreed lines using inter-simple sequence repeat (ISSR) amplification. *Mol. Breeding 1*:36-373.

Katan J. (1971). Symptomless carriers of the tomato *Fusarium* wilt pathogen. *Phytopathology 61*:1213–17.

Katan T., Di Primo P. (1999). Current status of vegetative compatibility groups in *Fusarium oxysporum. Phytoparasitica 27*: 51-64.

Kelly A., Alcala-Jimenez A.R., Bainbridge B.W., Heale J.B., Perez-Artes E., Jimenez-Diaz R.M. (1994). Use of genetic fingerprinting and random amplified polymorphic DNA to characterize pathotypes of *Fusarium oxysporum* f. sp. *ciceris* infecting chickpea. *Phytopathology 84*:1293–1298.

Kidwell M. G. (1992). Horizontal transfer *Curr. Opin. Genet. Dev 2*:868-873.

Kim H.-J., Choi Y.-K., Min B.-R. (2001). Variation of the Intergenic Spacer (IGS) Region of Ribosomal DNA among *Fusarium oxysporum* formae speciales. *The Journal of Microbiology*, December 2001, p.265-272.

Kim, Y., Hutmacher, R. B., Davis, R. M. (2005). Characterization of California isolates of *Fusarium oxysporum* f. sp. *vasinfectum. Plant Dis. 89*:366-372.

Kim D.H., Martyn R.D., Magill C.W. (1993a). Mitochondrial DNA (mt-DNA) relatedness among formae speciales of *Fusarium oxysporum* in the Cucurbitaceae. *Phytopathology 83*:91–97.

Kim D.H., Martyn R.D., Magill C.W. (1993b). Chromosomal polymorphism in *Fusarium oxysporum* f. sp. *niveum*. *Phytopathology 83*:1209–16.

Kiprop E.K., Baudoin J.P., Mwang'ombe A.W., Kimani P.M., Mergeai G., Maquet A. (2002). Characterization of Kenyan isolates of *Fusarium udum* from Pigeonpea [*Cajanus cajan* (L.) Millsp.] by cultural characteristics, aggressiveness and AFLP analysis *J. Phytopathol. 150*: 517-527.

Kistler H.C. (1997). Genetic diversity in the plant-pathogenic fungus. *Fusarium oxysporum*. *Phytopathology 87*:474–479.

Langin T., Capy P., Daboussi M. J. (1995). The transposable element *impala,* a fungal member of the *Tc1-mariner* superfamily *Mol. Gen. Genet 246*:19-28.

Larkin R.P., Hopkins D.L., Martin F.N. (1996). Suppression of *Fusarium* wilt of watermelon by nonpathogenic *Fusarium oxysporum* and other microorganisms recovered from a disease suppressive soil. *Phytopathology 86*:812–19.

Leslie J.F., Zeller K.A., Logrieco A., Mule G., Moretti, A., Ritieni A. (2004). Species diversity of and toxin production by *Gibberella fujikuroi* species complex strains isolated from native prairie grasses in Kansas. *Applied and Environmental Microbiology 70*: 2254-2262.

Lievens B., Rep M., Thomma B.P.H.J. (2008). Recent developments in the molecular discrimination of *formae speciales* of *Fusarium oxysporum. Pest Management Science Pest Manag Sci. (in Press).*

Lievens B., Claes L., Vakalounakis D.J., Vanachter A.C.R.C., Thomma B.P.H.J. (2007). A robust identification and detection assay to discriminate the cucumber pathogens *Fusarium oxysporum* f. sp. *cucumerinum* and f. sp. *radicis-cucumerinum*. *Environ. Microbiol 9*:2145–2161.

Lohe A. R., Moriyama E. N., Lidholm D. A., Hartl D. L.(1995). Horizontal transmission, vertical inactivation, and stochastic loss of *mariner*-like transposable elements *Mol. Biol. Evol 12*:62-72.

Ma Z.Q., Roder M., Sorrells M.E. (1996). Frequencies and sequence characteristics of di-, tri-, and tetra-nucleotide microsatellites in wheat. *Genome 39*: 123–130.

Maiwald M., Kappe R., Sonntag H. G. (1994). Rapid presumptive identification of medically relevant yeasts to the species level by polymerase chain reaction and restriction enzyme analysis. *J. Med. Vet. Mycol. 32*:115–122.

Majer D., Lewis B.G., Mithen R. (1998). Genetic variation among field isolates of *Pyrenopeziza brassicae*. *Plant Pathol. 47*:22-28.

Majer, D., Mithen R., Lewis B.G., Vos, P. , Oliver R.P.(1996). The use of AFLP fingerprinting for the detection of genetic variation in fungi. *Mycol. Res. 9*: 1107–1111.

Manulis S., Kogan N., Reuven M., Ben-Yephet Y. (1994). Use of the RAPD technique for identification of *Fusarium oxysporum* f. sp. *dianthi* from carnation. *Phytopathology 84*:98–101.

Marcon, M.J., Powell, D.A. (1987). Epidemiology, diagnosis, and management of *Malassezia furfur* systemic infection. *Diagn. Microbiol. Infect. Dis. 7*, 161–175.

Marois J.J., Dunn M.T., Papavizas G.C. (1983). Reinvasion of fumigated soil by *Fusarium oxysporum* f. sp. *melonis*, causal agent of wilt of muskmelon, *Cucumis melo var. reticulatus*. *Phytopathology 73*:680–84.

Mashiyama S., Sekiya T., Hayashi K. (1990). Screening of multiple DNA samples for detection of sequence changes. *Technique 2*:304–306.

Mbofung G.Y., Hong S.G., Pryor B.M. (2007). Phylogeny of *Fusarium oxysporum* f. sp. *lactucae* Inferred from Mitochondrial Small Subunit, Elongation Factor 1-α, and Nuclear Ribosomal Intergenic Spacer Sequence Data. *Phytopathology 97*:87-98.

McCreight J.D., Matheron M.E., Tickes B.R., Platts B. (2005). *Fusarium* wilt race 1 on lettuce. *HortScience 40*:529-531.

McDonald J. F. (1993). Evolution and consequences of transposable elements. *Curr. Opin. Genet. Dev. 3*:855–864.

McDonald B. A., Martinez P. J.(1990). DNA restriction fragment polymorphisms among *Mycosphaerella graminicola* (anamorph *Septoria tritici*) isolates collected from a single wheat field. *Phytopathology 80*:1368-1373.

Mes J.J., Haring M.A., Cornelissen B.J.C. (2000). *Foxy*: an active family of short interspersed nuclear elements from *Fusarium oxysporum*. *Mol. Gen. Genet. 263*:271–280.

Mes J., Weststeijn E.A., Herlaar F., Lambalk J.J.M., Wijbrandi J., Haring M.A., Cornelissen B.J.C. (1999). Biological and molecular characterization of *Fusarium oxysporum* f. sp. *lycopersici* divides race 1 isolates into separate virulence groups. *Phytopathology 89*:156–160.

Meyer W., Lieckfeldt E., Kuhls K., Freedman E.Z., Vilgalys R. (1993). Hybridization probes for conventional DNA fingerprinting can be used as single primers in the PCR to distinguish strains of Cryptococcus neoformans. *J. Clin. Microbiol. 31*:2274–2280.

Mitchell A., Cho S., Regier J.C., Mitter C., Poole R.W., Matthews M. (1997). Phylogenetic utility of elongation factor-1α in Noctuoidea (Insecta: Lepidoptera): the limits of synonymous substitution. *Molecular Biology and Evolution 14*: 381–390.

Migheli, Q., Briatore, E., Garibaldi, A. (1998). Use of random amplified polymrphic DNA to identify races 1, 2, 4,8 of *Fusarium oxysporum* f.sp. *dianthi* in Italy. *European Journal of Plant Pathology 104*: 49-57.

Milgroom M. G., Lipari S. E., Powell W. A. (1992). DNA fingerprinting and analysis of population structure in the chestnut blight fungus, *Cryphonectria parasitica Genetics 131*:297-306.

Mills R.E., Bennett E.A., Iskow R.C., Devine S.E. (2007). Which transposable elements are active in the human genome? *Trends in Genet. 23*:183–19.

Morgante M., Olivieri A.M. (1983). PCR-amplified microsatellites as markers in plant genetics. *The Plant J. 3*:175-182.

Mouyna I., Renard J.L., Brygoo Y. (1996). DNA polymorphism among *Fusarium oxysporum* f. sp. *elaeidis* populations from oil palm, using a repeated and dispersed 'Palm'. *Curr. Genet. 30*:174–180.

Mullis, K. B., Ehrlich H. A., Arnheim N., Saiki R. K. , Scharf S. J. (1987). Process for amplifying, detecting and/or cloning nucleic acid sequences. U.S. Pat. No. 4,683,195. 28 Jul. 1987. Assignee: Cetus Corporation, Emeryville, Calif., U.S.A.

Nagaoka T., Ogihara Y. (1997). Applicability of inter-simple sequence repeat polymorphisms in wheat for use as DNA markers in comparison to RFLP and RAPD markers. *Theor. Appl. Genet. 94*:597-602.

Namiki, F., T. Shiomi, T. Kayamura, T. Tsuge (1994). Characterization of the formae speciales of *Fusarium oxysporum* causing wilts of cucurbits by DNA fingerprinting with nuclear repetitive DNA sequences. *Appl. Environ. Microbiol. 60*, 2684–2691.

Nel B., Steinberg C., Labuschagne N., Viljoen A. (2006). Isolation and characterization of nonpathogenic *Fusarium oxysporum* isolates from the rhizosphere of healthy banana plants. *Plant Pathology 55*: 207–216.

Nelson P. E., Toussoun T. A., Cook R. J. (1981). *Fusarium*: diseases, biology, and taxonomy. Pennsylvania State University Press, University Park. 457pp.

Nelson P.E., Toussoun T.A., Marasas W.F.O. (1983). *Fusarium Species: An Illustrated Manual for Identification*. University Park, PA: Penn. State Univ. Press. 193 pp.

Nelson AJ, Elias KS, Are´ valo GE, Darlington LC, Bailey BA. (1997) Genetic characterization by RAPD analysis of isolates of Fusarium oxysporum f. sp. erytroxyli associated with an emerging epidemic in Peru. *Phytopathology 87*:1220–.1225

Nicholson, P. Jenkinson P., Rezanoor H.N., Parry D.W.(1993).Restriction fragment length polymorphism analysis of variation in *Fusarium* species causing ear blight of cereals. *Plant Pathology 42*: 905–914.

O'Dell M., Wolfe M. S., Flavell R. B., Simpson C. G., Summers R. W.(1989). Molecular variation in populations of *Erysiphe graminis* on barley, oats and rye *Plant Pathol. 38*:340-351.

O'Donnell K.(2000). Molecular phylogeny of the *Nectria haematococca- Fusarium solani* species complex. *Mycologia 92,*919-938.

O'Donnell K., Kisitler. H.C., Cigelnik. E., Ploetz. R.C. (1998a). Multiple evolutionary origins of the fungus causing panamadisetise of banana: concordant evidence from nuclear and mitochondrial gene genealogies. P. *Natl. Acad. Sci., USA 95*, 2044-2049.

O'Donnell K., Cigelnik E., Nirenberg H.I. (1998b). Molecular systematics and phylogeography of the *Gibberella fujikuroi* species complex. *Mycologia 90*: 456–493.

Okuda M., Ikeda K., Namiki F., Nishi K., Tsuge T. (1998). *Tfo1*: an *Ac*-like transposon from the plant pathogenic fungus *Fusarium oxysporum Mol. Gen. Genet. 258*:599-607.

Orita M., Iwahana H., Kanazawa H., Hayashi K., Sekiya T.(1989a). Detection of polymorphisms of human DNA by gel electrophoresis as single strand conformation polymorphisms. *Proc. Natl. Acad. Sci. USA 86*:2766–2770.

Orita M., Susuki Y., Sekiya T., Hayashi K. (1989b). Rapid and sensitive detection of point mutations and DNA polymorphisms using the polymerase chain reaction. *Genomics 5*:874–879.

Pasquali M., Acquadro A., Balmas V., Migheli Q., Garibaldi A., Gullino M. (2003). RAPD characterization of *Fusarium oxysporum* isolates pathogenic on *Argyranthemun frutescens* L. *J. Phytopathol. 151*:30–35.

Pasquali M., Acquadro A., Balmas V., Migheli Q., Garibaldi A., Gullino M.L. (2004a). Development of PCR primers for a new *Fusarium oxysporum* pathogenic on Paris daisy. *Eur. J. Plant Pathol. 110*:7–11.

Pasquali M., Marena L., Fiora E., Piatti P., Gullino M.L., Garibaldi A. (2004b). Real time PCR for the identification of a highly pathogenic group of *Fusarium oxysporum* f. sp. *chrysanthemi* on *Argyranthemum frutescens* L. *J. Plant Pathol. 86*:51–57.

Pasquali M., Dematheis F., Gullino M.L., Garibaldi A. (2007). Identification of race 1 of *Fusarium oxysporum* f. sp. *lactucae* on lettuce by inter-retrotransposon sequence-characterized amplified region technique. *Phytopathology 97*:987–996.

Peever TL, Ibañez A, Akimitsu K, Timmer LW (2002). Worldwide phylogeography of the citrus brown spot pathogen, *Alternaria alternata. Phytopathology 92*:794–802.

Purwantara A., Barrins J.M., Cozijnsen A.J., Ades P.K., Howlet B.J. (2000). Genetic diversity of the *Leptosphaeria maculans* species complex from Australia, Europe and North America using amplified fragment length polymorphism analysis, *Mycological Research 104*: 772–781.

Reed K.M., Hackett J.D., Phillips R.B. (2000). Comparative analysis of intra-individual and inter-species DNA sequence variation in *Salmonid* ribosomal DNA cistron. *Gene 249*, 115-125.

Restrepo S., Duque M., Tohme J., Verdier V. (1999). AFLP fingerprinting: an efficient technique for detecting genetic variation of *Xanthomonas axonopodis* pv. *Manihotis. Microbiology 145*: 107–114.

Robertson H. M., Lampe D. J. (1995*a*). Distribution of transposable elements in arthropods *Annu. Rev. Entomol. 40*:333-357.

Robertson H.M., Lampe D.J. (1995b). Recent horizontal transfer of a *mariner* element between Diptera and Neuroptera. *Mol. Biol. Evol. 12*: 850-862.

Roncero M.I.G., Hera C., Ruiz-Rubio M., Garca Maceira F.I, Madrid M.P., Caracuel Z.; Calero F., Delgado-Jarana J., Roldan-Rodrguez R., Martnez-Rocha A.L., Velasco C., Roa J., Martn-Urdiroz M., Cordoba D., Di Pietro A. (2003). *Fusarium* as a model for studying virulence in soilborne plant pathogens. *Physiol Mol Plant Pathol 62*:87-98.

Ruiz-Roldán MC, Di Pietro A, Huertas-González MD, Roncero MIG (1999). Two xylanase genes of the vascular wilt pathogen *Fusarium oxysporum* are differentially expressed during infection of tomato plants. *Mol. Gen. Genet. 261*:530–536.

Schneider RW. (1984). Effects of nonpathogenic strains of *Fusarium oxysporum* on celery root infection by *Fusarium oxysporum* f. sp. *apii* and a novel use of the Lineweaver-Burk double reciprocal plot technique. *Phytopathology 74*:646–53.

Sivaramakrishnan S., Kannan S., Singh S.D. (2002). Genetic variability of *Fusarium* wilt pathogen isolates of chickpea (*Cicer arietinum* L.) assessed by molecular markers. *Mycopathologia 155*: 171-178.

Snyder W.C., Hansen H.N. (1940). The species concept in *Fusarium. Am. J. Bot. 27*:64–67.

Spooner D.M., Tivang J., Nienhuis J., Miller J.T., Douches D.S., Contreras M.A. (1996). Comparison of four molecular markers in measuring relationships among wild potato relatives *Solanum* section *Etuberosum* (subgenus *Potatoe*). *Theoretical and Applied Genetics 92*: 532–540.

Suleman P., Straney D.C., Saleh A., Tohamy A.M., Madkour M. (1994). Variation in virulence, phytoalexin tolerance and polymorphism in DNA markers among isolates of *Fusarium oxysporum* f. sp. *lycopersici* (Abst.). *Phytopathology 84*:1075.

Tantaoui A., Ouinten M., Geiger J.P., Fernandez D. (1996). Characterization of a single clonal lineage of *Fusarium oxysporum* f. sp. *albedinis* causing bayoud disease of date palm in Morocco. *Phytopathology 86*, 787–92.

Taylor G.S. (1965). Studies on fungi in the root region. IV Fungi associated with the roots of *Phaseolus vulgaris*. *Plant Soil 22*:1–20.

Vakalounakis, D.J., Fragkiadakis, G.A. (1999). Genetic diversity of *Fusarium oxysporum* isolates from cucumber: differentiation by pathogenicity, vegetative compatibility and RAPD fingerprinting. *Phytopathology 89*:161–168.

Vakalounakis D.J., Doulis A.G., Klironomou, E. (2005). Characterization of *Fusarium oxysporum* f. sp. *radicis-cucumerinum* attacking melon under natural conditions in Greece. *Plant Pathology, 54*: 339-346.

Vos P., Hogers R., Bleeker M., Reijans M., van de Lee T., Hornes M., Frifiters A., Pot J., Peleman J., Kuiper M., Xabeau M. (1995). AFLP: A new technique for DNA fingerprinting. *Nucleic Acids Research 23*: 4407-4414.

Walton J.D. (1994). Deconstructing the cell wall. *Plant Physiol. 104*: 1113–1118.

Welsh, J., McClelland, M. (1990). Fingerprinting genomes using PCR with arbitrary primers. *Nucleic Acids Res. 18*, 7213-7218.

White T.J., Burns T., Lee S., Taylor J.(1990). Amplification and direct sequencing of fungal ribosomal RNA genes for phylogenetics, p. 315–324. *In* M. Innis et al. (ed.), *PCR protocols*. Academic Press, San Diego, Calif.

Williams J.G.K., Kubelik A.R., Livak K.J., Rafalski J.A., Tingey S.V. (1990). DNA polymorphisms amplified by arbitrary primers are useful as genetic markers. *Nucleic Acids Res. 18*:6531–6535.

Windels, C.E. (1992). *Fusarium*. p. 115–128. *In* L.L. Singleton et al.(ed.) *Methods for research on soilborne phytopathogenic fungi*. APS Press, St. Paul, MN.

Wong, J.Y. and Jeffries, P.l. (2006). Diversity of pathogenic *Fusarium* populations associated with asparagus roots in decline soils in Spain and the UK. *Plant Pathology 55*, 331-342.

Wollenweber H.W., Reinking O.A. (1935). *Die Fusarien. Ihre Beschreibung. Schadwirkung and Bekampfung*. Berlin: Paul Parey.

Wu K.S., Tanksley S.D. (1993). Abundance, polymorphism and genetic mapping of microsatellites in rice. *Mol. Gen. Genet. 241*: 225–235.

Wu K.S., Jones R., Danneberger L., Scolnik P.A. (1994). Detection of microsatellite polymorphism without cloning. *Nucleic Acids Res. 22*: 3257-3258.

Zabeau M., Vos P. (1993). Selective restriction fragment amplification: V general method for DNA fingerprinting. European Patent Application number: 92402629.7. Publication number 0534858A1.

Zanotti M. G. S., de Queiroz M. V, dos Santos J. K., Rocha R. B., de Barros E. G., Araujo E. F. (2006). Analysis of Genetic Diversity of *Fusarium oxysporum* f. sp. *phaseoli* Isolates, Pathogenic and Non-pathogenic to Common Bean (*Phaseolus vulgaris* L.). *J. Phytopathology 154,* 545 - 549.

Zamani M.R., Motallebi M., Rostamian A. (2004). Characterization of Iranian isolates of Fusarium oxysporum on the basis of RAPD analysis, virulence and vegetative compatibility. *J. Phytopathol. 152*:449–453.

Zeller K.A., Jurgenson J.E., El-Assiuty E.M., Leslie J.F. (2000). Isozyme and amplified fragment length polymorphisms (AFLPs) from *Cephalosporium maydis* in Egypt. *Phytoparasitica 28*: 121–130.

Zeller K.A., Summerell B.A., Bullock S., Leslie J.F. (2003). *Gibberella konza* (*Fusarium konzum*) sp. nov. from prairie grasses, a new species in the *Gibberella fujikuroi* species complex. *Mycologia 95*: 943-954.

Zietkiewics E., Rafalski A., Labuda D. (1994). Genome fingerprint by sequence repeat (SSR)-anchored polymerase chain reaction amplification. *Genomics 20*:176-183.

Yandell, D. W. (1991). SSCP analysis of PCR fragments. *Jpn. J. Cancer Res. 82*:1468–1469.

In: Current Advances in Molecular Mycology
Eds: Y. Gherbawy, R.L. Mach and M.K. Rai
ISBN 978-1-60456-909-4

Chapter VII

MORPHOLOGY AND MOLECULAR BIOLOGY OF *PHOMA*

L. Irinyi[1], A. K. Gade[2], A. P. Ingle[2], G. J. Kövics[1,*], M. K. Rai[2,&] and E. Sándor[1]

[1]Debrecen University, Faculty of Agriculture, Department of Plant Protection H-4015 Debrecen, P.O.Box 36, Hungary;
[2]Department of Biotechnology, Amravati University, Amravati-444 602, Maharashtra, India.

ABSTRACT

Phoma Sacc. emend. Boerema and G.J. Bollen (1975) is an ubiquitous genus, which has been reported from plants, soil, human beings, animals and air. The speciation of the genus *Phoma* was based on host-alone and later on, the trend was to study different species of *Phoma* in pure culture. On the basis of morphological studies on different culture media, several species were found to be identical. These studies have prompted many investigators to carry out morphological and cultural studies. The morphological criteria, such as diameter and colour of the colony, shape and size of pycnidia and pycnospores, formation of chlamydospores and pigmentation were considered for identification and differentiation of the species. However, these criteria are not always reliable and therefore, molecular studies were carried out. Molecular markers like RAPD and ITS-rDNA sequences were used for identification and differentiation of species of *Phoma* and to understand the evolutionary relationship among the species.

[*] Correspondence concerning this article should be addressed to: G.J. Kövics, Debrecen University, Faculty of Agriculture, Department of Plant Protection H-4015 Debrecen, P.O.Box 36, Hungary. E-mail: kovics@agr.unideb.hu; web: http://www.mtk.unideb.hu.

[&] Correspondence concerning this article should be addressed to: M.K. Rai, Department of Biotechnology, Amravati University, Amravati-444 602, Maharashtra, India. E-mail: mkrai123@rediffmail.comor pmkrai@hotmail.com.

The main goal of the present chapter is to discuss morphological and molecular identification and differentiation of selected species of *Phoma*.

Key words: *Phoma*, morphology, molecular biology, ITS-rDNA, tef1 sequences, β-tubulin.

INTRODUCTION

Many species of *Phoma* have been reported from diverse hosts/substrates ranging from plant, soil, water and air (Rai, 1981), marine environment (Sugano *et al.*, 1991), entomopathogenic (Narendra and Rao, 1974) and anthropophilic (Shukla *et al.*, 1984; Baker *et al.*, 1987; Zaitz *et al.*, 1997; Rishi and Font, 2003). The original genus concept of Saccardo (1880) was emended by Boerema and Bollen (1975) and after more than 40 years of taxonomical research the admitted *Phoma* species were arranged in nine Sections (Boerema *et al.*, 2004). *Phoma* can be easily differentiated from its allied genus *Phyllosticta* by formation of relatively smaller pycnidiospores. In addition, the later produces a hyaline sheath around pycnidiospores. A special appendage is also produced by the pycnospores, which can be visible in aquaceous condition, however this feature may disappear on dry herbarium materials. The fungus is characterized by the formation of single-celled, hyaline pycnospores borne inside a fruiting body referred to as pycnidia, which vary from globose, sub-globose to coalescence forms.

Up to 1980, the existing Indian species of *Phoma* were erected on the basis of host alone, and thus the importance of host specificity for the taxonomy of *Phoma* was over-emphasized. The assumption that each host genus or species was colonized by a specialized *Phoma* species prompted many mycologists to ignore morphological characters when erected new *Phoma* species. However, a morphological species may attack various host plants. For example, *P. exigua* has been reported by investigators on different hosts (Bhowmik and Singh, 1976; Raut, 1977; Kanaujia, 1979; Kamal and Singh, 1979; Wadje and Deshpande, 1981; Rao and Thirumalachar, 1981). Considering the host as the only criterion for identification, many new species have been erected from India (Tandon and Bilgrami, 1960; Bilgrami, 1963; Agarwal and Sahni, 1964; Dutta and Ghosh, 1965; Chandra and Tandon, 1965, 1966; Hasija, 1966; Pavgi and Singh, 1966; Shreemali, 1972, 1973, 1978; Jamaluddin *et al.*, 1975; Maiti *et al.*, 1978; Rai, J.N. and Misra, 1981; Agrawal and Misra, 1981). Such a way the criteria for identification of *Phoma* species should be inadequate, particularly in case of host identification difficulties because lack of floral parts, or when the fungus is grown on artificial media.

Boerema and his co-workers (1968) also studied some *Phoma* species from India and found that some newly described species are similar to the already existing species so later on these were reclassified and considered as synonyms.

For identification and delimitation of *Phoma* species which are suitable to study them in pure culture and an integrated approach based on cultural, morphological, physiological, determination of ontogeny, biochemical evaluation or in ambiguous cases molecular approaches should be applied (Rai, 1981; Monte *et al.*, 1991, Irinyi *et al.*, 2006a, 2006b, 2006c, 2007, 2008).

Lately, Boerema *et al.* (2004) published a modern monograph dealing with 223 specific and infraspecific taxa of *Phoma* Sacc. emend. Boerema & G.J. Bollen, with 1146 synonyms in various *Coelomycetes* genera.

Biotechnological importance of species of *Phoma* has been realized recently because some species are known to produce biologically active fungal metabolites (Baxter *et al.*, 1992; Pearce, 1997; Singh *et al.*, 1997; Deshmukh *et al.*, 2006). Among these, squalestatin production by *Phoma* sp. is worthy of note. A *Phoma* sp. (IMI 332962) known to produce the pharmaceutically active metabolites squalestatin-1 (S1) and squalestatin-2 (S2) (Baxter *et al.*, 1992). There are some antitumour agents produced by *Phoma* sp. (Singh *et al.*, 1997), particularly fusidienol-A, which is the second member of the fusidienol family of inhibitors to possess a novel tricyclic oxygen-containing heterocycle with a 7/6/6 ring system. Fusidienol-A was isolated from an unidentified species of fungal genus *Phoma* (MF 6118) collected from a vegetation sample. Singh *et al.* (1998) reported equisetin and phomasetin production by species of *Phoma*, which are useful against AIDS, equisetin derivatives from *Fusarium* or *Phoma* sp. (MF6070). This showed HIV virus integrase inhibitor activity. Due to its potential, the fungus can be used against AIDS. Sponga *et al.* (1999) reported antibiotic activities in microorganisms isolated from marine sediments. These microbes include *Phoma, Acremonium, Alternaria, Aspergillus, Cephalosporium, Chaetomium, Cladosporium, Geotrichum, Fusarium, Gliomastix, Humicola, Paecilomyces, Penicillium, Pestalotia, Plectosphaerella, Scopulariopsis, Stachybotrys,* and *Trichoderma* genera, the fungus exhibited antimycotic activities against *Enterococcus faecium*, *Escherichia coli,* and *Candida albicans*.

Phoma is also was being exploited for the control of weeds (Zhou *et al.*, 2005). Considering the importance of these species, they were realized to understand and study the morphology and their differentiation by using molecular markers.

The present chapter is aimed to describe some various morphological features and molecular markers for their identification and differentiation.

COMPARATIVE MORPHOLOGICAL AND CULTURAL STUDIES

Morphological characteristics must be studied *in vitro* in order to establish a more realistic and practical classification of *Phoma.* In pure culture, a large number of fungi previously assigned to several species actually represent a restricted number of morphological types. Examples are *Phoma exigua* Desm. (Boerema and Höweler, 1967; van der Aa and van Kesteren, 1971) and *Phoma macrostoma* Mont. (Boerema and Dorenbosch, 1970), which are known to develop widely on different hosts. Boerema (1970) stated that new *Phoma* species must be identifiable without the knowledge of substrate. This is possible in *Phoma*, if the morphology of pycnidia and conidia are considered with all other characters observable *in vitro* for species differentiation (habit, colony growth, crystal formation, pigmentation, and chlamydospores etc.).

Rai (1981) studied the Indian species of *Phoma*, and on the basis of morphological and cultural characters he formed 20 broad morphological groups (Rai 1985, 1986a, 1986b, 1987,

1989; Rai and Rajak, 1982a, 1982b, 1983a, 1983b, 1986a, 1986b, 1993; Rajak and Rai, 1982a, 1982b, 1983a, 1983b, 1984a, 1984b and 1985).

These groups include *P. pinodella* (Jones) Morgan-Jones & K.B. Burch, *P. medicaginis* Malbr. & Roum. var. *medicaginis*, *P. pomorum* Thüm., *P. herbarum* Westend., *P. exigua* Desm. var. *exigua*, *P. tropica* R. Schneid. & Boerema, *P. glomerata* (Corda) Wollenw. & Hochapfel, *P. sorghina* (Sacc.) Boerema *et al.*, *P. multirostrata* (Mathur *et al.*) Dorenb. & Boerema, *P. capitulum* V.H. Pawar *et al.*, *P. betae* A.B. Frank, *P. jolyana* Piroz. & Morgan-Jones var. *jolyana*, *P. fimeti* Brunaud, *P. chrysanthemicola* Hollós, *P. complanata* (Tode: Fr.) Desm., *P. destructiva* Plowr. var. *destructiva*, *P. eupyrena* Sacc., and *P. arachidicola* (Marasas *et al.,* 1974).

Conidial Ontogeny

Conidial ontogeny is one of the important criteria for genera and species differentiation. In *Phoma* conidia are produced by small conidiogenous cell inside pycnidia. Each parent cell can produce a series of conidia. Consequently, the new conidia take the place of older cells. The first conidium develops from a papilla at the top of the parent cell (Sutton and Sandhu, 1969). Boerema and Bollen (1975) have described conidial development in more details. The process is culminated by the formation of separation plate between the parent cell and the conidium. Some dissolved wall material remains surrounding the conidium, comprising a mucilaginous mass. A collarette is left on the parent cell when the conidium seceds. In some species, the conidia become 2-celled by an ingrowth from the lateral wall, which apparently from the beginning attains the thickness of the final septum. A pore remains in the center of the septum, which is associated with Woronin's bodies and membrane-bound plugs.

Conidia swell during germination. A new electron transparent layer can be distinguished inside the conidial wall that is continuous with the germ-tube emerges through the ruptured outer wall of the conidium.

To the study the conidial ontogeny, sophisticated equipments are required, and thus, cannot be used in routine mycological diagnostic work.

Biochemical Characteristics

A little work has been done on biochemical aspects of taxonomic criteria for species differentiation in *Phoma*. The NaOH spot test was introduced in *Phoma* taxonomy as a probe for checking 'E' metabolite (after *P. exigua*) near the growing margin of cultures with a drop of NaOH. The metabolite E-producing culture results within about 10 minutes in a greenish spot or ring (pigment α), which changes to red (pigment β) after one hour (Boerema and Höweler, 1967). Pigments are sometimes restricted to the cytoplasm or guttules in the hyphae, and usually diffused into the agar media are involved. These pigments are composed of anthraquinone components (pachybasin, chrysophanol, emodin, phomarin) (Bick and Rhee, 1966). Rajak and Rai (1983) tried cholesterol as a taxonomic marker or differentiation of species within the genus *Phoma*. They further reported that significant similarities were noted even among morphologically different species and vice versa, and owing to this reason, this cannot be recommended for identification and differentiation of species within the genus *Phoma*. Fluorescent metabolites may also be produced by some *Phoma* spp. (Boerema and Loerakker, 1985). Some species can produce very characteristic dendritic christals in the agar media, which may be brefeldin A, pinodellalide A and B, and radicin (Noordeloos *et al.*, 1993).

Integrated Approach of Speciation

Rai (1981) studied Indian species of *Phoma* in pure culture on malt-agar, oat-agar and rice-agar, and proposed a key to the identification and differentiation of different species of *Phoma*.

SELECTED *PHOMA* SPECIES

1) *P. arachidicola* Marasas *et al.* (1974)

Section *Phyllostictoides*
Teleomorph: Didymella arachidicola (Khokhr.) Taber *et al.*

Cultural Characteristics:

Colonies grey to dull-yellow, closely appressed at the margin, attaining a diameter of 5-5.5 cm on oat-meal agar in 14 days; pycnidia dark-brown, abundant, globose to subglobose to flask-shaped, pseudoparenchymatous, ostiolate, 100-250 μm; conidia hyaline, l-celled, mostly ovoid, few ellipsoid, ends obtuse, 4.8-8 x 2.5-3.9 μm; chlamydospores produced after 2 weeks, single or in groups, dark-brown, 2-to 3-celled, globose to subglobose, thick-walled, 7-15 μm.

Formation of chlamydospores single or in groups is a typical character of the fungus.

2) *P. betae*: Frank (1892)

Section *Pilosa*
Teleomorph: Pleospora betae (Berl.) Nevod.

Cultural Characteristics:

Colonies greenish-grey to black, attaining similar diameter on oat-agar and malt agar in 7 days; pycnidia black, abundant, globose to subglobose, pseudoparenchymatous, ostiolate, 74-150 μm; conidia hyaline, 1-celled, mostly ovoid on rice-meal agar and malt agar, ellipsoid on oat-meal agar, relatively broader, 6.5-7 x 3.2-4 μm; chlamydospores in chains, grey, relatively larger-celled.

P. betae characterized by relatively broad conidia (Boerema and Dorenbosch, 1973).

3) *P. capitulum* Pawar *et al.* (1967)

Section *Phoma*

Cultural Characteristics:

Colonies with cottony aerial mycelium, oval, attaining a diameter of 5 cm on oat-meal agar in 7-days; pycnidia 50-105 x 50-80 μm, usually in clusters of up to 20 specimens in a row, borne on radiating, dense, blackish hyphal strands, globose with 1-3 ostioles on a short neck, citrine to honey when young then olivaceous to olivaceous black, smooth, glabrous with thin walls, in concentric zones or sometimes in clusters after 7-14 days, mostly on agar, but also partly or entirely merged in the agar; exudate grey-saffron; conidiogenous cells 3-7 x 4-7 μm, phialidic, globose, thin-walled; conidia 3.2-4.4 x 2-3 μm, average 3.8 x 2.6 μm, broadly and shortly ellipsoidal with 1 or 2 guttules; NaOH spot test negative.

It is halophilic species and occurs in forest soil, saline soil, and marine environment.

4) *P. chrysanthemicola* Hollós (1907) [sensu lato]

Section *Peyronellaea*

Cultural Characteristics:

Colonies dark-grey with compact aerial mycelium, attaining a diameter of 3-3.5 cm on oat-meal agar in 7 days; pycnidia black, globose to subglobose, compound, parenchymatous, ostiolate, 88-250 μm; conidia hyaline, l-celled, mostly ovoid on rice-meal agar and malt agar, ellipsoid on oat-meal agar, 4-7.3 x 2.5-3.4 μm; orange-red discoloration of the medium appears. An important diagnostic character of this fungus is the occurrence of dark-brown to black pseudosclerotial, irregular masses (Dorenbosch, 1970). Rai (1998) overviewed the effect of different physical and nutritional conditions on the morphology and cultural characters.

Although Aa van and Vanev (2002) consider this species as a synonym of *Phyllosticta leucanthemi* Speg., Boerema *et al.* (2004) do not accept this status for the soil-borne fungus.

5) *P. complanata* (Tode: Fr.) Desm. (1851)

Section *Sclerophomella*

Cultural Characteristics:

Colonies cottony with much aerial mycelium, attaining a diameter of 6.7 cm on oat-meal agar in 7 days; pycnidia thick-walled, dark-yellow to black, globose, parenchymatous, 74-150 μm; conidia hyaline, 1- to 2-celled, 3.9-11 x 2-3.2 μm.

A typical identifying character of the fungus is formation of massive and thick-walled mature pycnidia (Boerema, 1976).

6) *P. destructiva* Plowr. (1881) var. *destructiva*

Section *Phoma*

Cultural Characteristics:

Colonies dark-green with much aerial mycelium, attaining a diameter of 5-6 cm on oat-meal agar in 7 days; pycnidia black, globose to subglobose, pseudoparenchymatous, ostiolate, 78-155 μm; conidia hyaline, 1-celled, mostly ellipsoid on oat-meal agar, ovoid on rice-meal agar and malt agar, 4.5-7.5 X 3.2-3.5 μm; chlamydospores in chains, single-celled, grey, non-guttulated or less guttulated.

Lately a new variety (*P. destructiva* var. *diversispora* Gruyter & Boerema apud de Gruyter, Boerema and van der Aa, 2002) was delimited and classified in Section *Phyllostictoides*. Then was generated *P. destructiva* var. *destructiva* (autonym).

7) *P. eupyrena* Sacc. (1879)

Section *Phoma*

Cultural Characteristics:

Colonies green with compact aerial mycelium, attaining a diameter of 4.2-4.8 cm on oat-meal agar in 7 days; pycnidia abundant, black, globose to subglobose, parenchymatous, 74-160 μm; conidia hyaline, 1-celled, mostly ellipsoid, few ovoid, guttulated, 3-7-6.2 x 2-3.5 μm; typically small, black chlamydospores produced in chains.

Formation of typically small chlamydospores in chains is stable character of this species (Dorenbosch, 1970). The chlamydospores of this fungus differ from *P. pinodella (P. medicaginis* var. *pinodella)* in their size. The former are relatively small in size and less guttulated.

8) *P. exigua* Desm. var. *exigua* (1849)

Section *Phyllostictoides*

Cultural Characteristics:

Colonies grey to black with irregularly scalloped margins, attaining a diameter of 6-7 cm on oat-meal agar in 7 days; pycnidia black, abundant, subglobose to globose, some coalesce to form irregular fructifications, ostiolate, parenchymatous, 62-290 μm; conidia hyaline, 1-2-celled, ovoid on malt agar and rice-meal agar, ellipsoid on oat-meal agar, 3.7-11.1 x 2-3.7 μm.

In spot test, the greenish-blue discolouration of the agar medium on application of a drop of NaOH takes place, which is an important diagnostic character. This greenish colour gradually becomes brown-reddish.

This fungus has numerous, about 150 synonyms (details see in Boerema *et al.*, 2004). Within the *P. exigua* species at least nine variety subtaxa (*varietas*, var.) was admitted (Boerema *et al.*, 2004) and confirmed based on molecular method (Abeln *et al.*, 2002).

9) *P. exigua* var. *heteromorpha* (Schulzer & Sacc.) Noordel. & Boerema (1987)

(Syn.: *P. exigua* var. *inoxydabilis* Boerema & Vegh apud Vegh, 1974)
Section *Phyllostictoides*

Cultural Characteristics:

Colonies black with irregularly scalloped margins, attaining a diameter of 4-5 (7) cm on oat-meal agar in 7 days; pycnidia black, abundant, globose to subglobose to flask-shaped, some coalesce to form irregular pycnidia, parenchymatous, ostiolate 67-286 μm; conidia differ from those of the type var. *exigua* by their extreme morphological variabiliy ('*heteromorpha*'), mostly ellipsoid, few ovoid, hyaline, 1-celled, 3.9-7.4 x 2.5-3.5 μm; lack of pigment production in this species is a characteristic feature for differentiation with *P. exigua* var. *exigua,* however some strains show slight positive reaction with NaOH (Boerema *et al.*, 2004).

Abeln *et al.* (2002) confirmed genetic differentiation of *P. exigua* varieties by means of AFLP fingerprints.

10) *P. fimeti* Brunaud (1889)

Section *Phoma*

Cultural Characteristics:

Colonies ashy to green with compact aerial mycelium, attaining a diameter of 1.7-3 cm on oat-meal agar in 7 days; pycnidia black, globose to subglobose, pseudoparenchymatous, ostiolate, 74-160 μm; conidia hyaline, 1-celled, mostly ovoid, few ellipsoid, ends obtuse, 2.9-5.2 x 2-3.2 μm; yellow discolouration of the medium on malt agar appears.

P. fimeti is well known for its slow rate of growth, and thus can easily be differentiated from other species. Van der Aa and Vanev (2002) suggested to classify this species as a synonym of *Phyllosticta epimedii* Sacc., but Boerema *et al.* (2004) rejected this taxonomic status for this soil fungus.

11) *P. glomerata* (Corda) Wollenw. & Hochapfel (1936)

Section *Peyronellaea*

Cultural Characteristics:

Colonies dull, dark-yellow, variable in appearance: flat with sparse and/or abundant aerial mycelium in sectors, dense and wooly places occur, olivaceous, olivaceous buff or dull green; reverse dark olivaceous to blackish, colonies relatively slow to fast growing, attaining a diameter of 4.5-7 cm on oat-meal agar in 7 days; pycnidia dark-yellow to black, usually solitary but sometimes coalescing, abundantly produced in concentric rings, globose to subglobose, some cylindrical, parenchymatous, ostiolate, (74) 100-160 (300) µm, frequently produce fertile micropycnidia, 20-50 µm diam.; conidia hyaline, 1-celled, mostly ellipsoid, some ovoid, 4-8.3 x 2.5-3.8 µm; NaOH spot-test the medium turns tea-brown. Chlamydospores and dictyochlamydospores (*Alternaria*-like) produced in chains. This species has numerous synonyms including three *Alternaria* because of multicellular, alternarioid chlamydospores.

12) *P. herbarum* Westend. (1852)

Section *Phoma*

Cultural Characteristics:

Colonies grey to green, massive mycelial mat, uniform rate of growth, attaining a diameter of 4-5 cm in 7 days on malt agar and oat-meal agar, in sectors; pycnidia grey to black, subglobose to globose, usually parenchymatous, sometimes with short neck, ostiolate, 74-150 (up to 350) µm; conidia hyaline, 1-celled, mostly ovoid on malt agar, ellipsoid on oat-meal agar, 3.7-8.0 x 2-3 µm; red pigments produced, which on application of NaOH changes bluish.

Sometimes pycnidial type strains of *P. herbarum* also occur, characterized by the production of abundant pycnidia and sparse mycelium. This creates confusion with *P. glomerata*. The latter can be differentiated from the former by the presence of chlamydospores and dictyochlamydospores.

At the 8th International Botanical Congress at Paris (1954) *P. herbarum* was selected as the type species of the 'form genus' *Phoma* (Boerema, 1964, 1970).

13) *P. jolyana* Pirozynski & Morgan-Jones var. *jolyana* (1968)

Section *Peyronellaea*

Cultural Characteristics:

Colonies greyish-black, attaining a diameter of 4-5 cm on oat-meal agar in 7 days; pycnidia black, globose to subglobose, compound, parenchymatous, ostiolate, 88-290 µm;

conidia hyaline, 1-celled, ellipsoid on oat-meal agar, ovoid on rice-meal agar and malt agar, few kidney-shaped, 3.7-8.5 x 2-3.8 μm; chlamydospores and dictyochlamydospores in chains, abundantly produced on malt agar, multicellular and laterally arranged.

This is a common soil-borne fungus in subtropical regions of Eurasia and Africa. Rai (1985) isolated from air, India. In Siberia and in the Sahara adapted varieties of the fungus occur (var. *circinata* and var. *sahariensis*).

14) *P. macrostoma* Mont. var. *macrostoma* (1849)

Section *Phyllostictoides*

Cultural Characteristics:

Colonies pinkish-red with much aerial mycelium, attaining a diameter of 4.5-7 (7.5) cm on oat-meal agar in 7 days; pycnidia black, globose to subglobose, pyriform, parenchymatous, swollen, 80-200 (300) μm; relatively wide ostiole up to 48 μm; conidial matrix salmon to flesh, conidia hyaline with some guttules, mainly 1-celled, some 2 or 3-celled mostly ellipsoid, a few ovoid, a few kidney-shaped, 7.5-14.2 x 2.5-3.7 μm. Applied a drop of NaOH on colonies on oat-meal agar reddish to purplish colour may occur.

The variety *macrostoma* can be easily distinguished from another variety, viz. *incolorata*, by lack of reddish-pigments in the hyphae (Boerema and Dorenbosch, 1970).

15) *P. macrostoma* var. *incolorata* (Horne) Boerema & Dorenb.

Section *Phyllostictoides*

Cultural Characteristics:

Colonies black, attaining a diameter of 7-7.5 cm on oat-meal agar in 7 days; pycnidia greyish-black, globose to subglobose, parenchymatous, ostiolate, 110-210 μm; conidial mass is white to buff or rosy buff; conidia hyaline, 1-2-celled, mostly ellipsoid, few ovoid, 4-11.9 X 2-3.5 μm; reddish pigments absent in hyphae. NaOH test negative.

This variety of *P. macrostoma* can be differentiated from variety *macrostoma* of the same species by absence of red pigments (Boerema and Dorenbosch, 1970).

16) *P. medicaginis* Malbr. and Roum (1886) var. *medicaginis*

Section *Phyllostictoides*

Cultural Characteristics:

Colonies irregular, sinuate outline, olivaceous or grey olivaceous at margin, buff at centre, attaining a diameter of 3.5-4 cm on oat-meal agar and 4-4.5 cm on malt agar in 7 days, sometimes with olivaceous sectors; formation of crystals on malt agar is occassionally (brefeldin A, Noordeloos *et al.*, 1993). The production of dendritic crystals differs strain by strain, can only be produced after some weeks of growth. Mycelium white, floccose, poorly developed; pycnidia grey to black, concentrically zonate, some in radial rows, globose to subglobose, parenchymatous, 80-200 µm; conidia hyaline, generally 1, rarely 2-celled, mostly ellipsoid, few ovoid, 4-8.7 x 1.5-3.3 µm, conidial mass whitish to pale pink; chlamydospores in chains, globose to subglobose, highly guttulated, occassionally produced in old culture.

P. medicaginis var. *medicaginis* differ from *P. pinodella* by growth pattern, chlamydospores produced only in old culture, and few crystals, only in cultures over 1-month old.

17) *P. pinodella* (Jones) Morgan-Jones & K.B. Burch (1988)

(syn.: *P. medicaginis* Malbr. and Roum. var. *pinodella* (Jones) Boerema apud Boerema, Dorenbosch & Leffring, 1965)

Section *Phyllostictoides*

Cultural Characteristics:

Colonies greenish or yellowish olivaceous, margin pale, flat, radially filamentous, attaining a diameter of 5-6.2 cm on oat-meal agar in 7 days, reverse on malt agar shows conspicous white, dendritic crystals; aerial mycelium practically absent; pycnidia grey to black, scattered, some in radial rows, globose to irregular, parenchymatous, (80) 96-175 (320) µm, sometimes micropycnidia occur near the apex of a pycnidia, 30-50 µm; conidia hyaline, 1-celled, sometimes 1-septate, mostly ellipsoid, few ovoid, variable in size, 4-8.7 x 2.5-3.5 µm; chlamydospores solitary or in chains, confluent, grey to black globose to subcylindrical, highly guttulated, 8-20 µm diameter.

Formation of crystals on malt-agar is one of the most important characteristic features of this fungus (pinodellalide A and B, Noordeloos *et al.*, 1993), however sometimes might missing. *Plurivorous* species isolated from wide range of plants, especially on *Leguminosae*.

18) *P. multirostrata* (Mathur *et al.*) Dorenb. & Boerema (1973) var. *multirostrata*

Section *Phoma*

Cultural Characteristics:

Colonies colourless to weak olivaceous, flat with little aerial mycelial mat, attaining a diameter of 6.5-7.2 cm on oat-meal agar in 7 days; colonies on malt agar olivaceous to olivaceous buff with felty to floccose, sometimes wooly aerial mycelium, reverse leaden grey to blackish; pycnidia in concentric rings, abundant, globose to subglobose, parenchymatous, several ostioles in a pycnidium with variously shaped necks, pycnidia always relatively large, more than 550 µm; conidia hyaline, 1-celled, ovoid on malt agar and rice-meal agar, ellipsoid on oat-meal agar with obtuse ends, sometimes one end acute and other end obtuse, guttulated,

4-7.2 x 2.2-3.1 μm; whitish to cream or rosy-buff conidial matrix produced. Chlamydospores common in old cultures, oblong to ellipsoidal, in chain or clustered, olivaceous with green guttules, 5-15 μm diameter.

The worldwide recorded thermotolerant fungus refers to soil isolates made in India. Numerous isolates differ by smaller pycnidia (mostly 150-300 μm diam.) and wide range in conidial dimensions which led to the differentiation of the vars *macrospora* Boerema (1986) and *microspora* (Allesch.) Boerema (1986). Intermediate variants also commonly occur. The descriptions of several isolates from India (Rai, 1998) mainly conform to var. *macrospora* (Boerema *et al.*, 2004).

19) *P. pomorum* Thüm. (1879) var. *pomorum*

Section *Peyronellaea*

Cultural Characteristics:

Colonies olivaceous to greyish-black, with much aerial mycelium, attaining a diameter of 4.5-6.5 cm on oat-meal agar and 5.5-7.5 cm on malt agar in 7 days; pycnidia black, globose to subglobose, scattered or coalesce to form irregular fructifications up to 1000 μm diameter with many ostioles, parenchymatous 73-225 μm, micropycnidia frequently occur; conidia hyaline, later becoming light brown, 1-celled, mostly ovoid, few ellipsoid, end obtuse, variable in shape and dimensions, 4-7.4 x 1.5-3.7 μm; conidial mass usually whitish to cream, later olivaceous brown; abundant production of single-celled chlamydospores in chains in combination with dictyochlamydospores, mostly terminal on mycelial branches, brown to black, 18-60 x 10-30 μm.

It can be distinguished from *P. glomerata* especially by its possession of chains of single chlamydospores as well as dictyochlamydospores. Rai (1998) referred about the effect of different physical and nutritional conditions on the morphology and cultural characters.

20) *P. sorghina* (Sacc.) Boerema *et al.* (1973)

Section *Peyronellaea*

Cultural Characteristics:

Colonies pink with much aerial mycelium, attaining a diameter of 5.5-7.2 cm on oat-meal agar in 7-days; aerial mycelium fluffy with greyish green or whitish with pinkish tinge, reverse reddish occassionally with needle-like chrystals; yellow discolouration of the medium in acidic condition below the colony; pycnidia grey, subglobose to flask-shaped, necked up to 80 μm, pseudoparenchymatous, 75-155 μm; conidial mass salmony; conidia variable in shape and dimensions, hyaline, 1-celled, mostly ellipsoid, few ovoid, 3.9-7.1 x 2-2.7 μm; chlamydospores abundant, uni- or multicellular, solitary or in chains, mostly globose, 8-35 μm, aseptates 5-15 μm. NaOH-test might be positive, green to red.

Occur worldwide as soil- and seed-borne fungus; often appear as a weak or secondary plant parasite about 80 host plant genera in the tropics and subtropics. A typical diagnostic character of this species is production of chlamydospores and dictyochlamydospores in chains, and red pigment in hyphae.

21) *P. tropica* Schneid. & Boerema (1975)

Section *Phoma*

Cultural Characteristics:

Colonies olivaceous to dull green, greasy with aerial mycelium, attaining a diameter of 4.5-6.2 cm on oat-meal agar in 7 days, reverse greenish olivaceous, light green margin; pycnidia dark-brown, globose, subglobose to flask-shaped, some coalesce to form irregular fruiting bodies, scattered or concentric rings on the agar, pseudoparenchymatous with 1-5 ostioles, 88-300 μm diameter; conidia hyaline, 1-celled, mostly ellipsoid, some ovoid, 3.5-6 x 2.2-2.6 μm, with distict guttules.

Thermotolerant saprophyte in European glasshouses on necrotic tissues of ornamental plants (Boerema *et al.*, 2004), repeatedly recorded from India (Rai, 1998).

MOLECULAR BIOLOGY OF *PHOMA*

In the middle of 90s, due to the advances in molecular and biochemical research of that time molecular markers were identified in *Phoma*. Some isozyme analyses were applied to distinct some morphologically identical *Phoma* species from each other (Kövics and Gruyter, 1995). Protein polymorphisms comparing to DNA polymorphisms is unfavourable, because protein electrophoresis assays the genotype indirectly, and a high proportion of the variation occurs at the DNA level may not be detectable, as it does not alter the amino acid composition of the protein. Similarly, some changes in amino acid composition do not change the electrophoretic mobility of the protein, and remain undetected, leading to different genotypes being assigned to the same allozyme allele.

DNA polymorphisms are based on differences in DNA sequences and have three enormous advantages over protein polymorphisms. The first is that the sequence differences are detected directly. The second advantage is that they occur in a genome at very high frequency, and finally, they are not subject to selection pressure, in case they do not affect the phenotype. But morphological characterization besides molecular tools will remain a basic and powerful key in the identification of *Phoma* species (Irinyi *et al.*, 2006a).

One of the most commonly used molecular techniques for assessing phylogenetic relationships is to evaluate the sequences of certain fungal DNA regions. Phylogenetic sequence comparisons concentrate on a comparison of the coding portions of the ribosomal genes and their RNA products, allowing discrimination at different taxonomic levels. Many phylogenetic works are based on the internally transcribed spacers (ITS), which are one of the most widely used molecular markers due to their high variability in nucleotide sequences.

According to Lutzoni *et al.*, (2004) 83.9% of fungal phylogenies are based exclusively on sequences from the ribosomal RNA tandem repeats. Because of it, there is a consequent trend toward inclusion of other gene loci in the data sets, gathered for phylogenetic analysis. Among these genes, protein-coding genes like β-tubulin and translation elongation factor (*tef1*) can contribute greatly to resolving deep phylogenetic relationships with high support and/or increase support for topologies inferred using ribosomal RNA genes.

Ribosomal DNA (rDNA) has long been used as a potential marker for phylogenetic studies (reviewed in Avise, 2004). rRNA genes are organized in clusters of tandemly repeated units, each of which consists of coding regions (18S, 5.8S, and 28S) (Gerbi, 1985) and 2 internal transcribed spacers (ITS) and intergenic spacer (formerly called as Non-Transcibed Spacer, NTS) region. While the coding regions are evolutionarily conserved and have been utilized for phylogenetic inferences for major phyla (reviewed in Hills and Dixon 1991), the 2 ITS regions are appropriate for detecting differences between co-specific individuals and are hence potentially useful markers to study the relationships of populations and closely related species in fungal, plant, and animal taxa due to their relatively rapid evolutionary rates (Baldwin, 1992; Schlötterer *et al.*, 1994; Mai and Coleman, 1997; Weekers *et al.*, 2001; Oliverio *et al.*, 2002; Chen *et al.*, 2000, 2002).

Many fungal taxonomy studies have applied ITS regions for resolving relationships at the genus and species level (Gardes and Bruns, 1993; Graser *et al.*, 1999; Shinohara *et al.*, 1999; Gottlieb and Lichtwardt, 2001; Nugent and Saville, 2004; Yli-Mattila *et al.*, 2004; Voglmayr and Yule, 2006, Morocko and Fatehi, 2007; Padamsee *et al.*, 2008; Takamatsu *et al.*, 2008).

Mendes-Pereira *et al.* (2003) used ITS sequences for studying the molecular phylogeny of *Leptosphaeria maculans – L. bilglobosa* species complex. Balmas *et al.* (2005) inferred phylogenetic relationships among isolates of *Phoma tracheiphila* on the basis of IT sequences as well as Fatehi *et al.* (2003) in *Ascochyta pinodes*-complex.

DNA sequences have been obtained from ITS, translation elongation factor and β-tubulin coding genes to resolve phylogenetic relationships among several *Phoma* species, since it has been shown that usage of multigene datasets can increase the resolution of molecular phylogenetic analyses.

A region of nuclear rDNA was obtained, containing the internal transcribed spacer regions 1 and 2 and the 5.8S rDNA (White *et al.*, 1990), which were studied (Irinyi *et al.*, 2006a, 2006b, 2006c, 2007, 2008) (Figure 1).

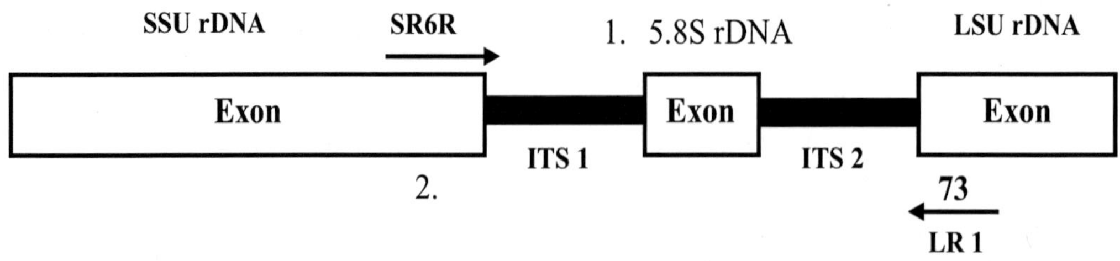

Figure 1. Schematic structure of ITS region in *Phoma* spp. and location of primers for phylogenetic analyses.

Translation elongation factor 1 subunit alpha (EF1=tefl) is part of the cytosolic EF1 complex, whose primary function is to promote the binding of aminoacyl-tRNA to the ribosome in a GTP-dependent process (Moldave, 1985). It is an essential component of the protein synthesis process in eukaryotes and archeabacteria. Complexed with GTP, it carries the aminoacyl-tRNA to the A site of the ribosome-mRNA-peptidyl-tRNA complex; upon hydrolysis of GTP it leaves the ribosome as EF-1-GDP.

Translation elongation factor 1 subunit alpha (EF1α) encoding gene (*tef1*) has a highly conserved sequence in eukaryotic organisms that has been suggested to have desirable properties for phylogenetic inference (Roger *et al.*, 1999). It is well-suited for determining phylogenetic relationships, due to its universal occurrence and presence typically as a single copy within the genome (Baldauf and Doolittle, 1997). It has been proven to be a useful gene to resolve phylogenetic relationships at species level as well as in deeper divergences (Roger *et al.*, 1999; Druzhinina and Kubicek, 2005). Knutsen *et al.* (2004) used *tef1* gene for phylogenetic analysis of *Fusarium poae*, *Fusarium sporotrichioides* and *Fusarium langsethiae* sepecies complex as well as Skovgaard *et al.* (2002) to assess genetic relatedness of *Fusarium oxysporum* complex isolated from pea.

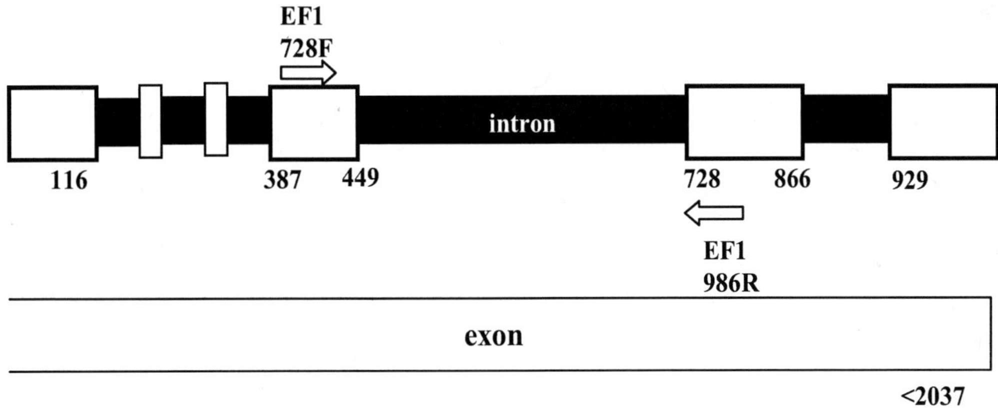

Figure 2. Schematic structure of *tef1* gene in *Phoma* spp. and location of primers for phylogenetic analyses.

However Tanabe *et al.* (2004) suggest that the *tef1* gene is unsuitable for resolving higher-level phylogenetic relationships within the Fungi.

Irinyi *et al.* (2006a, 2006b, 2007, 2008) used primer pair, which facilitates the PCR amplification of the large intron of *tef1* gene (Druzhinina and Kubicek, 2005) (Figure 2).

Like rRNA molecules, the tubulin proteins are involved in ancient eukaryotic functions as they are elementary subunits of the microtubules and can be preseumed to be ancient molecules and due to their universal occurrences they are very constant in function and therefore quite well conserved over considerable phylogenetic distances. Of the seven varieties of tubulin proteins (McKean *et al.*, 2001) α- and β-tubulins are the most abundant in the eukaryotic cell. Tubulin encoding genes, especially for β-tubulin (*tub1*), are receiving increasing attention in the investigation of evolutionary relationships at all levels: (i) in kingdomlevel phylogenetic analyses (Keeling and Doolittle 1996, Baldauf *et al.*, 2000), and

(ii) in studies of complex species groups within protists, animals, fungi and plants (Mages *et al.*, 1995, Keeling *et al.*, 1998, Schutze *et al.*, 1999, Ayliffe *et al.*, 2001, Edgcomb *et al.*, 2001; Yli-Mattila *et al.*, 2004).

β-tubulin has been determined to be a single copy gene in some genera of the Ascomycota (Byrd *et al.*, 1990; Neff *et al.*, 1983; Orbach *et al.*, 1986), but two highly divergent paralogs have been reported in others (May *et al.*, 1987; Panaccione and Hanau, 1990). Tubulin is well characterized in Ascomycota, but has not been used in phylogenetics of *Phoma.* Glass and Donaldson (1995) developed primer sets to amplify β-tubulin genes from filamentous Ascomycetes.

Several studies proved that β-tubulin at the nucleotide level can be suitable for phylogenetic studies at low taxonomic levels within the ascomycetes (Jong *et al.*, 2001; O'Donnell *et al.*, 1998; Schoch *et al.*, 2001). At the amino acid level partial β-tubulin sequences have been used by Landvik *et al.* (2001) to infer higher-level phylogenetic relationships in the ascomycetes, but their results suggest it is less useful than other genes at this level.

Voigt *et al.* (2005) used β-tubulin gene among other genes to analyze *Leptosphaeria maculans* (anamorph: *Phoma lingam*) species complex as well as Fatehi *et al.* (2003) to refer molecular relatedness within the '*Ascochyta pinodes*-complex'. Partial β-tubulin amino-acid sequences were used by Landvik *et al.* (2001) to assess higher-level phylogenetic relationships in the ascomycetes, but their results suggest it is less suitable than other genes at this level. According to Hansen *et al.* (2004) β-tubulin gene appears less useful at inter-generic level.

Obtaininig a part of the β-tubulin gene according to Glass and Donaldson, (1995) and O'Donnell and Cigelnik (1997) was applied (Irinyi *et al.*, 2008) (Figure 3).

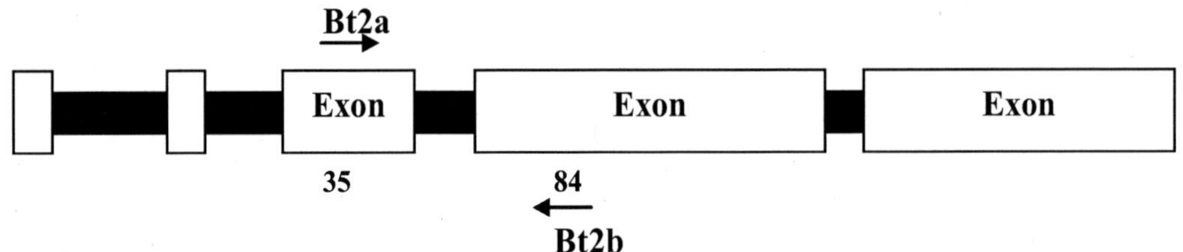

Figure 3. Schematic structure of β-tubulin gene in *Phoma* spp. and location of primers for phylogenetic analyses.

Twenty-two isolates of nine *Phoma* species were tested for phylogenetic analyses (Table 1). All isolates were identified morphologically according to Boerema *et al.* (2004) based on physiological and morphological characteristics.

Mycelium from isolates was transferred to 100 ml Erlenmeyer flasks containing 50 ml malt broth (2% malt exract). The cultures were grown at room temperature for 48 hours in the dark on a rotary shaker (125rpm). The mycelium was harvested by vacuum filtration. Total genomic DNA was extracted from freeze-dried mycelium and isolated using the E.Z.N.A.® TM Fungal DNA Isolation Kit (Omega Bio-tek, Inc., USA) according to the protocol (followed manufacturer instructions).

Primers used to amplify approximately 520bp of the ITS region containing the internal transcribed spacer regions 1 and 2 and the 5.8S rDNA are based on published composite sequences, SR6R and LR1 (White *et al.*, 1990) with the following amplification protocol: 3 min initial denaturing at 95°C, followed by 5 cycles of 1 min at 95°C, 1 min annealing at 50°C, 1 min at 72°C and 25 cycles of 1 min at 90°C, 1 min annealing at 50°C, 1min at 72°C and 15 min final extension at 72°C. The large intron (approx. 300bp) of the *tef1* gene was amplified by the EF1-728F and EF1-986R primer pair (Druzhinina and Kubicek, 2005) according to the following programme: 3 min initial denaturing at 95°C, followed by 5 cycles of 1 min at 95°C, 1 min annealing at 56°C, 1 min at 72°C and 25 cycles of 1 min at 90°C, 1 min annealing at 56°C, 1 min at 72°C and 15 min final extension at 72°C. Primers Bt2a and Bt2b (Glass and Donaldson, 1995) were used to amplify a 300bp fragment of β-tubulin gene and cycling conditions consisted of 3 min initial denaturing at 95°C, followed by 5 cycles of 1 min at 95°C, 1 min annealing at 58°C, 1 min at 72°C and 25 cycles of 1 min at 90°C, 1 min annealing at 58°C, 1min at 72°C and 15 min final extension at 72°C. PCR was carried out in a Primus (MWG Biotech) thermocycler. Amplification products were subjected to electrophoresis in a 0.7% agarose gel containing EtBr and visualized by UV illumination. The PCR products were purified by using YM-100 Microcon Centrifugal Filter Devices (Millipore). Purified amplification products were sequenced by MWG Biotech Company in Germany.

The obtained DNA sequences were aligned first with ClustalX (Thompson *et al.*, 1997) and manually checked for ambiguities and adjusted when necessary using Genedoc (Nicholas *et al.*, 1997). Single gaps were treated either as missing data or as the fifth base and multistate characters were treated as uncertain.

Phylogenetic analyses were performed in PAUP*4.0b (Swofford, 2002). Parsimony analysis (Kluge and Farris, 1969; Fitch, 1971) consisted of heuristic searches with 1000 random addition sequences and tree bisection-reconnection (TBR) branch swapping. All characters were equally weighted and alignment gaps were treated as missing date. Stability of clades was assessed with 1000 bootstrap replications. Phylogenetic trees were drawn by Treeview program (Page, 1996).

Table 1. Isolates of *Phoma* species

Isolate no.	Alternative isolate no.	Species identity	Host	Origin	Collector	GeneBank accession No.
D/035	BT-15	*Phoma pinodella*	*Glycine max*	Hungary	I. Walcz	**EU543973** **EU573015** **EU541416**
D/045	PD82/550	*P. pinodella*	*Hordeum vulgare*	Hungary	G.J. Köevics	**EU543971** **EU573025** **EU541417**
D/046	PD77/165 MYA-411	*P. pinodella*	*Pisum sativum*	Hungary	G.J. Köevics	**EU543972** **EU573024** **EU541419**
D/095	N.A.	*P. pinodella*	*P. sativum*	Hungary	L. Gergely	**EU543970** **EU573027** **EU541418**

Table 1. (Continued)

Isolate no.	Alternative isolate no.	Species identity	Host	Origin	Collector	GeneBank accession No.
D/159	CBS 318.90 PD 81/729	*P. pinodella*	*P. sativum*	Netherland s	M.E. Noordeloos	**EU595355 EU573028 EU595352**
D/054	MYA-406	*P. sojicola*	*G. max*	Hungary	G.J. Köevics	**EU543974 EU573023 EU541434**
D/056	CBS 567.97 PD97/2160	*P. sojicola*	*G. max*	Hungary	G.J. Köevics	**EU543976 EU573026 EU541433**
D/050	CBS 301.39	*Phyllosticta sojicola*	*G. max*	Germany	K. Böning	**EU595356 EU573029 EU595357**
D/075	N.A.	*P. exigua* var. *eigua*	*G. max*	Poland	G.J. Köevics	**EU543982 EU555533 EU541421**
D/077	N.A.	*P. exigua* var. *eigua*	*G. max*	Poland	G.J. Köevics	**EU543983 EU573010 EU541422**
D/063	Ph 58 MYA-408	*P. exigua* var. *eigua*	*Petroselinum crispum*	Poland	J. Marcinkowsk a	**EU543975 EU573012 EU541420**
D/145	N.A.	*P. exigua*	*Althaea officinalis*	Hungary	G. Nagy	– **EU573011 EU541425**
D/146	N.A.	*P. exigua*	*Althaea rosae*	Hungary	G. Nagy	**EU543984 EU573013 EU541427**
D/158	ICMP 15330	*P. exigua* var. *exigua*	*Agapanthus* sp.	New Zealand	M. Braithwaite	**EU543981 EU573008 EU541428**
D/157	ICMP 13336	*P. exigua*	*Cucurbita maxima*	New Zealand	P.G. Broadhurst	**EU543980 EU573007 EU541429**
D/071	PD 86/73	*P. exigua* var. *linicola*	*Linum usitatissimum*	Hungary	G.J. Koevics	**EU543979 EU573009 EU541423**
D/072	PD 75/907	*P. plurivora*	*Medicago sativa*	Australia	J. de Gruyter	**EU552929 EU573018 EU552932**
D/155	ICMP 6875	*P. plurivora*	*Pennisetum clandestinum*	New Zealand	P.R. Johnston	**EU552930 EU573019 EU552931**
D/034	AI-416	*P. glomerata*	*G. max*	Hungary	G.J. Koevics	**EU543969 EU573016 EU541424**
D/156	ICMP 15788	*P. glomerata*	*Yucca sp.*	New Zealand	C.F. Hill	**EU543968 EU573017 EU541426**

Table 1. (Continued)

Isolate no.	Alternative isolate no.	Species identity	Host	Origin	Collector	GeneBank accession No.
D/058	CBS 375.91 PD78/745	*P. eupyrena*	*Phaseolus vulgaris*	Netherlands	G. H. Boerema	**EU543977 EU573014 EU541415**
D/048	PD 76/1021	*P. foveata*	*Chenopodium quinoa*	Netherlands	G. H. Boerema	**EU543985 EU573021 EU541431**
D/044	PD 77/508	*P. multirostrata*	*Phylodendron* sp.	Netherlands	G. H. Boerema	**EU543986 EU573022 EU541430**
D/144	N.A.	*Ascochyta rabiei*	*Cicer arietinum*	Australia	N.A.	**EU595354 EU595358 EU595353**
D/160	CBS 581.83A	*Didymella rabiei*	*C. arietinum*	Syria	H.A. van der Aa, No. H.A.	**EU543978 EU573020 EU541432**

AI refers to Agrobotanical Institute, Tápiószele, Hungary;

BT refers to Fodder-plant Research Institute of Pannon University, Iregszemcse-Bicsérd, Hungary;

CBS refers to Centraalbureau voor Schimmelcultures, Utrecht, The Netherlands;

D refers to the culture collection of the Plant Protection Department of University of Debrecen, Hungary;

N.A. = data not available;

PD refers to Plantenziektenkundige Dienst; Dutch Plant Protection Service Collection, The Netherlands;

GenBank accession numbers are partial sequence data of the translation elongation factor-coding gene (*tef1*), 18S ribosomal RNA gene, partial sequence; internal transcribed spacer 1, 5.8S ribosomal RNA, internal transcribed spacer 2, complete sequence; and 26S ribosomal RNA gene, partial sequence, and partial sequence data of the β-tubulin coding gene.

Some Interesting Findings

Twenty-two isolates of nine *Phoma* species were compared. The morphological identification of the isolates was done following the descriptions of Boerema *et al.* (2004). The obtained results indicated that the microscopic and cultural characteristics of the concerned *Phoma* isolates fit to the identity of *Phoma* species given in Table 1.

Translation Elongation Factor

A 0.3 kb fragment of the large intron of the *tef1* gene from twenty-two isolates of nine *Phoma* species were amplified and sequenced (Irinyi *et al.*, 2006a, 2006b, 2007, 2008). In phylogenetic analyses of *tef1* fragments we involved other *Phoma* and *Ascochyta* species and *Leptosphaeria* species as outgroup; all were downloaded from GenBank maintained by the

NCBI (Table 2). *Didymella fabae* and *Didymella lentis* are the teleomorph of *Ascochyta fabae* and *Ascochyta lentis*, (Kaiser *et al.*, 1997).

Alignment of *tef1* (310bp) revealed 173 parsimony informative sites, 16 polymorphic sites and 121 sites are constant among all isolates (Figure 4).

Phoma species represented by more than one isolates constitute clades which are well separated from each other like *P. pinodella* and *P. exigua* var. *exigua* group, which prove that the *tef1* sequences are well suited for delineating phylogenetic relationships within the *Phoma* genus.

Table 2. Species involved in the phylogenetic analyses of *tef1* fragments

Species	Isolation code	Accession No.
Leptosphaerulina trifolii	WAC 6693	AY831543.1
Ascochyta pisi	AP2	DQ386494.1
teleomorf: *Didymella lentis* anamorf: *Ascochyta lentis* (Kaiser *et al.*, 1997)	SAT AL	AY831546.1
Ascochyta fabae f. sp. *viciae* (= *Ascochyta fabae*)	AV11	DQ386498.1
teleomorf: *Didymella lentis* anamorf: *Ascochyta lentis*	AL1	DQ386493.1
teleomorf: *Didymella fabae* anamorf: *Ascochyta fabae* (Kaiser *et al.*, 1997)	AF1	DQ386492.1
Phoma pinodella	CBS 318.90	AY831542.1
Phoma pinodella	WAC 7978	AY831545.1

The *Phoma* species are well separated from their closely related *Ascochyta* taxa. Most of *Phoma* species (*P. pinodella, P. exigua, P. glomerata, P. plurivora*) are well separated from the other tested *Phoma* species. Some *Phoma* species constitute clades but there are some species, which cannot be distinguished on the basis of *tef1* sequences (*P. pinodella* and *P. sojicola*, *Phyllosticta sojicola* and *P. exigua* as well as *P. foveata* and *P. multirostrata*).

ITS SEQUENCES

In the PCR reaction 0.5kb fragment of the rDNA gene containing the internal transcribed spacer regions 1 and 2 and the 5.8S regions were amplified and sequenced (Irinyi *et al.*, 2006a, 2006b, 2007, 2008).

In phylogenetic analyses of ITS region we involved other *Phoma* and *Ascochyta* species as well as *Didymella* and *Leptosphaeria* species as outgroup, all were downloaded from GenBank maintained by the NCBI (Table 3). The phylogenetic tree based on ITS sequences (Figure 5) is similar to that of tree based on *tef1* sequences.

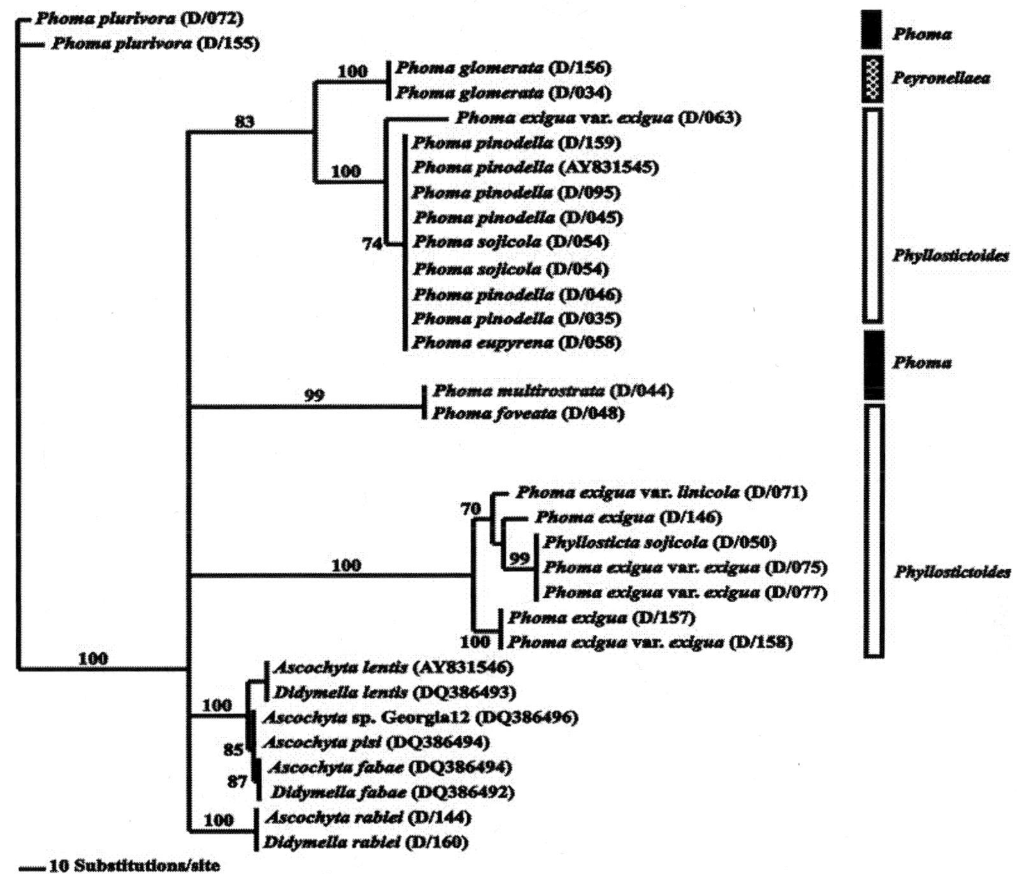

Figure 4. Phylogenetic relationships of *Phoma* strains inferred by the parsimony analysis of *tef1* sequences. The numbers above the lines represent the bootstrap values from 1000 bootstrap samples. The columns on the right side show the *Phoma* sections based on morphological characterization.

Alignment of ITS (454bp) revealed 32 parsimony informative sites, 5 polymorphic sites and 417 sites are constant among all isolates.

Species represented by more than one isolates constitute clades, which are well separated from each other. *Phoma* species are also well separated from their closely related *Ascochyta* taxa.

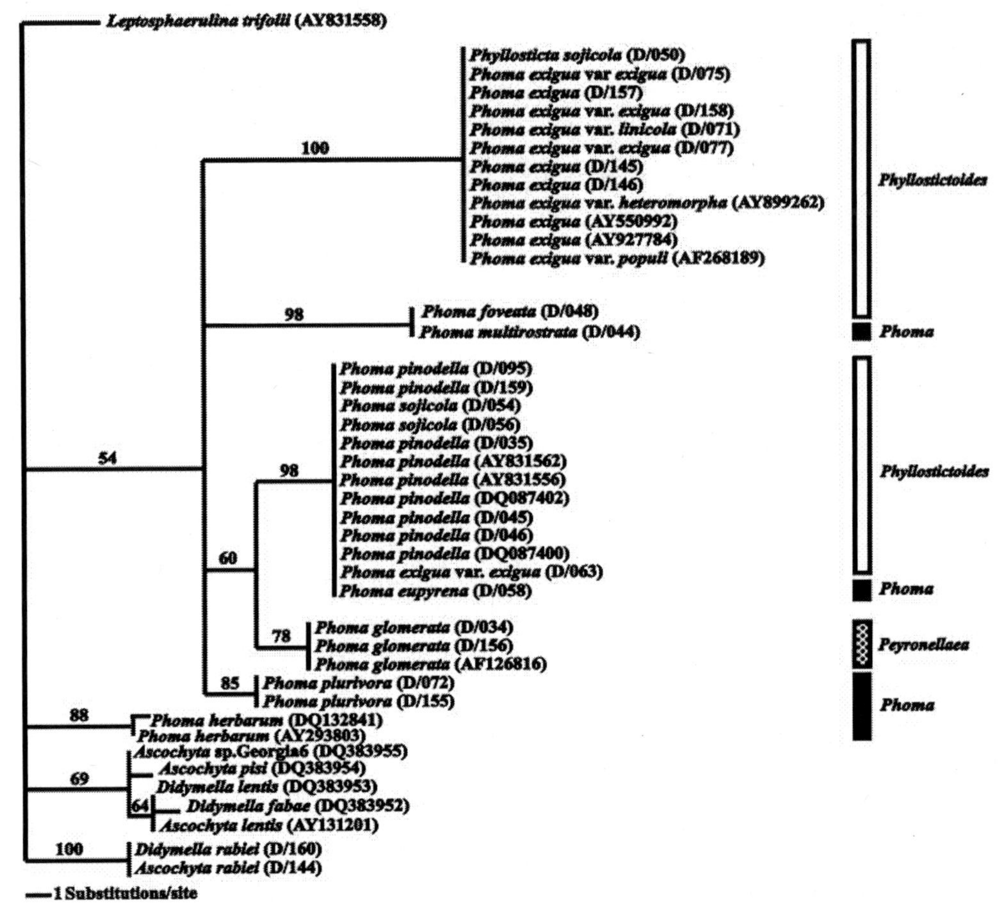

Figure 5. Phylogenetic relationships of *Phoma* strains inferred by the parsimony analysis of ITS sequences. The numbers above the lines represent the bootstrap values from 1000 bootstrap samples. The columns on the right side show the *Phoma* section based on morphological characterization.

β-Tubulin Sequences

In the PCR reaction 0.3kb fragment of the β-tubulin region was amplified and was subjected to phylogenetic analysis in which we involved other *Phoma* sequences downloaded from GenBank maintained by the NCBI (Irinyi *et al.*, 2008) (Table 4).

Alignment of β-tubulin (298bp) revealed 49 parsimony informative sites, 20 polymorphic sites and 229 sites are constant among all isolates.

Table 3. Species involved in the phylogenetic analyses of ITS fragments

Species	**Isolation code**	**Accession number**
Phoma exigua var. *heteromorpha*	N.A.	AY899262.1
Phoma exigua	CSL 20316964	AY550992.1
Phoma exigua var. *populi*	CBS 100167	AF268189.1
Phoma exigua	N.A.	AY927784.1
Phoma herbarum	N.A.	DQ132841.1
Phoma herbarum	ATCC 12569	AY293803.1
Phoma pinodella	VPRI 32177	DQ087402.1
Phoma pinodella	VPRI 32171	DQ087400.1
Phoma pinodella	WAC 7978	AY831556.1
Phoma pinodella	CBS 318.90	AY831562.1
Phoma glomerata	N.A.	AF126816.1
Ascochyta sp.	Georgia6	DQ383955.1
Ascochyta pisi	AP1	DQ383954.1
Ascochyta lentis	MU AL1	AY131201.1
Didymella lentis	AL1	DQ383953.1
Didymella fabae	AF1	DQ383952.1
Leptosphaerulina trifolii	WAC 6693	AY831558.1

N.A. = data not available.

The phylogenetic tree obtained by the analysis of β-tubulin sequences (Figure 6) is really similar to that of the two previous trees. Species represented by more than one isolate are also clustered in the same clades and they are well separated from each other. However *P. pinodella* and *P. sojicola*, *Phyllosticta sojicola* and *Phoma exigua* as well as *P. foveata* and *P. multirostrata* are also not distinguishable. *Ascochyta* isolates constitute well separated clades revealing that β-tubulin is also appropriate for inferring phylogenetic relationships in the genus.

Table 4. Species involved in the phylogenetic analyses of ITS fragments

Species	**Isolation code**	**Accession number**
Phoma pinodella	CBS 318.90	AY831517
Phoma pinodella	WAC 7978	AY831511
Phoma exigua	WAC 7988	AY831509
Phoma medicaginis	CBS 316.90	AY831518
Phoma medicaginis var. *medicaginis*	P3	DQ109962
Ascochyta lentis	SAT AL	AY831508
Leptosphaerulina trifolii	WAC 6693	AY831513

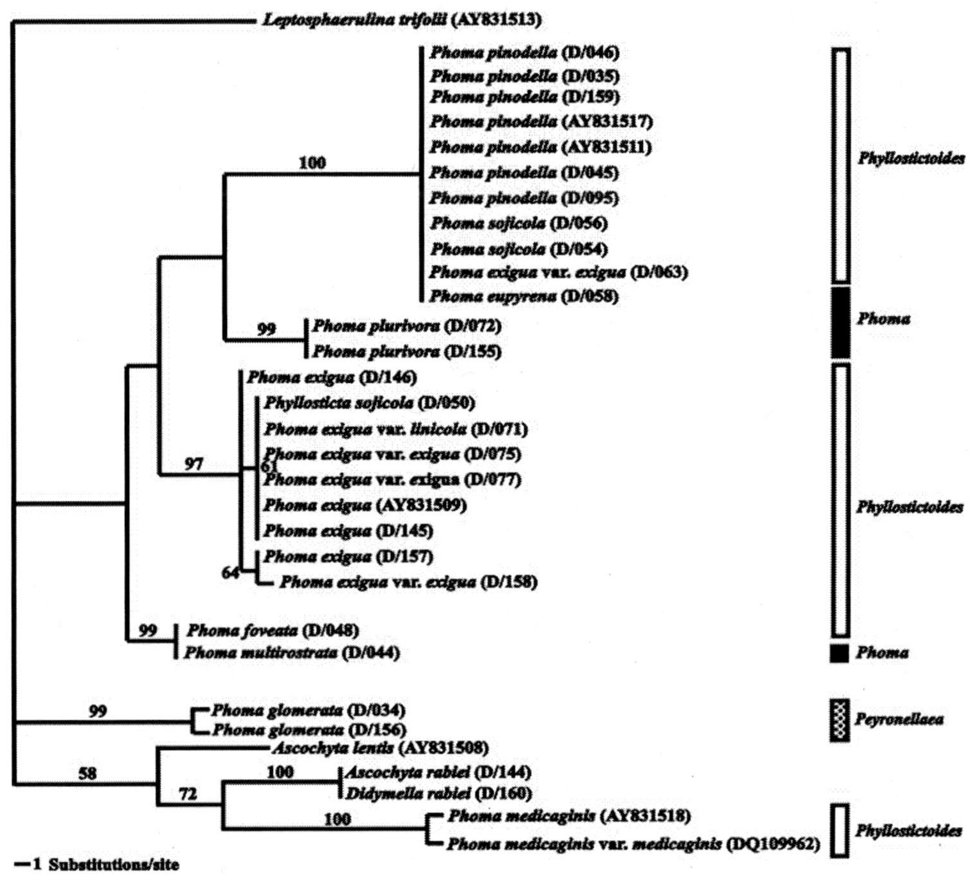

Figure 6. Phylogenetic relationships of *Phoma* strains inferred by the parsimony analysis of β-tubulin sequences. The numbers above the lines represent the bootstrap values from 1000 bootstrap samples. The columns on the right side represent the *Phoma* section based on morphological characterization.

Up to now, phylogenetic analyses within *Phoma* genus have only been used for defining phylogenetic relationships among isolates within one or closely related species (Fatehi *et al.*, 2003; Mendes-Pereira *et al.*, 2003 and Balmas *et al.*, 2005; Voigt *et al.*, 2005).

Previous publications (Irinyi *et al.*, 2006a, 2006b, 2006c, 2007, 2008) have been used *tef1*, ITS and β-tubulin sequences to resolve phylogenetic relationships within *Phoma* genus at higher taxonomic levels. These studies proved these phylogenetic markers are useful for defining *Phoma* species.

All *Phoma* species are well separated from their closely related *Ascochyta* taxa. As the identification of *Phoma* and *Ascochyta* genus based on morphological characteristics is often problematic, this new phylogenetic marker can be a useful tool for mycologists identifying an unknown species.

However, the phylogenetic tree does not support the traditional *Phoma* sections based on morphological characterization.

Conclusions

Phoma species have been reported ubiquitously from plants, soil, aerial and aquatic environment and from human and animals causing opportunistic infections. Moreover, some species of *Phoma,* are biotechnologically important, particularly for the production of pharmaceutically active fungal metabolites. Therefore, authentic identification is required for correct diagnosis. Some species of *Phoma* produces antibiotics, like squalestatin. Fusidienol-A is an antitumour agent produced by *Phoma* sp. Equisetin and phomasetin produced also by species of *Phoma, which* is useful against AIDS. Application of species of *Phoma* for biological control of weeds is another area of research, which warrants immediate attention.

The identification and delimitation of *Phoma*-like fungi based on morphological markers is time consuming and sometimes misapplied, therefore there is a pressing need to develop rapid molecular methods for their proper identification.

References

Aa, H.A. van der and Vanev, S (2002). A revision of the species described in *Phyllosticta.* Centralbureau voor Schimmelcultures, Utrecht.

Abeln, E.C.A., Stax, A.M., Gruyter, J. de, and Aa, H.A. van der (2002). Genetic differentiation of *Phoma exigua* varieties by means of AFLP fingerprints. *Mycol. Res. 106*(4): 419-427.

Agarwal, G.P. and Sahni, V.P. (1964). Fungi causing plant diseases at Jabalpur (M.P.). *Mycopath. et Mycol. Appl. 22*(2): 245-247.

Agrawal, S.C. and Misra, J.K. (1981). *Phoma lucknowensis* sp. nov. from Indian alkaline soils. *Curr. Sci. 50*(13): 600.

Ayliffe, M. A., Dodds, P. N., Lawrence, G. J. (2001). Characterisation of a beta-tubulin gene from *Melampsora lini* and comparison of fungal beta-tubulin genes. *Mycol. Res. 105*: 818–826.

Baker, J.G., Salkin, I.F., Forgacs, P., Haines, J.H., and Kemna, M.E (1987). First report of subcutaneous phaeohyphomycosis of the foot caused by *Phoma minutella. J. Cli. Microbiol. 25*(120): 2395-2397.

Baldauf, S.L. and Doolittle, W.F. (1997). Origin and evolution of slime molds (Mycetozoa). *Proc. Natl. Acad. Sci. USA. 94*: 12007-12012.

Baldauf, S. L., Roger, A. J., Wenk-Siefert, I., Doolittle, W. F. (2000). A kingdom-level phylogeny of eukaryotes based on combined protein data. *Science 290*: 972–977.

Baldwin, B.G. (1992). Phylogenetic utility of the internal transcribed spacers of nuclear ribosomal DNA in plants: an example from the Compositae. *Mol. Phylogenet. Evol. 1*: 3-16.

Balmas, V., Scherm, B., Ghignone, S., Salem, A.O.M., Cacciola, S.O., and Migheli, Q. (2005). Characterisation of *Phoma tracheiphila* by RAPD PCR, microsatellite-primed PCR and ITS rDNA sequencing and development of species primers for *in planta* PCR detection. *Eur. J. Plant Pathol. 111*: 235-247.

Baxter, A., Fitzgerald, B.J., Hutson, J.L., McCarthy, A.D., Motteram, J.M., Ross, B.C., Sapra, M., Snowden, M.A., Watson, N.S., and Williams, R.J. (1992). Squalestatin 1, a potent inhibitor of squalene synthase, which lowers serum cholesterol *in vivo*. *J. Biol. Chem. 267*(17): 11705-11708.

Bhowmik, T.P. and Singh, A. (1976). Occurrence of *Phoma exigua* Desm. on sunflower. *Curr. Sci. 45*: 776-777.

Bick, I.R.C. and Rhee, C. (1966). Anthraquinone pigments from *Phoma foveata* Foister. *Biochem. J. 98*: 112-116.

Bilgrami, K.S. (1963). Association of a new species of *Phoma* with *Pleospora herbarum* (Pers.) Rahb. *Curr. Sci. 32*: 174-175.

Boerema, G.H. (1964). *Phoma herbarum* Westend., type-species of the form-genus *Phoma* Sacc. *Persoonia 3*: 9-16.

Boerema, G.H. (1970). Additional notes on *Phoma herbarum. Persoonia 6*: 15-28.

Boerema, G.H. (1976). The *Phoma* species studied in culture by Dr R.W.G. Dennis. *Trans. Br. Mycol. Soc. 67*: 289-319.

Boerema, G.H. (1986). Mycologisch-taxonomisch onderzoek. Een subtropische bodemschimmel die zich ook thuis voelt in onze kassen. *Versl. Meded. Plantenziektenkundige Dienst Wageningen 164*: (Jaarb. 1985) 28-32.

Boerema, G.H. and Bollen, G.J. (1975). Conidiogenesis and conidial septation as differentiating criteria between *Phoma* and *Ascochyta. Persoonia 8*: 111-144.

Boerema, G.H. and Dorenbosch, M.M.J. (1970). On *Phoma macrostomum* Mont., an ubiquitous species on woody plants. *Persoonia 6*: 49-58.

Boerema, G.H and Dorenbosch, M.M.J. (1973). The *Phoma* and *Ascochyta* species described by Wollenweber and Hochapfel in their study on fruit rotting. *Studies in Mycol. 3*: 1-50.

Boerema, G.H. and Höweler, L.H. (1967). *Phoma exigua* Desm. and its varieties. *Persoonia 5*: 15-28.

Boerema, G.H., Dorenbosch M.M.J., and Kesteren, H.A. van (1968). Remarks on species of *Phoma* referred to *Peyronellaea* II. *Persoonia 5*: 201-205.

Boerema, G.H., Gruyter, J. de, Noordeloos, M.E., and Hamers, M.E.C. (2004). *Phoma* identification manual. CABI Publishing. CAB International Wallingford, Oxfordshire, UK

Boerema, G.H. and Loerakker, W.M. (1985). Notes on *Phoma* 2. *Trans. Brit. Mycol. Soc. 84*: 289-302.

Byrd, A.D., Schardl, C.L., Songlin, P.J., Mogen, K.L., Siegel, M.R. (1990). The β-tubulin gene of *Epichloë typhina* from perennial ryegrass (*Lolium perenne*). *Curr. Gen. 18*: 347–354.

Chandra, S. and Tandon, R.N. (1965). Two new leaf spot fungi. *Curr. Sci. 34*: 365-366.

Chandra, S. and Tandon, R.N. (1966). Three new leaf infecting fungi from Allahabad. *Mycopath. et Mycol. Appl. 29*: 273-276.

Chen, C.A., Yu, J.K., and Wei, N.W. (2000): Strategies for amplification by polymerase chain reaction of the complete sequence of nuclear large subunit ribosomal RNA-encoding gene in corals. *Mar. Biotechnol. 6*: 558-570.

Chen, C.A., Wallace, C.C., and Wolstenholme, J. (2002). Analysis of mitochondrial 12S RNA gene supports a two-clade hypothesis of the evolutionary history of scleractinian corals. *Mol. Phylogenet. Evol. 23*: 137-149.

Deshmukh, P., Rai, M.K., Kövics, G.J., Irinyi, L. and Sándor, E. (2006): *Phoma*s – can these fungi be used as biocontrol agents and sources of secondary metabolites? (A review). pp. 224-232. *In: 4th International Plant Protection Symposium at Debrecen University, Recent Developments of IPM. Proceedings*. Kövics, G.J. – Dávid, I. (Eds.). Debrecen University Centre for Agricultural Science, Faculty of Agriculture. 18-19 October, 2006, Debrecen, Hungary

Dorenbosch, M.M.J. (1970). Key to nine ubiquitous soil-borne *Phoma*-like fungi. *Persoonia 6*: 1-14.

Druzhinina, I. and Kubicek, C.P. (2005). Species concepts and biodiversity in *Trichoderma* and *Hypocrea*: from aggregate species to species cluster. *J. Zhejiang Univ. Sci.* 6B (2): 100-112.

Dutta, B.G. and Ghosh, G.R. (1965). Soil fungi from Orissa (India)-IV. Soil fungi of paddy fields. *Mycopath. et Mycol. Appl. 25*: 316-322.

Edgcomb, V.P., Roger, A.J., Simpson, A.G.B., Kysela, D.T., and Sogin, M.L. (2001). Evolutionary relationships among "jakobid" flagellates as indicated by alpha- and beta-tubulin phylogenies. *Mol. Biol. Evol. 18*: 514–522.

Fatehi, J., Bridge, P.D., Punithalingam, E. (2003). Molecular relatedness within the "*Ascochyta pinodes*"-complex. *Mycopathol. 156*: 317-327.

Fitch, W.M. (1971). Toward defining the course of evolution: Minimum change for a specific tree topology. *Syst. Zool. 20*: 406-416.

Gardes, M. and Bruns, T.D. (1993). ITS primers with enhanced specificity for basidiomycete-application to the identification of mycorrhizae and rusts. *Mol. Ecol. 2*: 113-118.

Gerbi, S.A. (1985). Evolution of ribosomal DNA. pp. 419-517. *In*: Molecular evolutionary genetics. Macintyre, R.J. (Ed.), Plenum, New York.

Glass, N.L. and Donaldson, G.C. (1995). Development of primer sets designed for use with the PCR to amplify conserved genes from filamentous Ascomycetes. *Appl. Env. Microbiol.* 1323-1330.

Gottlieb, A.M. and Lichtwardt, R.W. (2001). Molecular variation within and among species of Harpellales. *Mycologia 93*(1): 66-81.

Graser, Y., El Fari, M., Vilgalys, R., Kujipers, A.F.A., Hoog, G.S. de, Presber, W., and Tietz, H.J. (1999). Phylogeny and taxonomy of the family Arthrodermataceae (dermatophytes) using sequence analysis of the ribosomal ITS region. *Med. Mycol. 37*: 105-114.

Gruyter, J., Boerema, G.H., and Aa, H.A. van der (2002). Contributions toward a monograph of *Phoma* (Coelomycetes) – VI-2. Section *Phyllostictoides*: Outline of its taxa. *Persoonia 18* (1): 1-53.

Hansen, K., LoBuglio, K.F., and Pfister, D.H. (2004). Evolutionary relationships of the cup-fungus *Peziza* and Pezizaceae inferred from multiple nuclear genes: RPB2, β-tubulin, and LSU rDNA. *Mol. Phylogen. Evol. 36*: 1-23.

Hasija, S.K. (1966). Additions to the fungi of Jabalpur, M.P.-V. *Mycopath. et Mycol. Appl. 28*: 33-41.

Hillis, D.M. and Dixon, M.T. (1991). Ribosomal DNA: molecular evolution and phylogenetic inference. *Q. Rev. Biol. 66*: 411-453.

Irinyi, L., Kövics, G.J., Rai, M.K., and Sándor, E. (2006a). Studies of evolutionary relationships of *Phoma* species based on phylogenetic markers. pp. 99-113. *In: 4th International Plant Protection Symposium at Debrecen University, Recent Developments of IPM. Proceedings*. Kövics, G.J. – Dávid, I. (Eds.). Debrecen University Centre for Agricultural Science, Faculty of Agriculture. 18-19 October, 2006, Debrecen, Hungary

Irinyi, L., Kövics, G.J., and Sándor, E. (2006b). Classification of *Phoma* species using new phylogenetic marker. *Analele Universittății Din Oradea* Volume XII, Anul 12: 91-97.

Irinyi L., Kövics, G.J., and Sándor, E. (2006c). New phylogenetic marker for the classification of *Phoma* species. pp. 253-260. *In: "The 4th International Symposium on Natural Resources and Sustainable Development"*, 10-11 October, 2006, Oradea, Romania

Irinyi, L., Kövics, G.J., and Sándor, E. (2007). A phylogenetic study on different *Phoma* species. 15th International Congress of the Hungarian Society for Microbiology, July 18-20, 2007, Budapest. *Acta Microbiol. et Immunol. Hung. 54*: 51.

Irinyi, L., Kövics, G., Sándor, E. (2008). Bayesian módszer alkalmazása a *Phoma* taxonok filogenetikai vizsgálatában. (Application of Bayesian method in *Phoma* phylogenetics. pp. 4-12. In: *XVIII. Keszthelyi Növényvédelmi Fórum, Keszthely, 2008. január 30.-február 1., Pannon Egyetem, Keszthely (Proceedings of the 18th Keszthely Plant Protection Forum, Pannon University, Keszthely, Hungary)*

Jamaluddin, M., Tandon, P. and Tandon, R.N. (1975). A fruit rot of Aonla (*Phyllanthus emblica* L.) caused by *Phoma* sp. *Proc. Nat. Acad. Sci. India 45* (B): 75-76.

Jong, S.N., Lévesque, C.A., Verkley, G.J.M., Abeln, E.C.A., Rahe, J.E., Braun, P.G. (2001). Phylogenetic relationships among *Neofabraea* species causing tree cankers and bull's-eye rot of apple based on DNA sequencing of ITS nuclear rDNA, mitochondrial rDNA, and the β-tubulin gene. *Mycol. Res. 105*: 658-669.

Kaiser, W.J., Wang, B.C., and Rogers, J.D. (1997). *Ascochyta fabae* and *A. lentis*: Host specificity, teleomorphs (*Didymella*), Hybrid analysis, and taxonomic status. *Plant Dis. 81*: 809-816.

Kamal, S. and Singh, R.P. (1979). Fungi of Gorakhpur VII. *Indian J. Mycol. Pl. Pathol. 9*: 170-172.

Kanaujia, R.S. (1979). Notes on a new fungal disease of *Alocacia indica. Indian Phytopath. 32*: 463-464.

Keeling, P.J. and Doolittle, W.F. (1996). Alpha-tubulin from early-diverging eukaryotic lineages and the evolution of the tubulin family. *Mol. Biol. Evol. 13*: 1297-1305.

Keeling, P.J., Deane, J.A., and McFadden, G.I. (1998). The phylogenetic position of alpha- and beta-tubulins from the *Chlorarachnion* host and *Cercomonas* (Cercozoa). *J. of Eukaryotic Microbiol. 45*: 561-570.

Kluge, A.G. and Farris, J.S. (1969). Quantitative phyletics and the evolution of anurans. *Syst. Zool. 18*: 1-32.

Knutsen, A.K., Torp, M., and Holst-Jensen, A. (2004). Phylogenetic analyses of *Fusarium poae*, *Fusarium sporotrichioides* and *Fusarium langsethiae* species complex based on partial sequences of the tranlation elongation factor-1 alpha gene. *Int. J. Food Microbiol. 95*: 287-295.

Kövics, G.J. and Gruyter, J. de (1995). Comparative studies of esterase isozyme patterns of some *Phoma* species occurred on soybean. *DATE Tudományos Közleményei (Bulletin of Debrecen Agricultural University) 31*: 191-207.

Kövics, G.J., Gruyter, J. de, and Aa, H.A. van der (1999): *Phoma sojicola* comb. nov. and other hyaline-spored coelomycetes pathogenic on soybean. *Mycol. Res. 103*(8): 1065-1070.

Lutzoni, F., Kauff, F., Cox, C.J., McLaughlin, D., Celio, G., Dentinger, B., Padamsee, M., Hibbett, D., James, T.Y., Baloch, E., Grube, M., Reeb, V., Hofstetter, V., Schoch, C., Arnold, A. E., Miadlikowska, J., Spatafora, J., Johnson, D., Hambleton, S., Crockett, M., Shoemaker, R., Sung, G.H., Lucking, R., Lumbsch, T., O'Donnell, K., Binder, M., Diederich, P., Ertz, D., Gueidan, C., Hansen, K., Harris, R.C., Hosaka, K., Lim, Y.W., Matheny, B., Nishida, H., Pfister, D., Rogers, J., Rossman, A., Schmitt, I., Sipman, H., Stone, J., Sugiyama, J., Yahr, R., and Vilgalys, R. (2004). Assembling the fungal tree of life: progress classification and evolution of subcellular traits. *Am. J. Bot. 91*: 1446-1480.

Mages, W., Cresnar, B., Harper, J. F., Brüderlein, M., and Schmitt, R. (1995). *Volvox carteri* alpha-2-tubulin-encoding and beta-2-tubulin-encoding genes: regulatory signals and transcription. *Gene 160*: 47–54.

Mai, J.C. and Coleman, A.W. (1997): The internal transcribed spacer 2 exhibits a common secondary structure in green algae and flowering plants. *J. Mol. Evol. 44*: 258-271.

Maiti, S., Dinesh, C., Khatua and Chitreshwar, S. (1978). A new leaf-spot disease of betel vine. *Indian J. Mycol. Pl. Pathol. 8*(2): 217.

Marasas, W.F.O., Pauer, G.D., and Boerema, G.H. (1974). A serious leaf blotch disease of groundnuts (*Arachis hypogaea* L.) in southern Africa caused by *Phoma arachidicola* sp. nov. *Phytophylactica 6*: 195-200.

May, G.S., Tsang, M.L.S., Smith, H., Fidel, S., and Morris, N.R. (1987). *Aspergillus nidulans* β-tubulin genes are unusually divergent. *Gene 55*: 231–243.

McKean, P.G., Vaughan, S., and Gull, K. (2001). The extended tubulin superfamily. *J. Cell Sci. 114*: 2723-2733.

Mendes-Pereira, E., Balesdent, M.H., Brun, H. and Rouxel, T. (2003). Molecular phylogeny of the *Leptosphaeria maculans – L. biglobosa* species complex. *Mycol. Res. 107*(11): 1287-1304.

Moldave, K. (1985): Eukaryotic protein synthesis. *Annu. Rev. Biochem. 54*: 1109-1149.

Monte, E., Bridge, P.D., and Sutton, B.C. (1991). An integrated approach to *Phoma* systematics. *Mycopathologia 115*: 89-103.

Morgan-Jones, G. and Burch, K.B. (1988). Studies on the genus *Phoma*. X. Concerning *Phoma eupyrena*, an ubiquitous, soilborne species. *Mycotaxon 31*: 427-434.

Morocko, I. and Fatehi, J. (2007). Molecular characterization of strawberry pathogen *Gnomonia fragariae* and its genetic relatedness to other *Gnomonia* species and members of Diaporthales. *Mycol. Res. 111*: 603-614.

Narendra, D.V. and Rao, V.J. (1974). A new entomogenous species of *Phoma. Mycopath. et Mycol Appl. 54*: 135-140.

Neff, N.F., Thomas, J.H., GrisaW, P., Botstein, D. (1983). Isolation of the β-tubulin gene from yeast and demonstration of its essential function *in vivo*. *Cell 33*: 211–219.

Nicholas, K.B., Nicholas, H.B. Jr., and Deerfield, D.W. II. (1997). GeneDoc: Analysis and Visualization of Genetic Variation, *Embnew News*. *4*: 14.

Noordeloos, M.E., Gruyter, J. de, Eijk, G.W. van, and Roeijmans, H.J. (1993). Production of dendritic crystals in pure cultures of *Phoma* and *Ascochyta* and its value as a taxonomic character relative to morphology, pathology and cultural characteristics. *Mycol. Res. 97*: 1343-1350.

Nugent, K.G. and Saville, B.J. (2004). Forensic analysis of hallucinogenic fungi: a DNA-based approach. *Forensic Science International 140*: 147-157.

O'Donnell, K. and Cigelnik, E. (1997). Two divergent intragenomic rDNA ITS2 types within a monophyletic lineage of the fungus *Fusarium* are nonorthologous. *Mol. Phylogen. Evol. 7*(1): 103-116.

O'Donnell, K., Cigelnik, E., and Nirenberg, H.I. (1998). Molecular systematics and phylogeography of the *Gibberella fujikuroi* species complex. *Mycologia 90*: 465-493.

Oliverio, M., Cervelli, M., and Mariottini, P. (2002). ITS2 rRNA evolution and its congruence with the phylogeny of muricid neogastropods (Caenogastropoda, Muricoidea). *Mol. Phylogenet. Evol. 25*: 63-69.

Orbach, M.J., Porro, E.B., and Yanofsky, C. (1986). Cloning and characterization of the gene for b-tubulin from a benomyl-resistant mutant of *Neurospora crassa* and its use as a dominant selectable marker. *Mol. Cell. Biol. 6*: 2452-2461.

Padamsee, M., Matheny, P.B., Dentinger, B.T.M., and McLaughlin, D. (2008). The mushroom family Psathyrellaceae: Evidence for large-scalepolyphyly of the genus *Psathyrella*. *Mol. Phylogen. Evol. 46*: 415-429.

Page, R.D.M. (1996). TREEVIEW: An application to display phylogenetic trees on personal computers. *Comp. Appl. in the Biosci. 12*: 357-358.

Panaccione, D.G. and Hanau, R.M. (1990). Characterization of two divergent β-tubulin genes from *Colletotrichum graminicola*. *Gene 86*: 163-170.

Pavgi, M.S. and Singh, U.P. (1966). Parasitic fungi from north India. VII. *Mycopath. Mycol. Appl. 30*: 261-270.

Pawar, V.H., Mathur, P.N., and Thirumalachar, M.J. (1967). Species of *Phoma* isolated from marine soil in India. *Trans. Br. Mycol. Soc. 50*: 259-265.

Pearce, C. (1997). Biologically active fungal metabolites. *Adv. Appl. Microbiol. 44*: 1-80.

Rai, M.K. (1981). Studies on some Indian Sphaeropsidales with special reference to *Phoma* and related fungi. Ph.D. Thesis, University of Jabalpur, Jabalpur, Madhya Pradesh, India.

Rai, M.K. (1985). Taxonomic studies of species of *Phoma* isolated from air. *J. Econ. Tax. Bot. 7*(3): 645-647.

Rai, M.K. (1986a). Two new diseases of *Albizzia lebbek* and *Jasminum sambac*. *Acta Bot. Indica 14*: 238-239.

Rai, M.K. (1986b). A new report of *Phoma pomorum* Thüm. from India. *Geo-bios New Reports 5*: 157.

Rai, M.K. (1987). A first report of *Phoma herbarum* from tank water of Pune. *Geo-bios New Reporst 6*: 66-67.

Rai, M.K. (1989). *Phoma sorghina* infection in human beings. *Mycopath. 105*: 167-170.

Rai, M.K. (1998). The genus *Phoma* (Identity and Taxonomy). International Book Distributors, Dehra Dun, India.

Rai, J.N. and Misra, J.K. (1981). A new species of *Phoma* from Indian alkaline soil. *Curr. Sci. 50*: 377.

Rai, M.K. and Rajak, R.C. (1982a). A new leaf-spot disease of *Ailanthus excelsa. Curr. Sci. 51*(2): 98-99.

Rai, M.K. and Rajak, R.C. (1982b). A report of leaf-spot disease of *Holoptelea integrefolia* Planch caused by *Phoma sorghina. Indian J. Mycol. Pl. Pathol. 12*(2): 342.

Rai, M.K. and Rajak, R.C. (1983a). Studies on two species of *Phoma* from aquatic environment. *Geo-bios New Reports 2*(28): 135.

Rai, M.K. and Rajak, R.C. (1983b). A first report of two species of *Phoma* from tank water. *Geo-bios New Reports 2*(28): 136-137.

Rai, M.K. and Rajak, R.C. (1986a). *Phoma multirostrata* on three new hosts. *J. Econ. Tax. Bot.* 8(1): 229-230.

Rai, M.K. and Rajak, R.C. (1986b). A new disease of *Citrus medica* and the identity of its causal organism *Phoma exigua* var. *foevata* (Foister) Boerema. *Indian J. Mycol. Pl. Pathol. 17*(3): 320-321.

Rai, M.K. and Rajak, R.C. (1993). Distinguishing characteristics of selected *Phoma* species. *Mycotaxon 15*(8): 389-414.

Rajak, R.C. and Rai, M.K. (1982a). Species of *Phoma* from legumes. *Indian Phytopath. 35*(4): 609-611.

Rajak, R.C. and Rai, M.K. (1982b). Effect of different colours of light and temperature on the morphology of three species of *Phoma* in vivo. *Jabalpur Univ. J. 1*: 1-6.

Rajak, R.C. and Rai, M.K. (1983a). Effect of different factors on the morphology and cultural characters of 18 species and 5 varieties of *Phoma.* I. Effect of different media. *Bibl. Mycol. 91*: 301-317.

Rajak, R.C. and Rai, M.K (1983b). Cholesterol content in twenty species of *Phoma. Nat. Acad. Sci. Let. 6* (4): 113-114.

Rajak, R.C. and Rai, M.K. (1984a). Effect of different factors on the morphology and cultural characters of *Phoma.* II. Effect of different hydrogen-ion-concentrations. *Nova Hedwigia 40*: 299-311.

Rajak, R.C. and Rai, M.K. (1984b). A new leaf-spot disease of *Jasminum pubescence* caused by *Phoma herbarum. Ind. J. Mycol. Pl. Pathol. 14*(2): 173.

Rajak, R.C. and Rai, M.K. (1985). A key to the identification of species of *Phoma* in pure culture. *J. Econ. Tax. Bot. 7*(3): 588-590.

Rao, S. and Thirumalachar, U. (1981). *Phoma exigua* infecting brinjal leaves. *Indian Phytopath. 34*: 37.

Raut, J.G. (1977). *Phoma* leaf blight of sunflower. *Indian Phytopath. 30*: 269-270.

Rishi, K. and Font, R.L. (2003). Keratitis caused by an unusual fungus, *Phoma* species. *Cornea. 22*(2): 166-168.

Roger, A.J., Sandblom, O., Doolittle, W.F., Philippe, H. (1999). An evaluation of elongation factor 1 as a phylogenetic marker for eukaryots. *Mol. Biol. Evol. 16*: 218-233.

Saccardo, P.A. (1880). Conspectus generum fungorum Italiae inferiorum, nempe ad Sphaeropsideas, Melanconieas et Hyphomyceteas pertinentium, systemate sporologico dispositorum. *Michelia 2*: 1-38.

Schlötterer, C., Hauser, M., Haeseler, A. von, Tautz, D. (1994). Comparative evolutionary analysis of rDNA ITS regions in *Drosophila*. *Mol. Biol. Evol. 11*: 513-522.

Schoch, C.L., Crous, P.W., WingWeld, B.D., and WingWeld, M.J. (2001). Phylogeny of *Calonectria* based on comparisons of β-tubulin DNA sequences. *Mycol. Res. 105*: 1045-1052.

Schutze, J., Krasko, A., Custodio, M.R., Efremova, S.M., Muller, I. M., Muller, W.E.G. (1999). Evolutionary relationships of Metazoa within the eukaryotes based on molecular data from Porifera. *Proc. Royal Soc., London* B *266*: 63-73.

Shinohara, M.L., LoBuglio, K., and Rogers, S. (1999). Comparison of ribosomal DNA ITS regions among geographic isolates of *Cenococcum geophilum*. *Curr. Gen. 35*: 527-535.

Shreemali, J.L. (1972). Two new pathogenic fungi causing diseases on Indian medicinal plants. *Ind. J. Mycol. Pl. Pathol. 2*: 84-85.

Shreemali, J.L. (1973). Sphaeropsidean fungi obtained from Indian phylloclades. *Indian Phytopath. 26*: 56-62.

Shreemali, J.L. (1978). Two new species of *Phoma* from India. *Indian J. Mycol. Pl. Pathol. 8*: 220-221.

Shukla, N.P., Rajak, R.C., Agarwal, G.P., and Gupta, D.K. (1984). *Phoma minutispora* as a human pathogen, *Mykosen 27*: 255-258.

Singh, S.B., Ball, R.G., Zink, D.L., Monaghan, R.L., Polishook, J.D., Sanchez, M., Pelaez, F., Silverman, K.C., and Lingham, R.B. (1997). Fusidienol - A: a novel Ras farnesyl-protein-transferase inhibitor from *Phoma* sp. *J. Org. Chem. 62*(21): 7485-7488.

Singh, S.B., Zink, D.L., Goetz, M.A., Dombrowski, A.W., Polishook, J.D., and Hazuda, D.J.S. (1998). Equisetin and a novel opposite stereochemical homolog phomasetin, two fungal metabolites as inhibitors of HIV-1 integrase. *Biotechnol. Abst.* No. *98*: 06361.

Skovgaard, K., Bodker, L., and Rosendahl, S. (2002). Population structure and pathogenicity of members of the *Fusarium oxysporum* complex isolated from soil and root necrosis of pea (*Pisum sativum* L.). *FEMS Microbiol. Ecol. 42*: 367-374.

Sponga, F., Cavaletti, L., Lazzarini, A., Borghi, A., Ciciliato, I., Losi, D., and Marinelli, F. (1999). Biodiversity and potentials of marine-derived microorganisms. *J. Biotechnol. 70* (1-3): 65-69.

Sugano, M., Sato, A., Lijima, Y., Oshima, T., Furuya, K., Kuwano, H., Hata, T. and Hanzawa, H. (1991). Phomactin A: a novel PAF antagonist from a marine fungus *Phoma* sp. *J. Am. Chem. Soc. 113*(14): 5463-5464.

Sutton, B.C. and Sandhu, D.K. (1969). Electron microscopy and conidium development and secession in *Cryptosporiopsis* sp., *Phoma fumosa*, *Melanconium bicolor* and *M. apiocarpum*. *Can. J. Bot. 47*: 745-749.

Swofford, D.L. (2002). PAUP: Phylogenetic Analysis Using Parsimony (and other methods). Version 4b10. Sinauer Associates, Sunderland, MA

Takamatsu, S., Inagaki, M., Niinomi, S., Khodaparast, S.A., Shin, H.D., Grigaliunaite, B., Havrylenko, M. (2008). Comprehensive molecular phylogenetic analysis and evolution of the genus *Phyllactinia* (*Ascocmycota*: *Erysiphales*) and its allied genera. *Mycol. Res. 112*: 1-17.

Tanabe Y, Saikawa, M., Watanabe, M.M, and Sugiyama, J. (2004). Molecular phylogeny of Zygomycota based on EF-1α and RPB1 sequences: limitations and utility of alternative markers o rDNA. *Mol. Phylogen. Evol. 30*: 438-449.

Tandon, R.N., and Bilgrami, K.S. (1960). A new host for *Phyllosticta cycadina. Pass. Sci. Cul. 26*: 277-278.

Thompson, J.D., Gibson, T.J., Plewniak, F., Jeanmougin, F. and Higgins, D.G. (1997). The ClustalX windows interface: flexible strategies for multiple sequence alignment aided by quality analysis tools. *Nucleic Acids Res. 24*: 4876-4882.

Voglmayr, H. and Yule, C.M. (2006). *Polyancora globosa* sp. nov., an aeroaquatic fungus from Malaysian peat swamp forests. *Mycol. Res. 110*: 1242-1252.

Voigt K., Cozijnsen, A. J., Kroymann, J., Pöggeler, S., and Howlett, B.J. (2005). Phylogenetic relationships between members of the crucifer pathogenic *Leptosphaeria maculans* species complex as shown by mating type (MAT1-2), actin and β-tubulin sequences. *Mol. Phylogen. Evol. 37*: 541-557.

Wadje, S.S., and Deshpande, K.S. (1981). A new leaf spot of cotton caused by *Phoma exigua. Indian Phytopath. 33*: 126-127.

Weekers, P.H.H., Jonckheere, F.J. de and Dumont, H.J. (2001). Phylogenetic relationships inferred from ribosomal ITS sequences and biogeographic patterns in representative of the genus *Calopteryx* (Insecta: Odonata) of the West Mediterranean and adjacent West European zone. *Mol. Phylogen. Evol. 20*: 89-99.

White, T.J., Bruns, T., Lee, S., Taylor, J. (1990). Amplification and direct sequencing of fungal ribosomal RNA genes for phylogenetics. pp. 315-322. *In*: *PCR protocols.* A guide to methods and applications. Innis, M.A., Gelfand, D.H., Sninsky, J.J., and White, T.J. (Eds.) Academic Press, New York.

Yli-Mattila, T., Mach, R.L., Alekhina, I.A., Bulat, S.A., Koskinen, S., Kullnig-Gradinger, C.M., Kubicek, CP., and Klemsdal, S.S. (2004). Phylogenetic relationships of *Fusarium langsethiae* to *Fusarium poae* and *Fusarium sporotrichioides* as inferred by IGS, ITS, β-tubulin sequences and UP-PCR hybridization analysis. *Int. J. Food Microbiol. 95*: 267-285.

Zaitz, C., Heins, E.M., de Freitas, R.S., Arriagada, G.L., Ruiz, L., Totoli, S.A., Marques, A.C., Rezze, G.G., Muller, H., Valente, N.S and Lacaz, C.D. (1997). Subcutaneous phaeohyphomycosis caused by *Phoma cava.* Report of a case and review of the literature. *Rev. Inst. Med. Trop. Sao Paulo 39*(1): 43-48.

Zhou, L., Bailey, K.L., Chen, C.Y., and Keri, M. (2005). Molecular and genetic analyses of geographic variation in isolates of *Phoma macrostoma* used for biological weed control, *Mycologia 97*(3): 612-620.

In: Current Advances in Molecular Mycology
Eds: Y. Gherbawy, R.L. Mach and M.K. Rai

ISBN 978-1-60456-909-4

Chapter VIII

GENE REGULATION OF HYDROLASE EXPRESSION: LESSONS FROM THE INDUSTRIALLY IMPORTANT FUNGUS *TRICHODERMA REESEI* (*HYPOCREA JECORINA*)

A. R. Stricker and R. L. Mach*

Gene Technology, Gene Technology and Applied Biochemistry, Institute of Chemical Engineering, TU Wien, A-1060 Wien, Austria.

ABSTRACT

Trichoderma ssp. (*Hypocrea*) are saprophytic ascomycetes producing hydrolytic enzymes which are of broad industrial importance. This article gives a detailed insight in their various applications in different fields of industry. Moreover, it provides a synopsis on the research on the gene regulation of the hydrolase expression in *Trichoderma reesei* (*Hypocrea jecorina*) during the last decade. Information on the correlation of inducing compounds and activation of hydrolase-encoding genes is given. The role of some recently investigated transcription factors involved in the transcriptional regulation of expression of hydrolytic enzyme-encoding genes will be explained, in particular the function of the general hydrolase activator protein Xyr1 (Xylanase regulator 1). The mode of gene regulation will be exemplarily explained by the transcriptional regulation of the *xyn1* gene expression under inducing as well as repressing conditions. Finally, a future outline will point out potential applications of *Hypocrea* hydrolases in newly developing areas of the biotech-industry and highlight advantages of genetically engineered *Hypocrea* strains.

Key words: *Trichoderma reesei*, *Hypocrea jecorina*, hydrolases, Xylanase regulator 1 (Xyr1), *xyn1* gene expression, transcriptional regulation.

* Correspondence concerning this article should be addressed to: R.L. Mach, Institute of Chemical Engineering, TU Wien, Getreidemarkt 9/166/5/2, A-1060 Wien, Austria. E-mail: rmach@mail.zserv.tuwien.ac.at.

INTRODUCTION

The Industrially Important Fungus *Hypocrea*

Filamentous ascomycetes of the genus *Hypocrea* (anamorph *Trichoderma*) (Kuhls et al. 1996) (Figure 1) mainly act as saprophytes thereby degrading a wide variety of biopolymeric substrates such as cellulose and hemicelluloses, predominant components of plant material. Hemicellulose as a collective term summarizes a variety of heteropolysaccharides composed either of a backbone formed by xylose (xylans) or mannose and glucose (mannans, glucomannans) with additional side chain substitutes such as arabinose, galactose, and acetic or glucuronic acid. Hemicelluloses are largely water insoluble hence their hydrolysis sets a challenge for saprophytes. Complete degradation of hemicelluloses requires a large number of extracellular enzymes working in synergistic action to allow hydrolysis to smaller oligosaccharides and finally to the respective monomers.

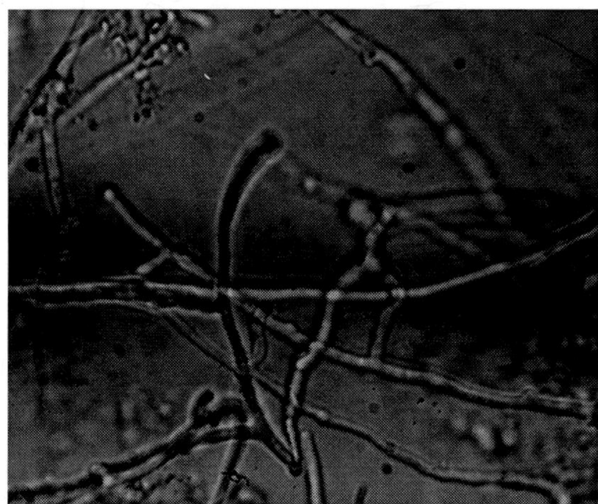

Figure 1. A light microscopic picture (400-fold magnification) of *Hypocrea jecorina* hyphes if the fungus is grown in a fermentation process on xylan (Stricker et al. 2007b).

A large number of hydrolytic enzymes secreted by *Hypocrea* have received broad industrial interest leading to widespread applications, such as in pulp and paper industry ((Buchert et al. 1998), and citations therein), in textile industry ((Galante et al. 1998b) and citations therein), and in food and feed production ((Galante et al. 1998a) and citations therein). It should be noted that *Hypocrea* has a remarkable secretory capacity since it is able to produce its native enzymes in amounts up to 100 g/L (Ward 2006). Detailed information on the numerous fields of applications of *Hypocrea ssp.* enzymes are listed in Table 1A and B. Consequently, this led to extensive investigations on the respective biochemical properties of these enzymes (e.g. (Biely et al. 1994; Poutanen et al. 1987; Tenkanen et al. 1992; Törrönen et al. 1992; Zeilinger et al. 1993)), their 3-dimensional structures (Törrönen et al. 1994; Törrönen and Rouvinen 1995), and the isolation and characterisation of the respective genes encoding them (Margolles-Clark et al. 1996a; Margolles-Clark et al. 1996b; Margolles-Clark et al. 1996c; Margolles-Clark et al. 1996d; Törrönen et al. 1992).

Table 1A. Examples of applications of *Hypocrea ssp.* enzymes

Application field	Enzyme/Protein	Application process/Benefits of enzyme usage
Pulp and paper	Cellulases	pulping
	β-Glucanases	
	Xylanases	
	β-Glucosidases	
	β-Xylosidase	
	Mannanases	
	β-Glucanases	improvement of drainage of recycled fibres
	Cellulases	
	β-Glucanases	improvement of pulp beatability
	Xylanases	delignification, bleaching
	Mannanases	
	β-Glucanase I	disturbing pitch depositions, increasement of runnability of paper machines
	Xylanases	de-inking from fibre surface
	Cellulases	
	Pectinases	
	Lipases	hydrolysis of soy-based ink carriers
Brewing	β-Glucanases	decreasing viscosity, better filterability
	α- and β-Amylase	mashing, liquefaction
Wine making	Pectinases	improving clarification, increasing terpene content, increasing wine quality/stability
	β-Glucanases	improving clarification/filtration
	Xylanases	degradation of neutral pectins associated to hemicellulases
	Cellulases	degradation of the grape cell wall
	β-Glycosidase	aroma development
Fruit juice production	Pectinases	liquefaction, clarification, extraction of valueable fruit components
	Cellulases	
	Xylanases	
	Pectinases	pulp and peel washing
	Cellulases	
Olive oil production	Cellulases	maceration, improving yield and quality (levels of tocopherol and polyphenols)
	Xylanases	
	Pectinases	
Bakery	Xylanases	conditioning of dough
	α-Amylase	improving crumb structure and loaf volume
Animal feed	Xylanases	degradation of non-starch polysaccarides, increasment of feed conversion rates
	β-Glucanases	
	Phytase	elimination of antinutritional compounds (myoinositolhexaphosphates)
	Amylases	supplementation of the animals digestive system
	Proteases	
Fuel ethanol	Cellulases	breakdown of corn fibers and stover
	Xylanases	
	α-Amylase	liquefaction
	Proteases	provision of nutrients for yeast
Textile industry	Cellulases	stonewashing, indigo dye redeposition
		biofinishing, defibrillation

Within the last 20 years a large number of hemicellulase-encoding genes have been cloned from *Hypocrea jecorina* (*Trichoderma reesei*) and other *Hypocrea spp.*, but only few of them (in particular *xyn1* and *xyn2*, encoding the two major Endo-β-1,4-xylanases I and II) received detailed attention concerning their regulation of gene expression (e.g. (Kubicek and Penttilä 1998; Mach et al. 1996; Mach et al. 1998; Margolles-Clark et al. 1997; Rauscher et al. 2006; Saloheimo et al. 2000; Würleitner et al. 2003; Zeilinger et al. 1996)).

Table 1B. Examples of heterologous protein expression in the host *Hypocrea*

Calf chymosin	expression of heterologous proteins under the *cbh1* promoter
Murine antibody Fab fragments	
Laccase	
Glucoamylase P	

A detailed understanding of such regulatory processes will not only assist in the improvement of enzyme production both by nutritional strategies and bioprocess design, but will also support efforts towards recombinant overproduction of these hydrolases or of heterologous proteins under the control of their promoters. Such strategies are necessary because the simple insertion of multiple copies of various hydrolytic genes into *Hypocrea* genomes did not substantially alter/improve the amount of enzymes produced (Kubicek-Pranz et al. 1991; Kubicek et al. 1993). During the last years the majority of the *in cis*-acting elements and some *in trans*-acting factors involved in the expression of these genes (*xyn1* and *xyn2*) have been identified and their functional characterization has been initiated (e.g. (Aro et al. 2002; Mach 2002; Mach and Zeilinger 2003; Stricker et al. 2006; Stricker et al. 2007b; Würleitner et al. 2003)). Nevertheless, neither the selective recognition of the different inducer molecules (xylose, xylobiose or sophorose) nor the respective role of corresponding *in trans*-acting factors probably involved in this process has been understood up to now. A detailed comprehension of the interplay between these inducer molecules and the respective regulatory proteins will not only serve as basis to develop strains expressing tailor-made enzyme cocktails, needed in several industrial applications, but should also lead to the design of artificial promoters with defined switching mechanisms to be used in heterologous protein production. Therefore, this should be a main focus of future research work.

Expression of *Hypocrea* Cellulases and Xylanases

Early studies establishing culture conditions for the production of xylanolytic activity reported abundant production when *Hypocrea* is cultivated on media containing cellulose, xylan or mixtures of plant polymers (reviewed by (Bisaria and Mishra 1989; Kubicek 1993; Zeilinger and Mach 1998)). Since these very potent natural inducing compounds cannot enter the fungal cell, it is generally believed that oligosaccharides released from the polymers and their derivatives function as the actual substances triggering induction of xylanase expression.

Whereas in *Aspergillus* the xylanolytic and cellulolytic system is strictly co-regulated via the inducer xylose (e.g. (Gielkens et al. 1999; Hasper et al. 2000)), enzymes participating in the respective *H. jecorina* hydrolytic complexes are not (see Figure 2). Their differential expression has been reported in several studies: Culturing *H. jecorina* on cellulose or xylan causes the formation of two specific xylanases (XYN I and XYN II) and one single unspecific endoglucanase (EGL I) (Senior et al. 1989). The disaccharide sophorose consists of two ß-1,2-linked glucose units and is the best to date known cellulase-inducing compound in *H. jecorina* which is concurrently considered as the natural inducer of cellulase formation (Mandels et al. 1962; Mandels and Reese 1960; Sternberg and Mandels 1979). If it is applied as an inducer, only one of both xylanases (XYN II) and Endoglucanase I can be detected, whereas xylobiose leads to the formation of both xylanases, and a ß-xylosidase, but not of Endoglucanase I (Hrmova et al. 1986; Margolles-Clark et al. 1997). Finally, *xyn1* transcription, in contrary to all other xylanolytic and cellulolytic enzyme-encoding genes in *H. jecorina,* is induced by xylose (Mach et al. 1996). More recently these data were confirmed by transcriptional analysis (Rauscher et al. 2006; Stricker et al. 2006). In addition, evidence for a different transcript formation pattern of these two xylanases with respect to glucose was provided (Zeilinger et al. 1996). It was demonstrated that *xyn2* transcript arises at a low basal level when the fungus is grown on glucose as sole carbon source. This basal level is elevated by the presence of xylan, xylobiose or sophorose. Simultaneous presence of glucose and xylan leads to a drop of transcription to the basal level, whereas induction of *xyn2* transcription by xylobiose is not affected by glucose (Würleitner et al. 2003). However, *xyn1* transcription which is induced by xylose is not affected by the simultaneous presence of glucose and xylose, whereas glucose alone completely represses its expression (Mach et al. 1996; Rauscher et al. 2006). A general model for the substrate recognition and induction mechanism of the major xylanolytic enzymes of *H. jecorina* implies that a low constitutive level of *xyn2* expression may initiate primarily a release of oligosaccharides from the xylan-backbone, thereafter triggering induction of Endoxylanases I and II and ß-xylosidase I. This model matches with the findings that all these genes are inducible by respective degradation products: *xyn1* by xylose (Mach et al. 1996) and *xyn2* and *bxl1* by xylobiose (Margolles-Clark et al. 1997; Zeilinger et al. 1996).

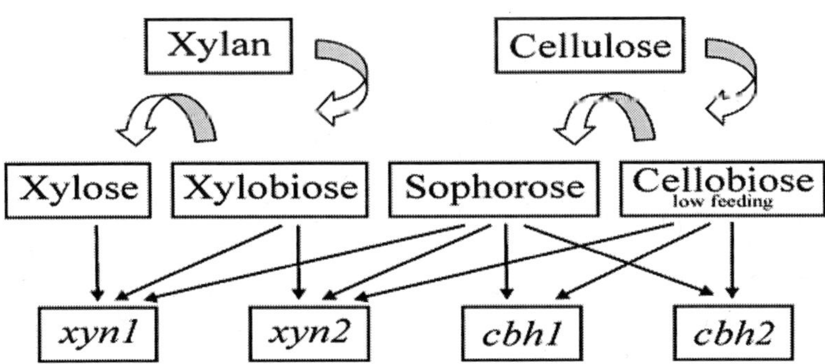

Figure 2. Induction pattern of four important hydrolytic enzyme-encoding genes. In *H. jecorina* biopolymeric substances as xylan and cellulose are broken down to the low molecular weight inducer molecules as cellobiose and sophorose or xylobiose and D-xylose, respectively. As known so far they act as the inducers for the *xyn1*, *xyn2*, *cbh1* or *cbh2* gene alone or in a rather sophisticated combination of these genes. For detailed insights please see for example: (Hrmova et al. 1986; Ilmén et al. 1997; Kubicek 1993; Margolles-Clark et al. 1997; Zeilinger et al. 1996).

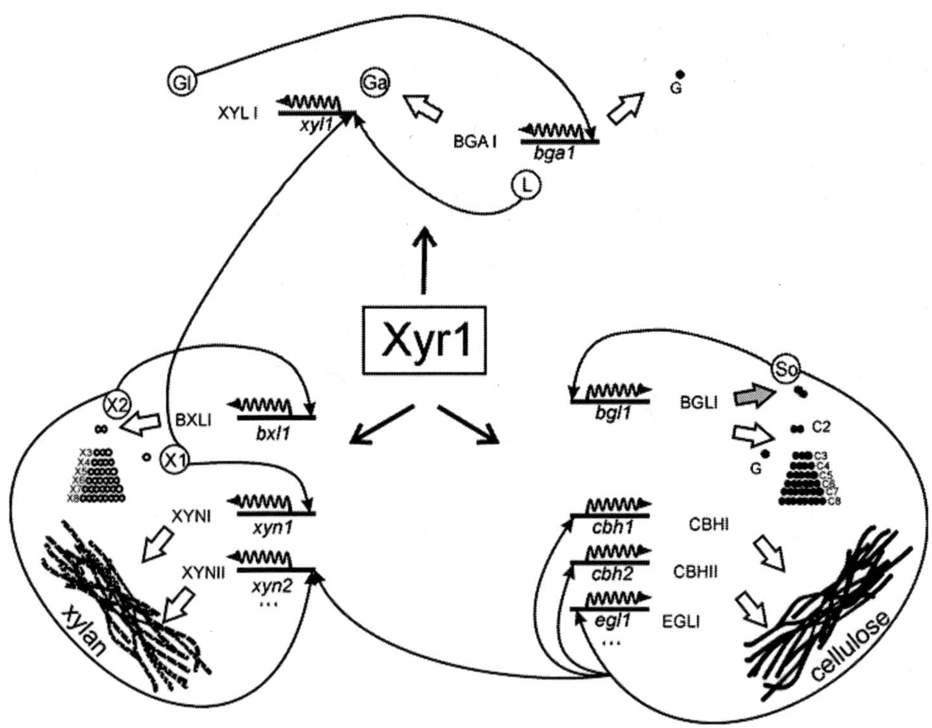

Figure 3. Schematic drawing of the central role of Xyr1 in the hydrolytic enzyme system involved in the degradation of xylan and cellulose as well as in the lactose metabolism of *H. jecorina*. Xyr1 acts as an *in trans*-acting factor (i) directly on the expression of xylanases (as XYNI, XYNII) and cellulases (as CBHI, CBHII, EGLI) as well as (ii) on the corresponding inducer (D-xylose, X1)-providing enzyme BXLI in the case of xylan utilization and on the corresponding inducer (sophorose, So)-providing enzyme BGLI in the case of cellulose utilization, and finally (iii) on the D-xylose reductase (XYLI) involved in the lactose (L) metabolism by converting galactose (Ga) into galactitol (Gl) which induces expression of β-galactosidase (BGAI). White arrows indicate enzymatic degradation of the respective biopolymer to oligo- or monomeric sugars (X1-X8 and G, C2-C8, Ga). Gray arrow indicates transglycosylation activity of BGLI. Thin black arrows indicate which inducers act on which gene.

As already mentioned strategies to influence hydrolase expression by manipulating (e.g. multi copying (Kubicek-Pranz et al. 1991; Kubicek et al. 1993) or knocking out (Seiboth et al. 1997)) respective hydrolase-encoding genes did not succeed. Regulation via the inducers is also restricted since broad usage of some of them would evoke considerable costs (for example α-sophorose costs about 1,400 €/g, and xylobiose about 3,000 €/g). Thus, the main focuses for improving the expression of hydrolases are on the understanding and appropriate modification of the function and mode of action of regulatory proteins. Recently, the main activator of hydrolytic enzymes-encoding genes in *H. jecorina*, Xyr1 (Xylanase regulator 1), which is the orthologue of the *Aspergillus niger* transactivator XlnR (van Peij et al. 1998), was identified (Stricker et al. 2006). A general model for the substrate recognition and the induction mechanisms of the major xylanolytic and cellulolytic enzymes as well as the lactose metabolism of *H. jecorina* involving the general *in trans*-acting activator Xyr1 was recently postulated (Figure 3) (Stricker et al. 2006; Stricker et al. 2007a). This model summarizes all discussed hydrolytic genes which are inducible by respective degradation and/or transglycosylation products derived from xylan and/or cellulose: *xyn1* by xylose (Mach et al. 1996), *xyn2* by xylobiose and sophorose (Zeilinger et al. 1996) and *bxl1* by xylobiose (Margolles-Clark et al. 1997), cellulases such as *cbh1*, *cbh2* and *egl1* (Ilmén et al. 1997) as well as *bgl1* (Fowler and Brown 1992) and *bgl2* (Saloheimo et al. 2002) by sophorose. The scheme points at the fact that in all cases Xyr1 - except for *bgl2* - is an indispensable transactivator (Stricker et al. 2006). Moreover, the involvement of Xyr1 in the lactose metabolism by directly influencing *xyl1* (D-xylose reductase 1-encoding) transcription (Stricker et al. 2007a) and by indirectly influencing *bga1* (β-galactosidase 1-encoding, (Seiboth et al. 2005)) transcription (Stricker et al. 2007a) providing the corresponding inducer galactitol (Fekete et al. 2007) is also given in Figure 3. Investigations on the function and the mode of action of such regulatory proteins are central aspects to understand and control hydrolase expression. Therefore, the next part of the article gives an introduction to recently isolated and characterized transcription factors participating in the hydrolase transcriptosome of *H. jecorina*.

Regulation of Hydrolytic Enzymes-Encoding Genes in *H. jecorina*

As mentioned above one important strategy for strain improvement is to find out which transcription factors regulate hydrolytic enzyme-encoding gene expression and how they exert their influence. In the following the recently obtained results for some transcription factors are summarized. It should be noted that further general transcription factors, also involved in xylanase and cellulase gene expression, as e.g. Cre1 (Carbon Catabolite Repressor Protein 1) or Hap 2/3/5 (CCAAT-binding protein) complex are not discussed in detail in this section. For more information see e.g.: (Cziferszky et al. 2002; Ilmén et al. 1996a; Mach et al. 1996; Strauss et al. 1995; Zeilinger et al. 2001).

Xyr1 (Xylanase Regulator 1) is the General Activator of Hydrolase Formation in *H. jecorina*

Comparative studies on the growth behavior of the different *xyr1* mutant strains (deleted and retransformed *xyr1* gene) on various carbon sources pointed at the strongly reduced ability of the *xyr1* deletion strain to utilize D-xylose and xylan. Transcriptional analysis of the *xyl1* gene as well as measurements of corresponding enzymatic activities gave evidence that Xyr1 takes part in control of the fungal D-xylose pathway, in particular in the regulation of the D-xylose reductase. It could be demonstrated that the uptake of D-xylose into the fungal cell is not importantly influenced in the Δ*xyr1*-strain ruling out the possibility of an inhibition of the D-xylose transport in a *xyr1* deletion strain (Stricker et al. 2006).

Moreover, Xyr1 was revealed to activate transcriptional regulation of genes encoding inducer-providing enzymes such as β-xylosidase BXLI and β-glucosidase BGLI, but not to be involved in the regulation of BGLII-encoding gene expression (Stricker et al. 2006).

In contrast to the diversity of inducers and induction mechanisms (described *vide supra*) it was recently demonstrated that transcriptional regulation of the major hydrolytic enzyme-encoding genes *xyn1*, *xyn2 cbh1*, *cbh2* and *egl1* is strictly dependent on Xyr1. Regulation of the respective genes via Xyr1 is not affected by the substances mediating induction (xylose, xylobiose, sophorose) and is indispensable for all modes of gene expression (Stricker et al. 2006). *Id est*, neither the Cre1-dependent de-repressed or induced transcriptional level of *xyn1* (Mach et al. 1996) nor the basal or induced transcriptional level of *xyn2* (Zeilinger et al. 1996) could be observed in a Δ*xyr1*-strain (Stricker et al. 2006). We have recently provided evidence that the corresponding *in cis*-acting elements occur as inverted repeats in both promoters with a spacing of 10 or 12 bases, respectively (Rauscher et al. 2006; Stricker et al. 2007b) (Figure 4). In case of the regulation of the *xyn1* expression it was shown that the induction-specific protein-DNA complex contains two Xyr1 proteins each contacting one of the inverted repeats (Rauscher et al. 2006). The corresponding inverted elements in the *xyn2* promoter are likewise bound by *in vitro* translation products of *xyr1,* whereas a mutation in one of both motifs fully abolishes binding (Stricker et al. 2007b). This finding strongly suggests that Xyr1 dimers can only contact a repeat of its binding element, nevertheless addition of sodium deoxycholate (interfering with protein-protein, but not with protein-DNA interactions (Baeuerle and Baltimore 1988; Kupfer et al. 1993)) leads to binding of Xyr1 as a monomer (Stricker, A. R., Mach, R. L., unpublished data). Summing up, we conclude that Xyr1 seems to dimerise before binding to its motifs but the dimerisation is not essential for binding.

Xyr1-motif (p*xyn1*)

Xyr1-motif (p*xyn2*)

Ace1 AGGCA(AA) (p*cbh1*, *H. jecorina*)
Ace2 GGCTAA (p*cbh1*, *H. jecorina*)
Xyr1 GGCTAA (p*xyn1*, *H. jecorina*)
XlnR GGCTAAA (p*xlnD*, *A. niger*, p*xlnA*, *A. tubingensis*)

Figure 4. Xyr1-binding motif within the *xyn1* promoter (p*xyn1*) and the *xyn2* promoter (p*xyn2*) of *H. jecorina*. Both motives are arranged as more or less exactly matching inverted repeats separated by 10 or 12 bp (N_{10}, N_{12}) respectively. The binding sites of the transcription factors Ace1, Ace2, Xyr1 and XlnR are given, together with the promoters of genes and species they have been deduced from.

Lactose is a disaccharide which actually is the predominant carbohydrate in milk and dairy products. Nevertheless, for several times lactose has been described as an activating substance for cellulase activity formation or for transcription of cellulase-encoding genes in *H. jecorina* (Ilmén et al. 1997; Margolles-Clark et al. 1997; Seiboth et al. 2004). But the mechanism of activation was not understood up to now. Latest studies proved that Xyr1 mediates the lactose induction signal and is vital for the transcriptional regulation of hydrolase expression on lactose (Stricker et al. 2007a). E.g. the transcription of the *xyn1* gene is inducible by lactose. In contrary, for the transcription of the *xyr1* gene itself on lactose a release from carbon catabolite repression (can occur if easily usable carbon sources as glucose are disposable) is responsible. Xyr1 influences the lactose metabolism of *H. jecorina* directly by regulating the *xyl1* transcription (*vide supra*, Figure 3) (Stricker et al. 2006; Stricker et al. 2007a). The *xyl1* gene product reduces galactose to galactitol and thereby catalyzes the first step of an alternative lactose metabolism pathway (Fekete et al. 2007; Pail et al. 2004; Seiboth et al. 2004). The longer known pathway is the so called Leloir pathway (Frey 1996) that starts with a galactokinase specifically metabolizing the α-anomer of D-galactose. Furthermore, Xyr1 acts indirectly on the lactose metabolism: as mentioned it activates transcription of *xyl1* whose gene product provides galactitol. Galactitol was described as an inducer of the *bga1* gene expression. The gene *bga1* encodes for a β-galactosidase, which converts lactose into D-galactose (see Figure 3), considered as inducer of cellulase-encoding gene expression (Karaffa et al. 2006). Thus, Xyr1 acts indirectly on *bga1* transcriptional regulation also involved in lactose metabolism (Figure 3).

As mentioned above Xyr1 is the essential activator for all levels of *xyn1* and *xyn2* transcription, obviously directly or indirectly receiving and mediating all different signals from the inducer molecules. In the past, not only evidence for the involvement of additional wide domain regulators (e.g. Hap2/3/5 and Cre1; (Mach et al. 1996; Rauscher et al. 2006; Würleitner et al. 2003)) could be demonstrated, but also strong indications for the participation of particularly specific transcription factors, notably Ace1 and Ace 2, were recently given (Rauscher et al. 2006; Stricker et al. 2007b).

Ace1 (Activator of Cellulase Genes 1) is a Specific Repressor of *xyn1* Gene Expression

In both xylanase promoters the Xyr1-binding elements closely resemble the consensus sequences of the recently cloned *H. jecorina* transcriptional regulator Ace1 (Figure 4) (Saloheimo et al. 2000). In contrast to the first report on this activator, Aro and coworkers (Aro et al. 2003) demonstrated that a Δ*ace1*-strain forms elevated levels of *xyn1* transcript on cellulose-based media. The distinct binding of Ace1 to an oligonucleotide corresponding to the part of the *xyn1* promoter which contains the inverted repeat of two Xyr1-binding elements was observed (Figure 4). Furthermore, electrophoretic mobility shift assays (EMSA) carried out with cell-free extracts prepared from *H. jecorina* mycelia grown on glucose resulted in a protein-DNA complex of lower mobility compared to cell-free extracts of *H. jecorina* Δ*ace1* (Rauscher et al. 2006). As summary, we recently deduced a competition between the Ace1 repressor and the Xyr1 activator for the Xyr1-binding elements (Rauscher et al. 2006). A similar mechanism has been shown for the regulation of the *alcA* gene of *A. niger* where the transcriptional activator AlcR competes with CreA for the corresponding, partially overlapping binding-sites (e.g. (Marmorstein et al. 1992; Marmorstein and Harrison 1994; Mathieu and Felenbok 1994; Narendja et al. 1999)).

As mentioned above the transcriptional regulation of *xyn1* and *xyn2* strongly differs according to their respective inducer molecules; we currently follow the working hypothesis that in addition to a general activator (Xyr1) specific repressors provoke these regulatory differences. A specific *xyn2*-repressing element (AGAA-box) has recently been identified (Würleitner et al. 2003) and potential candidate gene(s) encoding its (their) corresponding *in trans*-acting factor(s) have been cloned (Stricker, A. R., Mach, R. L., unpublished data, *vide infra,* Xrp1). In case of *xyn1* gene regulation, Ace1 acts as an antagonist of the Xyr1-driven *xyn1* gene transcription. This assumption is supported by the facts that *vice versa* neither the occurrence of a functional AGAA-like element in the *xyn1* promoter can be observed (Rauscher et al. 2006), nor a significant influence of Ace1 on the *xyn2* expression (Aro et al. 2003). The detailed interaction/competition of Ace1 and Xyr1 with each other and with their corporate binding element (right motif of the Xyr1-element) in the *xyn1* promoter remains to be investigated.

Ace2 (Activator of Cellulase Genes 2) is a Specific Regulator of *xyn2* Gene Expression

Using deletion analysis, Zeilinger et al. (1996) reported that a 55-bp fragment of the *xyn2* promoter contains all information necessary for the regulation of the *xyn2* gene expression. Further studies (Würleitner et al. 2003) using *in vitro* as well as *in vivo* strategies identified nucleotide sequences within the 55 bp, being essential for the binding of proteins and responsible for the *xyn2* gene regulation. The authors could demonstrate that a protein-DNA complex related to both basal transcription as well as induction of *xyn2* expression consists of at least the Hap2/3/5 complex (Zeilinger et al. 2001) and of Ace2 (Aro et al. 2001) contacting an undecameric motif 5'-GGGTAAATTGG-3' (Xylanase-activating element; XAE) (Würleitner et al. 2003). This induction-specific complex is counteracted by a yet only pre-characterised DNA-binding protein (complex) binding to a 5'-AGAA-3'-box located immediately upstream of XAE (*vide infra*, Xrp1). Applying *in vivo* footprinting we recently observed additional bases to be protected adjacent to XAE (Stricker et al. 2007b). Those bases are positioned within an inverted sequence repeat (spacing of 12 bp), each side closely resembling both the Xyr1- (Rauscher et al. 2006) as well as the Ace2-binding element (Aro et al. 2001) (Figure 4). Whereas in a previous study it was shown that binding domains of Ace2, but not of XlnR (van Peij et al. 1998) contact this motif (Würleitner et al. 2003), recent investigations revealed binding of Xyr1 (Stricker et al. 2007b). A probable explanation for this apparent contradiction could be a single amino acid exchange within the Zn-finger domain comparing Xyr1 and XlnR (Rauscher et al. 2006). Furthermore, it is noteworthy that only the Xyr1-binding elements in the *xyn1* promoter exactly match the XlnR-motif (and are *in vitro* bound by Xyr1 as well as by XlnR), whereas the according regions in the *xyn2* promoter are only similar but not identical (Figure 4). This observation is in strict accordance with the fact that Xyr1 binds to all corresponding motifs in both xylanase promoters. In contrary to this, Ace1 only contacts one part of the element in the *xyn1* but not in the *xyn2* promoter and Ace2 doing *vice versa* (Rauscher et al. 2006; Stricker et al. 2007b; Würleitner et al. 2003). EMSAs using *in vitro* translation products not only revealed that Ace2 binds to these Xyr1-elements in the *xyn2* promoter but also that this binding is abolished by the same mutations in the binding elements also essential for Xyr1-DNA interaction. In addition, both *in vitro* phosphorylation as well as dimerisation of Ace2 is essential for Ace2-DNA contact (Stricker et al. 2007b).

Anyway, our current working hypothesis is that an intense interplay between Ace2 and Xyr1, including several steps of phosphorylation and probably heterodimerization as well as the recruiting of additional proteins, is indispensable for the formation of an active xylobiose-dependent *xyn2* transcriptosome (Stricker et al. 2007b). Binding elements of Ace1 and Ace2 are quite similar (Figure 4); nevertheless their DNA-contact is dedicated only to one of the promoters of the two major xylanolytic enzyme-encoding genes (i.e. Ace1 to p*xyn1* and Ace2 to p*xyn2*). Therefore, we currently think that the different inducibility of the xylanase-encoding genes is caused by the fine tuning mediated by transcription factors as Ace1 and Ace2 besides the general activator Xyr1.

The interplay of these factors may not only involve acting together in the transcriptional regulation of a certain hydrolase-encoding gene but also directly influencing the regulation of gene expression of each other. The examination of the *xyr1* transcription levels clearly demonstrated a repressing effect of Ace2 on *xyr1* expression applying inducer molecules (xylobiose, sophorose). This could be affirmed by *in silico* analysis of the *xyr1* promoter revealing the presence of a GGCTAT-binding site 260 bp upstream of the *xyr1* coding sequence and further, three motifs closely resembling the GGCTAA-box (Stricker et al. 2007b). Finally, it remains to be investigated in detail if and how Ace2 and/or other transcriptional factors (e.g. Cre1) are playing a role in regulation of the *xyr1* transcript formation.

Aro et al. (2001) reported for the first time that xylanase 2 expression in a Δ*ace2*-strain is strongly reduced but not fully abolished. Recently, more detailed investigations revealed that Ace2 is responsible for the constitutive transcriptional level of *xyn2* expression. Moreover, the absence of Ace2 leads to a faster initial inducibility of *xyn2*, but the late transcript levels of the wild-type strain are only reached by half in a Δ*ace2*-strain. This regulatory influence of Ace2 on the *xyn2* expression can only be observed when the fungus is grown on xylan or induced by xylobiose but not if sophorose is applied (Stricker et al. 2007b). Again, this is in strict accordance with previous findings documenting that sophorose signals are not mediated via Ace2 (Aro et al. 2001). Summarizing, Ace2 is involved in the regulation of various aspects of *xyn2* gene expression such as the formation of its constitutive transcript level as well as in the induction by xylobiose. Ace2-DNA interaction studies currently provide evidence that this transcription factor needs to be phosphorylated and dimerised to contact its corresponding DNA-element. Above all, Ace2 most probably seems to interact with Xyr1 either in a competitive mode or even more likely as a heterodimerisation partner.

After all, it shall be noted that no orthologue of Ace2 is present in respective *Aspergillus* genomes, again strongly emphasising the significant differences in transcriptional regulation of hydrolytic genes compared to *Hypocrea*.

Xrp1 (Xylanase Repressor 1) is a Putative Novel Transcription Regulator of *xyn2* Gene Expression

The binding of a putative repressor protein to an AGAA-box within the *xyn2* promoter was shown by Würleitner and co-workers (Würleitner et al. 2003) by performing EMSAs using cell-free extracts of the mycelia of a *H. jecorina* wild-type strain grown on glucose. The same cell-free extracts were applied to an affinity chromatography using DNA oligonucleotides as a specific interaction matrix. Preliminary studies point at three additional DNA-binding proteins potentially involved in the *xyn2* gene regulation (Stricker, A. R., Mach, R. L., unpublished data). The potential regulatory protein-encoding genes are currently investigated by construction of corresponding deletion strains to determine the actual regulatory impact of the putative repressor proteins.

Transcriptional Regulation of *xyn1* Expression in *H. jecorina*

Deletion analysis of the *xyn1* 5'-regulatory sequences revealed a 214-bp fragment at –321 to –534 upstream of ATG to contain all crucial information for the xylanase 1 transcriptional regulation (Mach et al. 1996). Both *in vivo* and *in vitro* analyses demonstrated that an inverted repeat of two Cre1 consensus sequences within these 214 bp is responsible for tight glucose repression of *xyn1* transcription (Mach et al. 1996). Elimination of these two Cre1-sites led to a transcript formation on glucose comparable to that observed when the fungus was grown on the carbon source lactose. These data, together with the fact that inducibility by xylose is still given in this mutant strain, led to the consideration that *xyn1* expression is regulated by two – a repressing and an inducing – mechanisms. This assumption was further strengthened by observing the de-repressed *xyn1* expression in the Cre1-negative strain *H. jecorina* Rut C-30 (Ilmén et al. 1996b) when grown under repressing conditions (Mach et al. 1996). Recently, an investigation of the 214-bp fragment by *in vivo* footprinting analysis revealed a permanent protection of the CCAAT-box and also of an adjacent GGCTAA-motif (Rauscher et al. 2006) identical with the binding site of the *A. niger* xylanase activator XlnR (van Peij et al. 1998) responsible for the regulation of the expression of the major hydrolases and the D-xylose metabolism. In contrast to XlnR, in *H. jecorina* the binding site for the respective XlnR-orthologue, Xyr1, appears as an inverted repeat in the *xyn1* promoter (Rauscher et al. 2006), whereas in *A. niger* the corresponding binding sites mainly appear as single sites (van Peij et al. 1998).

As already discussed, the two major xylanases (XYN I and XYN II) of *H. jecorina* are simultaneously expressed during growth on xylan but respond differentially to low molecular-weight inducers. An *in vivo* footprinting analysis of the xylose-induced and the glucose-repressed mycelia of the *xyn1* promoter revealed three different nucleotide sequences (5'-GGCTAAATGCGACATCTTAGCC-3' on the non-coding strand; 5'-CCAAT-3' and 5'-GGGGTCTAGACCCC-3' (= double Cre1-site) on the coding strand, respectively) to be protected, hence, most probably to be bound by proteins. The protection of the Cre1-site is only observed under repressing conditions, whereas the protection of the two other motifs appears to be constitutive. An EMSA with heterologously expressed DNA-binding domains of the *H. jecorina* regulators Ace1 and Ace2 and the xylanase regulator Xyr1 suggests that Ace1 and Xyr1 but not Ace2 (shown to be involved in the transcriptional regulation of *xyn2, vide supra*) contact both GGCTAA-motifs. The *H. jecorina* transformants, containing correspondingly mutated versions of the *xyn1* promoter fused to the *A. niger goxA* gene as a reporter, revealed that the elimination of protein-binding to the left or the right GGCTAA-box, either strongly reduced or completely eliminated the induction of the *xyn1* transcript formation. The elimination of the Cre1-binding to its target released the basal transcriptional level from the glucose repression but did not influence the inducibility of the *xyn1* gene expression. The destruction of the CCAAT-box by insertion of a point mutation prevents the binding of the Hap2/3/5 complex *in vitro* and slightly antagonizes the loss of transcription of the *xyn1* gene caused by the mutation of the right GGCTAA-box. An EMSA with cell-free extracts from the wild-type strain and an *ace1* deletion strain prepared from mycelia grown under repressing and inducing conditions indicated a competition of Ace1 and Xyr1 for the right GGCTAA-box. These data support a proposed model for the *xyn1* regulation based on the interplay of Cre1 and Ace1 as a general and a specific repressor, Xyr1 as transactivator and Hap2/3/5. A schematic presentation of the *xyn1* transcriptional regulation under inducing (xylose) or repressing (glucose) conditions can be found in Figure 5.

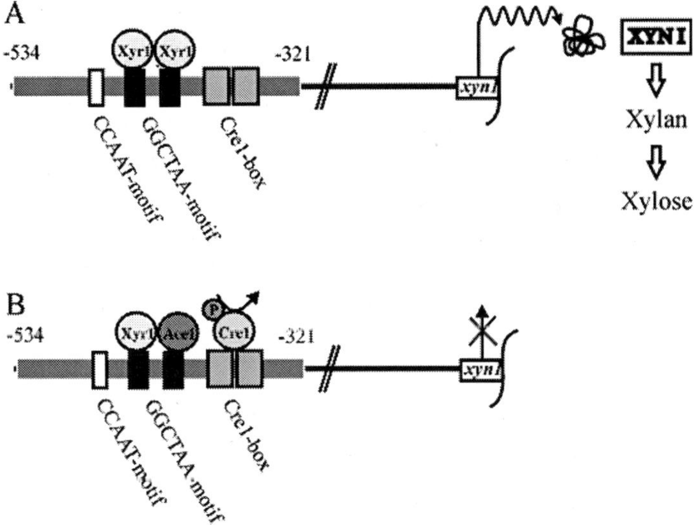

Figure 5. Architecture of the *xyn1* promoter of *H. jecorina*. (A) Binding of Xyr1 to the GGCTAA palindrome under inducing conditions leads to expression of the *xyn1* gene product, Xylanase I. (B) Binding of Ace1 - instead of Xyr1 - to the essential downstream part of the GGCTAA-motif and binding of Cre1 inhibits *xyn1* transcription under repressing conditions.

Future Prospectives and Biotechnological Relevance of the New Insights

As already mentioned promoter design and investigations on the architecture of the transcriptosomes of hydrolytic genes are indispensable pre-requisites for the production of defined enzyme mixtures matching the requirements of special industrial applications and also for the heterologous protein production. Besides that, the improvement of the yields of native enzyme production can be pressed ahead. Furthermore, it should be mentioned that xylitol and arabinitol count to the top 12 value added chemicals according to the reports of U.S. Department of Energy (Petersen, G. and Werpy, T., Pacific Northwest National Laboratory and the National Renewable Energy Laboratory, August 2004). Both alcohol sugars are hydrogenation products from xylose or arabinose (predominant brake-down products of xylan) respectively and can be easily converted from these sugars using nickel, rhodium or ruthenium catalysts with yields up to 99 %. Thus, a *Hypocrea* strain still able to degrade biopolymeric substances as xylan but with a blocked xylose metabolism would lead to D-xylose accumulation. Such strains, which can be designed on the basis of knowledge gained during the last years concerning the mechanisms of transcriptional regulation, would gain great importance for above mentioned purposes. Moreover, since xylose could serve as the starting material of precursor substances of important drugs, it has weighty impact on the drug industry.

REFERENCES

Aro N, Ilmén M, Saloheimo A, Penttilä M (2002) Ace1 is a repressor of cellulase and xylanase genes in *Trichoderma reesei*. G Vannacci, Sarrocco, S. (eds) *6th European Conference on Fungal Genetics*, Pisa, Italy

Aro N, Ilmén M, Saloheimo A, Penttilä M (2003) ACEI of *Trichoderma reesei* is a repressor of cellulase and xylanase expression. *Appl Environ Microbiol 69*: 56-65

Aro N, Saloheimo A, Ilmén M, Penttilä M (2001) ACEII, a novel transcriptional activator involved in regulation of cellulase and xylanase genes of *Trichoderma reesei*. *J Biol Chem 276*: 24309-14.

Baeuerle PA, Baltimore D (1988) Activation of DNA-binding activity in an apparently cytoplasmic precursor of the NF-kappa B transcription factor. *Cell 53*: 211-7

Biely P, Kremnicky L, Alfoldi J, Tenkanen M (1994) Stereochemistry of the hydrolysis of glycosidic linkage by endo-beta-1,4-xylanases of *Trichoderma reesei*. *FEBS Lett 356*: 137-40

Bisaria VS, Mishra S (1989) Regulatory aspects of cellulase biosnthesis and secretion *Crit Rev Biotechnol 9*: 61-103

Buchert J, Oksanen T, Pere J, Siika-aho M, Suurnäkki A, Viikari L (1998) Applications of *Trichoderma reesei* enzymes in the pulp and paper industry. In: GE Harman, CP Kubicek (eds) *Trichoderma & Gliocladium*, 2, Taylor & Francis Ltd., London, UK, pp 343-357

Cziferszky A, Mach RL, Kubicek CP (2002) Phosphorylation positively regulates DNA binding of the carbon catabolite repressor Cre1 of *Hypocrea jecorina* (*Trichoderma reesei*). *J Biol Chem 277*: 14688-94

Fekete E, Karaffa L, Kubicek CP, Szentirmai A, Seiboth B (2007) Induction of extracellular beta-galactosidase (Bga1) formation by D-galactose in *Hypocrea jecorina* is mediated by galactitol. *Microbiology 153*: 507-12

Fowler T, Brown RD, Jr. (1992) The *bgl1* gene encoding extracellular beta-glucosidase from *Trichoderma reesei* is required for rapid induction of the cellulase complex. *Mol Microbiol 6*: 3225-35

Frey PA (1996) The Leloir pathway: a mechanistic imperative for three enzymes to change the stereochemical configuration of a single carbon in galactose. *Faseb J 10*: 461-70

Galante YM, De Conti A, Monteverdi R (1998a) Application of *Trichoderma* enzymes in the food and feed industries. In: GE Harman, Kubicek, C.P. (eds) *Trichoderma & Gliocladium*, 2, Taylor & Francis, London, pp 327-342

Galante YM, De Conti A, Monteverdi R (1998b) Application of *Trichoderma* enzymes in the textile industry. In: GE Harman, Kubicek, C.P. (eds) *Trichoderma & Gliocladium*, 2, Taylor & Francis, London, pp 311-325

Gielkens MM, Dekkers E, Visser J, de Graaff LH (1999) Two cellobiohydrolase-encoding genes from *Aspergillus niger* require D- xylose and the xylanolytic transcriptional activator XlnR for their expression. *Appl Environ Microbiol 65*: 4340-5.

Hasper AA, Visser J, de Graaff LH (2000) The *Aspergillus niger* transcriptional activator XlnR, which is involved in the degradation of the polysaccharides xylan and cellulose, also regulates D-xylose reductase gene expression. *Mol Microbiol 36*: 193-200.

Hrmova M, Biely P, Vrsanska M (1986) Specificity of cellulase and ß-xylanase induction in *Trichoderma reesei* QM 9414. *Arch Microbiol 144*: 307-311

Ilmén M, Saloheimo A, Onnela ML, Penttilä ME (1997) Regulation of cellulase gene expression in the filamentous fungus *Trichoderma reesei. Appl Environ Microbiol 63*: 1298-306

Ilmén M, Thrane C, Penttilä M (1996a) The glucose repressor gene *cre1* of *Trichoderma*: isolation and expression of a full-length and a truncated mutant form. *Mol Gen Genet 251*: 451-60

Ilmén M, Thrane C, Penttilä M (1996b) The glucose repressor gene *cre1* of *Trichoderma*: isolation and expression of a full-length and a truncated mutant form. *Mol Gen Genet 251*: 451-60

Karaffa L, Fekete E, Gamauf C, Szentirmai A, Kubicek CP, Seiboth B (2006) D-Galactose induces cellulase gene expression in *Hypocrea jecorina* at low growth rates. *Microbiology 152*: 1507-14

Kubicek-Pranz EM, Gruber F, Kubicek CP (1991) Transformation of *Trichoderma reesei* with the cellobiohydrolase II gene as a means for obtaining strains with increased cellulase production and specific activity. *J Biotechnol 20*: 83-94

Kubicek CP (1993) From cellulose to cellulase inducers: facts and fiction. P Suominen, Reinikainen, T. (eds) *Proceedings of the 2nd Tricel Symposium on Trichoderma reesei cellulases and other hydrolases*, Espoo, Finland

Kubicek CP, Messner R, Gruber F, Mach RL, Kubicek-Pranz EM (1993) The *Trichoderma* cellulase regulatory puzzle: from the interior life of a secretory fungus. *Enzyme Microb Technol 15*: 90-99

Kubicek CP, Penttilä M, 1998. *Trichoderma & Gliocladium.* Taylor & Francis, London.

Kuhls K, Lieckfeldt E, Samuels GJ, Kovacs W, Meyer W, Petrini O, Gams W, Borner T, Kubicek CP (1996) Molecular evidence that the asexual industrial fungus *Trichoderma reesei* is a clonal derivative of the ascomycete *Hypocrea jecorina. Proc Natl Acad Sci U S A 93*: 7755-60

Kupfer SR, Marschke KB, Wilson EM, French FS (1993) Receptor accessory factor enhances specific DNA binding of androgen and glucocorticoid receptors. *J Biol Chem 268*: 17519-27

Mach R (2002) Regulation of gene expression in industrial fungi: the *Trichoderma* xylanase model G Vannacci, Sarrocco, S. (eds) *6th European Conference on Fungal Genetics*, Pisa, Italy

Mach RL, Strauss J, Zeilinger S, Schindler M, Kubicek CP (1996) Carbon catabolite repression of xylanase I (*xyn1*) gene expression in *Trichoderma reesei*. *Mol Microbiol 21*: 1273-81.

Mach RL, Zeilinger S (2003) Regulation of gene expression in industrial fungi: *Trichoderma*. *Appl Microbiol Biotechnol 60*: 515-22

Mach RL, Zeilinger S, Kristufek D, Kubicek CP (1998) Ca2+-calmodulin antagonists interfere with xylanase formation and secretion in *Trichoderma reesei*. *Biochim Biophys Acta 1403*: 281-9

Mandels M, Parrish FW, Reese ET (1962) Sophorose as an inducer of cellulase in *Trichoderma viride*. *J Bacteriol 83*: 400-8

Mandels M, Reese ET (1960) Induction of cellulase in fungi by cellobiose. *J Bacteriol 79*: 816-26

Margolles-Clark E, Ilmén M, Penttilä M (1997) Expression patterns of ten hemicellulase genes of the filamentous fungus *Trichoderma reesei* on various carbon sources. *J Biotechnol 57*: 167-179

Margolles-Clark E, Saloheimo M, Siika-aho M, Penttilä M (1996a) The alpha-glucuronidase-encoding gene of *Trichoderma reesei*. *Gene 172*: 171-2

Margolles-Clark E, Tenkanen M, Luonteri E, Penttilä M (1996b) Three alpha-galactosidase genes of *Trichoderma reesei* cloned by expression in yeast. *Eur J Biochem 240*: 104-11

Margolles-Clark E, Tenkanen M, Nakari-Setala T, Penttilä M (1996c) Cloning of genes encoding alpha-L-arabinofuranosidase and beta-xylosidase from *Trichoderma reesei* by expression in *Saccharomyces cerevisiae*. *Appl Environ Microbiol 62*: 3840-6

Margolles-Clark E, Tenkanen M, Soderlund H, Penttilä M (1996d) Acetyl xylan esterase from *Trichoderma reesei* contains an active-site serine residue and a cellulose-binding domain. *Eur J Biochem 237*: 553-60

Marmorstein R, Carey M, Ptashne M, Harrison SC (1992) DNA recognition by GAL4: structure of a protein-DNA complex. *Nature 356*: 408-14

Marmorstein R, Harrison SC (1994) Crystal structure of a PPR1-DNA complex: DNA recognition by proteins containing a Zn2Cys6 binuclear cluster. *Genes Dev 8*: 2504-12

Mathieu M, Felenbok B (1994) The *Aspergillus nidulans* CREA protein mediates glucose repression of the ethanol regulon at various levels through competition with the ALCR-specific transactivator. *Embo J 13*: 4022-7

Narendja FM, Davis MA, Hynes MJ (1999) AnCF, the CCAAT binding complex of *Aspergillus nidulans*, is essential for the formation of a DNase I-hypersensitive site in the 5' region of the *amdS* gene. *Mol Cell Biol 19*: 6523-31.

Pail M, Peterbauer T, Seiboth B, Hametner C, Druzhinina I, Kubicek CP (2004) The metabolic role and evolution of L-arabinitol 4-dehydrogenase of *Hypocrea jecorina*. *Eur J Biochem 271*: 1864-72

Poutanen K, Rättö M, Puls J, Viikari L (1987) Evaluation of different microbial xylanolytic systems. *J Biotechnol 6*: 49-60

Rauscher R, Würleitner E, Wacenovsky C, Aro N, Stricker AR, Zeilinger S, Kubicek CP, Penttilä M, Mach RL (2006) Transcriptional regulation of *xyn1*, encoding xylanase I, in *Hypocrea jecorina. Eukaryot Cell 5*: 447-56

Saloheimo A, Aro N, Ilmén M, Penttilä M (2000) Isolation of the *ace1* gene encoding a Cys(2)-His(2) transcription factor involved in regulation of activity of the cellulase promoter *cbh1* of *Trichoderma reesei. J Biol Chem 275*: 5817-25

Saloheimo M, Kuja-Panula J, Ylosmaki E, Ward M, Penttilä M (2002) Enzymatic properties and intracellular localization of the novel *Trichoderma reesei* beta-glucosidase BGLII (cel1A). *Appl Environ Microbiol 68*: 4546-53

Seiboth B, Hakola S, Mach RL, Suominen PL, Kubicek CP (1997) Role of four major cellulases in triggering of cellulase gene expression by cellulose in *Trichoderma reesei. J Bacteriol 179*: 5318-20

Seiboth B, Hartl L, Pail M, Fekete E, Karaffa L, Kubicek CP (2004) The galactokinase of *Hypocrea jecorina* is essential for cellulase induction by lactose but dispensable for growth on d-galactose. *Mol Microbiol* 51: 1015-25

Seiboth B, Hartl L, Salovuori N, Lanthaler K, Robson GD, Vehmaanpera J, Penttila ME, Kubicek CP (2005) Role of the *bga1*-encoded extracellular {beta}-galactosidase of *Hypocrea jecorina* in cellulase induction by lactose. *Appl Environ Microbiol 71*: 851-7

Senior DJ, Mayers PR, Saddler JN (1989) Production and purification of xylanases. *ACS Symp Ser 309*: 641-655

Sternberg D, Mandels GR (1979) Induction of cellulolytic enzymes in *Trichoderma reesei* by sophorose. *J Bacteriol 139*: 761-9

Strauss J, Mach RL, Zeilinger S, Hartler G, Stoffler G, Wolschek M, Kubicek CP (1995) Cre1, the carbon catabolite repressor protein from *Trichoderma reesei. FEBS Lett 376*: 103-7.

Stricker AR, Grosstessner-Hain K, Würleitner E, Mach RL (2006) Xyr1 (xylanase regulator 1) regulates both the hydrolytic enzyme system and D-xylose metabolism in *Hypocrea jecorina. Eukaryot Cell 5*: 2128-37

Stricker AR, Steiger MG, Mach RL (2007a) Xyr1 receives the lactose induction signal and regulates lactose metabolism in Hypocrea jecorina. *FEBS Lett 581*: 3915-20

Stricker AR, Trefflinger P, Aro N, Penttilä M, Mach RL (2007b) Role of Ace2 (Activator of Cellulases 2) in the *xyn2* transcriptosome in *Hypocrea jecorina. Fungal Genetics and Biology*

Tenkanen M, Puls J, Poutanen K (1992) Two major xylanases from *Trichoderma reesei. Enzyme Microb Technol 14*: 566-574

Törrönen A, Harkki A, Rouvinen J (1994) Three-dimensional structure of endo-1,4-beta-xylanase II from *Trichoderma reesei*: two conformational states in the active site. *Embo J 13*: 2493-501

Törrönen A, Mach RL, Messner R, Gonzalez R, Kalkkinen N, Harkki A, Kubicek CP (1992) The two major xylanases from *Trichoderma reesei*: characterization of both enzymes and genes. *Biotechnology (N Y) 10*: 1461-5.

Törrönen A, Rouvinen J (1995) Structural comparison of two major endo-1,4-xylanases from *Trichoderma reesei*. *Biochem J 34*: 847-856

van Peij NN, Visser J, de Graaff LH (1998) Isolation and analysis of *xlnR*, encoding a transcriptional activator co- ordinating xylanolytic expression in *Aspergillus niger*. *Mol Microbiol 27*: 131-42.

Ward M (2006) Improving secreted Enzyme Production by *Trichoderma reesei*. RL Mach, S Zeilinger (eds) 9th International Workshop on *Trichoderma* and *Gliocladium*, Vienna

Würleitner E, Pera L, Wacenovsky C, Cziferszky A, Zeilinger S, Kubicek CP, Mach RL (2003) Transcriptional regulation of *xyn2* in *Hypocrea jecorina*. *Eukaryot Cell 2*: 150-8

Zeilinger S, Ebner A, Marosits T, Mach R, Kubicek CP (2001) The *Hypocrea jecorina* HAP 2/3/5 protein complex binds to the inverted CCAAT-box (ATTGG) within the *cbh2* (cellobiohydrolase II-gene) activating element. *Mol Genet Genomics 266*: 56-63.

Zeilinger S, Kristufek D, Arisan-Atac I, Hodits R, Kubicek CP (1993) Conditions of formation, purification, and characterization of an alpha-galactosidase of *Trichoderma reesei* RUT C-30. *Appl Environ Microbiol 59*: 1347-53

Zeilinger S, Mach RL (1998) Xylanolytic enzymes of *Trichoderma reesei*: properties and regulation of expression. *Curr Topics Cer Chem 1*: 27-35

Zeilinger S, Mach RL, Schindler M, Herzog P, Kubicek CP (1996) Different inducibility of expression of the two xylanase genes *xyn1* and *xyn2* in *Trichoderma reesei*. *J Biol Chem 271*: 25624-9.

In: Current Advances in Molecular Mycology
Eds: Y. Gherbawy, R.L. Mach and M.K. Rai

ISBN 978-1-60456-909-4

Chapter IX

pH-Related Transcriptional Regulation of Lignocellulolytic Enzymes and Virulence Factors Genes in Fungi

Marcio José Poças-Fonseca[1,*], Thiago Machado Mello-de-Sousa[2] and Sérgio Martin Aguiar[2]

[1]Vienna University of Technology (TU Wien), Institute of Chemical Engineering, Gene Technology Group, Vienna, Austria;

[2]Department of Cellular Biology, Institute of Biological Sciences, University of Brasilia. 70.910-900, Brasilia-DF. Brazil.

Abstract

In several filamentous fungi and yeasts, the PacC zinc-finger transcription factor regulates gene expression in response to alkaline external pH. This response is dependent on a signal transduction cascade, in which the products of six *pal* genes take part. In the vast majority of species, PacC functional activation involves a two-step proteolytic process. Most recently, the participation of the multivesicular body complex and of the 26S proteasome was implicated in this activation. Functional PacC is capable of activating the transcription of genes which should be preferably expressed at alkaline pH, and to repress those which should be transcribed preferentially at acidic pH. The PacC system influences the production of plasma membrane permeases, exported enzymes, secondary metabolites - such as antibiotics and mycotoxins - and the pathogenicity against plants and humans. In this work, we summarize the most recent information concerning the molecular genetics of the PacC regulatory mechanism. Original data on the PacC effect over cellulases production by the thermophilic mould *Humicola grisea* var. *thermoidea*, and on the human pathogen *Paracoccidioides brasiliensis pac*C gene

* Correspondence concerning this article should be addressed to: Marcio José Poças-Fonseca, Vienna University of Technology (TU Wien), Institute of Chemical Engineering, Gene Technology Group. Getreidemarkt 9/166/5/2 A-1060. Vienna, Austria

structure and expression, are also presented. Pb*pacC* is the first transcription factor gene characterized in *P. brasiliensis* genome.

Key words: pH, transcription regulation, PacC, *Humicola grisea* var. *thermoidea*, *Paracoccidioides brasiliensis*.

INTRODUCTION

In several situations, microorganisms respond to alterations in the external environment by modulating their gene expression pattern. According to Espeso *et al.* (1993), this modulation can happen at two levels: 1) as an adaptative response in the intracellular metabolic pathways or 2) at the synthesis of exported molecules which could modify the environment.

Fungi are particularly versatile organisms, since they can grow at a wide range of different temperatures, pH values, osmolarity and ionic strength conditions. The ascomycete *Aspergillus nidulans*, for instance, can grow in culture media with pH varying from 2.5 to 10.5 (Rossi and Arst, 1990). Since this fungus permeases, secreted enzymes and exported metabolites normally lose their activity at extreme pH values, their synthesis is mainly regulated by the ambient pH (Espeso *et al.*, 1993).

In order to resist to variations of the external pH, microorganisms possess a homeostatic mechanism to control the intracellular pH variation. In this mechanism, the participation of a plasma membrane H^+-ATPase, which creates a protons electrochemical gradient, is essential (Portillo, 2000). Nonetheless, the only way to protect the molecules localized at the plasma membrane barrier, or beyond it, is to assure that they are synthesized only at the appropriate pH (van den Hombergh *et al.*, 1996).

In the last two decades, the scientific literature have been reporting the importance of the pH regulatory pathway for the expression of fungal enzymes of industrial interest, for the production of antibiotics, particularly the β-lactams, and for the biological processes involved in the virulence of fungi pathogenic to plants and humans. In this work, we present a general view of the conserved pH-driven gene regulation system in filamentous fungi and yeasts, focusing in enzymes production and in pathogenicity against humans. Recent results from our group, on the lignocellulolytic deuteromycete *Humicola grisea* var. *thermoidea* and the dimorphic human pathogen *Paracoccidioides brasiliensis*, are presented.

DEVELOPMENT

Several microorganisms are subject to pH-related gene expression (Hall *et al.*, 1995). Nonetheless, it is in the filamentous fungus *A. nidulans* that this biological process has been widely elucidated at the physiologic, genetic and molecular aspects, being considered a model organism for pH-driven gene regulation.

Caddick *et al.* (1986) verified that mutations in the *A. nidulans pacC* locus simulate fungal growth at alkaline pH: the mutants secrete, for instance, alkaline phosphatase at acidic pH culture medium. The same authors showed that mutations in the *palA*, *palB*, *palC*, *palE* and *palF* loci, which had been selected by Dorn (1965), reproduce the effect of growing at acid pH: even when cultivated at neutral or at basic pH, the mutants produce high levels of γ-aminobutyrate (GABA) permease, which presents an acidic optimum pH. Further characterization of these and of new mutants resulted in the proposition of the following model: loss-of-function mutations in the *pacC* ($pacC^{+/-}$) or in any of the *pal* loci lead to an acidity-mimicking phenotype, while gain-of-function mutations in the *pacC* locus ($pacC^{C}$) lead to alkaline-mimicking effects (Tilburn *et al*., 1995).

The *A. nidulans pacC* Model

Most of the knowledge concerning the molecular aspects of the *A. nidulans* pH regulatory system is due to the groups of Herbert N. Arst Junior (Imperial College London) and Miguel A. Peñalva (Centro de Investigaciones Biológicas CSIC, Madrid) and their colleagues. For detailed reviews, see Peñalva and Arst (2002), Arst and Peñalva (2003a), Arst and Peñalva (2003b) and Peñalva and Arst (2004).

In *A. nidulans*, the product of the *pacC* gene, the protein PacC, activates the transcription of genes which should be expressed in alkaline conditions and represses those whose expression should occur only in acidic conditions. This regulatory mechanism is triggered by a signal transduction pathway, involving the products of the *palA*, *palB*, *palC*, *palF*, *palH* and *palI* genes (Arst *et al*., 1994).

Tilburn *et al*. (1995) characterized the *A. nidulans pacC* gene. The structural analysis of the deduced protein, with 678 amino acids and 73 kDa, revealed typical features of transcription factors: the N-terminus presents three Cys_2His_2 zinc fingers domains; 2) the N-terminus also presents several alanine residues and an α-helix structure; 3) in the central region of the protein, a nuclear localization signal (*SKKR...KRRQ*) is found; 4) the protein presents several *S/TPXX* (*X* = any amino acid) motifs; 5) the C-terminus region is rich in glutamine and in acidic residues, such as aspartyl and glutamyl. Espeso *et al*. (1997) demonstrated that the three zinc fingers are the protein domains that bind to the DNA motif *5′-GCCARG-3′* in the promoters of the genes subject to PacC regulatory control. Only the second and the third zinc fingers domains interact directly with the DNA; the first domain functions by stabilizing the structure of the second one (Tilburn *et al*.,1995; Espeso *et al*., 1997).

It was initially believed that a single proteolytic cleavage of the C-terminus domain, which occurs in response to alkaline pH, would be sufficient to activate the PacC transcription factor (Orejas *et al*., 1995). Espeso *et al*. (2000) demonstrated that three internal domains of the protein interacted with each other, resulting in a "closed" conformation which prevented the proteolytic processing. Under alkaline conditions, the protein would adopt an "open" conformation, allowing the cleavage and resulting activation. The studies of Díez *et al*. (2002) were crucial to the elucidation of PacC processing and activation: these authors demonstrated that a "signalling protease" cleaves PacC in a so called *signalling protease box*, a region of 24 amino acids which is highly conserved among related proteins of different fungi. This cleavage removes about 180 residues of the C-terminus and generates an approximately 495 amino acids intermediate. This first cleavage only occurs upon alkaline ambient pH. A second protease, the "processing protease", would remove – in a pH-independent reaction- circa 245 C-terminal residues, generating a 250 amino acids N-terminal polypeptide which is the PacC functionally active form. Mingot *et al.* (2001) showed that the active and the intermediate forms are located in the nucleus, whereas the full length protein is cytoplasmatic. It was recently proposed that the "processing protease" actually corresponds to the proteasome complex (Hervás-Aguilar *et al.*, 2007). Figure 1 summarizes the two-step PacC activation mechanism.

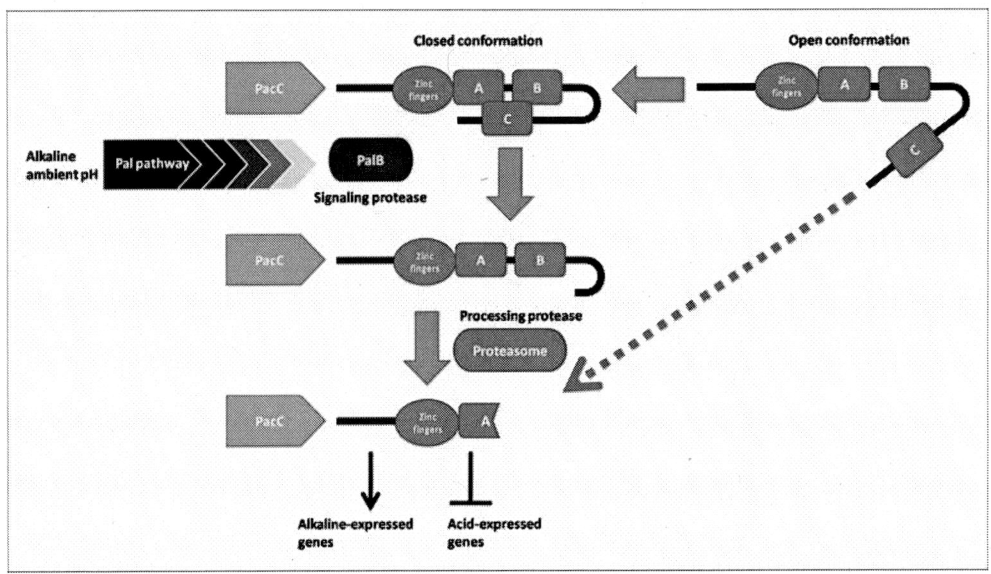

Figure 1. Schematic model for *Aspergillus nidulans* PacC functional activation (adapted from Peñalva, M. A. and Arst, H. N. Recent Advances in the Characterization of Gene Expression in Filamentous Fungi and Yeasts. Annu. Rev. Microbiol. 2004. 58: 425-451).

As it was shown by Espeso *et al.* (1997), for the isopenicillin-N-synthetase gene (*ipnA*) promoter, the alkaline pH-dependent PacC binding to the *5′-GCCARG-3′* motif is necessary and sufficient to activate the transcription. By analysing the γ-aminobutyrate (GABA) permease gene (*gabA*) promoter, Espeso and Arst (2000) demonstrated that a double PacC binding site overlaps the binding site for the IntA transcriptional activator, which is induced by certain amino acids. In this context, the authors verified that the GABA permease transcriptional repression under alkaline pH occurs because PacC competes with IntA for the binding to the promoter, thus avoiding the transcription induction. In another extremely elegant experiment, the same authors mutually exchanged the PacC/IntA recognition site with the IntA binding site of the acetamidase gene (*amdS*) promoter, which does not respond to pH variation. This change abolished *gabA* pH regulation and rendered *amdS* subject to it.

pacC Homologs and Mutations Effects

Homologs to the *A. nidulans pacC* gene were cloned and characterized for other filamentous fungi: *Aspergillus niger* (MacCabe *et al.*, 1996), *Penicillium chrysogenum* (Suarez and Peñalva, 1996), *Acremonium chrysogenum* (Schmitt *et al.*, 2001), *Sclerotinia sclerotiorum* (Rollins and Dickman, 2001), *Fusarium verticillioides* (Flaherty *et al.* 2003), *Fusarium oxysporum* (Caracuel *et al.*, 2003), *Ustilago maydis* (Aréchiga-Carvajal and Ruiz-Herrera, 2005), *Aspergillus giganteus* (Meyer *et al.*, 2005) and *Trichophytum rubrum* (Ferreira-Nozawa *et al.* 2006). DNA and protein sequence databanks present several other fungi putative *pacC* homologs, but data on gene cloning and characterization are still not available. *pacC* genes occur as single copy in all of these microorganisms.

The *pac*C genes open reading frames vary from 1,830 to 2,484 nucleotides (*F. oxysporum* and *U. maydis*, respectively) and they are interrupted by two to three intervening sequences. The *P. chrysogenum* gene presents just one intron, while for the *U. maydis* sequence no recognizable intron could be identified. The *F. verticillioides PAC1* gene presents three introns; nonetheless, the third intron is not always processed, leading to additional 25 amino acids residues in the C-terminus of the predicted protein. The physiological role of such an alternative splicing is not known. *pacC* genes intervening sequences present from 50 to 90 base pairs (bp), although for the genus *Aspergillus* this average number is higher (138 – 149 bp).

The region which codes for the zinc fingers, particularly for the second and third ones, is highly conserved amongst the genes mentioned above. The region which codes for the interacting domains (A, B and C, Figure 1), responsible for the maintenance of *A. nidulans* closed conformation (Espeso *et al.*, 2000), and the signalling protease recognition sequence (Díez *et al.*, 2002) can also be easily identified. The 3′ region of the genes is less conserved. A variable number (2 – 8) of binding sites for PacC itself is found in the 5′ upstream regions of all these genes, strongly suggesting that *pacC* is subject to autoregulation.

pacC null mutations in filamentous fungi seem to result in pleiotropic effects, although *pacC* is not essential for growth. In *A. nidulans*, besides the overexpression of enzymes that should only be produced in acidic environments, a poor growth and defective conidiation was reported (Tilburn *el al.*, 1995). In a murine experimental model for pulmonary aspergillosis, an *A. nidulans pacC* null mutant presented markedly attenuated virulence, while a mutant exhibiting a constitutively processed PacC phenotype was more virulent. These last data point out to the *pacC* regulatory pathway as a relevant virulence factor for opportunistic and pathogenic *Aspergilli* (Bignell *et al.*, 2005).

In the phytopathogen *S. sclerotiorum*, the functional deletion of the *pac1* locus led to growth inhibition at alkaline pH, aberrant development and maturation of sclerotia and to a dramatically reduced virulence, probably in function of the altered oxalic acid accumulation and endopolygalacturonase production (Rollins, 2003). On the other hand, in another plant pathogen, *F. oxysporum*, a loss-of-function *pacC* mutant resulted in an augmented virulence (Caracuel *et al.*, 2003). The authors propose that *pac*C could function as a negative regulator of virulence by preventing the expression of acid-expressed genes whose products would be important for plant infection. In *F. verticilliodes*, a disrupted *PAC1* gene drastically compromised the fungal growth at alkaline pH and increased the fumonisins production; in this view, *PAC1* would also act as a negative regulator for mycotoxins production (Flaherty *et al.*, 2003).

As discussed by Aréchiga-Carvajal and Ruiz-Herrera (2005), in the dimorphic phytopathogen *U. maydis* null *RIM101/pac*C mutants, a wide range of distinct phenotypic changes were observed: i- the dimorphism capacity was not affected, but mycelial morphology and septa distribution were different from the wild type; yeasts cells were also longer and some of them presented septa, suggesting some commitment of the morphogenesis process and cell cycle regulation. ii- the cell wall was more sensitive to the action of lytic enzymes. iii- abnormal polysaccharide and protein, particularly proteases, secretion. iv- a lower tolerance to ionic stress was detected. v- meiosis and mating were not affected and, interestingly, vi- there was no impairment in virulence.

In the human dermatophyte *T. rubrum*, the *pacC* gene disruption resulted in a decrease in conidiation, in the secretion of keratinolytic proteases and in the capacity to grow on human nails as the sole source of nutrition (Ferreira-Nozawa *et al.*, 2006).

In 2005, Meyer *et al.* presented some interesting data concerning the alkaline-dependent up-regulation of the gene that encodes for an antifungal protein (*afp*) in *A. giganteus*. This gene is only expressed at alkaline ambient and its promoter presents two consensus binding sites for the PacC transcription factor. These binding sites proved to be functional in an *in vitro* analysis. Curiously, the authors presented several experimental evidences showing that the alkaline activation of the *afp* promoter does not involve PacC, but probably a calcineurin signalling pathway.

Although the respective *pacC* genes have not yet been cloned and characterized, this pH-dependent transcription regulation system has been implicated in the expression of several other fungal genes. This assumption is based on the presence of putative PacC binding sites in the genes promoters and/or on the differential mRNA expression profile according to the culture medium pH. Some examples comprise: the *pgx2* gene, encoding a polygalacturonase from *Aspergillus awamori* (Nagai *et al.*, 2000); *papA*, encoding an aspartyl protease from the mycoparasite *Trichoderma harzianum* (Delgado-Jarana *et al.*, 2002); *aflR* and *pksA*, genes of the aflatoxin biosynthetic pathway, from *Aspergillus parasiticus* (Cary *et al.*, 2000; Ehrlich *et al.*, 2002); the *spr1* serine protease gene from the nematode pathogen *Monascroporium megalosporium* (Kanda *et al.*, 2006) and the *pnl* gene, coding for a pectin lyase from the phytopathogen *Penicillium occitanis* (Trigui-Lahiani and Gargouri, 2007).

The Transduction Signal Pathway

The proteolytic cleavage and resulting functional activation of the PacC protein depends on an alkaline pH-triggered signal transduction cascade, in which the products of the *palA*, *palB*, *palC*, *palF*, *palH* and *palI* genes are involved (Denison *et al.*, 1995; Orejas *et al.*, 1995).

The predicted product of the *palI* gene (PalI) presents four hydrophobic segments, compatible to the structure of transmembrane domains. It is possible that PalI represents a plasma membrane-associated sensor, related to the external pH detection (Denison *et al.*,

1998). PalI is homolog to *S. cerevisiae* RIM9, also a plasma membrane protein, related to the meiosis regulation and to invasive growth (Futai *et al.*, 1999).

The deduced PalH protein (*S. cerevisiae* homolog Rim21p) displays seven-transmembrane motifs and a long hydrophilic C-terminus; thus, it could also represent a plasma membrane sensor (Negrete-Urtasun *et al.*, 1999). Herranz *et al.* (2005) have assigned the product of the gene *palF* (Maccheroni *et al.*, 1995) as an arresting-like protein, capable of binding to seven-transmembrane receptors to regulate their function through phosphorylation and ubiquitination. It was shown that PalF (*S. cerevisiae* Rim8p) interacts with the hydrophilic C-terminus of PalH, which -in cooperation with PalI- would sensor the ambient pH (Herranz *et al.* 2005). These last authors also postulated that PalH could constitute a link between the ambient pH sensor(s) and the endocytic trafficking, from the plasma membrane to endosome, which have been more recently implicated in fungal pH-response regulatory pathway.

The PalB deduced protein (Denison *et al.*, 1995) is a calpain-like cysteine protease, whose homolog in *S. cerevisiae* is Rim13p (Lamb *et al.*, 2000). Recently, Peñas *et al.* (2007) have shown, by mutational analysis, that PalB is essential for the signalling proteolysis, which removes the negatively acting PacC C-terminus, but it is not required for the processing proteolysis, the final step for PacC functional activation (Figure 1). Interestingly, PalB levels do not depend on the ambient pH or on the Pal signal transduction pathway.

Negrete-Urtasun *et al.* (1997), by analysing the predicted protein from the *palA* gene, reported a nuclear localization signal (*RRKRK*) and a C-terminus extremely rich in proline. Vincent *et al.* (2003) demonstrated that PalA recognizes a protein-protein binding motif (*YPXL/I*) and that PacC presents two of such motifs flanking the signalling protease cleavage site. These motifs are conserved in all of the PacC homologs. In this view, PalA binding would be important to recruit or to facilitate the access of the signalling protease (PalB) to the full-length inactive PacC form. Xu and Mitchell (2001) had previously shown that *S. cerevisiae* and *C. albicans* present a PalA homolog, the Rim20p protein, which is essential for Rim101 (the PacC yeast homolog) processing.

The *palC* gene (Negrette-Urtasun *et al.*, 1999) was investigated by gene-specific mutation analysis (Tilburn *et al.*, 2005). The PalC protein is related to the *S. cerevisiae* endosomal sorting complexes required for transport (ESCRT). This complex participates in the recognition and traffic of endocytic vesicles to the multivesicular body (MVB) and then to the lysosome. ESCRT functions are related to the degradation of ubiquitinated proteins, recycling of plasma membrane components -such as receptors, ion channels and transporters- but also to the transport of lysosomal hydrolases precursors (for review, see Slagvold *et al.*, 2006; Williams and Urbé, 2007). Recently, Galindo *et al.* (2007) demonstrated that, upon alkaline conditions, PalC localizes to punctuate structures at or near the plasma membrane, and in this context, it could link the components responsible for sensing the environmental pH with those responsible for transcription activation or repression in the PacC pathway.

pH-Associated Gene Regulation in Yeasts

A pH-related gene regulation system, analogue to filamentous fungi *pacC*, was also described for the yeasts *Yarrowia lipolytica* (Lambert *et al.*, 1997; Gonzalez-Lopez *et al.*, 2002; Blanchin-Roland *et al.*, 2005), *S. cerevisiae* and *Candida albicans*. These last two, in which the *pacC* gene homologs are called "*RIM101*", will be addressed in this work. *S. cerevisiae RIM101* gene (initially called *RIM1*) was described by Su and Mitchell (1993) as a positive regulator of meiosis, since null mutants for this gene presented defects in meiosis, sporulation and invasive growth. Based on the similarity to *A. nidulans* PacC, Li and Mitchell (1997) suggested that the *S. cerevisiae* Rim101p protein could also take part in a pH-driven regulatory pathway. In the same work, the authors demonstrated that Rim101p is subject to C-terminal proteolytic cleavage, stimulated by alkaline pH and dependent on the products of the genes *RIM8*, *RIM9* and *RIM13* which, in this context, could be homologous to *A. nidulans pal* genes. These observations were confirmed by different subsequent studies.

Divergently from PacC activation, which involves a two-step proteolytic processing (Díez *et al.*, 2002), the Rim101p cleavage to the functional form depends on a single event (Xu and Mitchell, 2001). Another important difference to the *A. nidulans* PacC transcriptional factor, which can act both as transcriptional activator or repressor (vide supra), is that Rim101p seems to function exclusively as a repressor. Actually, it behaves as an indirect activator, by downregulating *NGR1*, a negative regulator of alkaline pH-induced genes, and *SMP1*, an inhibitor of invasive growth and sporulation related genes (Lamb and Mitchell, 2003).

Barwell *et al.* (2005) described a new *S. cerevisiae* Rim101p pathway component, the *DFG16* gene: it codes for a protein presenting seven membrane-spanning segments and a long hydrophilic C-terminus, resembling a G-protein-coupled receptor. *DFG16* is a homolog to *A. nidulans palH* and, according to the authors, its product could interact with Rim21p – another *palH* homolog, thus forming a heterodimeric environmental sensor that promotes Rim101p processing. Furthermore, *DFG16* function is conserved in *C. albicans*.

The dimorphic yeast *C. albicans* is a commensal microorganism of human mucosas, and, in immunocompromised individuals, it can cause severe opportunistic infections. As reviewed by Davis (2003), the capacity to change from the yeast to the hyphal form is crucial to *C. albicans* virulence and it is a pH-dependent process: an acidic environment favours the yeast growth, while an alkaline ambient promotes hyphal development. De Bernardis *et al.* (1998) demonstrated that two genes encoding for cell surface glycoproteins (*PHR1* and *PHR2*), involved in *C. albicans* morphogenesis, are divergently expressed according to the host niche pH and that their expression pattern is a virulence determinant. *PHR1* is expressed at alkaline pH and its deletion resulted in a mutant which was avirulent in a mouse model of systemic infection; however, the capacity to cause vaginal infection in rats remained unaffected. On the other hand, *PHR2* is expressed at acidic pH, and the null mutant of this gene proved to be virulent in the systemic infection model, but avirulent in the vaginitis induction experiments. Interestingly, heterozygous mutants presented an intermediate virulence phenotype, indicating a gene dosage effect.

Heinz *et al.* (2000) have cloned the *PHR1* and *PHR2* homologs of the *C. albicans* closely related species *Candida dubliniensis*, which is also a human pathogen. The pH-regulated mode of expression of the homolog genes was conserved in both species, further reinforcing the relevance of this regulatory pathway for *Candida* pathogenesis.

Kottom *et al.* (2001) identified the *PHR1* gene of *Pneumocystis carinii*, an opportunistic fungal pathogen frequently associated with severe pneumonia in patients with AIDS. This gene presents 37% homology to *C. albicans PHR1*/*PHR2* and it is only expressed at alkaline pH to take part in the maintenance of fungal cell wall integrity, thus acting as an important virulence factor.

The *C. albicans pac*C/*RIM101* homolog, initially designated *PRR2*, was cloned and characterized by Ramon *et al.* (1999). *C. albicans RIM101* null mutants lost the pH-dependent gene regulation pattern and exhibited defective hyphal development. As it was previously demonstrated for other organisms, the expression of *RIM101* itself was shown to be alkaline pH-driven and dependent on the product of the *PRR1* gene, an *A. nidulans palF* homolog (Porta *et al.*, 1999). The *C. albicans KER1* gene, encoding a plasma-membrane protein involved in cell aggregation, is also regulated by *RIM101*. A homozygous mutant strain for *KER1* presented defects in the cell wall composition and/or structure, leading to an attenuated virulence in a mouse systemic infection model (Galán *et al.*, 2004).

The ability to obtain iron from the environment is fundamental to human pathogens. Bensen *et al.* (2004) have shown that many genes possibly involved in *C. albicans* iron metabolism are pH regulated in a *RIM101*-dependent manner, indicating that the *RIM101* pathway is also important for adaptation to iron starvation. The involvement of PacC in the regulation of three *A. nidulans* genes, related to the biosynthesis and uptake of siderophores (low-molecular weight ferric iron chelators), was demonstrated by Eisendle *et al.* (2004).

The proteolytic processing of *C. albicans* Rim101p is mediated by the product of the *RIM13* gene, a calpain-like protease, homologous to *A. nidulans* PalB. *RIM13* null mutants are more sensitive to lithium chloride and to hygromycin B, characteristics often associated to cell wall defects (Li *et al.*, 2004). These authors also demonstrated that, besides the Rim13p-mediated cleavage, which occurs at alkaline pH, Rim101p is also proteolyticaly processed at acidic pHs, giving rise to a polypeptide that could possibly govern pH-independent gene regulation events.

Recently, Baek *et al.* (2006) showed that, besides acting as an inducer of alkaline-expressed genes, *C. albicans* Rim101p can also act as a direct repressor of the transcription of the *PHR2* gene at alkaline conditions. The same authors suggested that the Rim101p binding site in *C. albicans* genes promoters is *GCCAAGAA*, although slight context-dependent variations can occur.

By using an insertional mutagenesis strategy, with a gene disruption marker that permits the differentiation between homozygous from heterozygous mutants (the *UAU1* cassette, Enloe *et al.*, 2000), Davis *et al.* (2002) overcome the major problem for obtaining null mutants for a given gene in *C. albicans*: the fact that it is diploid and does not present a sexual cycle. This strategy allowed the identification of a new gene, *MDS3*, which is involved in alkaline pH gene regulation in a *RIM101*-independent manner. *MDS3* codes for a 1,383-residue protein with a kelch-like protein interaction domain. Loss of function mutants of this gene were defective in the expression of *HWP1* and *ECE1*, genes required for the hyphae formation and thus virulence, but not in the expression of all alkaline pH-dependent genes. In this view, Mds3p effects are not restricted to morphogenesis. By analysing *MDS3* and *RIM101* double null mutants, the authors could show that the two pathways act in parallel to regulate ion sensitivity and alkaline pH tolerance. Furthermore, in a mouse-tail-vein injection model, it was shown that Mds3p and Rim101p contributed independently to *C. albicans* virulence. It is worth commenting that the *S. cerevisiae* Mds3p homologs, Mds3p and Pmd1p, also promote a response to alkaline pH, according to the data of Davis *et al.* (2002).

The *C. albicans* ability to colonize a substrate surface, forming a biofilm, is considered a virulence trait. Richard *et al.* (2005), by analysing *MDS3* null mutants, concluded that this gene plays an important role in biofilm maturation, because its loss of function led to defective hyphae production. Curiously, *RIM101* is not required for biofilm formation.

As recently reviewed by Staib and Morschhäuser (2006), under oxygen and nutrient-limiting conditions or at low temperature, *C. albicans* and *C. dubliniensis* produce chlamydospores, which are large, spherical, thick-walled cells formed at the end of hyphal filaments. Although they have been found in clinical species, there is no experimental evidence that chlamydospores can play any role in infection. Yet, its biological function remains to be unveiled. Nobile *et al.* (2003) have shown that mutants for the *C. albicans* pH-response regulators *RIM13*, *RIM101* and *MDS3* produced fewer chlamydospores and that their development was delayed in comparison to the wild type yeast. Interestingly, chlamydospores are produced exclusively under acidic conditions, what suggests that a novel signal may activate the alkaline pH-response regulators.

Fungal Lignocellulolytic Systems and pH Regulation

Vegetal biomass annual production is estimated in about 155 billions of tons, (Rajaratham *et al.* 1992). In this view, microorganisms capable of acting on lignocellulose degradation are essential for the carbon cycle maintenance. As reviewed by Poças-Fonseca and Maranhão (2005), a wide diversity of filamentous fungi presents an elaborate and efficient enzymatic system for vegetal biomass decay. It is generally assumed that this substrate is formed mainly by 50% cellulose, 25% hemicellulose and 25% lignin.

Cellulose presents a simple chemical structure, being composed of 100 to 14,000 D-glucose residues, linked by beta-1,4 glycosidic bonds, forming linear homopolymers. The cellulose monomer is actually cellobiose, a glucose dimmer. On the other hand, the physical structure (reviewed by Béguin and Aubert, 1994) is rather complex: parallel linear chains are held together through intra and interchains hydrogen bonds and van der Waals forces, forming rigid and insoluble microfibrils. These microfibrils self assemble into highly crystalline fibers, among which some amorphous regions are found. Hemicellulose and lignin cooperate in maintaining the cellulose compact structure, protecting it from hydrolytic enzymes such as cellulases.

The physical structure and heterogeneity of the cellulosic substrate is reflected by the necessity of different enzymes in order to achieve its degradation. Using the *Trichoderma reesei* cellulolytic system as model, three classes of cellulases were identified: endoglucanases (EC 3.2.1.4), exoglucanases or cellobiohydrolases (EC 3.2.1.91) and beta-glucosidases (EC 3.2.1.21). These enzymes act synergistically to break down cellulose to glucose (Henrissat *et al.* 1985). It is commonly accepted that endoglucanases hydrolyze internal bonds of the cellulose chains, producing reducing and non-reducing ends from which cellobiohydrolases remove cellobiose residues; beta-glucosidases then cleave cellobiose and other cellooligosaccharides to glucose. Endoglucanases hydrolyze preferably amorphous, soluble and substituted celluloses, while cellobiohydrolases degrade crystalline cellulose. This model of hydrolysis is conserved amongst the vast majority of cellulolytic fungi studied to date (Poças-Fonseca and Maranhão, 2005).

Xylan is the principal component of hemicellulose. It is formed by a linear scaffold of ß-1,4-linked D-xylopyranose units, with side chains of arabinofuranose, glucoronic and methylglucoronic acids. Thus, on the contrary of cellulose, xylan is chemically heterogenous. Xylan degradation (reviewed by Polizeli *et al.*, 2005) is mainly achieved by endo-1,4-ß-xylanases (EC 3.2.1.8), which cleave the xylan internal ß-1,4 linkages, generating oligosaccharides, and by ß-xylosidases (EC 3.2.1.37), which hydrolyze these xylooligosaccharides to xylose units.

The production of enzymes involved in plant cell wall breakdown is regulated mainly at the transcriptional level (Aro *et al.*, 2005). Nonetheless, the involvement of the PacC pathway in the lignocellulolytic enzymes regulation has not been extensively studied.

Some interesting data have been obtained for xylanases regulation. Although the *A. nidulans* ß-xylosidase *xlnD* gene regulatory region presents several PacC binding sites, Pérez-Gonzáles *et al.* (1998) could not demonstrate that PacC was indeed involved in *xlnD* transcriptional regulation. Curiously, the *xlnA* and *xynB* endoxylanase genes from the same fungus present an opposite expression pattern in response to the time course of induction and ambient pH: while *xlnA*, encoding a neutral xylanase, is early expressed and exclusively at alkaline growth conditions, *xlnB* – which encodes an acidic xylanase - is expressed only after 4 hours of induction and mainly at acidic pH (MacCabe *et al.* 1998).

In *Penicillium purpurogenun*, expression analyses performed at different pH have shown that the endoxylanase genes *xynA* and *xynB* are clearly regulated by pH, despite the absence of PacC binding sites in the *xynB* promoter (Chávez *et al.*, 2002).

Recently, Tanaka *et al.* (2006), analyzing the expression of the *Aureobasidium pullulans xynII* endoxylanase gene by quantitative real time PCR, demonstrated that the transcription levels at pH 6.0 and 8.0 were respectively 8-fold and 22-fold higher than that at pH 2.7.

Our group have demonstrated that the thermophilic deuteromycete *Humicola grisea* var. *thermoidea*, is a potent cellulases producer (Azevedo *et al.* 1990, Poças-Fonseca *et al.* 1997, Poças-Fonseca *et al.* 2000), presenting a considerable potential for agricultural wastes bioconvertion processes (De-Paula *et al.* 2003; Rossi *et al.*, 2007). *H. grisea* var. *thermoidea* cellobiohydrolase *cbh1.1* and *cbh1.2* genes have been cloned and characterized (Azevedo *et al.* 1990; Poças-Fonseca *et al.* 1997).The *cbh1.1* promoter presents three consensus sites for PacC binding in the antisense strand, while the *cbh1.2* regulatory region also displays three PacC binding sites but in the sense strand. Preliminary assays with the *Escherichia coli lacZ* gene as reporter, in the heterologous system of *A. nidulans*, suggested that *cbh1.1* is not subject to pH regulation, while *cbh.1.2* transcription seems to be pH-driven in a PacC-dependent mechanism (Poças-Fonseca, 2000).

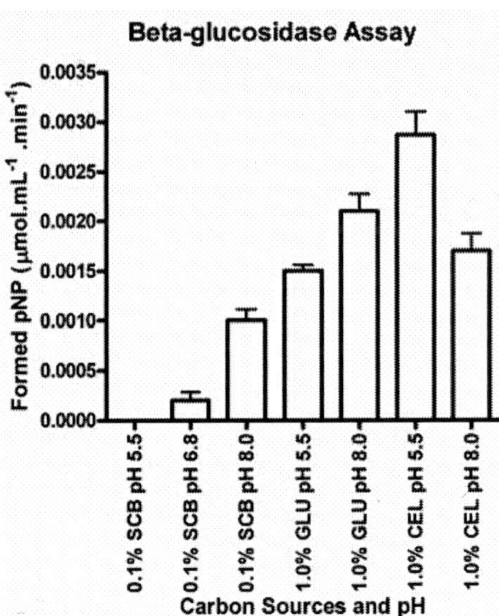

Figure 2. Beta-glucosidase activity in *Humicola grisea* var. *thermoidea* supernatants after cultivation on minimal medium (MM) supplemented with sugar cane bagasse (SCB), Glucose (GLU) or Cellobiose (CEL) at different pH values. 10^8 *H. grisea* spores were inoculated in 250 mL MM supplemented with 0.1% SCB, 1.0% GLU or 1.0% CEL. Culture media pH was adjusted to 5.5 with 30 mM citrate, to 6.8 with 2-(*N*-morpholino) ethansulfonic acid or to 8.0 with Tris-HCl. Cultures were incubated at 42 °C, under agitation for 12 h and filtered through filter paper. β-glucosidase activity was assayed by using a 1 mL reaction mixture containing 5 mM ρ-nitophenyl-β-D-glucopiranoside (ρNPG, Sigma), 50 mM acetate buffer (pH 6.0), and an appropriate dilution of culture supernatant. After 10 min of incubation at 42°C, the reaction was stopped by adding 1 mL of 1 M Na_2CO_3, and *p*-nitrophenol release was monitorated at A_{405}. One unit of β-glucosidase activity corresponded to the release of 1 μmol of ρ-nitrophenol min^{-1} under these conditions. The experiment was performed in triplicate.

More recently, we have studied the expression profile of the *H. grisea* var. *thermoidea* ß-glucosidases. As shown in Figure 2, total ß-glucosidase activity detected in *H. grisea* supernatants is strictly dependent on the culture medium carbon source and pH. When the minimal medium was supplemented with 0.1% sugar cane bagasse (SCB), no ß-glucosidase activity was detected after growing at pH 5.5. Nonetheless, an increased activity was observed upon culture medium alkalinization. The same result was noted when the fungus was grown on 1.0 % glucose (GLU). Interestingly, when 1.0% cellobiose (CEL) was employed as the sole carbon source, the ß-glucosidase activity was higher at an acidic pH (5.5) than at basic conditions (pH 8.0). This suggests that an interface between the carbon source and the pH-driven regulatory pathways may occur in certain growth conditions.

Figure 3 displays a total protein electrophoretical profile of *H. grisea* cultures supernatants after growing on complete medium buffered at acid, neutral and alkaline pH. The presence of differentially expressed proteins, according to the distinct pH values, corroborates our evidences supporting the existence of a pH regulatory pathway for *H. grisea* var. *thermoidea* transcriptional regulation.

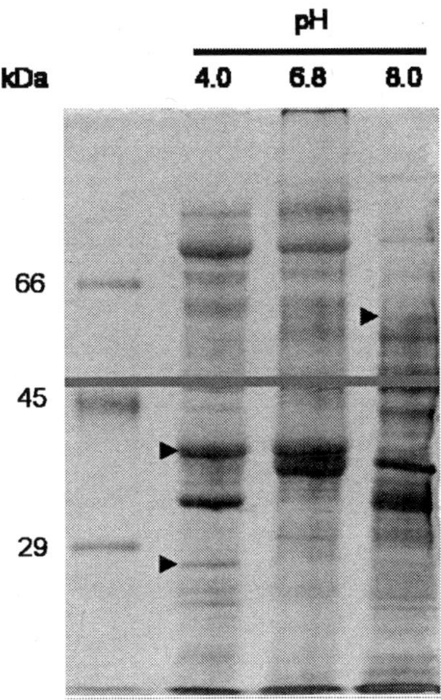

Figure 3. Electrophoretic analysis of *Humicola grisea* var. *thermoidea* supernatants after cultivation on complete medium at different pH values. Sodium dodecyl sulfate-polyacrylamide gel electrophoresis (SDS-PAGE) was performed using 15% polyacrylamide. Culture medium pH values are indicated. Molecular mass marker was purchased from Sigma. Gel was silver-stained by using the kit *plus one – Silver Staining Kit Protein* (Amersham). Differentially expressed proteins are indicated by arrows.

The *Paracoccidioides brasiliensis pacC* Gene

The dimorphic fungus *Paracoccidioides brasiliensis* is the etiologic agent of paracoccidioidomycosis, an important human systemic mycosis, endemic in Central and South America, particularly in Brazil. Infection is probably acquired by inhalation of airborne propagules derived from the mycelial saprophytic form of *P. brasiliensis* (Restrepo *et al.* 2001). In the lungs, the fungus undergoes a dimorphic transition, converting to the yeast form, which is an essential step for the establishment of the infection.

Structural Analysis

Aiming the isolation of a possible *pacC* ortholog from *P. brasiliensis*, genomic sequences of *pacC* homologs from *A. nidulans*, *A. niger*, *A. oryzae*, *S. cerevisiae* and *P. chrysogenum* (GenBank accession numbers: Z47081, X98417, AB035899, NC_001140 and PCU44726) were aligned for the design of PACF and PACR primers. Amplification of *P. brasiliensis* total DNA using these primers resulted in a 969 bp fragment, comprising the highly conserved *pacC* zinc fingers coding region. This amplicon was used as a homologous probe for screening a *P. brasiliensis* genomic library, allowing the identification of a *pacC* homologous sequence (Pb*pacC*, Figure 4A). The deduced translation product shares from

34.9% to 51.1% amino acid identity to other fungi PacC homologs (Figure 5), with highest identity observed with *T. rubrum*, also a human pathogen.

Four putative introns, displaying stop codons, 5' and 3' flanking sequences (5'-GT...AG-3') and the CTRAY motif, needed for the *lariat* formation (Turner, 1993), were identified in the Pb*pacC* sequence (Figure 4A). In this view, the *P. brasiliensis pacC* homolog presents the largest number of introns described to date in the scientific literature, considering this class of genes. *In silico* maintenance of the first three introns, together or individually, leads to a considerable loss of identity with PacC homologs.

Interestingly, the fourth putative intron does not contain stop codon in any reading frame. In this view, it could be subject to alternative splicing, generating additional 30 amino acids residues in the resulting protein. In order to test this hypothesis, primers flanking this intron were designed (Figure 4A). An RT-PCR experiment, using RNA samples extracted from the yeast form cultivated at different pH, suggests that the fourth intron is maintained in the mature mRNA (Figure 4B). In *F. verticillioides* grown at pH 8.4, two different *PAC1* transcripts, resulting from alternative splicing, were detected (Flaherty *et al.*, 2003). Considering this datum, the possibility of splicing of the Pb*pacC* fourth intron, under different physiological conditions, cannot be excluded.

Maintaining the fourth intron, translation of the 2,715 bp open reading frame (GenBank accession no. AF536981) results in a theoretical 706-amino-acid, 75-kDa protein. Most of the structural features described for PacC homologs, such as the three zinc fingers DNA binding domain, are present in PbPacC (Figure 5). The amino acid residues assumed as critical for DNA binding in *A. nidulans* (Lys125; Asp127; His128; Ser131; Arg153; Gln155; Asp156 and Lys158; reviewed by Peñalva and Arst, 2004) are conserved in the PbPacC second and third zinc fingers. The sequence *KKHVKT* (position 174) corresponds to the possible nuclear localization signal, superposed with the third zinc finger (Fernández-Martínez *et al.*, 2003). A glycine-rich region is present at the central region of the protein and could function as a flexible hinge (Suárez and Peñalva, 1996). The *S/TPXX* motif, frequently found in transcription factors (Suzuki, 1989), occurs 11 times in the *P. brasiliensis* PacC homolog. The N-terminus is extremely glutamine-rich (34 residues, including two stretches of 8 and 12 contiguous residues) in contrast to the N-termini of *Aspergilli* PacC homologs, which are particularly alanine-rich.

Pb*pacC* Expression Analysis

The expression profile of Pb*pacC* was evaluated by semiquantitative RT-PCR. RNA samples were collected after yeast 5-hour incubation in YPD media, adjusted to different pH values (5.0, 7.0 and 9.0); α-tubulin (α-*tub*) was used as the constitutive control (Figure 6). Our data show that Pb*pacC* transcripts considerably accumulate at alkaline growth conditions (pH 9.0), suggesting that its expression is pH-dependent, as verified for the homologs described to date.

Figure 4. DNA sequence and some structural features of the *P. brasiliensis pacC* gene. (A) Nucleotide and derived amino acid sequence of Pb*pacC* (GenBank accession no. AF536981). Zinc fingers region is shaded in grey. The four putative introns are indicated by lower-case letters; 5'-GT...AG-3' flanking nucleotides and the potential *lariat* motifs are underlined. Translation truncating stop codons (circled) are found in all the putative introns, except for the fourth one. Primers PACF (5'-CGCCACGTAGGGCGCAAG-3'), PACR (5'-GTGGGAGGTACGATAACTC-3'), PAC12 (5'-CTATGTCATTTGCGGAGGTT-3') and PAC13 (5'-CTTCTGGTGGTGCTGTTTTA-3') positions are shaded, with the direction being indicated by arrows. The derived amino acid sequence introduced by the putative fourth intron is boxed. (B) Amplification products from DNAse-treated total RNA extracted after yeast cultivation for 5 hours in YPD at pH 5.0; 7.0 (buffered with 50mM sodium acetate) or pH 9.0 (buffered with 20mM Tris-HCl). The RT-PCR 288 bp amplicon indicates that the fourth predicted intron is retained in the mature mRNA. RT-PCR assay performed without reverse transcriptase (RT) indicated the absence of genomic DNA contamination.

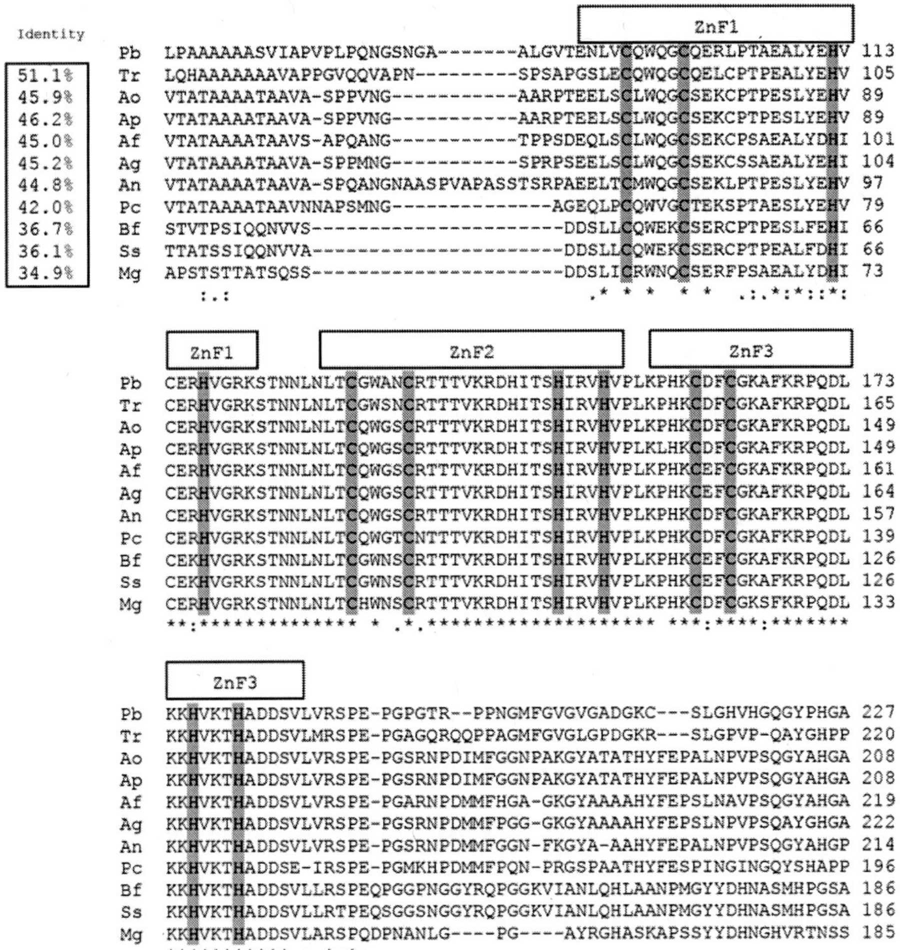

Figure 5. PacC partial amino acids sequence alignment for *P. brasiliensis* and other fungi related proteins. The zinc finger domains (ZnF) are indicated by boxes, with the essential Cys_2His_2 zinc-chelating conserved residues shaded. Abbreviations and accession numbers (GenBank) are as follows: Pb, *Paracoccidioides brasiliensis* (AF536981); Tr, *Trichophyton rubrum* (Q9C1A4); Ao, *Aspergillus oryzae* (Q9HFB3); Ap, *Aspergillus parasiticus* (Q96UW0); Af, *Aspergillus fumigatus* (Q4WY67); Ag, *Aspergillus giganteus* (Q5XL24); An, *Aspergillus nidulans* (Q00202); Pc, *Penicillium chrysogenum* (Q01864); Bf, *Botryotinia fuckeliana* (AAV54519.1); Ss, *Sclerotinia sclerotiorum* (Q9P413); Mg, *Magnaporthea grisea* (Q52B93).

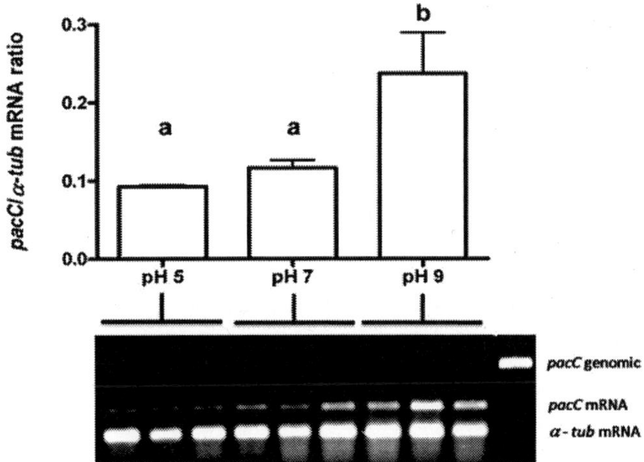

Figure 6. Semiquantitative expression analysis of *P. brasiliensis pacC* gene. Semiquantitative RT-PCR amplifications were realized with DNAse-treated total RNA extracted after yeast cultivation for 5 hours in YPD pH 5.0; 7.0 (buffered with 50mM sodium acetate) and pH 9.0 (buffered with 20mM Tris-HCl). The determination of the minimum number of PCR cycles, i.e. 23, was performed in order to make sure that the amplification was within the exponential range (Marone *et al.*, 2001). The zinc finger region spanning primers ZFpbF (5'-GGATCCGCCCTAGGTGTCACGGAGA-3') and ZFpbR (5'-CTCGAGCATTGGGTGGCCTGGTTC-3'), as well as the α-tubulin (*α-tub*) primers TUB4f (5'-TTCGTTGATCTGGACCCTTC-3') and TUB4r (5'-GGAGGGACGAGCAGTTATCA-3') were used in a multiplex reaction. Pb*pacC* gene relative expression was determined from the relative optical densities of the amplified bands after normalization to α-tubulin. The data are expressed as the mean ± S.E.M. from 3 experiments. Different letters denote statistical relevance, according to the analysis of variance (ANOVA), $P < 0.05$.

To our knowledge, Pb*pacC* is the first transcription factor gene characterized in *P. brasiliensis* genome.

Conclusion and Future Perspectives

In the last two decades, the pH-dependent regulatory pathway mediated by the PacC transcription factor has been elucidated at the biochemical, genetic and molecular level. Recent data concerning the functional activation of PacC indicate the participation of the multivesicular body complex and of the 26S proteasome. The specific role of the six *pal* genes, whose products constitute the signal transduction cascade which triggers pH-driven transcriptional response, has also been extensively studied.

The PacC regulatory system has been described for several distinct filamentous fungi and yeasts, and it seems to present a considerable degree of structural and functional conservation amongst these microorganisms. Biological processes, such as secondary metabolites production and pathogenicity against plants and humans, are markedly influenced by the environment pH. In this view, the studies on the PacC-mediated transcription regulation mechanism may result in important contribution to the optimization of the industrial processes for enzymes and antibiotics production, as well as to the control of agricultural pests and to the treatment of human mycosis.

In our group, we are particularly interested in evaluating the influence of the PacC system in the production of celullases and xylanases by the thermophilic mould *H. grisea* var. *thermoidea*, since this fungus presents a promising potential as an industrial microorganism for processes such as the bioconversion of agricultural wastes in biofuels, animal feed and fertilizers, amongst others (Rossi *et al.*, 2007).

We are also putting efforts in investigating the molecular mechanisms that underline *P. brasiliensis* human infection capacity. At the moment, we are working at the molecular cloning, structural and functional characterization of the Pb*pacC* promoter region, and at the evaluation of this gene influence on *P. brasiliensis* pathogenicity. In this view, we intend to examine the effect of the ambient pH in the expression of the possible virulence factors proposed by Tavares *et al.* (2007), such as cell metabolism, cell wall synthesis and maintenance, and oxidative stress genes.

Acknowledgments

M. J. Poças-Fonseca was supported by a post-doctoral fellowship from CAPES (Coordenação de Aperfeiçoamento de Pessoal de Nível Superior) – Brazil, which is gratefully acknowledged.

References

Aréchiga-Carvajal, E.T. and Ruiz-Herrera, J. (2005). The *RIM101/pacC* homologue from the basidiomycete *Ustilago maydis* is functional in multiple pH-sensitive phenomena. *Eukaryot Cell. 4*(6):999-1008.

Aro, N.; Pakula, T. and Penttilä, M. (2005). Transcriptional regulation of plant cell wall degradation by filamentous fungi. *FEMS Microbiol Rev. 29*(4):719-739.

Arst Jr., H. N. and Peñalva, M. A. (2003b). Recognizing gene regulation by ambient pH. *Fungal Genet. Biol. 40*:1-3.

Arst Jr., H. N. and Peñalva, M. A. (2003a). pH regulation in *Aspergillus* and parallels with higher eukaryotic regulatory systems. *TRENDS in Genetics. 19* (4): 224-231.

Arst Jr., H. N.; Bignell, E. and Tilburn, J. (1994). Two new genes involved in signalling ambient pH in *Aspergillus nidulans*. *Mol. Gen. Genet. 245*: 787-790.

Azevedo, M. O.; Felipe, M. S .S.; Astolfi-Filho, S. and Radford, A. (1990). Cloning, sequencing and homologies of the *cbh 1* (exoglucanase) gene of *Humicola grisea* var. *thermoidea*. *J. Gen. Microbiol. 136*: 2569-2576.

Baek, Y. U.; Martin, S. J. and Davis, D. A. (2006). Evidence for novel pH-dependent regulation of *Candida albicans* Rim101, a direct transcriptional repressor of the cell wall beta-glycosidase Phr2. *Eukaryot Cell. 5*(9):1550-1559.

Barwell, K. J.; Boysen, J. H.; Xu, W. and Mitchell, A. P. (2005). Relationship of DFG16 to the Rim101p pH response pathway in *Saccharomyces cerevisiae* and *Candida albicans. Eukaryot Cell. 4*(5):890-899.

Béguin, P. and Aubert, J. P. (1994). The biological degradation of cellulose. *FEMS Microbiol. Rev.* 13(1): 25-58, 1994.

Bensen, E. S.; Martin, S. J.; Li, M.; Berman, J. and Davis, D. A. (2004). Transcriptional profiling in *Candida albicans* reveals new adaptive responses to extracellular pH and functions for Rim101p. *Mol Microbiol. 54*(5):1335-1351.

Bignell, E.; Negrete-Urtasun, S.; Calcagno, A. M.; Haynes, K.; Arst Jr., H. N. and Rogers, T. (2005). The *Aspergillus* pH-responsive transcription factor PacC regulates virulence. *Mol Microbiol. 55*(4):1072-1084.

Blanchin-Roland, S.; Da Costa, G. and Gaillardin, C. (2005). ESCRT-I components of the endocytic machinery are required for Rim101-dependent ambient pH regulation in the yeast *Yarrowia lipolytica. Microbiology. 151*(11):3627-3637.

Caddick, M. X.; Brownlee, A.G. and Arst Jr., H. N. (1986). Regulation of gene expression by pH of the growth medium in *Aspergillus nidulans. Mol. Gen. Genet. 203*: 346-353.

Caracuel, Z.; Roncero, M. I.; Espeso, E. A.; González-Verdejo, C. I.; García-Maceira, F. I. and Di Pietro, A. (2003). The pH signalling transcription factor PacC controls virulence in the plant pathogen *Fusarium oxysporum. Mol Microbiol. 48*(3):765-779.

Cary, J. W.; Ehrlich, K. C.; Wright, M.; Chang, P. K. and Bhatnagar, D. (2000). Generation of aflR disruption mutants of *Aspergillus parasiticus. Appl Microbiol Biotechnol. 53*(6):680-684.

Chávez, R.; Schachter, K.; Navarro, C.; Peirano, A.; Aguirre, C.; Bull, P. and Eyzaguirre J. (2002). Differences in expression of two endoxylanase genes (*xynA* and *xynB*) from *Penicillium purpurogenum. Gene. 293*(1-2):161-168.

Davis, D. (2003) Adaptation to environmental pH in *Candida albicans* and its relation to pathogenesis. *Curr Genet. 44*(1):1-7.

Davis, D. A.; Bruno, V. M.; Loza, L.; Filler, S. G. and Mitchell, A. P.(2002). *Candida albicans* Mds3p, a conserved regulator of pH responses and virulence identified through insertional mutagenesis. *Genetics. 162*(4):1573-1581.

De Bernardis, F.; Mühlschlegel, F. A.; Cassone, A. and Fonzi, W. A. (1998). The pH of the host niche controls gene expression in and virulence of *Candida albicans. Infect Immun. 66*(7):3317-3325.

Delgado-Jarana, J.; Rincón, A. M. and Benítez, T. (2002). Aspartyl protease from *Trichoderma harzianum* CECT 2413: cloning and characterization. *Microbiology. 148*(5):1305-1315.

Denison, S. H.; Negrete-Urtasun, S.; Mingot, J. M.; Tilburn, J.; Mayer, W.A.; Goel, A.; Espeso, E. A.; Peñalva, M. A. and Arst Jr, H. N. (1998). Putative membrane components of signal transduction pathways for ambient pH regulation in *Aspergillus* and meiosis in *Saccharomyces* are homologous. *Mol. Microbiol. 30*(2): 259-264.

Denison, S. H.; Orejas, M. and Arst Jr., H. N. (1995). Signalling of ambient pH in *Aspergillus* involves a cysteine protease. *J. Biol. Chem. 270*: 28519-28522.

De-Paula, E. H., Poças-Fonseca, M. J. and Azevedo, M. O. (2003). The product of *Humicola grisea* var. *thermoidea cbh1.2* gene is the major expressed protein under induction by lignocellulosic residues. *World Journal of Microbiology & Biotechnology. 19*: 631-635.

Díez, E.; Alvaro, J.; Espeso, E. A.; Rainbow, L.; Suárez, T.; Tilburn, J.; Arst, H. N. Jr. e Peñalva, M. A. (2002). Activation of the *Aspergillus* PacC zinc finger transcription factor requires two proteolytic steps. *EMBO J. 21*(6):1350-1359.

Dorn, G. (1965). Genetic analysis of the phosphatases in *Aspergillus nidulans*. *Genet. Res. 6*: 13-26.

Ehrlich, K. C.; Montalbano, B. G.; Cary, J. W. and Cotty, P. J. (2002). Promoter elements in the aflatoxin pathway polyketide synthase gene. *Biochim Biophys Acta. 1576*(1-2):171-175.

Eisendle, M.; Oberegger, H.; Buttinger, R.; Illmer, P. and Haas, H. (2004). Biosynthesis and uptake of siderophores is controlled by the PacC-mediated ambient-pH Regulatory system in *Aspergillus nidulans*. *Eukaryot Cell. 3*(2):561-563.

Enloe, B.; Diamond, A. and Mitchell, A. P. (2000). A single-transformation gene function test in diploid *Candida albicans*. *J Bacteriol. 182*(20):5730-5736.

Espeso, E. A. and Arst Jr., H. N. (2000). On the mechanism by which alkaline pH prevents expression of an acid-expressed gene. *Mol. Cel. Biol. 20*(10): 3355-3363.

Espeso, E. A.; Roncal, T.; Diéz, E.; Rainbow, L.; Bignell, E.; Alvaro, J.; Suárez, T.; Denison, S. H.; Tilburn, J.; Arst Jr., H. N. and Peñalva, M. A. (2000). On how a transcription factor can avoid its proteolytic activation in the absence of signal transduction. *EMBO J. 19*(4): 719-728.

Espeso, E. A.; Tilburn, J., Arst Jr., H. N. and Peñalva, M. A. (1993). pH regulation is a major determinant in expression of a fungal penicillin biosynthetic gene. *EMBO J. 12*(10): 3947-3956.

Espeso, E. A.; Tilburn, J.; Sanches-Pulido, L.; Brown, C. V.; Valencia, A.; Arst Jr., H. N. and Peñalva, M. A. (1997). Specific DNA recognition by the *Aspergillus nidulans* three zinc finger transcription Factor PacC. *J. Mol. Biol. 274*: 466-480.

Fernández-Martínez, J.; Brown, C. V.; Díez, E.; Tilburn, J.; Arst Jr., H. N. (2003). Overlap of nuclear localisation signal and specific DNA binding residues within the zinc finger domain of PacC. *J. Mol. Biol. 334*:667-684.

Ferreira-Nozawa, M. S.; Silveira, H. C.; Ono, C. J.; Fachin, A. L.; Rossi, A. and Martinez-Rossi, N. M. (2006). The pH signaling transcription factor PacC mediates the growth of *Trichophyton rubrum* on human nail in vitro. *Med Mycol. 44* (7):641-645.

Flaherty, J. E.; Pirttilä, A. M.; Bluhm, B. H. and Woloshuk, C. P. (2003). *PAC1*, a pH-Regulatory Gene from *Fusarium verticillioides*. *Appl. Environ. Microbiol. 69* (9): 5222-5227.

Futai, E.; Maeda, T.; Sorimachi, H.; Kitamoto, K.; Ishiura, S. and Susuki, K. (1999). The protease activity of a calpain-like cysteine protease in *Saccharomyces cerevisae* is required for alkaline adaptation and sporulation. *Mol. Gen. Genet. 260*: 559-568.

Galán, A.; Casanova, M.; Murgui, A.; MacCallum, D. M.; Odds, F. C. Gow, N. A. and Martínez, J. P. (2004). The *Candida albicans* pH-regulated *KER1* gene encodes a lysine/glutamic-acid-rich plasma-membrane protein that is involved in cell aggregation. *Microbiology. 150*(8):2641-2651.

Galindo, A.; Hervás-Aguilar, A.; Rodríguez-Galán, O.; Vincent, O.; Arst, H. N. Jr.; Tilburn, J. and Peñalva, M. A. (2007). PalC, One of Two Bro1 Domain Proteins in the Fungal pH Signalling Pathway, Localizes to Cortical Structures and Binds Vps32. *Traffic. 8* (10):1346-1364.

Gonzalez-Lopez, C. I.; Szabo, R.; Blanchin-Roland, S. and Gaillardin, C. (2002). Genetic control of extracellular protease synthesis in the yeast *Yarrowia lipolytica. Genetics. 160*(2):417-427.

Hall, H. K.; Karem, K. L. and Foster, J. W. (1995). Molecular responses of microbes to environmental pH stress. *Adv. Microb. Physiol. 37*: 229-272.

Heinz, W. J.; Kurzai, O.; Brakhage, A. A.; Fonzi, W. A.; Korting, H. C.; Frosch, M. and Mühlschlegel, F. A. (2000). Molecular responses to changes in the environmental pH are conserved between the fungal pathogens *Candida dubliniensis* and *Candida albicans. Int J Med Microbiol. 290*(3):231-238.

Henrissat, B.; Driguez, H.; Viet, C. and Schülein, M. (1985). Synergism of cellulases from *Trichoderma reesei* in the degradation of cellulose. *Bio/Technology, 3*: 722-726. 1985.

Herranz, S.; Rodríguez, J. M.; Bussink, H. J.; Sãnchez-Ferrero, J. C.; Arst Jr., H. N.; Peñalva, M. A. and Vincent, O. (2005). Arrestin-related proteins mediate pH signaling in fungi. *PNAS. 102*(34):12141-12146.

Hervás-Aguilar, A.; Rodríguez, J. M.; Tilburn, J.; Arst Jr., H. N. and Peñalva, M. A (2007). Evidence for direct involvement of the proteasome in the proteolytic processing of the Aspergillus nidulans zinc-finger transcription factor PacC. *J Biol Chem.* In press.

Kanda, S.; Aimi, T.; Kano, S.; Ishihara, S.; Kitamoto, Y. and Morinaga, T. (2006). Ambient pH signaling regulates expression of the serine protease gene *(spr1)* in pine wilt nematode-trapping fungus, *Monacrosporium megalosporum. Microbiol Res.* In press.

Kottom, T. J.; Thomas, C. F. Jr. and Limper, A. H. (2001). Characterization of *Pneumocystis carinii PHR1*, a pH-regulated gene important for cell wall Integrity. *J Bacteriol. 183*(23):6740-6745.

Lamb, T. M. and Mitchell, A. P. (2003). The transcription factor Rim101p governs ion tolerance and cell differentiation by direct repression of the regulatory genes NRG1 and SMP1 in *Saccharomyces cerevisiae. Mol Cell Biol. 23*(2):677-686.

Lamb, T. M.; Xu, W.; Diamond, A.; Mitchell, A. P. (2000). Alkaline response genes of *Saccharomyces cerevisiae* and their relationship to the RIM101 pathway. *J Biol Chem. 276*(3):1850-1856.

Lambert, M.; Blanchin-Roland, S.; Le Louedec, F.; Lepingle, A. and Gaillardin, C. (1997). Genetic analysis of regulatory mutants affecting synthesis of extracellular proteinases in the yeast *Yarrowia lipolytica*: identification of a *RIM101/pacC* homolog. *Mol Cell Biol. 17*(7):3966-3976.

Li, M.; Martin, S. J.; Bruno, V. M.; Mitchell, A. P. and Davis, D. A. (2004). *Candida albicans* Rim13p, a protease required for Rim101p processing at acidic and alkaline pHs. *Eukaryot Cell. 3*(3):741-751.

Li, W. and Mitchell, A. P. (1997). Proteolytic activation of Rim1p, a positive regulator of yeast sporulation and invasive growth. *Genetics. 145*(1):63-73.

MacCabe, A. P.; Orejas, M.; Pérez-González, J. A. and Ramón, D. (1998). Opposite patterns of expression of two *Aspergillus nidulans* xylanase genes with respect to ambient pH. *J Bacteriol.180*(5):1331-1333.

MacCabe, A. P.; van den Hombergh, J. P. T. W.; Tilburn, J.; Arst Jr., H.N. and Visser, J. (1996). Identification, cloning and analysis of the *Aspergillus niger* gene *pac*C, a wide domain regulatory gene responsive to ambient pH. *Mol. Gen. Genet. 250*: 367-374.

Maccheroni Jr., W.; May, G. S.; Martinez-Rossi, N. M. and Rossi, A. (1997). The sequence of *palF*, an environmental pH response gene in *Aspergillus nidulans*. *Gene 194*: 163-167.

Marone, M.; Mozzetti, S.; De Ritis, D.; Pierelli, L. and Scambia, G. (2001). Semiquantitative RT-PCR analysis to assess the expression levels of multiple transcripts from the same sample. *Biol Proced Online. 3*:19-25.

Meyer. V.; Spielvogel, A.; Funk, L.; Tilburn, J.; Arst, H. N. Jr. and Stahl, U. (2005). Alkaline pH-induced up-regulation of the *afp* gene encoding the antifungal protein (AFP) of *Aspergillus giganteus* is not mediated by the transcription factor PacC: possible involvement of calcineurin. *Mol Genet Genomics. 274* (3):295-306.

Mingot, J. M.; Espeso, E. A.; Díez,, E. and Peñalva, M. A. (2001). Ambient pH signaling regulates nuclear localization of the *Aspergillus nidulans* PacC transcription factor. *Mol. Cell. Biol. 21*:1688-1699.

Nagai, M.; Ozawa, A.; Katsuragi, T.; Kawasaki, H. and Sakai, T. (2000). Cloning and heterologous expression of gene encoding A polygalacturonase from *Aspergillus awamori. Biosci Biotechnol Biochem.* 2000 Aug;64(8):1580-1587.

Negrete-Urtasun, S.; Denison, S. H. and Arst Jr., H. N. (1997). Characterization of the pH signal transduction pathway gene *palA* of *Aspergillus nidulans* and identification of possible homologs. *J. Bacteriol. 179*:1832-1835.

Negrete-Urtasun, S.; Reiter, W.; Diez, E.; Denison, S. H.; Tilburn, J.; Espeso, E. A.; Peñalva, M .A. and Arst Jr., H. N. (1999). Ambient pH signal transduction in *Aspergillus*: completion of gene characterization. *Mol. Microbiol. 33*(5): 994-1003.

Nobile, C. J.; Bruno, V. M.; Richard, M. L.; Davis, D. A. and Mitchell, A. P. (2003). Genetic control of chlamydospore formation in *Candida albicans*. *Microbiology. 149*(12):3629-3637.

Orejas, M.; Espeso, E.; Tilburn, J.; Sarkar, S.; Arst Jr., H. N. and Peñalva, M. A. (1995). Activation of the *Aspergillus* PacC transcription factor in response to alkaline ambient pH requires proteolysis of the carboxy-terminal moiety. *Genes Dev. 9*: 1622-1632.

Peñalva, M. A. and Arst Jr., H. N. (2002). Regulation of gene expression by ambient pH in filamentous fungi and yeasts. *Microbiol.Mol. Biol. Rev. 66*:426.46

Peñalva, M. A. and Arst Jr., H. N. (2004). Recent advances in the characterization of ambient pH regulation of gene expression in filamentous fungi ans yeast. *Annu. Rev. Microbiol. 58*:425.51.

Peñas, M. M.; Hervás-Aguilar, A.; Múnera-Huertas, T.; Reoyo, E.; Peñalva, M. A.; Arst Jr., H. N. and Tilburn, J. (2007). Further characterization of the signaling proteolysis step in the *Aspergillus nidulans* pH signal transduction pathway. *Eukaryot Cell. 6*(6):960-970.

Pérez-González, J. A.; van Peij, N. N.; Bezoen, A., MacCabe, A. P.; Ramón. D. and de Graaff, L. H. (1998). Molecular cloning and transcriptional regulation of the *Aspergillus nidulans xlnD* gene encoding a beta-xylosidase. *Appl Environ Microbiol. 64*(4):1412-1419.

Poças-Fonseca, M. J. (2000). Regulation of the Expression of the Cellobiohydrolase Genes from *Humicola grisea* var. *thermoidea.* PhD Thesis. Department of Cellular Biology. Institute of Biological Sciences. University of Brasilia – Brazil. 222 pp.

Poças-Fonseca, M. J. and Maranhão, A. Q. (2005). Diversity among Cellulolytic Fungi. In: Satyanarayana, T. and Johri, B. N. Eds. *Microbial Diversity: Current Perspectives and Potential Applications.* New Delhi, India, v. I, pp. 165-180.

Poças-Fonseca, M.J.; Lima, B.D.; Brígido, M.M.; Pereira, I.S.; Felipe, M.S.S.; Radford, A. and Azevedo, M.O. (1997). *Humicola grisea* var. *thermoidea cbh*1.2: a new gene in the family of cellobiohydrolases is expressed and encodes a cellulose-binding domain-less protein. *J. Gen. and Appl. Microbiol. 43*(2): 115-120, 1997.

Poças-Fonseca, M.J.; Silva-Pereira, I.; Rocha, B.B and Azevedo, M.O. (2000). Substrate-dependent differential expression of *Humicola grisea* var. *thermoidea* cellobiohydrolase genes. *Can. J. Microbiol. 46*: 749-752, 2000.

Polizeli, M; L.; Rizzatti, A. C.; Monti, R.; Terenzi, H. F.; Jorge, J. A. and Amorim, D. S. (2005). Xylanases from fungi: properties and industrial applications. *Appl Microbiol Biotechnol. 67*(5): 577-591.

Porta, A.; Ramon, A. M. and Fonzi, W. A. (1999). PRR1, a homolog of *Aspergillus nidulans palF*, controls pH-dependent gene expression and filamentation in *Candida albicans. J Bacteriol. 181*(24):7516-7523.

Portillo, F. (2000) Regulation of plasma membrane H^+-ATPase in fungi and plants. *Biochem. Biophys. Acta 1469*: 31-42.

Rajaratham, S.; Shashireka, M. N. and Bano, Z. (1992). Biopotentialities of the Basidiomacromycetes. *Adv. Appl. Microbiol. 37*: 233-361.

Ramon, A. M.; Porta, A. and Fonzi, W. A. (1999). Effect of environmental pH on morphological development of *Candida albicans* is mediated via the PacC-related transcription factor encoded by *PRR2. J Bacteriol. 181* (24):7524-7530.

Restrepo, A.; McEwen, J. G. and Castañeda, E. (2001). The habitat of *Paracoccidioides brasiliensis*: how far from solving the riddle? *Med Mycol. 39*(3):233-241.

Richard, M. L.; Nobile, C. J.; Bruno, V. M. and Mitchell, A. P. (2005). *Candida albicans* biofilm-defective mutants. *Eukaryot Cell. 4*(8):1493-1502.

Rollins, J. A. (2003). The *Sclerotinia sclerotiorum pac*1 gene is required for sclerotial development and virulence. *Mol. Plant-Microbe Interact. 16*:785.95

Rollins, J. A. and Dickman, M. B. (2001). pH sinaling in *Sclerotinia sclerotiorum*: Identification of *pac*C/R1M1 homolog. *Appl. Environ. Microbiol. 67*:75-81.

Rossi, A. and Arst Jr., H. N. (1990). Mutants of *Aspergillus nidulans* able to grow at extremely acidic pH acidify the medium less than wild type when grown at more moderate pH. *FEMS Microbiol. Lett. 66*: 51-54.

Rossi, M. S.; Poças-Fonseca, M. J. and Azevedo, M. O. (2007). Molecular Biology of Fungal Biodegradation: Cellulolytic activity over agricultural wastes (2007) In: (eds: R. Kuhad, C. and Singh, A. Eds. *Lignocellulose Biotechnology: Future Prospects.* I. K. International, New Delhi, India, pp 97-105.

Schmitt, E. K.; Kempken, R. and Kück, U. (2001). Functional analysis of promoter sequences of cephalosporin C biosynthesis genes from *Acremonium chrysogenum*: specific DNA-protein interactions and characterization of the transcription factor PACC. *Mol Genet Genomics. 265*(3):508-518.

Slagsvold, T.; Pattni, K.; Malerod, L. and Stenmark, H. (2006). Endosomal and non-endosomal functions of ESCRT proteins. *Trends Cell Biol. 16*(6):317-326.

Staib, P. and Morschhäuser, J. (2006). Chlamydospore formation in *Candida albicans* and *Candida dubliniensis* - an enigmatic developmental programme. *Mycoses. 50*(1):1-12.

Su, S. S. and Mitchell, A. P. (1993). Molecular characterization of the yeast meiotic regulatory gene *RIM1*. *Nucleic Acids Res. 21*(16):3789-3797.

Suarez, T. and Peñalva, M. A. (1996). Characterization of a *Penicillium chrysogenum* gene encoding a PacC transcription factor and its binding sites in the divergent *pcbAB-pcbC* promoter of the penicillin biosynthetic cluster. *Mol. Microbiol. 20*:529-540.

Suzuki, M. (1989). SPXX, a frequent sequence motif in gene regulatory proteins. *J Mol Biol. 207*(1):61-84.

Tanaka, H.; Muguruma, M. and Ohta, K. (2006). Purification and properties of a family-10 xylanase from *Aureobasidium pullulans* ATCC 20524 and characterization of the encoding gene. *Appl Microbiol Biotechnol.70*(2):202-211.

Tavares, A.; Silva, S. S.; Dantas, A.; Campos, E. G.; Andrade, R. V.; Maranhão, A. Q.; Brígido, M. M.; Passos-Silva, D. G.; Fachin, A. L.; Teixeira, S. M.; Passos, G. A.; Soares, C. M.; Bocca, A. L.; Carvalho, M. J.; Silva-Pereira I. and Felipe, M. S. (2007) Early transcriptional response of *Paracoccidioides brasiliensis* upon internalization by murine macrophages. *Microbes Infect. 9*(5):583-590.

Tilburn, J.; Sãnchez-Ferrero, J. C.; Reoyo, E.; Arst Jr., H. N. and Peñalva, M. A. (2005). Mutational analysis of the pH signal transduction component PalC of *Aspergillus nidulans* supports distant similarity to BRO1 domain family members. *Genetics. 171* (1):393-401.

Tilburn, J.; Sarkar, S.; Widdick, D. A.; Espeso, E. A.; Orejas, M.; Mungroo, Peñalva, M. A. and Arst Jr., H. N. (1995). The *Aspergillus* PacC zinc finger transcription factor mediates regulation of both acid- and alkaline-expressed genes by ambient pH. *EMBO J. 14*(4): 779-790.

Trigui-Lahiani, H. and Gargouri, A. (2007). Cloning, genomic organisation and mRNA expression of a pectin lyase gene from a mutant strain of *Penicillium occitanis*. *Gene. 388*(1-2):54-60.

Turner, G. (1993). Gene organization in filamentous fungi. In: Broda, P. M. A.; Oliver, S. G. and Sims, P. F. G. Eds. The Eukaryotic Genome Organization and Regulation. *Society for General Microbiology Symposium 50. Cambridge University Press*. pp. 107-125.

van den Hombergh, J. P. T. W.; MacCabe, A. P.; van den Vondervoort, P. J. I. and Visser, J. (1996). Regulation of acid phosphatases in an *Aspergillus niger pac*C disruption strain. *Mol. Gen. Genet. 251*: 542-550.

Vincent, O.; Rainbow, L.; Tilburn, J.; Arst Jr., H. N. and Peñalva, M. A. (2003). YPXL/I is a protein interaction motif recognized by *Aspergillus* PalA and its human homologue, AIP1/Alix. *Mol Cell Biol. 23*(5):1647-1655.

Williams, R. L. and Urbé, S. (2007). The emerging shape of the ESCRT machinery. *Nat Rev Mol Cell Biol. 8*(5):355-368.

Xu, W. and Mitchell, A. P. (2001). Yeast PalA/AIP1/Alix homolog Rim20p associates with a PEST-like region and is required for its proteolytic cleavage. *J Bacteriol. 183*(23):6917-6923.

In: Current Advances in Molecular Mycology
Eds: Y. Gherbawy, R.L. Mach and M.K. Rai

ISBN 978-1-60456-909-4

Chapter X

EXPLORING *ASCOCHYTA RABIEI* ON CHICKPEA AS A MODEL TO STUDY PATHOGENICITY FACTORS OF ASCOCHYTA BLIGHT OF COOL SEASON GRAIN LEGUMES

***W. Chen*[1] *and D. White*[2]**

[1]USDA-ARS, Grain Legume Genetics and Physiology Research Unit, Washington State University, Pullman, WA 99164, USA;

[2]Department of Plant Pathology, Washington State University, Pullman, WA 99164, USA.

ABSTRACT

Recent developments in molecular mycology have significantly advanced our understanding pathogenic process of fungal pathogens. In-depth studies in a number of model and well-studied pathosystems have revealed the interactions between host and pathogen at molecular levels. However, the disease of Ascochyta blight of cool season grain legumes has received little attention in this regard, despite the fact that Ascochyta blight is a serious disease of all cool season grain legume crops (chickpea, faba bean, lentil and pea). Although the pathogens that cause Ascochyta blight on these crops are of different species, they are closely related morphologically and phylogenetically. Therefore, investigation of the disease on one crop can serve as a model for other cool season grain legume crops. We have recently made important progress to provide the foundation and necessary tools to explore *A. rabiei* on chickpea as a model to study pathogenicity determinants of cool season grain legume diseases. An efficient protocol was optimized for transforming *Ascochyta rabiei*; A reproducible inverse PCR technique was developed for identification of insertion sites; Insertional mutants with altered pathogenicity were identified, and the DNA sequences adjacent to insertion sites were determined; A phage DNA library of *A. rabiei* was constructed; And probes for specific genes- an acyl-CoA ligase (*cps*1) and a polyketide synthase gene (*pks*1) - with potential of being general pathogenic determinants in *A. rabiei* were developed, and positive library clones were identified. The positive clones containing the specific genes or the insertion sites will be useful for either ectopic complementation tests or targeted mutagenesis.

Key words: Ascochyta blight, pathogenicity determinants, fungal transformation, inverse-PCR.

INTRODUCTION

Technological advances in molecular biology have accelerated our understanding of fundamental biological questions in many biological disciplines including mycology and plant pathology (Oliver and Osbourn, 1995). Recent developments in mutagenesis, genome sequencing and microarray have helped enhance our understanding of complex interactions between fungal pathogens and plants. An understanding of the host-pathogen interactions at molecular levels has emerged in several model and well studied pathosystems. However, the diseases collectively called Ascochyta blight of cool season grain legumes (pea, chickpea, lentil and faba bean) caused by a group of related fungi have received little attention. A common and often devastating disease among all cool season grain legumes is the fungal disease Ascochyta blight. Different species are involved in causing the disease on different cool season food legume crops [*Ascochyta rabiei* (Pass.) Labrousse on chickpea, *A. fabae* Speg. on faba bean, *A. lentis* Vassiliesky on lentil, and *A. pisi* Lib., *Mycosphaerella pinodes* (Berk. & Blox) Vestergr. and *Phoma medicagenisis* Malbr. & Roum. var *pinodella* (Jones) Boerema on pea] (Nene et al., 1988; Peever, 2007; Taylor and Ford, 2007). Despite the economical importance of Ascochyta blight in cool season grain legume production, the mechanisms of pathogenicity of the pathogens are still not well understood. This chapter is aimed at reviewing our recent research progress in exploring the chickpea-*Ascochyta rabiei* pathosystem as a model of studying pathogenic mechanisms of Ascochyta blight of cool season grain legumes.

Among the Ascochyta pathogens of cool season grain legumes, *Ascochyta rabiei* on chickpea is probably the most intensively studied pathosystem. *Ascochyta rabiei* causing chickpea ascochyta blight is known to produce phytotoxin solanopyrones through the polyketide synthesis pathway (Alam et al., 1989; Köhl et al., 1991), and hydrolytic or cell-wall degrading enzymes (Tenhaken and Barz, 1991; Tenhaken et al., 1997). The phytotoxins play a role in causing blight (Chen and Strange, 1991; Kaur, 1995). Hydrolytic enzymes are considered necessary for fungal nutrition and to facilitate spatial spread of fungi (Walton, 1994). *A. rabiei* was first transformed with the protoplast/PEG protocol with a GUS reporter gene for observing the infection process (Köhler et al., 1995), and later transformed with *Agrobacterium*-mediated transformation (AMT) for studying pathogenicity factors (White and Chen, 2006; Morgensen et al., 2006). However, currently little information is available about the genetic determinants of pathogenicity of the *Ascochyta* pathogens. Knowledge of pathogenic determinants will allow us to develop novel or more effective measures in managing the disease.

Two approaches (targeted gene disruption and insertional mutagenesis) are available to investigate pathogenic determinants of fungal pathogens. In targeted gene disruption, target genes could be previously reported pathogenicity genes in other pathosystems. By comparing the pathogenicity of the mutant with that of the wild type, we can assess the role of the gene in infection. Insertional mutagenesis is to generate random and tagged mutations within the pathogen genome. The modern technique of choice is of either restriction-enzyme-mediated integration (Oliver and Osbourn, 1995; Kahmann and Basse, 1999) or *Agrobacterium*-mediated transformation (Michielse et al., 2005). This approach does not require a priori knowledge of gene function, and it involves generation of a library of random mutations, screening the library for altered phenotypes or pathogenicity, and characterization of disrupted genomic regions. This approach is powerful in identification of previously unknown pathogenicity factors.

A number of previously reported conserved fungal virulence factors could be explored in *A. rabiei.* Lu et al (2003) described a general fungal virulence factor (an acyl-CoA ligase *cps*1) in several plant-pathogenic ascomycetes. Disruption of the *cps*1 homolog in several plant pathogens produced no observable changes in growth phenotype, but showed reduced virulence. Production of melanin has also been shown to be a virulence factor in some pathogenic fungi (Henson et al., 1999; Kawamura et al., 1999). *A. rabiei* produces melanin through the 1,8-dihydroxynaphthalene pathway via polyketide synthesis (Chen et al., 2004b). Thus, polyketide synthases could potentially be pathogenicity factors in *A. rabiei* through their involvement in melanin biosynthesis or in the synthesis of phytotoxin solanapyrones (Hohl et al., 1991).

In exploring *A. rabiei* as a model for studying other Ascochyta pathogens, important research progress has been made. We have optimized a transformation protocol, developed a technique to identify insertion sites, identified and characterized tagged mutants with reduced or lost pathogenicity, developed gene-specific probes, constructed a phage library of the *A. rabiei* genome, and isolated clones containing potential pathogenicity factors through screening the library. The research provides foundation and necessary tools for further assessing the roles of the respective genes in causing Ascochyta blight.

Materials and Methods

Biological Materials

Microbial strains used in this study included *Escherichia coli* strain DH10B (Invitrogen, Carlsbad, CA), *A. tumefaciens* strain AGL-1, and *A. rabiei* strain AR628. Plasmids pCAMBIA1300 (Cambia, Melbourne, AU), pGEM-T (Promega, Madison, WI), pII99 (Namiki et al. 2001), pSH75 (Kimara and Tsuge 1993), kindly provided by Takashi Tsuge, and plasmids pDJW2, pDJW5 (White and Chen 2006) were maintained in *E. coli.*

Fungal Transformation

An optimized transformation procedure was described by White and Chen (2006). Briefly, cultures of *A. tumefaciens* strain AGL-1 containing plasmid pCAMBIA1300, pDJW2, or pDJW5 were co-cultured with conidia of *A. rabiei* strain AR628 on dialysis or nylon membranes on IM agar (ca. 200 μl/22 cm membrane). Following co-cultivation, the membranes containing co-cultivate were transferred and inverted onto either PDA or V8 Agar plates containing hygromycin B, chloramphenicol, and cefotaxime. *A. rabiei* transformants that arose on the selection medium 4-10 days post transfer were immediately isolated and transferred to fresh selection plates followed by transfer to APDA or V8 plates containing timentin to eliminate bacteria. Transformants were maintained on PDA or V8 agar with or without hygromycin selection.

Characterization of Transformants

Two methods of detecting T-DNA (PCR and Southern hybridization) were used to confirm transformants. PCR detection of T-DNA was done by amplifying an internal 854 bp region of the *hph* gene of T-DNA using primers hph-F (5'-GAGCCTGACCTATTGCAT CTC-3') and hph-R (5'- CCGTCAACCAAGCTCTGATAG-3'). Southern hybridizations were performed to determine the frequency and randomness of T-DNA integration. The DNA probe for Southern analysis was generated by PCR using the hph-F and hph-R primers with the PCR DIG Probe Synthesis Kit (Roche, Indianapolis, IN). Aqueous hybridizations were performed at 65°C overnight and detected using the DIG Luminescent Detection Kit (Roche) according to the manufacture's instructions.

Detection and Analysis of T-DNA Insertion Sites

The inverse PCR procedure (Gardner et al. 2005) was used to identify genomic DNA sequences flanking T-DNA insertions (White and Chen, 2006). DNA was digested separately with XhoI or with SacI for amplifying, respectively, the right and left boarders of T-DNA. After ligation and PCR with primers LB5IP (5'-AGTCGTTTACCCAGAATGCACAG GTACACT-3') and RB5IP (5'-CTTGCACAAATTGGATGTCCATCTTCGAAAC-3') using Elongase (Invitrogen), products were separated on 1% agarose gels (NuSeive, IBC BioExpress). Reactions yielding single products visible on agarose gels were either sequenced directly using primer LB5IP or RB5IP after clean-up using the Exo-Sap-It Kit (USB Cleveland, OH) or cloned into plasmid pGEM-T Easy and then sequenced using primer T7 or SP6 at the Washington State University Bioinformatics Core laboratory, Pullman, WA. Assembled sequences were compared to each other to verify that each contained a unique region of the *A. rabiei* genome and translated in all six reading frames for comparison to the GenBank database as well as the *Stagonospora nodorum* genome (http://www.broad.mit.edu). The *S. nodorum* genome was selected because for the fungi with genomes available *S. nodorum* is the closest phylogenetically related to *A. rabiei* (Peever *et al.*, 2007).

Pathogenicity Assay

Transformants were screened for their pathogenicity using a mini-dome technique as described by Chen et al. (2005). The transformants were always compared with wild-type strains AR19 (pathotype I) and AR628 (pathotype II) in the pathogenicity assays on standard differential chickpea cultivars Dwelley and Spanish White. Disease severity was assessed according to the 1-to-9 rating scale (Chen *et al.*, 2004a). The transformants that showed reduced pathogenicity in the first assay were tested again in a second assay. Transformants that showed reduced or lost pathogenicity in at least two pathogenicity assays were chosen for further analysis.

Figure 1. Mini-dome pathogenicity assay of transformants of *Ascochyta rabiei* on chickpea cultivar Spanish White. Left: non-inoculated control plants; Middle: Plants inoculated with wild type, parental strain AR628; Right: Plants inoculated with transformant ArW519.

Development of Gene-Specific Probes for *A. Rabiei*

Specific probes were developed for genes that could be potential virulence factors in *A. rabiei*. A portion (562 bp) of the gene encoding the polyketide synthase (*pks*1) based on *Glarea lozoyensis* (Zhang *et al.*, 2003) was amplified with the primers pksF2 (5'CACTACCACTGCCGTCGCAT-3') and pksR2 (5'-TAGACTTGACCATGCCACTGCA-3'). A portion (683 bp) of the acyl-CoA ligase (*cps*1) based on *Cochliobolus heterostrophus*

(Lu *et al.*, 2003) was amplified with primers cpsF (5'-GGGACAAGAGCAACCTCTA-3') and cpsR (5'-TGGTAGTTGTATGCAGC-3'). PCR products were cloned into the pGEM-T Easy vector and sequenced using the M13F and M13R primers. The sequences were used to compare with previously published gene sequences, and were used as probes in Southern hybridization in screening a genomic library of *A. rabiei.*

Construction and Screening of *Ascochyta Rabiei* Lambda Phage Library

High molecular weight DNA of *A. rabiei* strain AR628 was partially digested with the restriction enzyme ApoI and size fractionated on a 0.8% low melting point agarose gel. Fragments corresponding to 7-10 kb were isolated and treated with agarase enzyme. The fractionated *A. rabiei* DNA was mixed with phage arms and ligated in the presence of T4 ligase. Ligated arms were packaged using Gigapack III packaging extracts at 25°C for 2 hours followed by chloroform extraction. The recombinant (clear plaques) and non-recombinant (blue plaques) were screened in the presence of 5-bromo-4-chloro-3-indolyl-beta-D-galactoside (X-gal) and isopropylthio-beta-D-galactosidase (IPTG). Ten randomly selected recombinant plaques were used for plasmid rescue using ExAssist® helper phage and the *E. coli* host strain SOLR under ampicillin selection. Recovered plasmid DNA was digested with ApoI enzyme and separated in 1% agarose to estimate average insertion size in the Phage library. The genome coverage of this primary *Ascochyta* library was estimated based on the average size of the insert DNA in the recombinant phage and the number of recombinants of the library.

To isolate clones from the library that contain either *pks*1 or *cps*1 homologs, probes were constructed using PCR with the *pks* or *cps* primers. Approximately 80,000 plaques were transferred from NZY agar to nylon membranes (Amersham) and probed sequentially, first with the *cps*1 probe, then with the *pks*1 probe. Single plaques that hybridized with each probe were recovered from the corresponding NZY plate and *in vivo* excision reactions were performed to rescue phagemid DNA. Recovered phagemid DNA was used as template for PCR with the corresponding primer pairs used to generate the probe, and were also digested with ApoI to estimate the insert size by agarose gel electrophoresis.

To isolate clones from the phage library that contain DNA adjacent to the T-DNA insertion sites, approximately 80,000 plaques were screened with probes generated from inverse-PCR products. Probes from the transformants were first mixed together for the primary hybridization and detection screen. Phage from positive plaques in the primary screen was harvested and pooled in SM buffer to make an enriched phage stock for infecting *E. coli* XL-1 Blue cells. Plaques generated from the enriched phage stock were transferred to nylon membranes and screened with individual probes. Phagemids were recovered by *in vivo* excision from phage collected from three plaques identified by each probe and analyzed by restriction digestion and sequencing.

RESULTS

Identification and Characterization of Transformants

The optimal conditions for efficient transformation of *Ascochyta rabiei* using *Agrobacterium*-mediated transformation have been determined. Hygromycin B resistance (*hph*) expressed by the *Aspergillus nidulans trp*C promoter was found to be superior to geneticin resistance (*nptII*) for selecting transformants. Co-cultivation on solid media for 72 hours was optimal for generating transformants of *Ascochyta rabiei* (White and Chen, 2006).

A wide range of variation in colony morphology, growth rate, and conidial production was observed among transformants. Examples include lost ability to produce conidia and constitutive production of black mycelium. Generally transformants produced less conidia under antibiotic selection than in the absence of selection. The wild-type strain AR628 consistently produced about 3.7×10^7 conidia per plate, 63 transformants produced about 10% of conidia of the wild type, seven transformants produced about 1.5% of conidia of the wild type. Two transformants produced 5 times more conidia than the wild type.

Approximately 800 transformants were screened for pathogenicity in this study. The transformants that lost ability to produce conidia were not screened because the screening procedure uses conidia as inoculum (Chen *et al.*, 2004a). Most of the transformants screened were about equally virulent as the parental wild-type strain, producing disease scores above 6. To date, 21 of the transformants produced significantly lower disease severity (score <4 on a 1-9 rating scale) than that of the wild-type in at least two independent pathogenicity assays. Nine of the 21 transformants plus two transformants that lost ability to produce conidia were selected for further characterization.

Characterization of Insertion Sites

Southern hybridization of digested transformant DNA probed with the hygromycin-resistance gene (*hph*) showed single hybridization bands of various sizes, confirming that the T-DNA was integrated into the genome of *A. rabiei* and that each transformant contained a single insertion. Inverse PCR amplified single products from transformants, ranging in size from 850 to 2500 bp. Sequences adjacent to the insertion sites determined from the inverse-PCR products were first compared among themselves, and comparison showed that two pairs of the 11 transformants were identical in insertion locations. This reduced the number of characterized transformants from 11 to 9. These two pairs were probably originated from the same insertion events because they were from the same transformation plates. This result also indicates that the inverse-PCR technique for identifying insertion sites in transformants is reliable and reproducible.

The sequences flanking the T-DNA insertion site from each of the nine transformants were used as queries in tBLASTx searches of the GenBank database as well as the genome database of *S. nodorum*. DNA recovered from two of the transformants shared a high degree of similarity with known proteins while sequence from another transformant shared significant similarity with a hypothetical protein of *A. nidulans* (Table 1). The translated

DNA (576 bp) from a transformant shared 71% identity (91/128 aa) with the kinesin of *C. heterostrophus* (accession AY230433). Translated DNA from transformant ArW540 (440 bp) shared 66% identity (86/130 aa) with the transposase protein of the *S. nodorum* transposon *molly*. Three additional sequences shared minimal sequence similarity with proteins in the database as indicated by low E values (Table 1). The remaining three sequences did not have any similarity to known proteins. In searching the *S. nodorum* genome, sequences of three transformants shared significant similarity to translated regions (hypothetical proteins) of the genome, while sequences of the remaining six transformants did not have any similarity with any translated region of the *S. nodorum* genome (data not shown).

Table 1. Pathogenic and molecular characterization of insertion sites in selected transformants of *Ascochyta rabiei*

Strain	Disease score[a]	Inverse PCR[b]	Sequence length[c]	GenBank tBLAST results (Accession #)	E value
ArW8	3.5±0.5	1900	576	*C. heterostrophus* kinesin (AY230433)	$1e^{-46}$
ArW247/ArW251	NT[d]	500	175	*A. nidulans* hypothetical protein (XM_659106)	$3e^{-10}$
ArW519	2.0±0.5	1600	433	*M. musculus* p21 activated kinase (AK08851)	0.096
ArW520/ ArW525	1.3±0.3	850	786	*A. erytherum* put. transcript. rep. (AY62365)	0.051
ArW522	1.2±0.3	2200	742	No significant similarity	--
ArW524	2.5±0.9	1500	755	No significant similarity	--
ArW529	1.0±0	2100	818	No significant similarity	--
ArW540	1.8±0.6	2500	440	*S. nodorum* transposon *molly* (AJ488502)	$3e^{-34}$
ArW541	1.2±0.3	2300	619	*A. oryzae* cDNA, contig sequence: (AoEST1849)	6.6
Ar628 (WT)	8.0±0.8	--[e]		--	--

[a]Pathogenicity score (± standard deviation, n=3) on chickpea cultivar Dwelley using a 1-9 rating scale (1= healthy, no disease and 9 = dead plant);
[b]Size in base pairs of total product of inverse PCR;
[c]Sequence length used in tBlast searches;
[d]Not tested due to lack of conidia;
[e]Not Applicable.

Construction and Screening of Genomic Library

The constructed phage library consisted of 1.7 x 10^6 recombinants containing *A. rabiei* DNA. The average insert size of the recombinants was about 6,300 bp. Thus, this DNA library contains more than 10,000 MB of *A. rabiei* DNA. Assuming a genome size of 40 MB for *A. rabiei* (Akamatsu and Peever, 2005), this library would provide more than 250X coverage.

Both the polyketide synthase (*pks*1) and the acyl-CoA ligase (*cps*1) genes were detected in *A. rabiei* using PCR amplification and sequence confirmation. The amplified *pks*1 DNA fragment had 81% (455/562) identity to the *Bipolaris oryzae* polyketide synthase gene (accession AB176546). The amplified *cps*1 fragment was 82% (560/683) identical to the *cps*1 gene (accession AF332878) of *Cochliobolus heterostrophus*. The two sequences from *Ascochyta rabiei* were deposited into GenBank and assigned the accession numbers EF092313 (*ARcps*1) and EF092314 (*ARpks*1). Positive clones for both genes were identified in the phage library (White and Chen, 2007).

A mixture of the nine probes of the transformants was initially used to screen approximately 80,000 plaques. Positive plagues were pooled and enriched for secondary screening. Then individual probes were used to screen the enriched phage stock. Each probe in the secondary screening was exposed to approximately 30,000 plaques generated from the enriched phage stock. Phagemid DNAs were rescued from three random plaques identified by each single probe and in each case the three recovered phagemids contained the same sized-DNA inserts. It was assumed that the three clones represented the same region of *A. rabiei* genomic DNA and only one clone of the three was needed for further analysis.

CONCLUSIONS AND FUTURE LINE OF RESEARCH

Essential molecular tools have been developed to provide the foundation to explore *Ascochyta rabiei* on chickpea as a model to study pathogenic mechanisms of Ascochyta blight of cool season grain legumes. An efficient protocol was optimized for transforming *Ascochyta rabiei*; A reproducible inverse-PCR techniques was developed for identification of insertion sites in interested transformants; Insertional mutants with altered pathogenicity were identified through *in vivo* pathogenicity assays, and the DNA sequences adjacent to insertion sites were determined; A phage DNA library of *A. rabiei* was constructed with sufficient coverage for isolating single-copy genes. Hybridization with either gene-specific probes or probes generated from random insertion sites of transformants always identified positive clones in the library, proving its utility in isolating other genes; And probes for specific genes (*cps*1 and *pks*1) with potential of being general pathogenic determinants in *A. rabiei* were developed, and positive library clones were identified through Southern hybridization. The positive clones containing the specific genes or the insertion sites will be useful for either ectopic complementation tests or targeted mutagenesis.

Insertion in one of the transformants appears to be within a known fungal gene. The T-DNA insertion in transformant ArW8 has disrupted a kinesin-like gene. Kinesins play important roles in the transport of cell organelles, polarized growth, and secretion (Schoch *et al.*, 2003), and the kinesins of the yeast *Schizosaccharomyces pombi* as well as the corn smut fungus *Ustilago maydis* have been studied extensively (Steinberg and Fuchs, 2004; Straube *et al.*, 2006). However, this is the first report of a kinesin-like gene potentially being involved in plant pathogenesis. Its role remains to be confirmed and its mechanisms in pathogenesis are not clear.

Southern hybridization and sequence analyses showed that T-DNA insertions in *A. rabiei* were random, and diverse sequences in insertion sites were found, indicating several potential genes were disrupted. Pathogenesis is a complex biological process involving diverse factors. Pathogenesis of the necrotrophic pathogen *A. rabiei* is predicted to involve a number of processes including attachment and penetration of host plant tissue, as well as production and secretion of extracellular enzymes and phytotoxins, and each process is likely controlled by several genes. A mutation in any gene involved in these processes could result in altered pathogenicity.

We found that several of the sequences recovered from the transformants had no significant matches either in the GenBank database or within the *S. nodorum* genome. This inability to detect any known sequences with significant similarity could be due to the limited length of query sequences available, to the fact that they are unique pathogenicity factors in *A. rabiei*, or to the fact that small deletions of genomic DNA occurred during T-DNA integration events (Bundock and Hooykaas, 1996), resulting in the loss of a coding region flanking the insertion.

We developed gene-specific probes to isolate a polyketide synthase gene and an acyl-CoA ligase gene from the *A. rabiei* library. These genes were selected because they were shown to be conserved pathogenicity factors in other pathosystems (Kawamura *et al.*, 1999, Lu *et al.*, 2003). The results also demonstrated the completeness and usefulness of the phage DNA library. Both probes identified positive plaques in the library despite the fact only a portion (80,000 plaques) of the library was exposed to the probes. Thus, this library should be useful for isolating other genes of interest and it will be a valuable resource available to the scientific community for studying *A. rabiei* or other related plant pathogens.

Two approaches need to be taken to unequivocally demonstrate the roles of the identified potential pathogenicity determinants in *A. rabiei*. Ectopic complementation tests will be one approach to prove the role of the genes disrupted in the random insertional mutants. A second shuttle vector carrying the *nptII* gene for geneticin resistance expressed by the *A. nidulans trpC* promoter has been created for delivering library DNA via AMT to hygromycin-resistant transformants, and selection on hygromycin and geneticin has been shown to be stable (unpublished). Additionally, *A. rabiei* is heterothallic (Trapero-Casas and Kaiser, 1992). Thus segregation analysis could also be employed. Co-segregation of hygromycin resistance and lack of pathogenicity would prove the role of mutated genes in pathogenesis.

Targeted mutagenesis specifically on the *cps* and *pks* genes to create knockout mutants will be another approach to study pathogenicity determinants of *A. rabiei*. To create deletions in the *pks* and *cps* library fragments, a short region of each clone has been removed by restriction digest and replaced with the *trpC-hph* antibiotic resistance cassette. Disruption cassettes containing library clones in a markerless T-DNA shuttle vector are being constructed for delivery into *A. rabiei* wild-type strains via AMT. Integration can occur at the genomic site of interest (homologous recombination) or at other sites (illegitimate recombination), which would be distinguished by PCR or Southern hybridization. Since transformation may induce unexpected changes in chromosome structure or complement, we shall be prudent to evaluate the phenotypes including pathogenicity of a number of independent transformants including those that have not undergone disruption at the gene of interest.

Significant advances have been made in understanding genetic factors of pathogenicity in a number of phytopathogenic fungi (e.g., Gilbert *et al.*, 2004; Talbot 2004). However, little information is available about pathogenicity determinants in *Ascochyta* spp causing Ascochyta blight of cool season grain legumes. Using *A. rabiei* as a model for Ascochyta and other closely related plant pathogens, our research showed the feasibility of and provided necessary tools for studying pathogenicity determinants in Ascochyta blight pathogens of cool season grain legumes. A detailed knowledge of pathogenic determinants of *A. rabiei* and of chickpea resistance response (Cho *et al.*, 2005; Coram and Pang, 2006) will be invaluable in developing our understanding of the interaction between *A. rabiei* and chickpea, and in devising novel or more effective measures in managing the disease. We anticipate that the information will also be applicable to Ascochyta blight of other cool-season grain legumes.

REFERENCES

Akamatsu, H. and Peever, T.L. (2005). Molecular karyotypes of the phytopathogenic fungus *Ascochyta rabiei* and related legume-infecting *Ascochyta* spp. XXIII Fungal Genetics Conference. *Fungal Genetics Newsletter 52*(Supplement): 204.

Alam, S.S., Bilton, J.N., Slawin, A.M.Z., Williams, D.J., Sheppard, R.N. and Strange, R.N. (1989). Chickpea blight – Production of the phytotoxins solanapyrone A and C by *Ascochyta rabiei. Phytochemistry 28*: 2627-2630.

Bundock, P. and Hooykaas, P.J.J. (1996). Integration of *Agrobacterium tumefaciens* T-DNA in the *Saccharomyces cerevisiae* genome by illegitimate recombination. *Proceedings of National Academy of Science USA 93*: 15272-15275.

Chen, W., Coyne, C., Peever, T. and Muehbauer, F.J. (2004a). Characterization of chickpea differentials for Ascochyta blight and identification of resistance sources for *Ascochyta rabiei. Plant Pathology 53*: 759-769.

Chen, W., McPhee, K.E. and Muehlbauer, F.J. (2005). Use of a mini-dome bioassay and grafting to study resistance of chickpea to Ascochyta blight. *J Phytopathology 153*: 579-587.

Chen, W., Sharma, K.D. and Wheeler, M.H. (2004b). Demonstration of the 1,8-dihydroxynaphthalene melanin pathway in *Ascochyta rabiei. Inoculum 55*: 11(Abstract).

Chen, Y.M. and Strange, R.N. (1991). Synthesis of the solanapyrone phytotoxin by *Ascochyta rabiei* in response to metal cations and development of a defined medium for toxin production *Plant Pathology 40*: 401-407.

Cho, S., Chen, W. and Muehlbauer, F.J. (2005). Constitutive expression of the flavanone 3-hydroxylase gene related to pathotype-specific ascochyta blight resistance in *Cicer arietinum* L. *Physiological and Molecular Plant Pathology 67*:100-107.

Coram, T.E. and Pang, E.C.K. (2006). Expression profiling of chickpea genes differentially regulated during a resistance response to *Ascochyta rabiei. Plant Biotechnology Journal 4*:647-666.

Gilbert, M.J., Soanes, D.M. and Talbot, N.J. (2004). Functional genomic analysis of the rice blast fungus *Magnaporthe grisea*. In: Arora DK & Khachatourians GG (eds.), *Applied Mycology and Biotechnology Volume 4: Fungal Genomics*. Elsevier Science: Amsterdam, Netherlands, pp. 331-352.

Henson, J., Butler, M. and Day, A. (1999). The dark side of the mycelium: Melanins and phytopathogenic fungi. *Annu Rev Phytopath 37*: 447-471.

Höhl, B., Weidemann, C., Höhl, U. and Barz, W. (1991). Isolation of solanapyrones A, B and C from culture filtrates and spore germination fluids of *Ascochyta rabiei* and aspects of phytotoxin action. *J Phytopath 132*: 193-206.

Kahmann, R. and Basse, C. (1999). REMI (Restriction enzyme mediated integration) and its impact on the isolation of pathogenicity genes in fungi attacking plants. *Europ J Plant Pathol 105*: 221-229.

Kaur, S. (1995). Phytotoxicity of solanapyrones produced by the fungus *Ascochyta rabiei* and their possible role in blight of chickpea (*Cicer arietinum*). *Plant Sci 109*: 23-29.

Kawamura, C., Tsujimoto, T. and Tsuge, T. (1999). Targeted disruption of a melanin biosynthesis gene affects conidial development and UV tolerance in the Japanese pear pathotype of *Alternaria alternata. Mol Plant-Microbe Interact 12*: 59-63.

Köhler, G., Linkert, C. and Barz, W. (1995). Infection studies of *Cicer arietinum* (L.) with GUS-(*E. coli* β-*glucuronidase*) transformed *Ascochyta rabiei* strains. *J Phytopathol 143*: 589-595.

Lu, S., Kroken, S., Lee, B.N., Robbertse, B., Churchill, A., Yoder, O.C. and Turgeon, B. (2003). A novel class of gene controlling virulence in plant pathogenic ascomycete fungi. *Proceed Nationl Acad Sci USA 100*: 5980-5985.

Michielse, C.B, Hooykaas, P.J.J., Cees, A., van-den Hondel, J. and Ram, A.F.J. (2005). *Agrobacterium* mediated transformation as a tool for functional genomics in fungi. *Curr Genet 48*: 1-17.

Morgensen, E., Challen, M. and Strange, R.N. (2006). Reduction in solanapyrone phytotoxin production by *Ascochyta rabiei* transformaned with *Agrobacterium tumefaciens. FEMS Microb Lett 255*: 255-261.

Nene, Y.L., Hanounik, S.B., Qureshi, S.H. and Sen, B. (1988). Fungal and bacterial foliar diseases of pea, lentil, faba bean and chickpea. In: R. J. Summerfield (Ed.), *World crops: Cool season food legumes*. Kluwer Academic Publishers: Dordrecht, Netherlands. pp. 577–589.

Oliver, R. and Osborun, A. (1995). Molecular dissection of fungal phyopathogenicity. *Microbiology 141*: 1-9.

Peever, T.L. (2007). Role of host specificity in the speciation of Ascochyta pathogens of cool season food legumes. *Europ J Plant Pathol 119*: 119-126.

Peever, T.L., Barve, M.P., Stone, L.J. and Kaiser, W.J. (2007). Evolutionary relationships among Ascochyta species infecting wild and cultivated hosts in the legume tribes Cicereae and Vicereae. *Mycologia 99*: 59-77.

Schoch, C.L., Aist, J.R., Yoder, O.C. and Turgeon, B.G. (2003). A complete inventory of fungal kinesins in representative filamentous ascomycetes. *Fungal Genet Biol 39*: 1-15.

Steinberg, G. and Fuchs, U. (2004). The role of microtublues in cellular organization and endocytosis in the plant pathogen *Ustilago maydis*. *J Microscopy 214*: 114-123.

Straube, A., Hause, G., Fink, G. and Steinberg, G. (2006). Conventional kinesin mediates microtubule-microtubule interactions *in vivo*. *Mole Biolo Cell 17*: 907-916.

Talbot, N.J. (Ed.) (2004). *Plant-Pathogen Interactions*. Blackwell Publishing: Oxford, UK

Taylor, P. and Ford, R. (2007). Diagnostics, genetic diversity and pathogenic variation of Ascochyta blight of cool season food and feed legumes. *European Journal of Plant Pathology 119*: 127-133.

Tenhaken, R. and Barz, W. (1991). Characterization of pectic enzymes from the chickpea pathogen *Ascochyta rabiei. Verlag der Zeitschrift fur Naturforschung 46c*: 51-57.

Tenhaken, R., Arnemann, M., Kohler, G. and Barz, W .(1997). Characterization and cloning of cutinase from *Ascochyta rabiei. Verlag der Zeitschrift fur Naturforschung 52c*: 197-208.

Trapero-Casas, A. and Kaiser, W.J. (1992). Development of *Didymella rabiei,* the telomorph of *Ascochyta rabiei*, on chickpea straw. *Phytopathology 82*: 1261-1266.

Walton, J. (1994). Deconstructing the plant cell wall. *Plant Physiology 104*: 1113-1118.

White, D. and Chen, W. (2006a). Genetic transformation of *Ascochyta rabiei* using *Agrobacterium*-mediated transformation. *Current Genetics 49*: 272-280.

White, D. and Chen, W. (2006b). Construction of a lambda phage library of the chickpea blight pathogen *Ascochyta rabiei* genome. *International Chickpea and Pigeon Pea Newsletter 13*: 1-2.

White, D. and Chen, W. (2007). Towards identifying pathogenic determinants of the chickpea pathogen *Ascochyta rabiei. European Journal of Plant Pathology 119*: 3-12

Zhang, A., Lu, P., Dahl-Roshak, A.M., Paress, P.S., Kennedy, S., Tkacz, J.S. and An, Z. (2003). Efficient disruption of a polyketide synthase gene (*pks*1) required for melanin synthesis through *Agrobacterium*-mediated transformation of *Glarea lozoyensis. Molecular Genetics and Genomics 268*: 645-55.

In: Current Advances in Molecular Mycology
Eds: Y. Gherbawy, R.L. Mach and M.K. Rai
ISBN 978-1-60456-909-4

Chapter XI

Diversity, Genotypic Identification, Ultrastructural and Phylogenetic Characterization of Zygomycetes from Different Ecological Habitats and Climatic Regions: Limitations and Utility of Nuclear Ribosomal DNA Barcode Markers

Kerstin Hoffmann[1], Sabine Telle[1], Grit Walther[2], Martin Eckart[1], Martin Kirchmair[3], Hansjörg Prillinger[4], Adam Prazenica[5], George Newcombe[5], Franziska Dölz[1], Tamás Papp[6], Csaba Vágvölgyi[6], G. Sybren deHoog[2], Lennart Olsson[7] and Kerstin Voigt[1,*]

[1]University of Jena, Institute of Microbiology, 07743 Jena, Germany;
[2]Centraalbureau voor Schimmelcultures, 3584 CT Utrecht, The Netherlands;
[3]University of Innsbruck, Institute of Microbiology, 6020 Innsbruck, Austria;
[4]University of Natural Resources and Applied Life Sciences Vienna, Department of Biotechnology, Institute of Applied Microbiology, 07743 Vienna, Austria;
[5]University of Idaho, College of Natural Resources, Department of Forest Resources, Moscow, Idaho 83844-1133, USA;
[6]University of Szeged, Faculty of Sciences, Department of Microbiology, 6726 Szeged, Hungary;
[7]University of Jena, Institute of Systematic Zoology and Evolutionary Biology, 07743 Jena, Germany.

[*] Correspondence concerning this article should be addressed to: Kerstin Voigt, Friedrich-Schiller-Universität Jena, Institut für Mikrobiologie, Pilz-Referenz-Zentrum, Neugasse 24, D-07743 Jena, Germany. Phone: +49 3641 949321; Fax: +49 3641 949322; E-mail: kerstin.voigt@uni-jena.de.

Abstract

In addition to the morphological criteria traditionally used in taxonomy, molecular markers, which could be useful as DNA barcodes, are of increasing interest for the identification of fungi. The success of unequivocal identification relies on a rational selection of potential marker genes. The search for a candidate that is universally applicable for fungal identification is currently the subject of much debate. The aim of this study is to investigate the reliability of genes from the ribosomal RNA cluster, small subunit (SSU) 18S and large subunit (LSU) 28S ribosomal DNA (rDNA), internal transcribed spacer (ITS) 1 and 2 interrupted by 5.8S rDNA as identification markers. For this purpose a total of 98 fungal strains isolated from different habitats in diverse geographic regions in the northern hemisphere were examined. The fungi investigated comprise 68 mucoralean, two mortierellalean and one kickxellalean zygomycete, as well as several ascomycetes and basidiomycetes, which serve as outgroups for the comparison. Phylogenetic reconstruction and analysis of a dataset encompassing a total of 115 new sequences (Acc nos. EU484190-EU484303, EF589886) including ITS1-5.8S-ITS2 nucleotide sequences of morphologically well-defined type strains suggest that the accessibility of type strains which can serve as references are an essential prerequisite for obtaining reliable results.

Key words: Internal transcribed spacer 1 & 2, 5.8S, 18S (SSU), 28S (LSU) rDNA; Mucorales; DNA barcoding; phylogeny; reference and type strains.

1. Introduction

1.1. The Fascinating World of Fungi

The number of fungal species described so far ranges between 72,000 and 120,000 (Hawksworth and Rossman, 1997; Hawksworth, 2001), but the total number of fungal species is estimated to be at least 1.5 million (Hawksworth, 1991; Kirk *et al.*, 2001; Hawksworth, 2001). The gap between the number of species described and the estimated total number of fungi nevertheless indicates that a very large number of undescribed fungi exists all over the world, many of which may be found in associations with plants, insects, animals or as lichen-forming fungi, particularly in the tropics (Hawksworth, 2001). But also hidden species (cryptic species), hitherto believed to belong to species already described, are discovered, for example in the dikaryomycotan genera *Armillaria* (Pegler, 2000), *Trichoderma* (Gams and Bissett, 1998), *Letharia* (Kroken and Taylor, 2001) or *Fusarium* (e.g. Baayen *et al.*, 2000; O`Donnell, 2000; O`Donnell *et al.*, 2000). On the other hand large intra- and interspecific variations in phenotypic, physiological, biological and ecological traits often hamper identification, if it tries to take all properties into account, as shown in *Fusarium* (Gherbawy *et al.*, 2001) or in yeasts and yeast phases of dimorphic fungi (Barnett *et al.*, 1990; Boekhout *et al.*, 2000; Prillinger *et al.*, 1990a, b; 1991 a, b; Messner *et al.*, 1994; Schweigkofler *et al.*, 2002).

1.2. Taxonomy Versus a Natural System of Fungi

Traditionally four major groups (phyla) are distinguished within the kingdom Fungi: Chytridiomycota, Zygomycota, Ascomycota and Basidiomycota. However, it has become evident that this traditional scheme does not reflect the phylogenetic relationships among fungi and the classification of the fungi is still in flux. The monophyletic sister groups Asco- and Basidiomycota are well-characterized phyla at the molecular phylogenetic level (Sugiyama, 1998; Berbee and Taylor, 2001; van de Peer *et al.*, 2000). Because phylogenetic analyses reveal that both phyla share a hypothetic common ancestor they were treated as the subphyla Asco- and Basidiomycotina and combined into the phylum Dikaryomycota (as used in Tehler *et al.*, 2000) or even into the subkingdom Dikarya (as used in James *et al.*, 2006 and established in Hibbett *et al.*, 2007).

The most important changes to phylogenetic relationships and classification in recent years concern groups traditionally embedded in the phyla Zygomycota and Chytridiomycota, for example to establish the phyla Glomeromycota (Schüßler *et al.*, 2001), Blastocladiomycota (James *et al.*, 2007), and Neocallimastigomycota (Hibbett *et al.*, 2007).

Constant changes of terminologies and taxonomic ranks applied to the fungal groups indicate a lack of consensus in classification. Major advances towards a modernization of classification were obtained by Kirk *et al.* (2001) and McLaughlin *et al.* (2001a, 2001b). Databases which are available online at Index Fungorum (www.indexfungorum.org), GenBank (www.ncbi.nlm.nih.gov/taxonomy), MycoBank (www.mycobank.org), Myconet (www.fieldmuseum.org/myconet), the Tree of Life web project (www.tolweb.org), Assembling the Fungal Tree of Life (AFTOL, http://aftol.org) and Deep Hyphae (http://ocid.nacse.org/research/deephyphae) provide a powerful tool for the generalization of mycological taxonomy and are easy to access. Although there is still no finalized congruency in all these classification projects, a consensus classification based on monophyletic groups was proposed by Hibbett *et al.* (2007).

1.3. The outsider Position of Zygomycetes: the Morphological Tradition meets Molecular Phylogenetics

The phylum Zygomycota is traditionally divided into Zygomycetes and Trichomycetes at the level of classes, and in Asellariales, Dimargaritales, Endogonales, Entomophthorales, Harpellales, Kickxellales, Mortierellales, Mucorales and Zoopagales at the level of orders (White *et al.*, 2006). In contrast, if the monophyla-based classification system as proposed by Hibbett *et al.* (2007) is applied, the polyphyletic phylum Zygomycota is dissolved and its monophyletic groups are subdivided into the Mucoromycotina, Kickxellomycotina, Zoopagomycotina and Entomophthoromycotina. The orders Mucorales, Endogonales and Mortierellales are assigned to the subphylum Mucoromycotina. Kickxellales, Dimargaritales, Harpellales and Asellariales are assigned to the Kickxellomycotina.

Nevertheless, bearing in mind that there is a taxonomical revolution ongoing, the terminology of the traditional Zygomycota will be kept during this study, because the phylogeny-based affiliation of its monophyletic clades to taxa of higher ranks is still controversial if compared with different deep-level gene genealogies (Sugiyama, 1998; Voigt and Wöstemeyer, 2001; Schüßler *et al.*, 2001; Tanabe *et al.*, 2005; James *et al.*, 2006).

Members of the Zygomycota generally reproduce asexually by the formation of aplanospores such as sporangiospores, arthrospores or chlamydospores and sexually by the formation of zygospores. Sporangiospores can be located in multi-spored sporangia, few-spored sporangiola or merosporangia (Benjamin, 1979; Benny *et al.*, 2001). Zygomycetes are ubiquitously distributed worldwide. The class includes common saprobionts in soil and dung as well as, facultative or obligate parasites (Benjamin, 1979). They also inhabit many important niches with impact on humans as plant pathogens, agents of storage decay, and as pathogens on animals and man (e.g. Benjamin, 1959; Hesseltine and Ellis, 1966; Holliday, 1980; Humber *et al.*, 1988; Michailides and Spotts, 1990; Ogawa *et al.*, 1992; Ribes *et al.*, 2000; Voigt *et al.*, 1999a; Wolf, 1917). But they also possess industrial importance as biocatalysts for the production of steroids, organic acids, beta carotenes and in various food fermentation processes (e.g. Certik and Shimizu, 1999; Hachmeister and Fung, 1993; Hesseltine and Ellis, 1973).

Since the references cited above reflect only a modest representation of the rapidly growing studies and research on fungal systematics and phylogenetics, there is a pressing need for a consensus on the classification and identification of new species within the zygomycetes. One way to achieve this goal is to combine phenotypical and molecular information with emphasis on molecular data, which are (more) independent from external influences. Getting informative phenotypic data is often highly problematic due to a lack of sufficient distinguishing morphological characters, large intra-specific variability of the present characters and their dependence upon physiological growth parameters (Schipper, 1973, Zycha *et al.,* 1969). On the other hand precise morphological identification is essential to ensure the correct assignment of molecular data to a fungal microorganism. In combination with DNA barcoding, which applies DNA sequences as taxon-specific molecular markers, it is possible to achieve a more precise delimitation of species and subspecies, and to identify morphologically identical species. A necessary requirement for a reliable DNA-barcode based species identification is a broad range of specimens and a well-defined taxonomical species description (Meyer and Pauley, 2005). Barcodes are not suitable to define species but there are useful to assign unidentified specimens to a known species or to elucidate new or cryptic species corresponding to tedious groups requiring multilocus-analyses (see the ′triage tool′ concept proposed by Schindel and Miller, 2005). Many zygomycetes are opportunistic pathogens causing mucor- or entomophthoromycoses, whose treatment requires a fast and reliable diagnosis (Voigt *et al.*, 1999a). Isolation and cultivation of fungi is time-consuming and labour intensive. The morphological identification is affected by atypical growth of fungal hyphae in human tissue in comparison to their growth on the appropriate artificial media. Moreover, the optimal growth conditions for reliable morphological observations are often unknown (Benny, 1995; Schipper, 1973; Scholer *et al.*, 1983; Weitzman *et al.,* 1995). In this respect consensus classifications as proposed by Hibbett *et al.* (2007), accessible culture-collections (CBS, DSMZ, ATCC, ARS-NRRL, PRZ etc.) providing reliably maintained type strains, as well as reliable sequence information on type strains used as references and available from public databases (GenBank, EMBL etc.) are important prerequisites for the precise molecular identification of fungal species. The number of species might be underestimated due to variable ranges of sequence identities. Thus, the control of the reliability of fungal organisms, which are already identified or freshly isolated from environmental samples, with such comprehensive databases allows the assignment to new species (e.g. Wuczkowski *et al.*, 2003).

In this study the reliability of morphological and molecular markers for the identification of fungi, the influence of the geographic origin and the species delimitation borders of DNA sequence identity percentages are reviewed and discussed. Furthermore an overview of DNA barcode-based identification is provided which gives a short introduction to the utility, limitations, applications and experimental strategy of DNA barcode markers with emphasis on nuclear ribosomal DNA applied on a large range of fungi. Evidence is provided that DNA barcoding really works. Due to the complexity DNA barcoding has gained in recent years, a comprehensive DNA barcoding study would go beyond the possibilities of the present review and rather an impulse for the intensification of DNA barcoding projects at the regional, national, European and international levels is intended.

2. Experimental Strategies, Methods and Results

2.1. Preliminary Methodological Remarks

Ninety-nine fungal strains belonging to the phyla Zygomycota (92), Ascomycota (6) and Basidiomycota (1; Table 1) were characterized based on nucleotide sequence analyses (Table 2) and morphological criteria (Table 3). Ten strains represent type, neo-, iso- or syntype strains and were used for comparison using DNA barcodes based on ITS1-5.8S-ITS2. Seventy-seven fungal strains were freshly isolated from nature and represent new environmental isolates from different substrates of diverse climatic regions of the northern hemisphere (Table 1). These strains are deposited at the Fungal Reference Centre (University Jena) and available online upon request at http://www.prz.uni-jena.de. The dataset was augmented with twenty strains from the Centraalbureau voor Schimmelcultures (CBS, Utrecht, The Netherlands, at http://cbs.knaw.nl) to verify the Blast searches with well-studied references, preferably from type strain material (Table 1).

All strains were cultivated at room temperature on complex media like malt extract medium (30g/l) supplemented with 5g/l yeast extract or on supplemented minimal medium (SUP; Wöstemeyer, 1985). For isolation of fungi from natural substrates SMA (synthetic *Mucor* agar; Hesseltine, 1954), SNA (synthetic nutrient-poor agar; Nirenberg, 1981), soil extract agar and horse manure agar (see instructions in the CBS catalogue) or SUP agar were used. The content of colony forming units on each substrate was estimated after incubation of dilution series of up to the 10^8 dilutions on all five types of solid media. Morphological investigations as shown in figures 3-6 were performed by light microscopic visualisation (Axiophot, Zeiss, Germany) and as shown in figures 7-10 by scanning electron microscopy (Philips XL 30 ESEM, The Netherlands and Emitech K500 sputter coater, England) as described in Voigt and Olsson (2008).

The extraction of genomic DNA and the amplification of marker genes by the polymerase chain reaction were carried out following the protocols of Einax and Voigt

(2003). SSU rDNA fragments were amplified using forward primer NS1 (White *et al.*, 1990) and reverse primers NS41 or NS8Z (O`Donnell *et al.*, 1998), whereas the LSU rDNA was amplified using the primer pair NL1 and NL4 (O`Donnell, 1993). The ITS regions 1 and 2 including 5.8S rDNA was amplified with the primer pair ITS1 and ITS4 (White *et al.*, 1990). Purification, cloning and sequencing of the amplicons were done as previously described by Hoffmann *et al.* (2007). A total of 115 new nucleotide sequences were generated and deposited under the accession numbers EF589886 and EU484190-EU484303 in GenBank at http://ncbi.nlm.nih.gov/. The accession numbers of 93 new sequences are listed in Table 2. We also provide the lengths of the sequenced fragments including the best hits during BLAST searches (provided as an online tool by the National Center for Biotechnology Information [NCBI] at http://www.ncbi.nlm.nih.gov) as indicated by the accession numbers of reference sequences, and the relative and absolute sequence identities as indicated by the number of identical base pairs. The number of corresponding sequences from type and non-type material presently available in GenBank is also indicated in Table 2.

Additionally, nineteen nucleotide sequences of SSU rDNA and twenty-two nucleotide sequences of LSU rDNA (as listed in Table 4) were retrieved from GenBank and used as references in the phylogenetic analyses shown in figures 11 and 12, respectively.

Alignments were carried out in ClustalX version 1.83 (Higgins and Sharp, 1988, 1989; Thompson *et al.*, 1997) and controlled manually using BioEdit version 7.0.3 (Hall, 1999). Phylogenetic reconstructions were conducted using PAUP*4.0b10 (Swofford, 1998) for the 18S rDNA and the 28S rDNA datasets. The 18S rDNA alignment consists of 49 taxa and the 28S rDNA alignment of 53 taxa with 1211 and 800 characters, respectively. The neighbor-joining trees (Saitou and Nei, 1987) shown in figures 11 and 12 were inferred from logdet/paralinear distances with uncorrected P distance measure. Both phylograms were displayed and printed in TreeView version 1.6.6. (Page, 1996). Bootstrap supports (BS) (Felsenstein, 1985; 50% majority rule) were obtained by 1000 replicates using logdet/paralinear distances. The aligned data matrices and the phylogenetic trees are available from TreeBase at http://www.treebase.org/treebase (study accession no. S2040, matrix accession nos. M3822-M3823).

Figure 1. Ascoma of a representative of the *Morchella elata* group in Idaho, which mycelia and sporangia of the zygomycetous mycoparasite *Mucor hiemalis* f. *corticola* FSU3008 were recovered from. (Image: Adam Prazenica).

Table 1. Fungal strains used in this study. FSU…Friedrich Schiller University Jena, Germany; CBS…Centraalbureau voor Schimmelcultures Utrecht, The Netherlands

strain collection numbers		geographic origin	substrate	isolated by
FSU	additional			
Mucorales, Zygomycota				
497		South-Island	mud near geyser	K. Voigt
499		Hafnarfjördhur, South-Island	dung of horse	K. Voigt
500		Hafnarfjördhur, South-Island	dung of horse	K. Voigt
501		Hafnarfjördhur, South-Island	dung of horse	K. Voigt
514		Rotmoostal Ötztal, Tirol, Austria	unknown	M. Kirchmair
697		Michigan, USA	unknown	P.Sebek
2005		Rotmoostal Ötztal, Tirol, Austria	unknown	M. Kirchmair
2572		Nägelstedt, Germany	dung of horse	K. Voigt
2666		Jena, Germany	Ericaceae	E. Kothe
2797	RA1-02	Rothwald, Austria	soil, leaf litter	E. Metzger
2798	RA1-21	Rothwald, Austria	soil, leaf litter	E. Metzger
2799	RA1-26	Rothwald, Austria	soil, leaf litter	E. Metzger
2800	RA1-41	Rothwald, Austria	soil, leaf litter	E. Metzger
2801	RA1-49	Rothwald, Austria	soil, leaf litter	E. Metzger
2802	RB1-01	Rothwald, Austria	soil, leaf litter	E. Metzger
2803	RB1-02	Rothwald, Austria	soil, leaf litter	E. Metzger
2804	RB1-25	Rothwald, Austria	soil, leaf litter	E. Metzger
2805	RB1-32	Rothwald, Austria	soil, leaf litter	E. Metzger
2806	RB1-45	Rothwald, Austria	soil, leaf litter	E. Metzger
2807	RB1-53	Rothwald, Austria	soil, leaf litter	E. Metzger
2808	RB1-54	Rothwald, Austria	soil, leaf litter	E. Metzger
2809	SA1-13	Bad Sauerbrunn, Austria	soil, leaf litter	E. Metzger
2810	SA1-25	Bad Sauerbrunn, Austria	soil, leaf litter	E. Metzger
2811	SA1-30	Bad Sauerbrunn, Austria	soil, leaf litter	E. Metzger
2812	SA1-31	Bad Sauerbrunn, Austria	soil, leaf litter	E. Metzger

Table 1. (continued 1)

strain collection numbers		geographic origin	substrate	isolated by
FSU	additional			
2813	SB1-05	Bad Sauerbrunn, Austria	soil, leaf litter	E. Metzger
2814	SB1-07	Bad Sauerbrunn, Austria	soil, leaf litter	E. Metzger
2815	SB1-15	Bad Sauerbrunn, Austria	soil, leaf litter	E. Metzger
2816	SB1-27	Bad Sauerbrunn, Austria	soil, leaf litter	E. Metzger
2817	ZM10	nationalpark Danube, meadow, Lobau, Austria	meadow of Danube river, soil under *Salix alba*	M. Wuczkowski
2818	ZM11	nationalpark Danube, meadow, Lobau, Austria	meadow of Danube river, soil under *Salix alba*	M. Wuczkowski
2819	ZM12	nationalpark Danube, meadow, Lobau, Austria	meadow of Danube river, soil under *Salix alba*	M. Wuczkowski
2820	ZM13	Mannswoerth, Austria	meadow of Danube river, soil under *Salix alba*	M. Wuczkowski
2821	ZM14	Mannswoerth, Austria	meadow of Danube river, soil under *Salix alba*	M. Wuczkowski
2822	ZM15	Mannswoerth, Austria	meadow of Danube river, soil under *Salix alba*	M. Wuczkowski
2824	SB2-20	unknown	meadow of Danube river	unknown
2825	ZM18	nationalpark Danube, meadow, Lobau, Austria	meadow of Danube river	M. Wuczkowski
3008		Idaho, USA	ascocarp of *Morchella elata*	A. Prazenica
3846	CID285	Idaho, USA	endophyte of *Centaurea stoebe*	G. Newcombe
4726		Innsbruck, Tirol, Austria	damp concrete	M. Kirchmair
4754		Innsbruck, Tirol, Austria	potting soil, *Capsicum sp.* (Solanaceae)	M. Kirchmair
4755		Geisenheim, Germany	soil of vineyard	M. Kirchmair
4756		Geisenheim, Germany	soil of vineyard	M. Kirchmair
6154		Crimea	dung of cow	S. Telle
6155		Crimea	dung of cow	S. Telle
6156		Crimea	dung of cow	S. Telle
6157		Crimea	dung of horse	S. Telle
6158		Crimea	dung of horse	S. Telle
6159		Crimea	dung of bat	S. Telle
6160		Crimea	dung of bat	S. Telle
6161		Crimea	dung of bat	S. Telle
6162		Crimea	dung	S. Telle
6163		Crimea	dung	S. Telle
6164		Crimea	dung	S. Telle

Table 1. (continued 2)

strain collection nos.		geographic origin	substrate	isolated by
FSU	additional			
6165		Crimea	dung	S. Telle
6166		Crimea	dung	S. Telle
6167		Crimea	dung	S. Telle
6168		Crimea	dung	S. Telle
6169		Crimea	dung	S. Telle
6170		Germany	agaric	S. Telle
6171		Germany	agaric	S. Telle
6172		Germany	agaric	S. Telle
6173		Germany	agaric	S. Telle
6174		Germany	agaric	S. Telle
6175		Germany	agaric	S. Telle
6176		Germany	basidiocarb of agaric fungi	S. Telle
6177		Germany	dung of horse	S. Telle
6178		Germany	dung of horse	S. Telle
6179		Germany	dung of cow	S. Telle
Kickxellales, Zygomycota				
2823		nationalpark Danube, meadow, Lobau, Austria	meadow, soil under *Populus x canescens*	M. Wuczkowski
Mortierellales, Zygomycota				
2735		unknown	unknown	A. Zeuner
2736		unknown	unknown	A. Zeuner
Eurotiales, Ascomycota				
503		Hafnarfjördhur, South-Island	dung of horse	K. Voigt
506		Hafnarfjördhur, South-Island	dung of horse	K. Voigt
1257		Moskau?	unknown	G. Arnold
2667		Göttingen, Germany	unknown	Boritzki
Hypocreales, Ascomycota				
2858		Germany	3M potassium chloride	J. Voigt
2883		Germany	leakage water, waste disposal site	H. Bindara

Table 1. (continued 3)

strain collection nos.		geographic origin	substrate	isolated by
FSU	**additional**			
Polyporales, Basidiomycota				
2734		unknown	unknown	A. Zeuner
complementary CBS strains				
620	CBS 109.16[1]	unknown	unknown	unknown
757	CBS 148.22[2]	unknown	unknown	M.B.Church
775	CBS 136.28[3]	unknown	dung of horse	unknown
796	CBS 137.28[4]	unknown	dung of horse	unknown
869	CBS 811.69[5]	near Frederiksseter, Norway	meadow soil	M.A.A. Schipper
6019	CBS 127.08[6]	unknown	unknown	unknown
6020	CBS 110.17[7]	Switzerland	unknown	A. Lendner
6022	CBS 257.28[8]	Taiwan	Formosan peka	R. Nakazawa
6025	CBS 329.47[9]	unknown	unknown	unknown
6027	CBS 285.55[10]	Den Haag, Netherlands	tempeh	M.B. Schol-Schwarz
-	CBS 101.08[11]	unknown	unknown	O. Hagem
-	CBS 106.08[12]	unknown	unknown	unknown
-	CBS 112.07[13]	Netherlands	unknown	F.A.F.C. Went
-	CBS 114.08[14]	Geneve, Jussy, Switzerland	soil	A. Lendner
-	CBS 169.25[15]	unknown	*Pyrus communis* (Rosaceae), decaying fruit	C. Wehmer
-	CBS 201.65[16]	Michigan, USA	unknown	P.Sebek
-	CBS 260.68[17]	Basel, Switzerland	unknown	M.A.A. Schipper
-	CBS 311.52[18]	Germany	soil	G. Linnemann
-	CBS 444.65[19]	Wyoming, Medicine Bow, Libby Flats, USA	soil	G. Rall
-	CBS 115583[20]	Rothamsted, UK, England	wallpaper	H. Kwasna

[1] *Mucor mucedo*, [2] *Rhizopus oryzae*, [3-4] *Chaetocladium brefeldii*, [5] *Chaetocladium jonesii*, [6-10] *Rhizopus oryzae*, [11] Type of *Absidia glauca*, [12] *Absidia spinosa* var. *spinosa* identified by A. Lendner (describer of the species *Absidia spinosa*), [13] Type of *Rhizopus oryzae*, [14] Type of *Mucor genevensis*, [15] Neotype of *Mucor piriformis*, [16] Neotype of *Mucor hiemalis* f. *hiemalis*, [17] Type of *Mucor racemosus* f. *racemosus*, [18] Syntype of *Mortierella parvispora*, [19] Type of *Zygorhynchus moelleri*, [20] Isotype of *Absidia repens*

Table 2. Associated information for each fungal strain with respect to the nucleotide sequences obtained (type of gene, length and Genebank accession number of the gene sequences), results of BLAST searches (Genebank accession number of the best hit, length and percentage of matching base pairs [bp]), the number of available reference sequences in Genebank (for the type strain as well as the total number of sequences, respectively), and identities of the type strains. Relative identities [in percent] of ITS sequences below or equal to 97 % are indicated in bold. Variance to reference sequences higher than 6 % indicates new genera. Variances ranging between 3 % and 6 % indicate new species. FSU...Friedrich Schiller University Jena, Germany; CBS...Centraalbureau voor Schimmelcultures, Utrecht, The Netherlands; T...type strain, NT... neotype strain, IT... isotype strain; 1...reference sequences total/type strain refers to the final identity of the strain (Table 3); 2...if sequences of the type strain are not available the CBS number is printed in grey.

strain no.	sequence information			BLAST results			reference sequences	type strain[2]
				best hit	identity			
	gene	bp	Acc.no.	Acc.no	bp	%	total/ type[1]	
Mucorales, Zygomycota								
FSU 497	ITS	552	EU484190	DQ118992, *Mucor hiemalis* f. *hiemalis* (NT)	552/553	99	2/1	CBS 201.65 NT
	18S rDNA	1153	EU484191	AF113428, *Mucor hiemalis* f. *hiemalis* (NT)	1152/1155	99	1/1	
FSU 499	ITS	545	EU484192	AY213659, *Mucor racemosus* f. *racemosus* (T)	544/546	99	7/2	CBS 260.68 T
FSU 500	ITS	544	EU484193	AJ878775, *Mucor racemosus* f. *sphaerosporus*	539/544	99	1/0	CBS 115.08 IT
	18S rDNA	1155	EU484194	AF113430, *Mucor racemosus* f. *racemosus* (T)	1153/1155	99	0/0	
FSU 501	ITS	549	EU484195	AY243943, *Mucor circinelloides* f. *circinelloides* (NT)	543/549	98	10/2	CBS 195.68 NT
	18S rDNA	1155	EU484196	AF113430, *Mucor racemosus* f. *racemosus* (T)	1149/1155	99	0/0	
FSU 514	ITS	546	EU484197	AJ878785, *Zygorhynchus moelleri*	546/546	100	1/0	CBS 444.65 T
	18S rDNA	1155	EU484198	AF157132, *Dicranophora fulva*	1141/1157	98	0/0	
FSU 697	ITS	558	EU484200	AY213659, *Mucor racemosus* f. *racemosus* (T)	557/558	99	7/2	CBS 260.68 T
FSU 2005	ITS	569	EU484205	AY997097, *Umbelopsis ramanniana*	541/570	**94**	18/?	?
	18S rDNA	1154	EU484206	DQ322627, *Umbelopsis ramanniana*	1151/1154	99	2/?	
FSU 2572	ITS	609	EU484207	AF412289, *Mucor mucedo*	586/609	**96**	1/0	CBS 640.67 NT
FSU 2666	ITS	558	EU484208	AY213659, *Mucor racemosus* f. *racemosus* (T)	556/558	99	7/2	CBS 260.68 T
FSU 2797	28S rDNA	698	AY706219	AF113468, *Mucor hiemalis* f. *hiemalis* (NT)	695/698	99	30/1	CBS 201.65 NT
FSU 2798	28S rDNA	698	AY706220	AF113468, *Mucor hiemalis* f. *hiemalis* (NT)	694/698	99	30/1	CBS 201.65 NT
FSU 2799	28S rDNA	698	AY706221	AF113468, *Mucor hiemalis* f. *hiemalis* (NT)	693/698	99	30/1	CBS 201.65 NT
FSU 2800	28S rDNA	698	AY706222	AF113468, *Mucor hiemalis* f. *hiemalis* (NT)	694/698	99	30/1	CBS 201.65 NT

Table 2. (continued 1)

strain no.	sequence information			BLAST results			reference sequences	type strain[2]
				best hit	identity			
	gene	bp	Acc.no.	Acc.no	bp	%	total/ type[1]	
FSU 2801	28S rDNA	698	AY706223	AF113468, *Mucor hiemalis* f. *hiemalis* (NT)	693/698	99	30/1	CBS 201.65 NT
FSU 2802	28S rDNA	698	AY706224	AF113468, *Mucor hiemalis* f. *hiemalis* (NT)	695/698	99	30/1	CBS 201.65 NT
FSU 2803	28S rDNA	698	AY706225	AF113468, *Mucor hiemalis* f. *hiemalis* (NT)	691/698	99	30/1	CBS 201.65 NT
FSU 2804	28S rDNA	698	AY706226	AF113468, *Mucor hiemalis* f. *hiemalis* (NT)	691/698	99	30/1	CBS 201.65 NT
FSU 2805	28S rDNA	698	AY706227	AF113468, *Mucor hiemalis* f. *hiemalis* (NT)	691/698	99	30/1	CBS 201.65 NT
FSU 2806	28S rDNA	698	AY706228	AF113468, *Mucor hiemalis* f. *hiemalis* (NT)	692/698	99	30/1	CBS 201.65 NT
FSU 2807	28S rDNA	698	AY706229	AF113468, *Mucor hiemalis* f. *hiemalis* (NT)	691/698	99	30/1	CBS 201.65 NT
FSU 2808	28S rDNA	698	AY706230	AF113468, *Mucor hiemalis* f. *hiemalis* (NT)	962/698	99	30/1	CBS 201.65 NT
FSU 2809	28S rDNA	694	AY706231	AF113448, *Absidia repens*	630/681	**92**	1/0	CBS 115583 IT
FSU 2810	28S rDNA	700	AY706232	AF113447, *Absidia glauca*	690/700	98	3/0	CBS 101.08 T
FSU 2811	28S rDNA	680	AY706233	AF157193, *Helicostylum elegans*	677/682	99	1/?	?
FSU 2812	28S rDNA	700	AY706234	AF113447, *Absidia glauca*	690/700	98	3/0	CBS 101.08 T
FSU 2813	28S rDNA	694	AY706235	AF113448, *Absidia repens*	628/681	**92**	1/0	CBS 115583 IT
FSU 2814	28S rDNA	694	AY706236	AF113448, *Absidia repens*	630/681	**92**	1/0	CBS 115583 IT
FSU 2815	28S rDNA	694	AY706237	AF113448, *Absidia repens*	629/681	**92**	1/0	CBS 115583 IT
FSU 2816	28S rDNA	694	AY706238	AF113448, *Absidia repens*	630/683	**92**	1/0	CBS 115583 IT
FSU 2817	28S rDNA	700	AY706239	AF157172, *Absidia glauca*	698/700	99	3/0	CBS 101.08 T
FSU 2818	28S rDNA	700	AY706240	AF157172, *Absidia glauca*	700/700	100	3/0	CBS 101.08 T
FSU 2819	28S rDNA	698	AY706241	AF113468, *Mucor hiemalis* f. *hiemalis* (NT)	688/701	98	30/1	CBS 201.65 NT
FSU 2820	28S rDNA	698	AY706242	AF113468, *Mucor hiemalis* f. *hiemalis* (NT)	687/701	98	30/1	CBS 201.65 NT
FSU 2821	28S rDNA	700	AY706243	AF157172, *Absidia glauca*	699/700	99	3/0	CBS 101.08 T
FSU 2822	28S rDNA	655	AY706244	AB363778, *Zygorhynchus exponens*	626/660	**94**	?/?	?
FSU 2824	28S rDNA	694	AY706246	AF113448, *Absidia repens*	630/682	**92**	1/0	CBS 115583 IT
FSU 2825	28S rDNA	623	AY706247	AB250187, *Rhizopus oryzae* (T)	623/623	100	10/2	CBS 112.07 T

Table 2. (continued 2)

strain no.	sequence information			BLAST results			reference sequences	type strain[2]
				best hit	identity			
	gene	bp	Acc.no.	Acc.no	bp	%	total/ type[1]	
FSU 3008	ITS	553	AY633568	AY243950, *Mucor hiemalis* f. *corticola*	550/553	99	1/0	?
	18S rDNA	519	AY633567	AF113428, *Mucor hiemalis* f. *hiemalis* (NT)	519/519	100	0/0	
	β-tubulin	1226	AY633571	AY944791, *Mucor mucedo*	773/805	96	0/0	
	actin	807	AY633569, AY633570	AJ287174, *Mucor hiemalis* f. *hiemalis* (NT)	1102/1241	88	0/0	
FSU 3846	ITS	348	EF589886	AJ608958, *Mucor fragilis*	338/349	96	3/0	?
FSU 4726	ITS	749	AY944890	AJ877962, *Absidia repens* (IT)	531/544	98	5/1	CBS 115583 IT
	18S rDNA	1156	EU484209	AF113410, *Absidia repens*	1131/1154	98	1/0	
FSU 4754	ITS	525	EU484210	AB193546, *Umbelopsis isabellina*	517/532	97	7/?	?
	18S rDNA	1156	EU484211	AF157166, *Umbelopsis isabellina*	1154/1156	99	2/?	
FSU 4755	ITS	611	EU484212	AF346409, *Cunninghamella elegans*	608/611	99	7/?	?
	18S rDNA	1158	EU484213	AF113422, *Cunninghamella elegans*	1151/1158	99	5/?	
FSU 4756	ITS	543	EU484214	AY944888, *Absidia spinosa* var. *spinosa*	531/542	97	1/0	?
	18S rDNA	1157	EU484215	AF113410, *Absidia repens*	1125/1157	97	0/0	
FSU 6154	ITS	548	EU484221	AF412286, *Mucor circinelloides* f. *circinelloides*	548/548	100	10/2	CBS 195.68 NT
	18S rDNA	1155	EU484222	AF113430, *Mucor racemosus* f. *racemosus* (T)	1147/1156	99	0/0	
FSU 6155	ITS	538	EU484223	DQ119016, *Syncephalastrum racemosum*	428/491	87	1/0	CBS 213.78 T
	18S rDNA	1151	EU484224	X89437, *Syncephalastrum racemosum*	1135/1151	98	1/0	
FSU 6156	ITS	536	EU484225	DQ119033, *Rhizopus oryzae*	535/536	99	>20/4	CBS 112.07 T
	18S rDNA	1146	EU484226	AB250174, *Rhizopus oryzae*	1145/1146	99	3/1	CBS 112.07 T
FSU 6157	ITS	543	EU484227	AJ271061, *Mucor circinelloides* f. *lusitanicus*	541/543	99	1/0	?
	18S rDNA	1155	EU484228	AF113427, *Mucor circinelloides* f. *lusitanicus*	1152/1155	99	2/0	
FSU 6158	ITS	548	EU484229	AF474242, *Mucor fragilis*	545/548	99	3/?	?
	18S rDNA	1155	EU484230	AF113430, *Mucor racemosus* f. *racemosus* (T)	1150/1156	99	0/0	
FSU 6159	ITS	537	EU484231	DQ990328, *Rhizopus oryzae*	537/537	100	>20/4	CBS 112.07 T
	18S rDNA	1146	EU484232	AB250164, *Rhizopus oryzae* (T)	1146/1146	100	3/1	

Table 2. (continued 3)

strain no.	sequence information			BLAST results			reference sequences	type strain[2]
				best hit	identity			
	gene	bp	Acc.no.	Acc.no	bp	%	total/ type[1]	
FSU 6160	ITS	536	EU484233	DQ119033, *Rhizopus oryzae*	536/536	100	>20/4	CBS 112.07 T
	18S rDNA	1146	EU484234	AB250174, *Rhizopus oryzae*	1146/1146	100	3/1	
FSU 6161	ITS	545	EU484235	AF412290, *Mucor plumbeus*	545/545	100	7/?	?
FSU 6162	ITS	549	EU484236	DQ118987, *Mucor circinelloides* f. *circinelloides*	549/549	100	10/2	CBS 195.68 NT
FSU 6163	ITS	537	EU484237	DQ641279, *Rhizopus oryzae*	537/537	100	>20/4	CBS 112.07 T
FSU 6164	ITS	547	EU484238	AF474242, *Mucor fragilis*	546/547	99	3/?	?
FSU 6165	ITS	537	EU484239	DQ990328, *Rhizopus oryzae*	537/537	100	>20/4	CBS 112.07 T
	18S rDNA	1146	EU484240	AB250164, *Rhizopus oryzae* (T)	1146/1146	100	3/1	
FSU 6166	ITS	545	EU484241	AF412290, *Mucor plumbeus*	545/545	100	7/?	?
	18S rDNA	1156	EU484242	AF548078, *Mucor plumbeus*	1115/1117	99	1/0	
FSU 6167	ITS	548	EU484243	AF412286, *Mucor circinelloides* f. *circinelloides*	547/548	99	10/2	CBS 195.68 NT
	18S rDNA	1155	EU484244	AF113430, *Mucor racemosus* f. *racemosus* (T)	1151/1156	99	0/0	
FSU 6168	ITS	545	EU484245	AF412290, *Mucor plumbeus*	545/545	100	7/?	?
	18S rDNA	1156	EU484246	AF548078, *Mucor plumbeus*	1117/1117	100	1/0	
FSU 6169	ITS	548	EU484247	AF412286, *Mucor circinelloides* f. *circinelloides*	548/548	100	10/2	CBS 195.68 NT
	18S rDNA	1155	EU484248	AF113430, *Mucor racemosus* f. *racemosus* (T)	1151/1156	99	0/0	
FSU 6170	ITS	549	EU484249	DQ119006, *Rhizomucor variabilis* (T)	499/547	**91**	5/1	CBS 103.93 T
	18S rDNA	1154	EU484250	AF113435, *Rhizomucor variabilis*	1148/1154	99	1/0	
FSU 6171	ITS	552	EU484251	AY243951, *Mucor hiemalis* f. *luteus* (T)	551/554	99	2/2	CBS 244.35 T
FSU 6172	ITS	558	EU484252	AY243949, *Mucor hiemalis* f. *hiemalis*	557/558	99	2/1	CBS 201.65 NT
FSU 6173	ITS	553	EU484253	DQ118992, *Mucor hiemalis* f. *hiemalis* (NT)	553/553	100	2/1	CBS 201.65 NT
	18S rDNA	1155	EU484254	AF113428, *Mucor hiemalis* f. *hiemalis* (NT)	1154/1155	99	1/1	
FSU 6174	ITS	549	EU484255	AB193540, *Umbelopsis ramanniana*	546/550	99	18/?	?
	18S rDNA	1154	EU484256	DQ322627, *Umbelopsis ramanniana*	1151/1154	99	2/?	
FSU 6175	ITS	561	EU484257	AJ876491, *Absidia glauca*	557/561	99	11/0	CBS 101.08 T
	18S rDNA	1155	EU484258	AF157118, *Absidia glauca*	1146/1155	99	2/0	

Table 2. (continued 4)

strain no.	sequence information			BLAST results			reference sequences	type strain[2]
				best hit	identity			
	gene	bp	Acc.no.	Acc.no	bp	%	total/ type[1]	
FSU 6176	ITS	659	EU484259	AY243943, *Mucor circinelloides* f. *circinelloides* (NT)	319/358	**89**	0/0	CBS 114.08 T
	18S rDNA	1154	EU484260	AF113428, *Mucor hiemalis* f. *hiemalis* (NT)	1134/1158	97	0/0	
FSU 6177	ITS	544	EU484261	AF412290, *Mucor plumbeus*	544/545	99	7/?	?
FSU 6178	ITS	545	EU484262	AF412290, *Mucor plumbeus*	545/545	100	7/?	?
FSU 6179	ITS	553	EU484263	DQ118992, *Mucor hiemalis* f. *hiemalis* (NT)	553/553	100	2/1	CBS 201.65 NT
Kickxellales, Zygomycota								
FSU 2823	18S rDNA	1150	EU484264	AF007539, *Martensiomyces pterosporus*	1028/1190	86	0/0	?
	28S rDNA	649	AY706245	AF031066, *Martensiomyces pterosporus*	446/541	82	0/0	
Mortierellales, Zygomycota								
FSU 2735	ITS	557	EU484265	AJ878780, *Mortierella hyalina*	540/567	95	0/0	?
FSU 2736	ITS	554	EU484266	EF031107, *Mortierella sp.*	525/554	94	0/0	CBS 311.52 ST
Eurotiales, Ascomycota								
FSU 503	ITS	493	EU484267	AJ005677, *Penicillium roquefortii*	493/493	100	6/0	CBS 221.30 NT
FSU 506	ITS	496	EU484268	AF033473, *Penicillium echinulatum* (IT)	496/496	100	3/1	CBS 317.48 IT
Eurotiales, Ascomycota								
FSU 1257	ITS	494	EU484269	DQ339570, *Penicillium dipodomyicola*	494/494	100	>20/?	?
				DQ339557, *Penicillium griseofulvum*	494/494	100		
				AY371615, *Penicillium dipodomyis*	494/494	100		
				AF514301, *Penicillium urticae*	494/494	100		
FSU 2667	ITS	491	EU484270	AY373897, *Penicillium brevicompactum*	491/491	100	44/3	CBS 257.29 NT
Hypocreales, Ascomycota								
FSU 2858	ITS	547	AY633563	DQ682584, **Hypocreales sp.**	538/538	100	>20/?	?
	18S rDNA	511	AY633562	AY567009, *Acremonium strictum* (T)	425/425	100	2/2	CBS 346.70 T
	28S rDNA	564	AY633564	AY138483, *Acremonium strictum* (T)	562/564	99	5/2	CBS 346.70 T
FSU 2883	ITS	475	AY633561	DQ094534, *Fusarium solani*	471/478	98	>20/?	?
	28S rDNA	563	AY633560	AY097316, *Fusarium solani*	560/563	99		

Table 2. (continued 5)

strain no.	sequence information			BLAST results			reference sequences	type strain[2]
				best hit	identity			
	gene	bp	Acc.no.	Acc.no	bp	%	total/ type[1]	
Polyporales, Basidiomycota								
FSU 2734	ITS	828	EU484271	AF267644, AF267656, *Trichaptum abietinum*	677/695	97		?
complementary CBS strains								
CBS 109.16	ITS	598	EU484199	AF412289, *Mucor mucedo*	575/575	100	1/0	CBS 640.67 NT
CBS 148.22	ITS	564	EU484201	DQ119031, *Rhizopus oryzae* (T)	561/571	98	>20/4	CBS 112.07 T
CBS 136.28	ITS	596	EU484202	AF412290, *Mucor plumbeus*	483/533	90	0/0	?
CBS 137.28	ITS	596	EU484203	AF412290, *Mucor plumbeus*	483/533	90	0/0	?
CBS 811.69	ITS	597	EU484204	AF412290, *Mucor plumbeus*	482/549	87	0/0	?
CBS 127.08	ITS	564	EU484216	DQ119031, *Rhizopus oryzae* (T)	564/564	100	>20/4	CBS 112.07 T
CBS 110.17	ITS	564	EU484217	DQ119031, *Rhizopus oryzae* (T)	564/564	100	>20/4	CBS 112.07 T
CBS 257.28	ITS	564	EU484218	DQ119031, *Rhizopus oryzae* (T)	563/564	99	>20/4	CBS 112.07 T
CBS 329.47	ITS	564	EU484219	AB181317, *Rhizopus oryzae*	564/564	100	>20/4	CBS 112.07 T
CBS 285.55	ITS	564	EU484220	DQ119031, *Rhizopus oryzae* (T)	557/564	98	>20/4	CBS 112.07 T
CBS 101.08	ITS	562	EU484272	AY944879, *Absidia glauca*	557/573	97	11/0	CBS 101.08 T
CBS 106.08	ITS	551	EU484273	AY944888, *Absidia spinosa* var. *spinosa*	550/551	99	1/0	?
CBS 112.07	ITS	584	EU484274	DQ119031, *Rhizopus oryzae* (T)	567/567	100	>20/4	CBS 112.07 T
CBS 114.08	ITS	602	EU484275	AY243943, *Mucor circinelloides* f. *circinelloides* (NT)	319/358	89	0/0	CBS 114.08 T
CBS 169.25	ITS	618	EU484276	AJ278359, *Mucor piriformis*	603/621	97	1/0	CBS 169.25 NT
CBS 201.65	ITS	553	EU484277	DQ118992, *Mucor hiemalis* f. *hiemalis* (NT)	553/553	100	2/1	CBS 201.65 NT
CBS 260.68	ITS	546	EU484278	AY213659, *Mucor racemosus* f. *racemosus* (T)	546/546	100	7/2	CBS 260.68 T
CBS 311.52	ITS	549	EU484279	EF031107, *Mortierella sp.*	506/529	95	0/0	CBS 311.52 ST
CBS 444.65	ITS	546	EU484280	AJ878785, *Zygorhynchus moelleri*	546/546	100	1/0	CBS 444.65 T
CBS 115583	ITS	664	EU484281	AY944890, *Absidia repens*	660/675	97	5/1	CBS 115583 IT

Table 3. Final identity of the fungal strains according to morpho- and genotyping based on morphological and molecular parameters (following the system of Zycha *et al.* 1969), respectively. FSU...Friedrich Schiller University Jena, Germany; CBS...Centraalbureau voor Schimmelcultures, Utrecht, The Netherlands

strain no.	morphological identity	final identity
Mucorales, Zygomycota		
FSU 497	*Mucor hiemalis* f. *hiemalis*	*Mucor hiemalis* f. *hiemalis*
FSU 499	*Mucor racemosus*	*Mucor racemosus* f. *racemosus*
FSU 500	*Mucor racemosus* f. *sphaerosporus*	*Mucor racemosus* f. *sphaerosporus*
FSU 501	*Mucor circinelloides* f. *circinelloides*	*Mucor circinelloides* f. *circinelloides*
FSU 514	*Zygorhynchus moelleri*	*Zygorhynchus moelleri*
FSU 697	*Mucor racemosus* f. *racemosus*	*Mucor racemosus* f. *racemosus*
FSU 2005	*Umbelopsis ramanniana*	**unknown Mucorales**
FSU 2572	*Mucor mucedo*	**new species of *Mucor***
FSU 2666	*Mucor racemosus* f. *racemosus*	*Mucor racemosus* f. *racemosus*
FSU 2797	*Mucor hiemalis* f. *hiemalis*	*Mucor hiemalis* f. *hiemalis*
FSU 2798	*Mucor hiemalis* f. *hiemalis*	*Mucor hiemalis* f. *hiemalis*
FSU 2799	*Mucor hiemalis* f. *hiemalis*	*Mucor hiemalis* f. *hiemalis*
FSU 2800	*Mucor hiemalis* f. *hiemalis*	*Mucor hiemalis* f. *hiemalis*
FSU 2801	*Mucor hiemalis* f. *hiemalis*	*Mucor hiemalis* f. *hiemalis*
FSU 2802	*Mucor hiemalis* f. *hiemalis*	*Mucor hiemalis* f. *hiemalis*
FSU 2803	*Mucor hiemalis* f. *hiemalis*	*Mucor hiemalis* f. *hiemalis*
FSU 2804	*Mucor hiemalis* f. *hiemalis*	*Mucor hiemalis* f. *hiemalis*
FSU 2805	*Mucor hiemalis* f. *hiemalis*	*Mucor hiemalis* f. *hiemalis*
FSU 2806	*Mucor hiemalis* f. *hiemalis*	*Mucor hiemalis* f. *hiemalis*
FSU 2807	*Mucor hiemalis* f. *hiemalis*	*Mucor hiemalis* f. *hiemalis*
FSU 2808	*Mucor hiemalis* f. *hiemalis*	*Mucor hiemalis* f. *hiemalis*
FSU 2809	*Absidia repens*	*Absidia* sp.*
FSU 2810	*Absidia glauca*	*Absidia glauca*
FSU 2811	*Helicostylum elegans*	*Helicostylum elegans*
FSU 2812	*Absidia glauca*	*Absidia glauca*
FSU 2813	*Absidia repens*	*Absidia* sp.*
FSU 2814	*Absidia repens*	*Absidia* sp.*
FSU 2815	*Absidia repens*	*Absidia* sp.*
FSU 2816	*Absidia repens*	*Absidia* sp.*
FSU 2817	*Absidia glauca*	*Absidia glauca*
FSU 2818	*Absidia glauca*	*Absidia glauca*
FSU 2819	*Mucor hiemalis*	*Mucor hiemalis* f. *hiemalis*
FSU 2820	*Mucor hiemalis*	*Mucor hiemalis* f. *hiemalis*
FSU 2821	*Absidia glauca*	*Absidia glauca*
FSU 2822	*Rhizopus sp.*	**unknown Mucorales**
Mucorales, Zygomycota		
FSU 2824	*Absidia repens*	*Absidia* sp.*
FSU 2825	*Rhizopus oryzae*	*Rhizopus oryzae*
FSU 3008	*Mucor hiemalis* f. *corticola*	*Mucor hiemalis* f. *corticola*
FSU 3846	*Mucor fragilis*	**new species of *Mucor***
FSU 4726	*Absidia repens*	*Absidia repens*
FSU 4754	*Umbelopsis isabellina*	Umbelopsis sp.*
FSU 4755	*Cunninghamella elegans*	*Cunninghamella elegans*
FSU 4756	*Absidia spinosa* var. *spinosa*	*Absidia* sp.*

Table 3. (Continued)

strain no.	morphological identity	final identity
FSU 6154	*Mucor circinelloides* f. *circinelloides*	*Mucor circinelloides* f. *circinelloides*
FSU 6155	*Syncephalastrum racemosum*	**unknown Mucorales**
FSU 6156	*Rhizopus oryzae*	*Rhizopus oryzae*
FSU 6157	*Mucor circinelloides*	*Mucor circinelloides* f. *lusitanicus*
FSU 6158	*Mucor hiemalis*	*Mucor fragilis*
FSU 6159	*Rhizopus oryzae*	*Rhizopus oryzae*
FSU 6160	*Rhizopus oryzae*	*Rhizopus oryzae*
FSU 6161	*Mucor plumbeus*	*Mucor plumbeus*
FSU 6162	*Mucor circinelloides* f. *circinelloides*	*Mucor circinelloides* f. *circinelloides*
FSU 6163	*Rhizopus oryzae*	*Rhizopus oryzae*
FSU 6164	*Mucor hiemalis*	*Mucor fragilis*
FSU 6165	*Rhizopus oryzae*	*Rhizopus oryzae*
FSU 6166	*Mucor plumbeus*	*Mucor plumbeus*
FSU 6167	*Mucor circinelloides* f. *circinelloides*	*Mucor circinelloides* f. *circinelloides*
FSU 6168	*Mucor plumbeus*	*Mucor plumbeus*
FSU 6168	*Mucor plumbeus*	*Mucor plumbeus*
FSU 6169	*Mucor circinelloides* f. *circinelloides*	*Mucor circinelloides* f. *circinelloides*
FSU 6170	*Mucor hiemalis*	**unknown Mucorales**
FSU 6171	*Mucor hiemalis* f. *luteus*	*Mucor hiemalis* f. *luteus*
FSU 6172	*Mucor hiemalis*	*Mucor hiemalis* f. *hiemalis*
FSU 6173	*Mucor hiemalis*	*Mucor hiemalis* f. *hiemalis*
FSU 6174	*Umbelopsis ramanniana*	*Umbelopsis ramanniana*
FSU 6175	*Absidia glauca*	*Absidia glauca*
FSU 6176	*Mucor genevensis*	Mucor genevensis
FSU 6177	*Mucor plumbeus*	*Mucor plumbeus*
FSU 6178	*Mucor saturninus*	*Mucor saturninus*
FSU 6179	*Mucor hiemalis* f. *hiemalis*	*Mucor hiemalis* f. *hiemalis*
Kickxellales, Zygomycota		
FSU 2823	*Coemansia aciculifera*	*Coemansia aciculifera*
Mortierellales, Zygomycota		
FSU 2735	*Mortierella minutissima*	*Mortierella minutissima*
FSU 2736	*Mortierella parvispora*	*Mortierella parvispora*
Eurotiales, Ascomycota		
FSU 503	*Penicillium sp.*	*Penicillium roquefortii*
FSU 506	*Penicillium sp.*	*Penicillium echinulatum*
FSU 1257	*Penicillium brevicompactum*	*Penicillium sp.*
FSU 2667	*Penicillium brevicompactum*	*Penicillium brevicompactum*
Hypocreales, Ascomycota		
FSU 2858	Hypocreales	*Acremonium-like* Hypocreales
FSU 2883	*Fusarium sp.*	*Fusarium solani*
Polyporales, Basidiomycota		
FSU 2734	*Trichaptum sp.*	*Trichaptum* sp.*
complementary CBS strains		
CBS 109.16	*Mucor mucedo*	*Mucor mucedo*
CBS 148.22	*Rhizopus oryzae*	*Rhizopus oryzae*
CBS 136.28	*Chaetocladium brefeldii*	*Chaetocladium brefeldii*
CBS 137.28	*Chaetocladium brefeldii*	*Chaetocladium brefeldii*
CBS 811.69	*Chaetocladium jonesii*	*Chaetocladium jonesii*
CBS 127.08	*Rhizopus oryzae*	*Rhizopus oryzae*
CBS 110.17	*Rhizopus oryzae*	*Rhizopus oryzae*

Table 3. (Continued)

strain no.	morphological identity	final identity
CBS 257.28	*Rhizopus oryzae*	*Rhizopus oryzae*
CBS 329.47	*Rhizopus oryzae*	*Rhizopus orzae*
CBS 285.55	*Rhizopus oryzae*	*Rhizopus oryzae*
CBS 101.08	*Absidia glauca*	*Absidia glauca*
CBS 106.08	*Absidia spinosa* var. *spinosa*	*Absidia spinosa* var. *spinosa*
CBS 112.07	*Rhizopus oryzae*	*Rhizopus oryzae*
CBS 114.08	*Mucor genevensis*	*Mucor genevensis*
CBS 169.25	*Mucor piriformis*	*Mucor piriformis*
CBS 201.65	*Mucor hiemalis* f. *hiemalis*	*Mucor hiemalis* f. *hiemalis*
CBS 260.68	*Mucor racemosus* f. *racemosus*	*Mucor racemosus* f. *racemosus*
CBS 311.52	*Mortierella parvispora*	*Mortierella parvispora*
CBS 444.65	*Zygorhynchus moelleri*	*Zygorhynchus moelleri*
CBS115583	*Absidia repens*	*Absidia repens*

* due to a lack of reference sequences a final species delimitation is impossible. Classification to a new species is likely. This applies to *Absidia* spp. FSU2809, FSU2813-2816, FSU2824, FSU4756, *Umbelopsis* sp. FSU4754 and *Trichaptum* sp. FSU 2734.

2.2. Identification of Fungi using Morphological Characters (Morphotypization)

Classical approaches to identify fungi range from comparisons with the fossil record (Hawksworth *et al.*, 1995), the use of physiological and biochemical markers (Bridge, 1985; Paterson and Bridge, 1994; Rath *et al.*, 1995), such as the composition of the cell wall (Bartnicki-Garcia, 1970, 1987) and isoenzyme patterns (Fuhrmann *et al.*, 1990; Kohn, 1992; Maxson and Maxson, 1990), over metabolic criteria (Le´John, 1974; Vogel, 1964) to observations on the ultrastructure (e.g. Fuller, 1976; Heath, 1980; Kimbrough, 1994). All these parameters are used in comparison with macro- and micromorphological characters to obtain criteria, which are taxon-specific. Traditional classification and identification of Zygomycota depends mainly on the morphology of asexual (sporangia, sporangiola) and sexual (zygospores) reproduction structures, the branching order of the sporangiophores and remaining vegetative hyphae, the occurrence of septae, and the conditions for growth in relation to the morphology of mitospores (Benny, 1995). Morphological criteria for identification can be easily observed using any microscope and following appropriate keys and descriptions, which are widely available. A comprehensive compendium on the major groups traditionally studied by mycologists is provided by the constantly updated ´Ainsworth & Bisby`s Dictionary of the Fungi´ (9th ed: Kirk, *et al.*, 2001). Special keys and descriptions to Zygomycota are provided by Benjamin (1979), Benny (1982), Benny *et al.* (2001) or Zycha *et al.* (1969), where different orders like the Mucorales, Mortierellales, Kickxellales, Endogonales and Dimargaritales are described. For some families or genera of the Zygomycota a diverse number of separate keys exists e.g. for the genus *Mucor* (Schipper, 1973, 1975, 1976, 1978; Watanabe, 1994), *Rhizopus* (Schipper, 1984; Zheng, *et al.*, 2007), *Absidia* (Hesseltine and Ellis, 1961, 1964, 1966; Ellis and Hesseltine, 1965, 1966; Schipper,

1990), *Zygorhynchus* (Schipper, 1986) and Thamnidiaceae (Benny and Benjamin, 1975, 1976).

Table 3 lists the identity of our freshly isolated fungi. The decision about final species identities depends upon morphology as investigated using light and scanning electron microscopy (Figures 3-10) compared to a molecular identification based on rDNA barcodes. Problems regarding final identity and morphological identity are discussed under point 3.5.

2.3. Identification of Fungi using Molecular Data (Genotypization)

Since molecular tools like e.g. DNA hybridization, randomly amplified polymorphic DNA (RAPD), the polymerase chain reaction, DNA sequencing or restriction fragment length polymorphisms (RFLP) are available, molecular systematics is developing rapidly and provides new insights into fungal evolution. The advantage of molecular techniques lies in their universal applicability. Detailed and comprehensive applications of molecular methods are reviewed by e.g. Bruns *et al.* (1991), Hibbett (1992), Kohn (1992), Maresca and Kobayashi (1994), and Weising *et al.* (1995). The polymerase chain reaction (PCR) can be a very useful tool in cases where only a few cells (e.g. spores) with genetic material are available or where a microbial community is investigated, which may even independent from cultivation in a laboratory (Blackwood *et al.*, 2005; Ganley and Newcombe, 2006). A specific DNA sequence can be targeted for this purpose with PCR (Anderson *et al.*, 2003; Kennedy and Clipson, 2003). This method is also useful in cases where rapid and reliable identification of fungal organisms is essential, e.g. the diagnosis and monitoring of fungal infections and antifungal therapy in humans (Mattner *et al.*, 2004; Mayr *et al.*, 2004; Nyilasi *et al.*, 2008) or plants (Casimiro *et al.*, 2004). It is also possible to obtain PCR products for DNA sequence analyses from extinct organisms (Pääbo, 1989). With the availability of universal primers the access to e.g. specific fungal sequences is easy (e.g. Sandhu *et al.*, 1995; Vilgalys and Hester, 1990; White *et al.*, 1990). Hence analysis of molecular data obtained by DNA sequencing provides a direct and fast way to estimate genetic variation within specific genes among diverse organisms, and is therefore a useful tool for inferring phylogenetic relationships. Relationships based on molecular data can also confirm morphological markers traditionally used for phylogeny (e.g. Garnica *et al.*, 2007). For the investigation of deep phylogenetic branches at the level of orders and above (classes, phyla, or kingdoms), sequences of more conserved coding regions like the nuclear small subunit (SSU) ribosomal DNA are used. Orders and family structures can also be investigated phylogenetically using the nuclear large subunit (LSU) rDNA. Due to faster evolutionary rates in the internal transcribed spacer regions (ITS) these sequences are generally suitable for uncovering relationships between species and subspecies (reviewed by Hwang and Kim, 1999) but bearing in mind that there are also some exceptions (Skouboe *et al.*, 1999). To classify the fungal organisms shown in table 1, nucleotide sequences from different fungal species, to which the unknown species are related (or identical), were included as reference sequences (Table 4) in the phylogenetic analyses (Figure 11 and 12) regarding results from morphological identification and BLAST searches as described under points 3.4 and 4.

Table 4. Reference sequences retrieved from GenBank for the phylogenetic analyses shown in figures 11 and 12. FSU…Friedrich-Schiller-University Jena, Germany; CBS…Centraalbureau voor Schimmelcultures, Utrecht, The Netherlands

18S rDNA organisms	Acc.no.	28S rDNA organisms	Acc.no.
Mucorales, Zygomycota		**Mucorales, Zygomycota**	
Absidia glauca	AF157118	*Absidia coerulea*	AF113443
Absidia repens	AF113410	*Absidia glauca*	AF113447
Cunninghamella elegans	AF113422	*Absidia glauca*	AF157172
Mucor hiemalis f. *hiemalis*	AF113428	*Absidia repens*	AF113448
Mucor plumbeus	AF548078	*Dicranophora fulva*	AF157186
Mucor racemosus f. *racemosus*	AF113430	*Helicostylum elegans*	AF157193
Rhizomucor variabilis	AF113435	*Mucor hiemalis* f. *hiemalis*	AF113468
Rhizopus oryzae	AB250164	*Rhizomucor variabilis*	AF113476
Rhizopus oryzae	AF548078	*Rhizpous oryzae*	AB250187
Syncephalastrum racemosum	X89437	*Rhizpous oryzae*	AY213624
Umbelopsis ramanniana	DQ322627	*Thamnidium elegans*	AF157217
Umbelopsis isabellina	AF157166	*Zygorhynchus exponens*	AB363778
Zygorhynchus heterogamus	AF157170	*Zygorhynchus heterogamus*	AF157224
Kickxellales, Zygomycota		**Kickxellales, Zygomycota**	
Dipsacomyces acuminosporus	AF007534	*Coemansia reversa*	AY546689
Kickxella alabastrina	AF007537	*Dipsacomyces acuminosporus*	AF031065
Martensiomyces pterosporus	AF007539	*Linderina pennispora*	AF031063
Hypocreales, Ascomycota		*Martensiomyces pterosporus*	AF031066
Acremonium strictum	U43968	**Mortierellales, Zygomycota**	
Acremonium strictum	AY567009	*Mortierella chlamydospora*	AF157197
Acremonium kiliense	U43973	*Mortierella multidivaricata*	AF157198
		Mortierella polycephala	AF113464
		Hypocreales, Ascomycota	
		Acremonium strictum	AY138483
		Fusarium solani	AY097316

3. IDENTIFICATION OF FUNGI IN COMMON LABORATORY PRACTICE

3.1. Identification of Species designated to the phyla Ascomycota and Basidiomycota

In some cases morpho- and genotyping identification give conflicting species designations. Under these circumstances identification below the genus level is tedious. The strain FSU1257 was identified as *Penicillium sp.* by morphological parameters, but its ITS sequences were 100% identical to sequences from *P. dipodomyicola, P. griseofulvum, P. dipodomyis* and *P. urticae*. The inadequacy of ITS sequences for identification of *Penicillium* is a known problem (Skouboe *et al.*, 1999). Here supplementary DNA barcodes become necessary, e.g. gene encoding the nuclear intergenic spacer (IGS) rRNA or protein-coding genes which provide highly resolutive introns like the genes of beta tubulin or translation elongation factor 1 alpha. The ascomycete FSU2858 was identified as an *Acremonium-like* ascomycete within the order Hypocreales. 18S and 28S rDNA indicate an identity as *Acremonium strictum* with 100% and 99%, respectively. Although there are more than 20 ITS sequences in the GenBank database of the NCBI, none occurred to be the best hit in a BLAST search. The best hit (100%) matched an unidentified hypocrealean species (Table 2).

It is known that the majority of the described fungi (about 98%) belong to the Asco- and Basidiomycota (James *et al.*, 2006). Therefore, searching public data bases makes it obvious that the emphasis of mycological research concerns those two fungal phyla (more than 100,000 entries in PubMed database, which is more than twenty-fold that of the entries for Zygomycota). Within the ′Taxonomy′ section at NCBI there are more than 50 times more entries for Asco- and Basidiomycota than for Zygomycota. Because many species of Asco- and Basidiomycota are of industrial relevance, such as in food production and drug discovery, of clinical interest regarding pathogens and antimicrobial agents (e.g. Mayr *et al.*, 2004), species differentiation often depends on small differences in physiology e.g. volatile metabolites (Larsen and Frisvad, 1995; Karlshøj and Larsen, 2005).

3.2. Ascomycetes meet Zygomycetes: Barcoding the Fungal Partner in the Symbiotic or Parasitic Interaction between *Mucor* spp. and Morels (*Morchella Elata*)

Zygomycetes are known to be effective facultative mycoparasites such as on ascomycetous fungi. Figure 1 shows an ascoma of the morel *Morchella elata* potentially covered with a species of the zygomycetous order Mucorales. This isolate was deposited in the Fungal Reference Centre Jena as FSU3008 (Table 1) and identified as *Mucor hiemalis f. corticola* by nucleotide sequence comparisons based on nuclear SSU and ITS1-5.8S-ITS2 rDNA and genes encoding actin and beta-tubulin (Table 2) in combination with light and scanning electron microscopy (Figure 2).

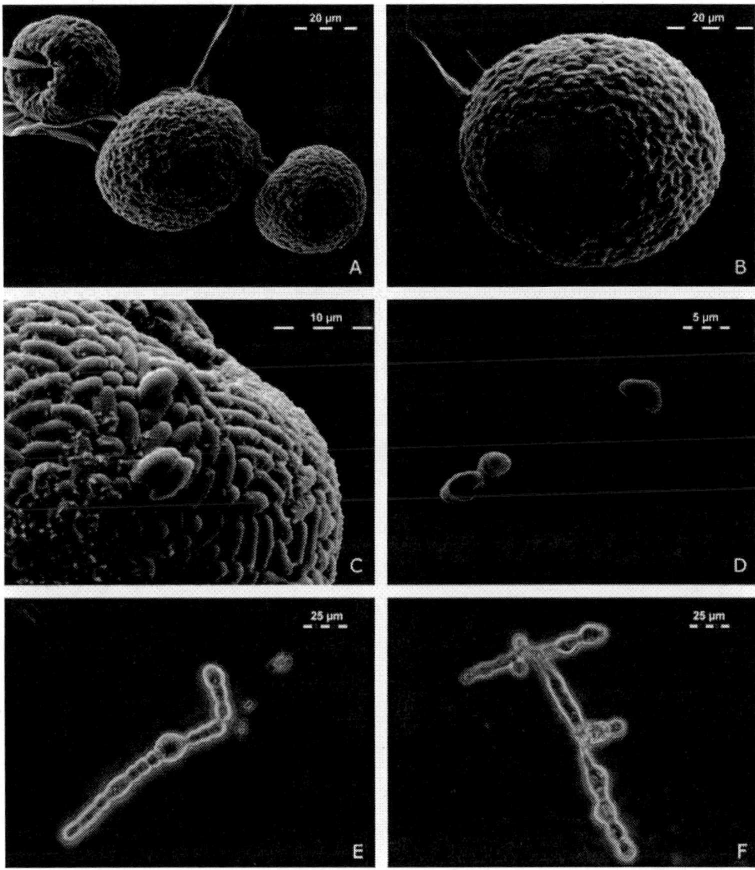

Figure 2. Scanning electron and light microphotographs of *Mucor hiemalis* f. *corticola* FSU3008. Scale bars indicate 20 micrometers for A and B; 10 micrometers for C; 5 micrometers for D; and 25 micrometers for E and F, respectively.

3.3. Barcoding the Fungal Partner of the Interaction between *Mucor* spp. and *Centaurea Stoebe*

Apart from the interaction of *Mucor* spp. with other fungi, also green plants were observed to host an endophytic *Mucor*, isolate CID285 (deposited in the Fungal Reference Centre as FSU3846, Table 1). Endophytic fungi are known to influence growth, competition, and protection of an invasive plant. Based on nuclear ITS1-5.8S-ITS2 rDNA sequence comparisons this isolate shows 96% sequence identity to *Mucor fragilis* (Acc no. AJ608958). Unfortunately, no type strain of *M. fragilis* is available, which could easily be used as reference sequence. But also the *M. fragilis* isolate CBS236.35 deviates from CID285 by 3.4%. Comparing all accessible ITS barcodes of *Mucor*, the most closely related taxon is *Mucor circinelloides* f. *janssenii* with an ITS-sequence deviation of 2.9-3.2%. Based on the number of nucleotide substations CID285 could represent a new species which should be verified by a multilocus analysis.

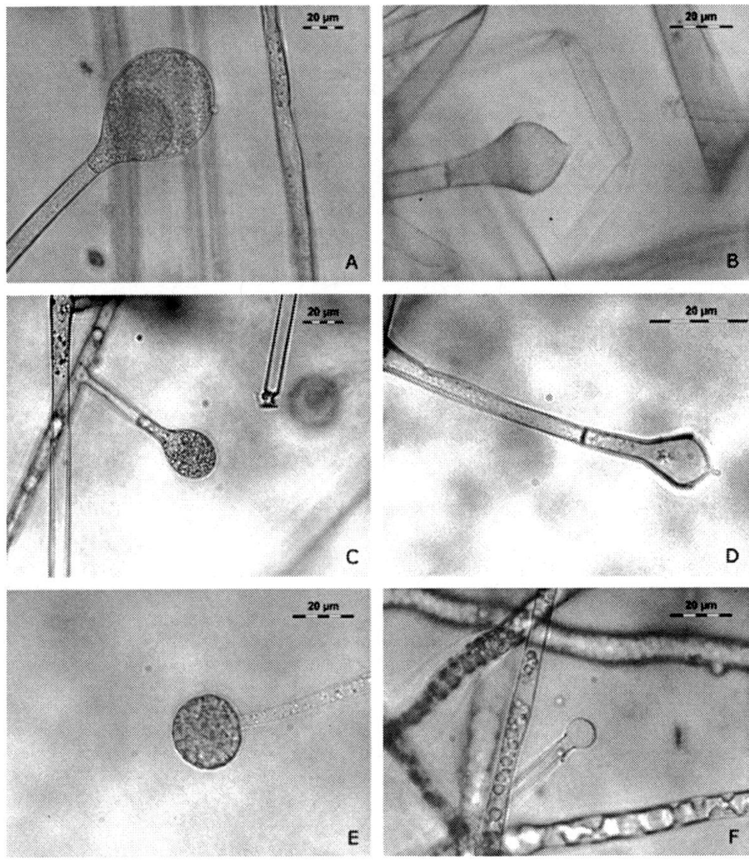

Figure 3. Ligth microscopic images of different multi-spored sporangial morphologies of columellate mucoralean fungi. *Absidia glauca* FSU6175: A – pyriform sporangium, B - columella; *Absidia repens* FSU4726: C – primary sporangium, D - secondary sporangium; unknown Mucorales FSU2005: E - sporangium, F – columella. Scale bars indicate 20 micrometers for A-D, and 10 micrometers for E and F.

3.4. Zygomycete Communities in Different Climatic Regions assessed by Mycelial Isolation, Morpho- and Genotyping via Nuclear Ribosomal DNA Barcodes

Morphotyping based on colony- and micromorphological as well as ultrastructural characters (Figures 2-10) belong to a common practice in fungal identification. Morphological parameters like the presence of a columella (Figures 3-7) and rhizoids (Figure 6) or characteristics of the sporangiophor such as appendices (e.g. apophysis; Figure 9), the development of multi-spored sporangia (Figures 3-9), few-spored sporangiola or merosporangia (Figure 5) and the morphology of the zygospore (Figures 4, 5, 10) are useful morphological markers for identification. Few of these criteria are synapomorphies and determine phylogenetic monophyla (see chapter: Voigt *et al.* within this book). But identification using solely morphological parameters depends often on cultivation of the fungus on different media and observation over a long period of time, because the morphotypic structures used to appear sequentially. Therefore, morphotyping is time-consuming, labour-intensive, prone to variations due to physiological conditions and requires an experienced eye. Comparisons of DNA sequences of common genes became a rapid method of fungal identification in molecular biological laboratories, where gene amplification via Polymerase Chain Reaction (PCR) is well-established and DNA sequencing equipment is easily accessible.

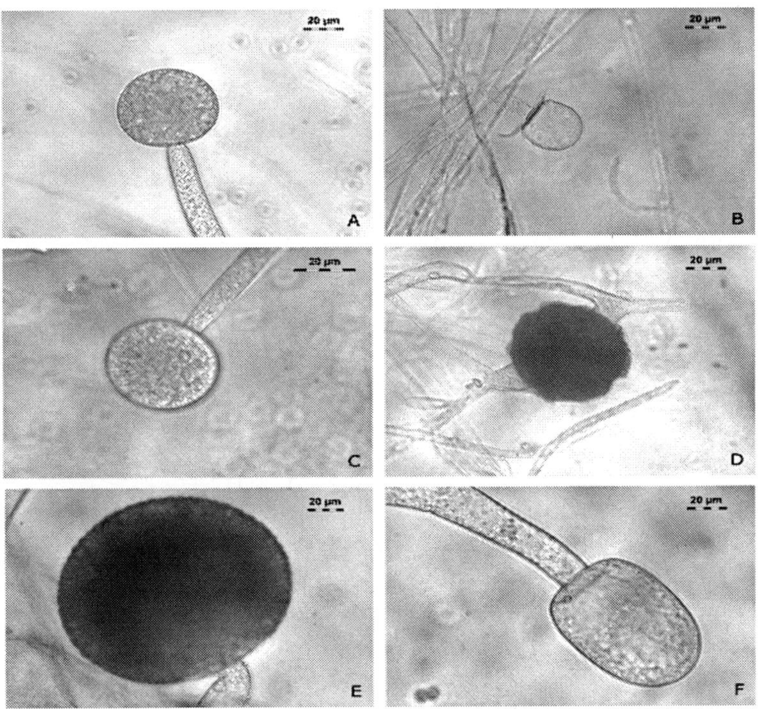

Figure 4. Ligth microscopic images of characteristic micromorphological structures from different *Mucor* spp. *Mucor fragilis* FSU6158: A - sporangium, B - columella; *Mucor genevensis* FSU6176: C – asexually formed sporangium, D – sexually formed zygospore; *Mucor hiemalis* f. *hiemalis* FSU6179: E - sporangium, F - columella. Scale bars indicate 20 micrometers, respectively for A-F.

A useful tool to identify fungal organisms solely by their DNA barcode sequences, is to search for similarities with other sequences stored in various databases like at the National Center for Biotechnology Information (NCBI). If the closest neighbour species have been gathered the reconstruction of phylogenetic trees (Figures 11, 12) based on nucleotide sequence alignments (Figures 14, 16) might be effective. The establishment of sequence similarity matrices (Figures 13, 15, 17-18) give deep insights into gene structure, the presence or absence of deletions and insertions, which may be of molecular taxonomic value.

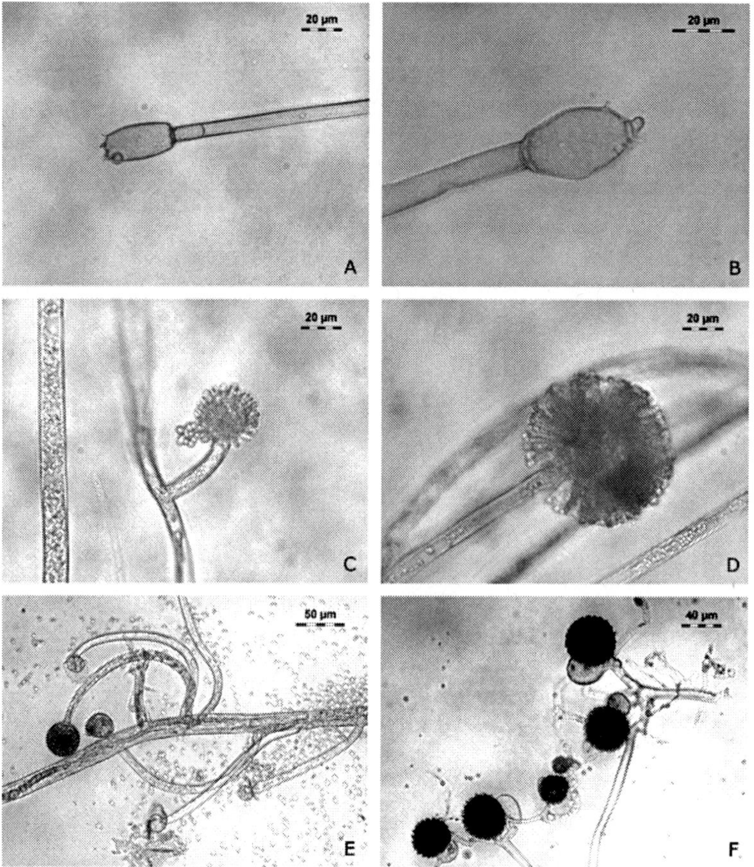

Figure 5. Ligth microscopic images of characteristic micromorphological structures from different mucoralean fungi. *Mucor plumbeus* FSU6161: A - columella; *Mucor plumbeus* FSU6177: B - columella; unknown Mucorales FSU6155: C - young merosporangia, D - matured merosporangia; *Zygorhynchus moelleri* FSU514: E – asexually formed multi-spored sporangia, F – sexually formed zygospores. Scale bars indicate 20 micrometers for A-D; 50 micrometers for E and 40 micrometers for F.

Authentic strains and type strains, including neotype, isotype and syntype strains are unique organisms connected with a specific name and a specific description, because it is recommended to identify unknown isolates by such references. Our BLAST searches revealed similarities of 98 to 100% for 18S and 28S rDNA sequences with only a few exceptions (Table 2). Knowing that nuclear ribosomal DNA sequences of the 18S and 28S rRNA genes are very useful to uncover higher level phylogenetic relationships (White *et al.*, 2006, Woese, 1987; Woese *et al.*, 1990; Tehler *et al.*, 2000), it is not remarkable that some fungi are misidentified with high similarity values because of the non-existence of appropriate, highly resolving sequences in the database. This bias of nuclear 18S rDNA and ITS for estimating fungal biodiversity in environmental samples was also observed by Anderson *et al.* (2003). Such cases are for example *Mucor racemosus* f. *sphaerosporus* FSU500, which was identified as *M. racemosus* f. *racemosus* based on 18S rDNA with 99% identical sequence residues, because no sequences are available for the forma/subspecies *'sphareosporus'*. For *Mucor circinelloides* f. *circinelloides* also no 18S rDNA sequences are accessible and therefore FSU501, FSU6154, FSU6167 and FSU6169 were misidentified as *M. racemosus* f. *racemosus*. This general problem of missing data is also responsible for the false molecular identities of *Zygorhynchus moelleri* (FSU514), *Mucor fragilis* (FSU6158), *Mucor genevensis* (FSU6176), *Mucor hiemalis* f. *corticola* (FSU3008) and *Absidia spinosa* var. *spinosa* (FSU4756) based on 18S (Table 2).

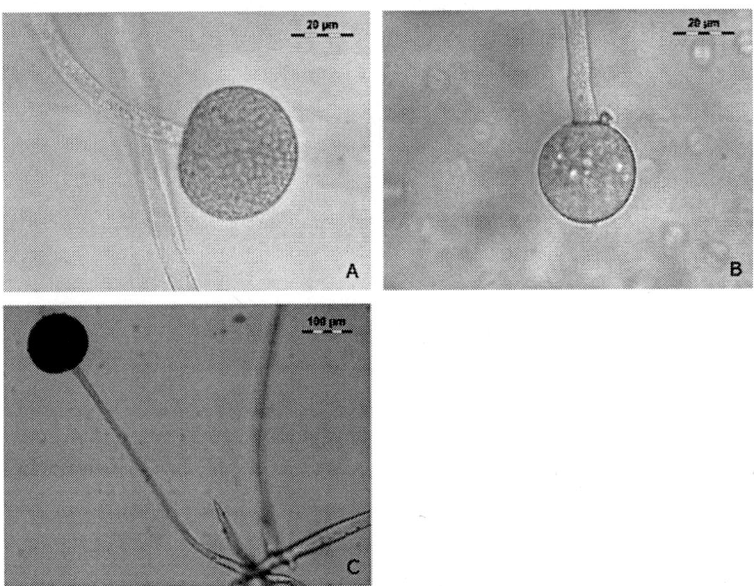

Figure 6. Ligth microscopic images of rhizoid-forming mucoraceaous fungi. unknown Mucorales FSU6170: A - sporangium; *Rhizopus oryzae:* B - columella, C - sporangia with rhizoids. Scale bars indicate 20 micrometers for A and B; 100 micrometers for C.

To avoid such misidentifications due to high sequence similarities between different subspecies, species or genera, it is necessary to rely on sequences with high similarities at the species or subspecies level but not at the genus level, such as sequences of the internal transcribed spacer regions 1 and 2 (reviewed by Hwang and Kim, 1999).

Comparing the sequences of ITS 1 and 2 including 5.8S rDNA of the misidentified species, BLAST searches resulted in 97 to 100% identity for all *Mucor* species, *Zygorhynchus moelleri*, *Umbelopsis isabellina* and *Absidia spinosa.* For example, species of *Mucor circinelloides* were frequently identified as *Mucor racemosus* with high sequence similarities based on 18S rDNA because of missing sequences. At the phenotypic level both species can be easily distinguished by their spores, chlamydospores, their maximal growth temperature and the often recurved sporangiophores of *M. circinelloides*. ITS sequence comparisons show that *Mucor racemosus* is clearly different from *Mucor circinelloides* with approximately 9% sequence dissimilarities. Also two formae of *Mucor circinelloides*, namely f. *circinelloides* and f. *lusitanicus*, can be distinguished by 4% dissimilarity from each other. Keeping in mind that the threshold for species delimitation is below or equal to 3% (Table 2) both of these formae may represent distinct species.

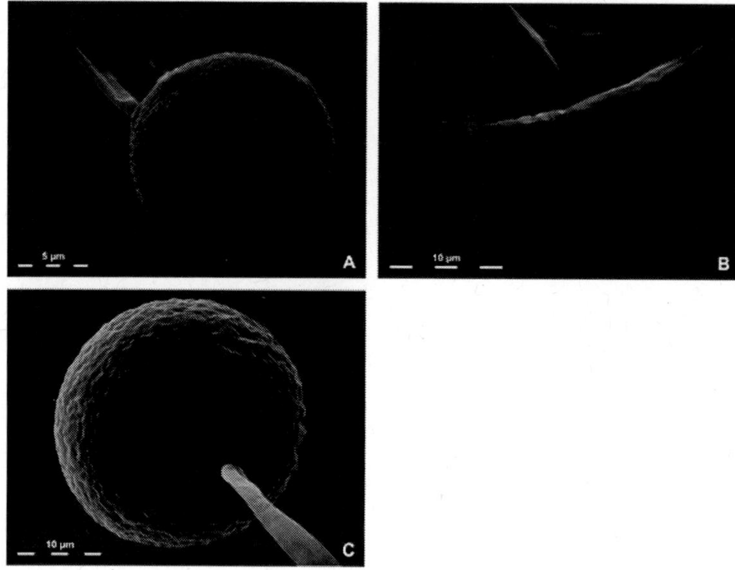

Figure 7. Scanning electron microscopic images of acolumellate, globose sporangia from different *Mortierella* spp. A and B - *Mortierella minutissima* FSU2735 and C - *M. parvispora* FSU2736. Scale bars indicate 5 micrometers for A and 10 micrometers for B and C.

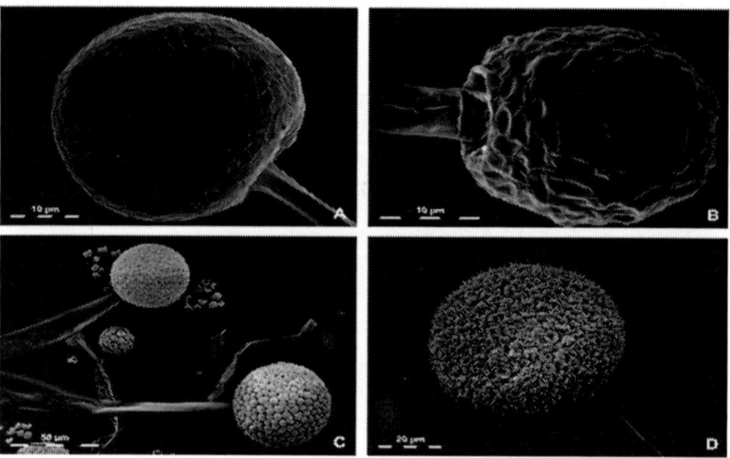

Figure 8. Scanning electron microscopic images of columellate, globose sporangia from different *Mucor* spp. A - *Mucor genevensis* FSU6176, B - *Mucor hiemalis* f. *hiemalis* FSU6179, C and D - *Mucor plumbeus* FSU6177. Scale bars indicate 10 micrometers for A and B, 50 micrometers for C, and 20 micrometers for D.

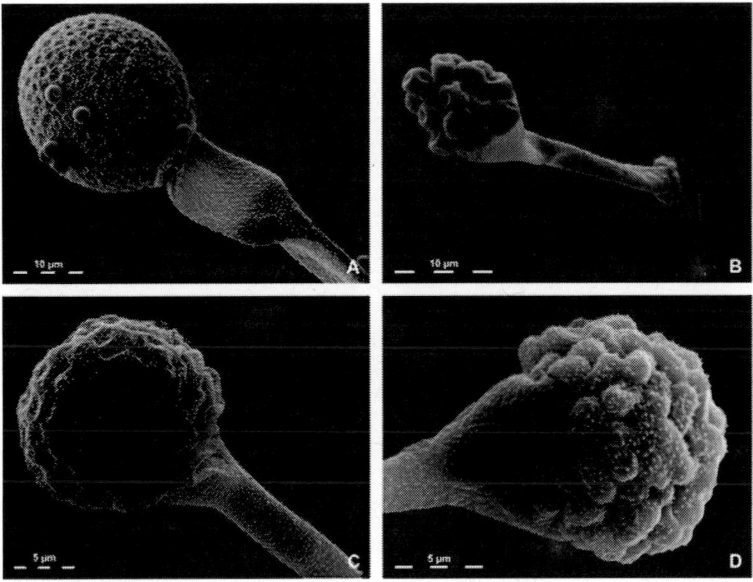

Figure 9. Scanning electron microscopic images of collumellate, apophysate, pyriform sporangia from different *Absidia* spp. A - *Absidia glauca* FSU6175, B - *A. repens* FSU4726, C and D - *A.spinosa* FSU4756. Scale bars indicate 10 micrometers for A – B and 5 micrometers for C - D.

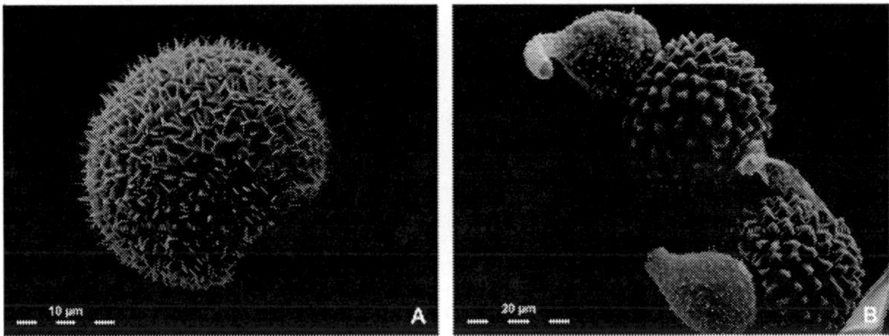

Figure 10. Scanning electron microscopic images of *Zygorhynchus moelleri* FSU514: A - sporangium and B - zygospores. Scale bars indicate 10 and 20 micrometers, respectively for A and B.

3.5. The Power of Morphological and Physiological Data for the Identification of the Members of the Zygomycota

Because informations about morphological and physiological characters of nearly all known species exists, identification of unknown organisms is traditionally performed using such data over the last decades. Another reason for using this type of data is that only about five percent of all known fungal species in Index Fungorum (http://www.indexfungorum.org; currently 424303 records online) are present in GenBank (a total of 22810 fungi in NCBI Taxonomy; as of January 24, 2008).

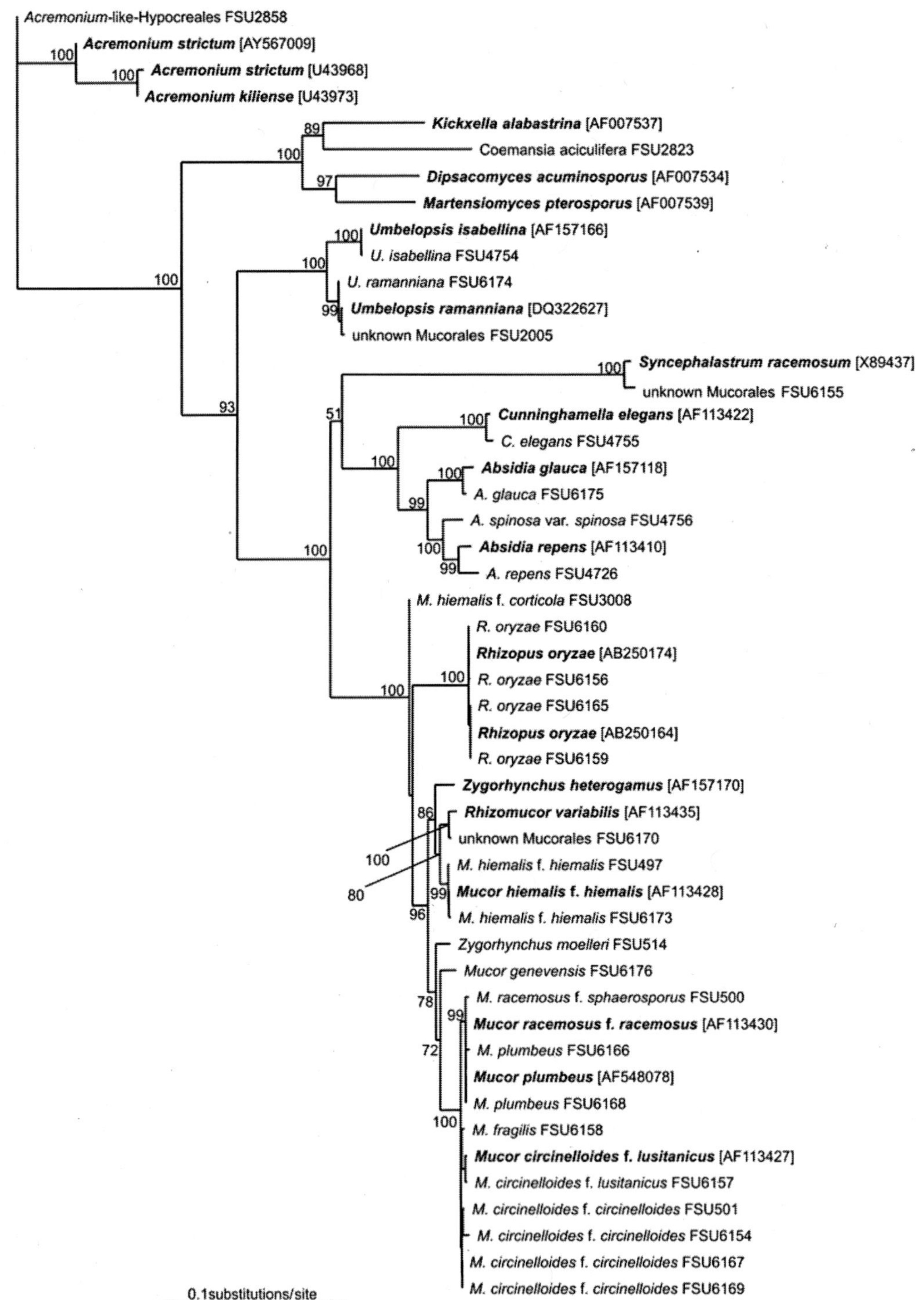

Figure 11. Neighbor-joining tree based on 1211 aligned characters of nuclear SSU (18S) rDNA nucleotide sequences from 49 taxa. Bootstrap proportions are given above the branches. Scale bar indicates nucleotide substitutions per site.

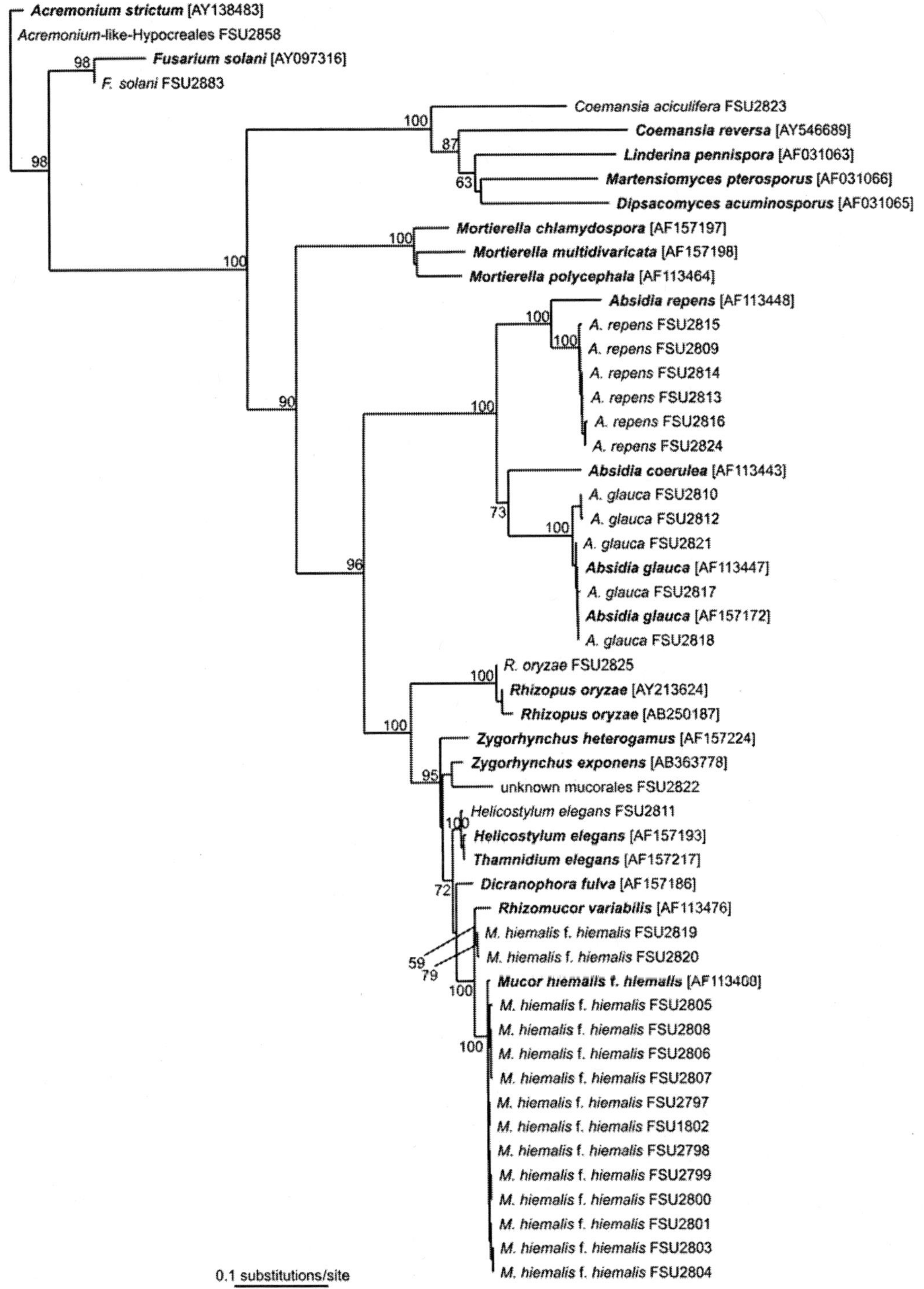

Figure 12. Neighbor-joining tree based on 800 aligned characters of nuclear LSU (28S) rDNA nucleotide sequences from 53 taxa. Bootstrap proportions are given above the branches. Scale bar indicates nucleotide substitutions per site.

Some species that lack sequences in GenBank are *Coemansia aciculifera*, *Mortierella parvispora, Mortierella minutissima*, and *Mucor saturninus*. *Coemansia aciculifera* (FSU2823) was misidentified using 18S and 28S rDNA sequences as *Martensiomyces pterosporus* (Table 2), which is also a member of the order Kickxellales just like *Coemansia.* However, these two species are clearly identifiable by their sporocladia, which are umbel-like for *Martensiomyces* and consisting of whorls for *Coemansia* (Zycha *et al.*, 1969). *Mortierella minutissima* FSU2735 and *M. parvispora* FSU2736 can be distinguished by e.g. their spore size from *M. hyalina* (Zycha *et al.*, 1969), which was a BLAST result from ITS sequences (Table 2). *Mucor saturninus* FSU6178 was identified as *Mucor plumbeus* with 100% based on ITS data (Table 2), but these species differ clearly in e.g. their sporangiophores and zygospores (Schipper, 1978).

Misidentification through ITS sequences even on the family level has occurred for *Chaetocladium* species (FSU775, FSU796, FSU869) with the best BLAST hits for *Mucor plumbeus* with 87% to 90% sequence similarities (Table 2). Here also no appropriate ITS sequences are available in the database. The Mucoraceae show typical columellate many-spored sporangia whereas the Chaetocladiaceae possess columellate unispored sporangiola.

Sometimes distinguishing different species depends only on slight phenotypical differences (e.g. colour of colonies and spores, spore size and morphology, growth temperature). Therefore the morphologically obtained identity of *Mucor fragilis* (FSU6158, FSU6164) and *Mucor hiemalis* f. *luteus* (FSU6171) differs from the final identity (Table 3) which is a combined decision based on molecular and, although partially, also on morphological data.

4. Problems Concerning the Molecular Identification of Zygomycota

4.1. Incomplete Sequence Data Bases

In the case of the homothallic *Mucor genevensis* (FSU6176) morphological identification is clear despite the formed zygospores (Figure 4 C-D; Zycha *et al.*, 1969). However, molecular comparisons yield no satisfying results regarding 18S rDNA and ITS sequences (Table 2). New ITS sequences generated from the type strain CBS114.08 (Acc. no. EU484275) show only 91% similarity to FSU6176. The 'low' percentage of identity between the new environmental isolate and the type strain is mainly due to a major insertion/deletion in the ITS1 region (Figure 16).

4.2. Sequence Deviations Caused by Different Geographical Regions

Another interesting aspect is that different isolates of a species can differ in their sequences because they are members of geographically distinct populations. One example is *Absidia repens.* The isotype CBS115583 was isolated in England. All new environmental isolates in this study, FSU4726, FSU2809, FSU2813, FSU2814, FSU2815, FSU2816 and

FSU2824 were isolated in Austria. All other strains with sequences available in GenBank originated from the American continents, such as CBS101.32 and CBS102.32 that were isolated by A.F. Blakeslee in Venezuela. A clinical specimen was presumably isolated in Minnesota (Hall *et al.*, 2003) and the neotype NRRL1336 was isolated from soil of unknown origin, but isolated by Blakeslee, who did a lot of his research in America. This strain also shows strong mating reactions with his tester strain NRRL1339, which he isolated from Venezuela (Hesseltine and Ellis, 1966). A morphological character common to all isolates is the formation of two types of sporangia. Firstly, sporangiophores and sporangia which are typical for *Absidia* and secondly, in older cultures, abundant shorter sporangiophores with small and sometimes few-spored sporangia, so called secondary sporangia (Figures 3 C-D, 9 B). Comparing the sequences of 28S rDNA (Figure 13) and ITS (Figure 15) the species from Europe differ markedly from the american ones with 8% dissimilarity for 28S rDNA and only 58%-59% identity for ITS. This large difference is mainly due to a 103bp insertion/deletion at the beginning of ITS1 (Figure 14), but each continental population shows higher sequence identity among each other with up to 100% for 28S rDNA. Therefore, it is important to include information about the geographic origin within the strain records, to verify the results with multi-locus analyses and, eventually to consider the erection of new taxa.

Strain	Geographic origin	NRRL1336	14849A*	FSU 2809	FSU 2813	FSU 2814	FSU 2815	FSU 2816	FSU 2824
NRRL1336	(South-)America	ID							
14849A*	Minnesota	0,992	ID						
FSU 2809	Austria	0,921	0,875	ID					
FSU 2813	Austria	0,918	0,875	0,994	ID				
FSU 2814	Austria	0,921	0,875	0,997	0,997	ID			
FSU 2815	Austria	0,918	0,875	0,994	0,994	0,994	ID		
FSU 2816	Austria	0,918	0,875	0,994	0,994	0,994	0,994	ID	
FSU 2824	Austria	0,920	0,877	0,995	0,995	0,995	0,995	0,998	ID

Figure 13. Sequence identity matrix comparing the nuclear LSU sequences from different isolates of *Absidia repens* originating from different geographical regions. *Note that the clinical isolate 14849A, AY234881 is a shorter sequence and the percentage identity refers in this case to shortened sequences of the other isolates. Abbreviations: FSU...Friedrich Schiller University Jena, Germany; NRRL... Agricultural Research Service (ARS) Culture Collection, Peoria, Illinois, USA.

4.3. Unsolved Problems

The interpretation of sequence comparisons is somewhat difficult for two other Mucorales that were misidentified based on rDNA: *Umbelopsis ramanniana* FSU2005 and *Syncephalastrum racemosum* FSU6155, with only 94% and 87% similarity to database sequences (Table 2). *Umbelopsis ramanniana* FSU2005 also shows considerable differences to *Umbelopsis ramanniana* FSU6174 with 8% dissimilarity. However, morphological identification gave clear results (Figures 3 E-F and 5 C-D). Due to missing additional information about the geographic origin of all isolates involved in this study, possibilities such as a geographical dependence of this problem could not be investigated. Another thought could be that one of the investigated species is a cryptic one, which remains distinguishable on a molecular but not on a morphological basis.

```
CBS115583   1 GAAATGCTGGGAAGCCTCCGGGTGGACCTAACTTTTTTCTACTGTGCACT 50
FSU4726     1 GAAATGCTGGGAAGCCTCCGGGTGGACCTAACTTTTTTCTACTGTGCACT 50
CBS101.32   1 -------------------------------------------------- 50
CBS102.32   1 -------------------------------------------------- 50
NRRL1336    1 -------------------------------------------------- 50

CBS115583   51 GTTTTTTAGGGGGTTGCTTGGGAAGGGATTCGTTTCTTCCCTTGATGTTT 100
FSU4726     51 GTTTTTTAGGGGGTTGCTTGGGAAGGGATTCGTTTCTTCCCTTGATGTTT 100
CBS101.32   51 -------------------------------------------------- 100
CBS102.32   51 -------------------------------------------------- 100
NRRL1336    51 -------------------------------------------------- 100

CBS115583   101 GGGGGAATTTTATTATTCCCCCTTCATGGGAAAGTTTTACTACTTTCCCC 150
FSU4726     101 GGGGGAATTTTATTATTCCCCCTTCATGGGAAAGTTTTACTACTTTCCCC 150
CBS101.32   101 ---AAAATGCGGCCGGTTCTCTTTCGGGAGGATTGGTCAACAGATTTAAT 150
CBS102.32   101 ---AAAATGCGGCTGCCTCTCCT---GTAGAGGTGGTCAACAGATTTAAT 150
NRRL1336    101 ---AAAATGCGGCTGGCTCTCTTT--GGAGGGTTGGTCAACAGATTTAAT 150

CBS115583   151 TTCTCCCACCCTGGGTAAAGCCCTTTTTCCTT-----TGGGAGAATCCGG 200
FSU4726     151 TTCTCCCACCCTGGGTAAAGCCCTTTTTCCTT-----TGGGAGAATCCGG 200
CBS101.32   151 TCTGTGCACTGTTTTTAATTGGGGGTTTTCTTGAAAAAGGGAGCCTCCTG 200
CBS102.32   151 TCTGTGCACTGTTTTTAATTGGGGGTTTCCTT-----TGGGAGCCTCCTG 200
NRRL1336    151 TCTGTGCACTGTTTTTAATTGGGGGTTTCCTT-----AGGGAGCCTCCTG 200

CBS115583   201 TTTGCCCAGTTGAATTCCCCTTCTTTCATAGGGGGGGGGTTTTCAAGTTT 250
FSU4726     201 TTTGCCCAGTTGAATTCCCCTTCTTTCATAGGGGGGGGGTTTTCAAGTTT 250
CBS101.32   201 CCCTGGGTATTGCTCTTTTTCCTTTGGGAAGAAATCAGCTTGCCCTATTA 250
CBS102.32   201 CCCTGGGTATTGCTCTTTTTCCTTTGGGAAGAAATCAGCTTGCCCTATTA 250
NRRL1336    201 CCCTGGGTATTGCTCTTTGTCCTTTGGGAAGAAATCAGCTTGCCCTATTA 250
```

Figure 14. Partial alignment of ITS sequences from different isolates of *Absidia repens* originating from different geographical regions.

strain	Geographic origin	CBS115583	FSU 4726	CBS101.32	CBS102.32	NRRL1336
CBS115583	UK	ID				
FSU 4726	Austria	1,000	ID			
CBS101.32	Venezuela	0,579	0,579	ID		
CBS102.32	Venezuela	0,591	0,591	0,933	ID	
NRRL1336	(South-)America	0,591	0,591	0,944	0,976	ID

Figure 15. Sequence identity matrix comparing ITS sequences of different isolates of *Absidia repens* from different geographical regions. Abbreviations: FSU...Friedrich Schiller University Jena, Germany; CBS...Centraalbureau voor Schimmelcultures, Utrecht, The Netherlands; NRRL... Agricultural Research Service (ARS) Culture Collection, Peoria, Illinois, USA.

```
CBS114.08  1   AATAATTTATATGATAGTAA------------------------------   50
FSU 6176   1   AATAATTTATATGATAGCAAAAAAAAAAAAAAAAAAAAAACTGAGCCTTT   50

CBS114.08  51  --------------------------TTTTTACTATTAATATTTTTTTTA  100
FSU 6176   51  TCGGGCGCTTTTTTTCTTTTTTTCTTTTTTTACTATTAATATTTTTTTTA  100

CBS114.08  101 CTGTGAAACGTATTATTATACCTGACGCTTTGAGGAATGCCTAGCCCGCT 150
FSU 6176   101 CTGTGAAACGTATTATTATACCTGACGCTTTGAGGAATGCCTAGCCCGCT 150
```

Figure 16. Partial alignment of the ITS1 region comparing *Mucor genevensis* CBS114.08 and *M. genevensis* FSU6176. 91% of the total ITS region is identical except for a major insertion/deletion within ITS1.

4.4. Consequences, Threshold Values and Defining new Species

In two cases morphological and molecular identification was very contradictory. FSU2822 was identified as some kind of *Rhizopus* based on morphology, but the best sequence comparison of 28S rDNA with 94% identity was with *Zygorhynchus exponens*. Both species are morphologically very distinct. FSU6170, morphologically identified as *Mucor hiemalis* shows sequence similarities with 99% (18S rDNA) and 91% (ITS) to *Rhizomucor variabilis*. Both potential species are very similar morphologically, and phylogenetically *Rhizomucor variabilis* seems to be very closely related to *Mucor* species (Voigt *et al.*, 1999a)

In summary, problems occurring in identification of Zygomycota, such as a lack of appropriate sequences, geographical differences of subpopulations and discrepancies between phenotype and genotype, suggest that a comprehensive revaluation should be considered.

Perhaps a critical threshold value for ITS sequences can help to draw a line between reliable and doubtful species identification. In our cases we assume this threshold to be at least 97% (but better 98%) sequence identity, which should be validated by multi-locus barcoding as proposed by Taylor *et al.* (2000). Therefore, strains FSU2572 and FSU3846 represent new species of the genus *Mucor* and FSU2005, FSU2822, FSU6155 and FSU6170 require deeper investigations to clarify the justification for erecting a new species or subspecies (Table 3). Failing to assign an unknown fungal isolate to a possible new species can create cryptic species which will complicate any further investigations. A known general problem is the estimation, that in GenBank up to 20% of the fungal deposits are not named correctly. Also, the quality of a sequence depends strongly on the correct identification of the underlying organism (Bridge *et al.*, 2003).

5. PHYLOGENY

The evolutionary distances obtained from molecular data are often displayed as cladogram or phylogenetic tree. A large variety of different algorithms is available for constructing such trees. We conducted phylogenetic reconstructions using PAUP*4.0b10 (Swofford, 1998) for the 18S rDNA and the 28S rDNA datasets after aligning the sequences in ClustalX version 1.83 (Higgins and Sharp, 1988, 1989; Thompson *et al.*, 1997) and manually controlling in BioEdit version 7.0.3 (Hall, 1999). According to the best BLAST hit, both datasets were complemented with nineteen additional reference sequences for 18S and twenty-two additional reference sequences for 28S rDNA, retrieved from GenBank (Table 4). If available, sequences from type strains were preferred. In the neighbor-joining trees obtained from 18S rDNA (Figure 11) and 28S rDNA sequences (Figure 12) the outgroup Ascomycota is clearly separated from the ingroup Zygomycota (BS=100%). In the outgroup the new isolate *Acremonium*-like Hypocreales FSU2858 is distinct from *Acremonium strictum* which was the direct BLAST hit for 18S and 28S rDNA sequence comparisons (Table 2). This distinctness was supported by an analysis of ITS sequences (chaper 3.1.). The Zygomycota are divided into the orders Kickxellales (BS=100%) and Mucorales (BS=93%) in the 18S rDNA phylogeny into and Kickxellales (BS=100%), Mortierellales (BS=100%) and Mucorales (BS=96%) in the 28S rDNA phylogeny.

Coemansia aciculifera FSU2823 belongs to the order Kickxellales, which is the sister group to a group containing Mortierellales and Mucorales (BS=90%). Although there is not a single sequence for this species in GenBank, it could easily be distinguished from other Kickxellelales through a morphological comparison (see point 3.5.). The Mortierellales are distinguished from the Mucorales by non-apophysate sporangia and a lacking columella (Figure 7 A-C). Within the order Mucorales the family Umbelopsidaceae appears as monophyletic basal sister lineage to the Mucoraceae (each BS=100%; Figure 11). *Umbelopsis ramanniana* FSU6174 and *U. isabellina* FSU4754 were verified by sequence identities of 99%, and 97% for ITS sequences and 99% (in both species) for 18S rDNA sequences, respectively. Although the identity of FSU2005 is still not known, it is in close vicinity to the other species of *Umbelopsis* in the *Umbelopsis*-clade, and resembles the typical non-apophysate sporangia and the rudimentary columella shown in figures 3 E and F.

Concentrating on the 28S rDNA phylogenetic tree, species of the genus *Absidia* form a monophyletic group (BS=100%) within the Mucoraceae comprising *Absidia repens*, *A. coerulea,* and *A. glauca*. All *A. repens* isolates from Europe are clearly separated from *A. repens* NRRL1336 with 100% BS supporting its geographic distinctness (see point 4.2.).

All isolates of *Mucor hiemalis* f. *hiemalis* are identical with 98% to 99% to the sequence of the neotype strain of this species Both strains of *Mucor hiemalis* f. *hiemalis* with an identity of 98% are separated from the other strains of *M. hiemalis* f. *hiemalis* and show 97% sequence identity to *Rhizomucor variabilis*, but the bootstrap support is very weak (59%). Both species, *Mucor hiemalis* f. *hiemalis* and *Rhizomucor variabilis* form well supported sister groups with BS=100% in the 28S rDNA phylogeny and 80% BS in the 18S rDNA tree. These relationships suggest that FSU6170 is closely related to these forms.

Although the 18S sequences are 99% identical between *Rhizomucor variabilis, Mucor hiemalis* f. *hiemalis* and the isolate FSU6170, there are differences between the ITS sequences. FSU6170 shows 87% identity to *Rhizomucor variabilis* and 84% to *Mucor hiemalis. Rh. variabilis* and *M. hiemalis* themselves are identical to 85%. Due to morphological similarities (Figures 4 E–F, 6 A and 8 B) this isolate was primarily described as *Mucor hiemalis*. However, based on the discrepancies mentioned above regarding phenotype and genotype, the final identity is still unknown. In the polyphyletic genus *Mucor* the species *M. racemosus*, *M. plumbeus*, *M. fragilis* and *M. circinelloides* form a well supported (BS=100%) subclade based on 18S rDNA sequence data (Figure 11) with 99 to 100% sequence identity. They could be clearly identified morphologically (e.g. the flattened columella seen in figure 4 B, and the prolonged columella with appendages seen in figures 5 A-B) or using ITS sequence similarities (Figures 17, 18). The species most closely related to this *Mucor* subclade are the homothallic fungi *Mucor genevensis* (Figures 4 C-D, 8 A) and *Zygorhynchus moelleri* (Figures 5 E-F, 10 A-B) with BS values 72% and 78%, respectively. The isolates of *Rhizopus oryzae* form a monophyletic group (BS=100%) with a sequence identity of 99%-100%, based on 18S rDNA and ITS data. A typical sporangium directly opposite to the rhizoids is shown in figure 6 C.

	AY213659	AY243943	FSU 6154	FSU 6154	AJ271061
Mucor racemosus f. *racemosus* CBS260.68 AY213659	ID				
Mucor circinelloides f. *circinelloides* CBS195.68 AY243943	0,905	ID			
Mucor circinelloides f. *circinelloides* FSU 6154	0,903	0,987	ID		
Mucor circinelloides f. *lusitanicus* FSU 6157	0,910	0,956	0,961	ID	
Mucor circinelloides f. *lusitanicus* CBS277.49 AJ271061	0,908	0,956	0,961	0,996	ID

Figure 17. Sequence identity matrix comparing ITS sequences of *Mucor racemosus* f. *racemosus* CBS260.68 , *M. circinelloides* f. *circinelloides* CBS195.68 and FSU6154 and *M. circinelloides* f. *lusitanicus* CBS277.49 and FSU6157. FSU…Friedrich Schiller University Jena, Germany; CBS…Centraalbureau voor Schimmelcultures, Utrecht, The Netherlands.

	FSU 6158	FSU 6164	FSU 6171	FSU 6172	FSU 6173	AF254944	AY243951	AY243949	AF474242
M. fragilis FSU 6158	ID								
M. fragilis FSU 6164	0,994	ID							
M. hiemalis f. *luteus* FSU 6171	0,761	0,760	ID						
M. hiemalis f. *hiemalis* FSU 6172	0,743	0,745	0,824	ID					
M. hiemalis f. *hiemalis* FSU 6173	0,734	0,734	0,806	0,931	ID				
M. h. f. *luteus* CBS244.35 AF254944	0,749	0,747	0,983	0,818	0,799	ID			
M. h. f. *luteus* CBS 243.35 AY243951	0,761	0,759	0,994	0,828	0,809	0,978	ID		
M. h. f. *hiemalis* CBS 242.35 AY243949	0,741	0,743	0,823	0,998	0,933	0,816	0,826	ID	
M. fragilis AF474242	0,994	1,000	0,760	0,745	0,734	0,747	0,759	0,743	ID

Figure 18. Sequence identity matrix comparing ITS sequences of *Mucor fragilis* FSU6158, FSU6164, *M. hiemalis* f. *hiemalis* FSU6172, FSU6173 and *M. hiemalis* f. *luteus* FSU6171 with reference material of *M. fragilis* GenBank acc.no AF474242, *M. hiemalis* f. *hiemalis* CBS242.35 and *M. hiemalis* f. *luteus* CBS243.35 and CBS244.35. FSU...Friedrich Schiller University Jena, Germany; CBS...Centraalbureau voor Schimmelcultures, Utrecht, The Netherlands.

General Conclusions

The Zygomycetes is a fascinating and versatile group of true fungi (Mycota) harbouring immense biotechnological potential. The Zygomycota is one of the most diverse fungal phyla whose traditional classification based on phenotypic characteristics is changing by assigning polyphyletic related groups to new phyla and subphyla comprising well supported monophyletic groups (Hibbett *et al.*, 2007). Clearly defined structures in classification and generally accepted concepts are essential for communication within a growing community of scientists with varying backgrounds.

The identification of unknown fungi must not depend on single methods; it should be supported by several different techniques. Simple morphology for example is still not standardized with a stable terminology. It is also highly prone to the investigator´s personal subjective opinion and some characters depend upon environmental conditions. Morphology is furthermore not suitable for the precise species identification of microscopic fungi that develop exclusively vegetative hyphae without any specific distinguishing structures (e.g. hyphal appendices, asexual or sexual sporulation propagules). Even the occurrence of clamp connections in dikaryotic hyphae of Basidiomycota allow solely delimitation of phyla rather than the identification of species without the development of a basidioma. The possibility to analyze molecular sequence data with PCR technology has revolutionized the taxonomy of fungi (White *et al.*, 1990). Nevertheless, a phylogeny based on sequence information devoid of information on morphology, physiology or ecology and will work only for well studied species. The exploration of new species depends on the combination of all of these criteria. The reconstruction of phylogenetic relationships can be used to erect taxonomic systems, which are essential for the organization of biological information and for a deeper understanding of the reconstructed evolution (Wheeler, 2004; Fenchel and Finlay, 2006).

The decision about which molecular markers are used for the reconstruction of phylogenetic relationships should be considered carefully. The markers need to fulfil certain requirements for phylogenetic comparison, such as being present in all taxa of interest, and they should not be affected by horizontal gene transfer events. Furthermore, they should be diverse enough to distinguish single species, like the internal transcribed spacer 2 region, which also reveals useful information about higher-level relationships based on secondary structure (Coleman, 2007).

Comparing the sequenced material with already existing data deposited in data bases can lead to misidentification, either due to highly conserved sequences or simply because of false annotation of the material. The first problem can easily be compensated by comparing different genomic sequences. However, mislabelled material is a general problem and a precise identification scheme has to cope with varying species delineations. Species concepts are widely discussed and usage is very different between the different fungal phyla. A challenging hypothesis is that, at least for plants and fungi, two organisms belong to distinct species with 93% reliability once they differ in at least one compensatory base change (CBC) in the internal transcribed spacer 2 region (Müller *et al.*, 2007). Moreover, geographical differences in phenotype and genotype should be considered carefully. Describing fungi solely based on morphological characters make them seem to be globally distributed, but comparing molecular data can reveal several endemic species. These differences in geographic range of fungi depending on the method of species recognition can be due to the fact that the rate of morphological change is slower than the rate of genetic change for organisms with less elaborate development and fewer cells (Taylor *et al.*, 2006).

Because a broad range of different genes and sequence fragments can be used for the characterization of an organism, a common problem is to find appropriate sequences that can complement the data produced in the lab. Generating DNA barcodes can be a helpful alternative. DNA barcodes are short genetic sequences from a standard part of the genome. For nearly all animals this region is part of a gene encoding the mitochondrial cytochrome C oxidase 1 (CO1). However, in plants CO1 evolves too slowly to be used as a barcode and another region will soon be proposed (http://www.dnabarcodes.org). Because only one case study concerning fungi is available at the Consortium for the Barcode of Life (www.bolinfonet.org/casestudy), there is no fixed genomic region for barcoding fungi. Although CO1 is proposed and has proven useful (Seifert *et al.* 2007), other proposals exist, such as using repetitive sequences (Healy *et al.*, 2004; Wise *et al.*, 2007) or ITS sequences (Summerbell *et al.*, 2007).

In the taxonomy browser of Barcode of Life Data systems (BOLD at http://www.boldsystems.org/views/taxbrowser) more then the half of the fungal species belong to the Basidiomycota (1108 specimens). The Eumycota (718 specimens) comprises 592 Ascomycota, 1 member of the Sordariomycetes, 10 members of the Pezizomycetes, 110 members of the Leotiomycetes and 3 members of the Zygomycetes, namely *Smittium culisetae, Mortierella verticillata* and *Rhizopus oryzae.*

Conclusions

1. If reliable threshold values of sequence similarities and dissimilarities are defined, ITS barcoding represents a powerful tool in the identification of zygomycetes and distinction at the species and intraspecies level.
2. Thresholds of ITS-based similarity indices could be determined for species and genus delimitations. Variance to reference sequences higher than 6 % indicates new genera, ITS-sequence deviations ranging between 3 % and 6 % indicate new species. A similarity less or equal to 3% indicates identity to a particular species as determined by a well-defined reference species. These values are maximal ones and may rather be corrected to smaller numbers.
3. Molecular barcoding relies on a wide collection of morphologically well-defined reference strains. Its major task is species identification but also help to define ′white spots′ where new or cryptic species can be hypothesized. Due to a lack of morphological markers especially the ′lower′ fungi, such as zygo- and chytridiomycetes, harbour vast majorities of cryptic species.
4. Based on our experiences with molecular identification of microscopic fungi the following guidelines for doing BLAST searches aiming at molecular spcies identification of fungi can be defined:
   - Priority on highly variable ITS over SSU, LSU and conserved protein-coding genes such as genes encoding actin and beta-tubulin (as shown in Hoffmann *et al.*, 2007).
   - BUT, like other regions of the nuclear ribosomal DNA cluster also the ITS exists in multiple copies causing difficulties in sequence homogeneities (Voigt *et al.*, 1999b). Variable ITS sequence entities within a single genome make a direct sequencing impossible. Therefore, additional supplementary barcodes other than ITS are required.
   - The multi-copy nature of most molecular barcodes makes a multi-gene barcoding necessary for species identification. Even the protein-coding gene barcodes are often moderately repetitive due to gene duplication (e.g. beta-tubulin; Einax and Voigt, 2003).
   - BLAST – searches rely on the access to comparable molecular barcodes of type/isotype/neotype or authentic strains providing reference information.
   - If such strains are not available the search for morphologically safely identified strains will be essential.
   - Since the data entries in public sequence databases rely on accuracy of the submitter, the species description lines are sometimes misleading.
   - Thus, barcoding without morphological background will never be successful which is accordance to Taylor *et al.* (2000) and Schindel and Miller (2005).

Future Line of Research

Strain identification on the basis of morphological characters in combination with sequence analysis of LSU 28S rDNA, SSU 18S rDNA and internal transcribed spacer regions 1 and 2, including 5.8S rDNA is a powerful tool in taxonomy and systematics of fungi. A unified and sensitive species reference data base system including morpho- and genotypes of type strain material is required to provide safe reference entries and sufficient reference points for the efficient interpretation of the upcoming enormous amounts of barcoding data in future perspectives. Ideally a very broad ranged sampling should include specimens from different types of substrates and diverse geographic regions as well as known species of a genus. The identification of cryptic species, which show morphologies equal to reference species but deviating DNA barcodes, will have a major rule in the taxonomy of basal fungi.

In order to support the organization and analysis of barcode data there will be an increasing importance for a centralized open accessible resources, which unites individual data bases and serves as online workbench for the DNA barcode community. The Barcode of Life Data Systemes (BOLD: http://www.barcodinglife.org/; Ratnasingham and Hebert, 2007) fulfil this requirement by providing a repository for barcode records, by storing specimen data and images as well as sequences and trace files and by monitoring the number of barcode sequence records and species coverage. In future this facility should become more and more crucial as an efficient interface for submitting barcode records to GenBank and as the identification engine based on the current barcode library within the barcoding community.

Acknowledgements

This work was supported by the Deutsche Forschungsgemeinschaft (VO 772/4-1, 2 & 3 to KV) and by the Thüringer Ministerium für Wissenschaft, Forschung und Kunst to KH. The study could not have been carried out without the inspiring interaction with colleagues from the European Consortium for the Barcoding of Life (ECBoL). We express our gratitude to Mrs. Gisela Baumbach for excellent assistance with strain maintenance. The authors thank Anke Zeuner for help with fungal isolation and the undergraduate students in the four week lab course ('Großpraktikum') during the winter semesters of 2006/07 and 2007/08 for their technical contributions to the molecular identification of some dubious fungal strains. KV and LO thank Katja Felbel and Nadine Piekarski for their help with scanning electron microscopy. The authors provided no information concerning the existence of conflicting or dual interests.

REFERENCES

Anderson, I.C., Campbell, C.D. and Prosser, J.I. (2003). Potential bias of fungal 18S rDNA and internal transcribed spacer polymerase chain reaction primers for estimating fungal biodiversity in soil. *Environ. Microbiol. 5*: 36-47.

Baayen, R.P., O`Donnell, K., Bonants, P.J.M., Cigelnik, E., Kroon, L.P.N.M., Roebroeck, E.J.A. and Waalwijk, C. (2000). Gene genealogies and AFLP analyses in the *Fusarium oxysporum* complex identify monophyletic and nonmonophyletic formae speciales causing wilt and rot disease. *Phytopathology 90*: 891-900.

Barnett, J.A., Payne, R.W. and Yarrow, D. (1990). Yeasts: characteristics and identification. Cambridge University Press, Cambridge, UK.

Bartnicki-Garcia, S. (1970). Cell wall composition and other biochemical markers in fungal phylogeny, pp. 81–103. *In* J. B. Harbone (ed.), *Phytochemical phylogeny.* Academic Press, Ltd., London, United Kingdom.

Bartnicki-Garcia, S. (1987). The cell wall in fungal evolution. *In*: Rayner, A.D.M., Brasier, C.M. and Moore, D. (eds.), *Evolutionary biology of the fungi.* Cambridge University Press, New York, N.Y., pp. 389–403.

Benjamin, R.K. (1959). The merosporangiferous Mucorales. *Aliso 4*: 321-433.

Benjamin, R.K. (1979). Zygomycetes and their spores. *In*: Kendrick, B. (ed.). *The whole fungus the sexual-asexual synthesis.* National Museum of Natural Science, National Museums of Canada and The Kananaskis Foundation. Ottawa, Canada., pp. 573-616.

Benny, G.L. (1982). Zygomycetes. *In*: Parker, S.P. (Ed.). *Synopsis and Classification of Living Organisms.* Vol. I. McGraw-Hill Book Company, Inc., New York., pp. 184-195.

Benny, G.L. (1995). Classical morphology in zygomycete taxonomy. *Can. J. Bot. 73* (Suppl. 1): 725–730.

Benny, G.L. and Benjamin, R.K. (1975). Observations on Thamnidiaceae (Mucorales). New taxa, new combinations, and notes on selected species. *Aliso 8*: 301-351.

Benny, G.L. and Benjamin, R.K. (1976). Observations on Thamnidiaceae (Mucorales). II. *Chaetocladium, Cokeromyces*, *Mycotypha,* and *Phascolomyces*. *Aliso 8*: 391-424.

Benny, G.L., Humber, R.A. and Morton, J.B. (2001). The Zygomycota: Zygomycetes. *In*: McLaughlin, D.J., McLaughlin, E.G. and Lemke, P.A. (eds.). *The Mycota.* Vol. 7A. Systematics and Evolution. Springer-Verlag, Berlin, Germany., pp. 113-146.

Berbee, L.M. and Taylor, J.W. (2001). Fungal molecular evolution: gene trees and geologic time. *In*: McLaughlin, D.J., McLaughlin, E.G. and Lemke, P.A. (eds.). *The Mycota.* Vol. 7B. Systematics and Evolution. Springer-Verlag, Berlin, Germany., pp. 229-245.

Blackwood, C.B., Oaks, A. and Buyer, J.S. (2005). Phylum- and class-specific PCR primers for general microbial community analysis. *Appl. Environ. Microbiol. 71*: 6193-6198.

Boekhout, T., Fell, J.W., Fonseca, A., Prillinger, H.J., Lopandic, K. and Roeijmans, H. (2000). The basidiomycetous yeast *Rhodotorula yarrowia* comb. nov. *Antonie van Leeuwenhoek 77*: 355-358.

Bridge, P.D., Roberts, P.J., Spooner, B.M. and Panchal, G. (2003). On the unreliability of published DNA sequences. *New Phytol. 160*: 43-48.

Bridge, P.D. (1985). An evaluation of some physiological and biochemical methods as an aid to the characterization of species of *Penicillium* subsection Fasciculata. *J. Gen. Microbiol. 131*: 1887–1895.

Bruns, T.D., White, T.J. and Taylor, J.W. (1991). Fungal molecular systematics. *Annu. Rev. Ecol. Syst. 22*: 525–564.

Casimiro, S., Moura, M., Zé-Zé, L., Tenreiro, R. and Monteiro, A.A. (2004). Internal transcribed spacer 2 amplicon as a molecular for identification of *Peronospora parasitica* (crucifer mildew). *J. Appl. Microbiol. 96*: 579–587.

Certik, M. and Shimizu, S. (1999). Biosynthesis and regulation of microbial polyunsaturated fatty acid production. *J. Biosci. Bioeng. 87*: 1-14.

Coleman, A.W. (2007). Pan-eukaryote ITS2 homologies revealed by RNA secondary structure. *Nuc. Acid. Res. 35*: 3322-3329.

Einax, E. and Voigt, K. (2003). Oligonucleotide primers for the universal amplification of beta-tubulin genes facilitate phylogenetic analyses in the regnum Fungi. *Org. Divers. Evol. 3*: 185-194.

Ellis, J.J. and Hesseltine, C.W. (1965). The genus *Absidia*: globose-spored species. *Mycologia 57*: 222-235.

Ellis, J.J. and Hesseltine, C.W. (1966). Species of *Absidia* with ovoid sporangiospores. II. *Sabouraudia 5*: 59-77.

Felsenstein J. (1985). Confidence limits of phylogenies: An approach using the bootstrap. *Evolution 39*: 783-791.

Fenchel, T. and Finlay, B.J. (2006). The diversity of microbes: resurgence of the phenotype. *Phil. Trans. R. Soc. B 361*: 1965–1973.

Fuhrmann, B., Roquebert, M.F., Lebreton, V. and van Hoegaerden, M. (1990). Immunological differentiation between *Penicillium* and *Aspergillus* taxa. *In*: Samson, R.A. and Pitt, J.I. (ed.), *Modern concepts in Penicillium and Aspergillus classification.* Plenum Press, New York, N.Y., pp. 423–432.

Fuller, M.S. (1976). Mitosis in fungi. *Int. Rev. Cytol. 45*: 113–153.

Gams, W. and Bissett, J. (1998). Morphology and identification of *Trichoderma*. *In*: Kubicek, C.P. and Harman, G.E. (eds.), *Trichoderma* and *Gliocladium*. *Basic Biology, Taxonomy and Genetics*. Taylor and Francis, London, UK, pp. 3-34.

Ganley, R.J. and Newcombe, G. (2006). Fungal endophytes in seeds and needles of *Pinus monticola*. *Mycol. Res. 110*: 318-327.

Garnica, S., Weiss, M., Walther, G. and Oberwinkler, F. (2007). Reconstructing the evolution of agarics from nuclear gene sequences and basidiospore ultrastructure. *Mycol. Res. 111*: 1019-1029.

Gherbawy, Y.A.M.H., Adler, A. and Prillinger, H. (2001). Genotypic identification of *Fusarium subglutinans, F. proliferantum* and *F. verticillioides* strains isolated from maize in Austria. *Egyptian J. Biol. 3*: 37-46.

Hachmeister, K.A. and Fung, D.Y. (1993). Tempeh: a mold-modified indigenous fermented food made from soybeans and/or ceral grains. *Crit. Rev. Microbiol. 19*: 137-188.

Hall, T.A. (1999). BioEdit: a user-friendly biological sequence alignment editor and analysis program for Windows 95/98/NT. *Nucl. Acids Symp.* Ser. *41*: 95-98.

Hall, L., Wohlfiel S. and Roberts G.D. (2003). Experience with the MicroSeq D2 large-subunit ribosomal DNA sequencing kit for identification of commonly encountered clinically important yeast species. *J. Clin. Microbiol. 41*: 5099-5102.

Hawksworth, D.L. (1991). The fungal dimension of biodiversity: magnitude, significance, and conservation. *Mycol. Res. 95*: 641-655.

Hawksworth, D.L., Kirk, P.M., Sutton, B.C. and Pegler, D.N. (1995). Ainsworth and Bisby's dictionary of the fungi, 8th ed. International Mycological Institute, Egham, United Kingdom.

Hawksworth, D.L. and Rossman, A.Y. (1997). Where are all the undescribed fungi? *Phytopathology 87*: 888-891.

Hawksworth, D.L. (2001). The magnitude of fungal diversity: the 1.5 million species estimate revisited. *Mycol. Res. 105*: 1422-1432.

Healy, M., Reece, K., Walton, D., Huong, J., Shah, K. and Kontoyiannis, D.P. (2004). Identification to the species level and differentiation between strains of *Aspergillus* clinical isolates by automated repetitive-sequence-based PCR. *J. Clin. Microbiol. 42*: 4016-4024.

Heath, I.B. (1980). Variant mitoses in lower eukaryotes: indicators of the evolution of mitosis? *Int. Rev. Cytol. 64*: 1–80.

Hesseltine, C.W. (1954). The section *Genevensis* of the genus *Mucor*. *Mycologia 64*: 358-366.

Hesseltine, C.W. and Ellis, J.J. (1961). Notes on Mucorales, especially *Absidia*. *Mycologia 53*: 406-426.

Hesseltine, C.W. and Ellis, J.J. (1964). The genus *Absidia: Gongronella* and cylindrical-spored species of *Absidia*. *Mycologia 56*: 568-601.

Hesseltine, C.W. and Ellis, J.J. (1966). Species of *Absidia* with ovoid sporangiospores. I. *Mycologia 58*: 761-785.

Hesseltine, C.W. and Ellis, J.J. (1973). Mucorales. *In*: Ainsworth, G.C., Sparrow, F.K., and Sussman, A.S. (Eds.), The Fungi: An Advanced Treatise. Vol. 4B: A Taxonomic Review with Keys: Basidiomycetes and Lower Fungi. Academic Press, London, pp. 187-217.

Hibbett, D.S. (1992). Ribosomal RNA and fungal systematics. *Trans. Mycol. Soc. Jpn. 33*: 533–556.

Hibbett, D.S., Binder, M., Bischoff, J.F., Blackwell, M., Cannon, P.F., Eriksson, O.E., Huhndorf, S., James, T., Kirk, P.M., Lücking, R., Lumbs H.T., Lutzoni, F., Matheny, P.B., McLaughlin, D.J., Powell, M.J., Redhead, S., Schoch, C.L., Spatafora, J.W., Stalpers, J.A., Vilgalys, R., Aime, M.C., Aptroot, A., Bauer, R., Begerow, D., Benny, G.L., Castlebury, L.A., Crous, P.W., Dai, Y.C., Gams, W., Geiser, D.M., Griffith, G.W., Gueidan, C., Hawksworth, D.L., Hestmark, G., Hosaka, K., Humber, R.A., Hyde, K.D., Ironside, J.E., Kõljalg, U., Kurtzman, C.P., Larsson, K.H., Lichtwardt, R., Longcore, J., Miadlikowska, J., Miller, A., Moncalvo, J.M., Mozley-Standridge, S., Oberwinkler, F., Parmasto, E., Reeb, V., Rogers, J.D., Roux, C., Ryvarden, L., Sampaio, J.P., Schüssler, A., Sugiyama, J., Thorn, R.G., Tibell, L., Untereiner, W.A., Walker, C., Wang, Z., Weir, A., Weiss, M., White, M.M., Winka, K., Yao, Y.J. and Zhang, N. (2007). A higher-level phylogenetic classification of the Fungi. *Mycol Res. 111*: 509-547.

Higgins, D.G. and Sharp, P.M. (1988). CLUSTAL: a package for performing multiple sequence alignment on a microcomputer. *Gene 73*: 237-244.

Higgins, D.G. and Sharp, P.M. (1989). Fast and sensitive multiple sequence alignments on a microcomputer. *Comput. Appl. Biosci. 5(2)*: 151-153.

Hoffmann, K., Discher, S. and Voigt, K. (2007). Revision of the genus *Absidia* (Mucorales, Zygomycetes): based on physiological, phylogenetic and morphological characters: Thermotolerant *Absidia* spp. form a coherent group, the Mycocladiaceae fam. nov. *Mycol. Res. 111*: 1169-1183.

Holliday, P. (1980). Fungus diseases of tropical crops. Cambridge University Press, Cambridge.

Humber, R.A., Brown, C.C. and Kornegay, R.W. (1988). Equine zygomycosis caused by *Conidiobolus lamprauges*. *J. Clin. Microbiol. 27*: 573-576.

Hwang, U.-W. and Kim, W. (1999). General properties and phylogenetic utilities of nuclear ribosomal DNA and mitochondrial DNA commonly used in molecular systematics. *The Korean Journal of Parasitology 37*: 215-228.

James, T.Y., Kauff, F., Schoch, C.L., Matheny, P.B., Hofstetter, V., Cox, C., Celio, G., Gueidan, C., Fraker, E., Miadlikowska, J., Lumbsch, H.T., Rauhut, A., Reeb, V., Arnold, E.A., Amtoft, A., Stajich, J.E., Hosaka, K., Sung, G-H., Johnson, D., O'Rourke, B., Crockett, M., Binder, M., Curtis, J.M., Slot, J.C., Wang, Z., Wilson, A.W., Schüßler, A., Longcore, J.E., O'Donnell, K., Mozley-Standridge, S., Porter, D., Letcher, P.M., Powell, M.J., Taylor, J.W., White, M.M., Griffith, G.W., Davies, D.R., Humber, R.A., Morton, J., Sugiyama, J., Rossman, A.Y., Rogers, J.D., Pfister, D.H., Hewitt, D., Hansen, K., Hambleton, S., Shoemaker, R.A., Kohlmeyer, J., Volkmann-Kohlmeyer, B., Spotts, R.A., Serdani, M., Crous, P.W., Hughes, K.W., Matsuura, K., Langer, E., Langer, G., Untereiner, W.A., Lücking, R., Büdel, B., Geiser, D.M., Aptroot, A., Diederich, P., Schmitt, I., Schultz, M., Yahr, R., Hibbett, D.S., Lutzoni, F., McLaughlin, D., Spatafora, J. and Vilgalys, R. (2006). Reconstructing the early evolution of the fungi using a six-gene phylogeny. *Nature 443*: 818–822.

James, T.Y., Letcher, P.M., Longcore, J.E., Mozley-Strandridge, S.E., Porter, D., Powell, M.J., Griffith, G.W. and Vilgalys, R. (2006) ['2007']. A molecular phylogeny of the flagellated fungi (Chytridiomycota) and description of a new phylum (Blastocladiomycota). *Mycologia 98*: 860-71.

Karlshøj, K. and Larsen, T.O. (2005). Differentiation of species from the *Penicillium roqueforti* group by volatile metabolite profiling. *J. Agric. Food Chem. 53*: 708-715.

Kennedy, N. and Clipson, N. (2003). Fingerprinting the fungal community. *Mycologist 17*: 158–164.

Kimbrough, J.W. (1994). Septal ultrastructure and ascomycete systematics. *In*: Hawksworth, D.L. (ed.). *Ascomycete systematics: problems and perspectives in the nineties*. Plenum Press, New York, N.Y., pp. 127–141.

Kirk, P.M., Cannon, P.F., David, J.C. and Stalpers, J.A. (2001). *Ainsworth & Bisby's Dictionary of the fungi*. 9th edn., CABI Publishing, Wallingford, United Kingdom.

Kohn, L.M. (1992). Developing new characters for fungal systematics: an experimental approach for determining the rank of resolution. *Mycologia 84*: 139–153.

Kroken, S. and Taylor, J.W. (2001). Species boundaries in *Letharia*. *Mycologia 93*: 38-53.

Larsen, T.O. and Frisvad, J.C. (2005). Characterization of volatile metabolites from 47 *Penicillium*-Taxa. *Mycol. Res. 99*: 1153-1166.

Le'John, H. B. (1974). Biochemical parameters of fungal phylogenetics. *Evol. Biol. 7*: 79–125.

Maresca, B. and Kobayashi, G.S. (1994). *Molecular biology of pathogenic fungi*. Telos Press, New York, N.Y.

Mattner, F., Weissbrodt, H. and Strueber, M. (2004). Two case reports: fatal *Absidia corymbifera* pulmonary tract infection in the first postoperative phase of a lung transplant patient receiving voriconazole prophylaxis, and transient bronchial *Absidia corymbifera* colonization in a lung transplant patient. *Scand. J. Infect. Dis. 36*: 312–314.

Maxson, L.R. and Maxson, R.D. (1990). Proteins II. Immunological techniques. *In*: Hillis, D.M. and Moritz, C. (ed.). *Molecular systematics*. Sinauer Associates, Sunderland, Mass., pp. 127-155.

Mayr, A., Kirchmair, M., Rainer, J., Rossi, R., Kreczy, A., Tintelnot, K., Dierich, M.P. and Lass-Flörl, C. (2004). Chronic paracoccoidioidomycosis in a femal patient in Austria. *Eur. J. Clin. Microbiol. Infect. Dis. 23*: 916-919.

McLaughlin, D.J., McLaughlin, E.G. and Lemke, P.A. (eds.) (2001a). *The Mycota*. Vol. VII. Part A. Systematics and Evolution. Springer Verlag, Berlin.

McLaughlin, D.J., McLaughlin, E.G. and Lemke, P.A. (eds.) (2001b). *The Mycota*. Vol. VII. Part B. Systematics and Evolution. Springer Verlag, Berlin.

Messner, R., Prillinger, H., Altmann, F., Lopandic, K., Wimmer, K., Molnár, O. and Weigang, F. (1994). Molecular characterization and application of random amplified polymorphic DNA analysis of *Mrakia* and *Sterigmatomyces* species. *Int. J. Sys. Bacteriol. 44*: 694-703.

Meyer, C.P. and Paulay, G. (2005). DNA barcoding: error rates based on comprehensive sampling. *PLoS BIOLOGY 3*: 2229-2238.

Michailides, T.J. and Spotts, R.A. (1990). Postharvest diseases of pome and stone fruits caused by *Mucor piriformis* in the Pacific northwest and California. *Plant Dis. 74*: 537-543.

Müller, T., Philippi, N., Dandekar, T., Schultz, J. and Wolf, M. (2007). Distinguishing species. *RNA 13*: 1469-1472.

Nirenberg, H. (1981). A simplified method for identifying *Fusarium* spp. occurring on wheat. *Can. J. Bot. 59*: 1599-1609.

Nyilasi, I., Papp, T., Csernetics, Á., Krizsán, K., Nagy, E. and Vágvölgyi, C. (2008). High-affinity iron permase (*FTR1*) gene sequence-based molecular identification of clinical important Zygomycetes. *Clin. Microbiol. Infect. 14*: 393-397.

O'Donnell, K. (1993). *Fusarium* and its near relatives. *In*: Reynolds, D.R. and Taylor, J.W. (eds.). *The fungal holomorph: mitotic, meiotic and pleomorphic speciation in fungal systematics*. CAB International, Wallingford, United Kingdom. pp. 225–233.

O'Donnell, K., Cigelnik, E. and Benny, G.L. (1998). Phylogenetic relationships among Harpellales and Kickxellales. *Mycologia. 90*: 624-639.

O'Donnell, K. (2000). Molecular phylogeny of the *Nectria haematococca – Fusarium solani* species complex. *Mycologia 92*: 919-938.

O`Donnell, K., Kistler, H.C., Tacke, B.K. and Casper, H.H. (2000). Gene genealogies reveal global phylogeographic structure and reproductive isolation among lineages of *Fusarium graminearum*, the fungus causing wheat scab. *Proc. Natl. Acad. Sci. USA 97*: 7905-7910.

Ogawa, J.M., Sonada, R.M. and English, H. (1992). Postharvest diseases of tree fruit. *In*: Kumar, J., Chaube, H.S., Singh, U.S. and Mukhopadhyay, A.N. (eds.). *Plant diseases on international importance*. Vol. III, Diseases of fruit crops. Prentice Hall, Englewood Cliffs, pp. 405-422.

Pääbo, S. (1989). Ancient DNA: extraction, characterization, molecular cloning, and enzymatic amplification. *Proc. Natl. Acad. Sci. U.S.A. 86*: 1939-1943.

Page R.D.M. (1996). TreeView: an application to display phylogenetic trees on personal computers. *Computational and Applied Biosciences 12*: 357-358.

Paterson, R.R.M. and Bridge, P.D. (1994). *Biochemical techniques for filamentous fungi*. CAB International, Wallingford, United Kingdom.

Pegler, D.N. (2000). Taxonomy, nomenclature and description of *Armillaria*. *In*: Fox, R.T.V. (ed.). Armillaria Root Rot: biology and control of honey fungus. *Intercept, Andover*, pp. 81-93.

Prillinger, H., Dörfler, C., Laaser, G., Eckerlein, B. and Lehle, L. (1990a). Ein Beitrag zur Systematik und Entwicklungsbiologie höherer Pilze: Hefe-Typen der Basidiomyceten. Teil I: Schizosaccharomycetales, *Protomyces*-Typ. *Z. Mykol. 56*: 219-250.

Prillinger, H., Dörfler, C., Laaser, G. and Hauska, G. (1990b). Ein Beitrag zur Systematik und Entwicklungsbiologie höherer Pilze: Hefe-Typen der Basidiomyceten. Teil III: *Ustilago*-Typ. *Z. Mykol. 56*: 251-278.

Prillinger, H., Deml, G., Dörfler, C., Laaser, G. and Lockau, W. (1991a). Ein Beitrag zur Systematik und Entwicklungsbiologie höherer Pilze: Hefe-Typen der Basidiomyceten. Teil II: *Microbotryum*-Typ. *Bot. Acta 104*: 5-17.

Prillinger, H., Laaser, G., Dörfler, C. and Ziegler, K. (1991b). Ein Beitrag zur Systematik und Entwicklungsbiologie höherer Pilze: Hefe-Typen der Basidiomyceten. Teil IV: *Dacrymyces*-Typ, *Tremella*-Typ. *Sydowia 53*: 170-218.

Rath, A.C., Carr, C.J. and Graham, B.R. (1995). Characterization of *Metarrhizium anisopliae* strains by carbohydrate utilization (API 50 CH). *J. Invertebr. Pathol. 65*:152–161.

Ratnasingham, S. and Hebert, P.D.N. (2007). BOLD: The barcoding of life data system (http://www.barcodinglife.org*). Molecular Ecology Notes 7*: 355-364.

Ribes, J.A., Vanover-Sams, C.L. and Baker, D.J. (2000). Zygomycetes in human disease. *Clin. Microbiol. Rev. 13*: 236-301.

Saitou, N. and Nei, M. (1987). The neighbor-joining method: a new method for reconstructing phylogenetic trees. *Molecular Biology and Evolution 4*: 406-425.

Sandhu, G.S., Kline, B.C., Stockman, L. and Roberts, G.D. (1995). Molecular probes for diagnosis of fungal infections. *J. Clin. Microbiol. 33*: 2913–2919.

Schindel, D.E. and Miller S.E. (2005). DNA barcoding a useful tool for taxonomists. *Nature 435*: 17

Schipper, M.A.A. (1973). A study on variability in *Mucor hiemalis* and related species. *Stud. Mycol. (Baarn) 4*: 1-40.

Schipper, M.A.A. (1975). *Mucor mucedo, Mucor flavus* and related species. *Stud. Mycol. (Baarn) 10*: 1-33.

Schipper, M.A.A. (1976). On *Mucor circinelloides, Mucor racemosus* and related species. *Stud. Mycol. (Baarn) 12*: 1-40.

Schipper, M.A.A. (1978). On certain species of *Mucor* with a key to all accepted species. *Stud. Mycol. (Baarn) 17*: 1-52.

Schipper, M.A.A. (1984). A revision of the genus *Rhizopus*. I. The *Rhizopus stolonifer*-group and *Rhizopus oryzae*. *Stud. Mycol. (Baarn) 25*: 1-19.

Schipper, M.A.A. (1986). Notes on *Zygorhynchus* species. *Persoonia 13*: 97-105.

Schipper, M.A.A. (1990). Notes on Mucorales - I Observations on *Absidia*. *Persoonia 14*: 133-149.

Scholer, H.J., Müller, E. and Schipper, M.A.A. (1983). Mucorales. *In*: Howard D.H. (ed.). *Fungi pathogenic for humans and animals*. A. Biology. Marcel Dekker, Inc., New York, N.Y., pp. 9–59.

Schüßler, A., Schwarzott, D. and Walker, C. (2001). A new fungal phylum, the Glomeromycota: phylogeny and evolution. *Mycol. Res. 105*: 1413-1421.

Schweigkofler, W., Lopandic, K., Molnár, O. and Prillinger, H. (2002). Analysis of phylogenetic relationships among Ascomycota with yeast phases using ribosomal DNA sequences and cell wall sugars. *Org. Divers. Evol. 2*: 1-17.

Seifert, K.A., Samson, R.A., Dewaard, J.R., Houbraken, J., Lévesque, C.A., Moncalvo, J.M., Louis-Seize, G. and Hebert, P.D. (2007). Prospects for fungus identification using CO1 DNA barcodes, with *Penicillium* as a test case. *Proc. Nat. Acad. Sci. U.S.A. 104*: 3901-3906.

Skouboe, P., Frisvad, J.C., Taylor, J.W., Lauritsen, D., Boysen, M. and Rossen, L. (1999). Phylogenetic analysis of nucleotide sequences from the ITS region of terverticillate *Penicillium* species. *Mycol Res 103*: 873-881.

Sugiyama, J. (1998). Relatedness, phylogeny, and evolution of the fungi. *Mycoscience 39*: 487-511.

Summerbell, R.C., Moore, M.K., Starink-Willemse, M. and Van Iperen, A. (2007). ITS barcodes for *Trichophyton tonsurans* and *T. equinum*. *Med. Mycol. 45*: 193-200.

Swofford, D.L. (1998). PAUP*: Phylogenetic Analysis using Parsimony (*and Other Methods). Version 4. Sinauer Associates, Sunderland, MA

Tanabe, Y., Watanabe, M.M. and Sugiyama, J. (2005). Evolutionary relationships among basal fungi (Chytridiomycota and Zygomycota): Insights from molecular phylogenetics. *J Gen. Appl. Microbiol. 51*: 267-276.

Taylor, J.W., Jacobson, D.J., Kroken, S., Kasuga, T., Geiser, D.M., Hibbett, D.S. and Fisher, M.C. (2000). Phylogenetic species recognition and species concepts in fungi. *Fungal Genetics and Biology 31*: 21–32.

Taylor, J.W., Turner, E., Townsend, J.P., Dettman J.R. and Jacobson, D. (2006). Eucaryotic microbes, species recognition and the geographic limits of species: examples from the kingdom Fungi. *Phil. Trans. R. Soc. B. 361*: 1947-1963.

Tehler, A., Farris, J.S., Lipscomb, D.L. and Kallersjo, M. (2000). Phylogenetic Analyses of the Fungi Based on Large rDNA Data Sets. *Mycologia 92*: 459-474.

Thompson, J.D., Gibson, T.J., Plewniak, F., Jeanmougin, F. and Higgins, D.G. (1997). The Clustal X windows interface: flexible strategies for multiple sequence alignment aided by quality analysis tools. *Nucleic Acids Res. 25*: 4876-4882.

van de Peer, Y., Baldauf, S.L., Doolittle,W.F. and Meyer, A. (2000). An updated and comprehensive rRNA phylogeny of (crown) eukaryotes based on rate calibrated evolutionary distances. *Journal of Molecular Evolution 51*: 565-576.

Vilgalys, R. and Hester, M. (1990). Rapid genetic identification and mapping of enzymatically amplified ribosomal DNA from several *Cryptococcus* species. *J. Bacteriol. 172*: 4238–4246.

Vogel, H.J. (1964). Distribution of lysine pathways among fungi: evolutionary implications. *Am. Nat. 98*: 435–446.

Voigt, K., Cigelnik, E. and O`Donnell, K. (1999a). Phylogeny and PCR identification of clinical important Zygomycetes based on nuclear ribosomal-DNA sequence data. *J. Clin. Microbiol. 37*: 3957-3964.

Voigt, K., Matthäi, A. and Wöstemeyer, J. (1999b) Phylogeny of zygomycetes – a molecular approach towards systematics of Mucorales. *Cour. Forsch.-Inst. Senckenberg 215*: 207-213.

Voigt, K. and Wöstemeyer, J. (2001). Phylogeny and origin of 82 zygomycetes from all 54 genera of the Mucorales and Mortierellales based on combined analysis of actin and translation elongation factor EF-1alpha genes. *Gene 270*: 113-120.

Voigt, K. and Olsson, L. (2008). Molecular phylogenetic and scanning electron microscopic analyses places the Choanephoraceae and the Gilbertellaceae in a monophyletic group within the Mucorales (Zygomycetes, Fungi). *Acta Biol. Hun. 59(3)*: in press.

von Arx, J.A. (1983). On Mucoraceae s. str. and other families of the Mucorales. *Sydowia 35*: 10-26.

Watanabe, T. (1994). Two new species of homothallic *Mucor* in Japan. *Mycologia 86*: 691-695.

Weising, K., Nybom, H., Wolff, K. and Meyer W. (1995). *DNA fingerprinting in plants and fungi*. CRC Press, Inc., Boca Raton, Fl.

Weitzman, I., Whittier, S., McKitrick, J.C. and Della-Latta, P. (1995). Zygospores: the last word in identification of rare and atypical zygomycetes isolated from clinical specimens. *J. Clin. Microbiol. 33*: 781–783.

Wheeler, Q.D. (2004). Taxonomic triage and the proverty of phylogeny. *Phil. Trans. R. Soc. B. 359*: 571-583.

White, T.J., Bruns, T., Lee, S. and Taylor, J. (1990). Amplification and direct sequencing of fungal ribosomal RNA genes for phylogenetics. *In*: *PCR protocols: a guide to methods and applications*. Innis, M.A., Gelfand, D.H., Sninsky, J.J., White, T.J. (eds.), Academic Press, San Diego., pp. 315-322.

White, M.M., James, T.Y., O`Donnell, K., Cafaro, M.J., Tanabe, Y. and Sugiyama, J. (2006). Phylogeny of the Zygomycota based on nuclear ribosomal sequence data. *Mycologia 98*: 872-884.

Wise, M.G., Healy, M., Reece, K., Smith, R., Walton, D., Dutch, W., Renwick, A., Huong, J., Young, S., Tarrand, J. and Kontoyiannis, P.D. (2007). Species identification and strain differentiation of clinical *Candida* isolates using the DiversiLab system of automated repetitive sequence-based PCR. *J. Med. Microbiol. 56*: 778-787.

Woese, C.R. (1987). Bacterial evolution. *Microbiol. Rev. 51*: 221–271.

Woese, C.R., Kandler, O. and Wheelis, M.L. (1990). Towards a natural system of organisms: proposal for the domains *Archaea, Bacteria*, and *Eucarya. Proc. Natl. Acad. Sci. USA 87*: 4576-4579.

Wolf, F.A. (1917). A squash disease caused by *Coanephora cucurbitarum. J. Agric. Res. 8*: 319-327.

Wöstemeyer, J. (1985). Strain dependent variation in ribosomal DNA arrangement in *Absidia glauca. Eur. J. Biochem. 146*: 443-448.

Wuczkowski, M., Druzhinina, I., Gherbawy, Y., Klug, B. and Prillinger, H. (2003). Species pattern and genetic diversity of *Trichoderma* in mid-european, primeval floodplain-forest. *Microbiol. Res. 158*: 125-133.

Zheng, R.Y., Chen, G.Q., Huang, H. and Liu, X.Y. (2007). A monograph of *Rhizopus. Sydowia 59*: 273-372.

Zycha, H., Siepmann, R. and Linnemann, G. (1969). Mucorales. Eine Beschreibung aller Gattungen und Arten dieser Pilzgruppe. J. Cramer, Lehre.

In: Current Advances in Molecular Mycology
Eds: Y. Gherbawy, R.L. Mach and M.K. Rai

ISBN 978-1-60456-909-4

Chapter XII

REVISION OF THE FAMILY STRUCTURE OF THE MUCORALES (MUCOROMYCOTINA, ZYGOMYCETES) BASED ON MULTIGENE-GENEALOGIES: PHYLOGENETIC ANALYSES SUGGEST A BIGENERIC PHYCOMYCETACEAE WITH *SPINELLUS* AS SISTER GROUP TO *PHYCOMYCES*

***K. Voigt*[1,*], *K. Hoffmann*[1], *E. Einax*[1,&], *M. Eckart*[1], *T. Papp*[2], *C. Vágvölgyi*[2] *and L. Olsson*[3]**

[1]University of Jena, Institute of Microbiology, Fungal Reference Centre, 07743 Jena, Germany;

[2]University of Szeged, Faculty of Sciences, Department of Microbiology, 6726 Szeged, Hungary;

[3]University of Jena, Institute of Systematic Zoology and Evolutionary Biology, 07743 Jena, Germany.

ABSTRACT

Zygomycetes encompass microscopic fungi forming zygospores in sexual interactions. The most prominent and largest order is the Mucorales comprising saprotrophic and facultatively parasitic species. The traditional family system of the Mucorales deviates substantially from molecular phylogenies. Multi-gene genealogies

---

* Correspondence concerning this article should be addressed to: Kerstin Voigt, Friedrich-Schiller-Universität Jena, Institut für Mikrobiologie, Pilz-Referenz-Zentrum, Neugasse 24, D-07743 Jena, Germany. Phone: +49 3641 949321; Fax: +49 3641 949322; E-mail: kerstin.Voigt@uni-jena.de.

& present address: Thüringer Tierseuchenkasse, Victor-Goerttler-Str. 4, 07745 Jena, Germany.

based on Bayesian inference analyses of the exonic genes for actin (*act*), beta-tubulin (*btub*) and translation elongation factor 1 alpha (*tef*), the nuclear genes for the small (18S) and large (28S) subunit ribosomal RNA (comprising 811, 1168, 1112, 871, 402 characters, respectively) of twenty-seven genera of the Mucorales shows that the traditional family structure is highly unnatural at the level of groups with monophyletic origin. Four families can be considered to be monophyletic; the Umbelopsidaceae, the Phycomycetaceae, the Absidiaceae and the Choanephoraceae including *Gilbertella*. The establishment of a natural system suggests *Spinellus* as the closest phylogenetic relative to *Phycomyces*, which converts the former monogeneric family Phycomycetaceae to becoming bigeneric. The Mucoraceae and the Thamnidiacaeae, both representing the largest families, are polyphyletic and therefore highly unnatural. Special emphasis was given to the determination of micromorphological and ultrastructural apomorphies derived by light and scanning electron microscopy. Furthermore, phylogenetic reconstructions based on a combined profile distance analysis of small and large subunit rDNA (1431 and 387 characters, respectively) of seventy-two taxa suggest a close phylogenetic relationship between the Mucorales and the Entomophthorales. The Mortierellales, which has been recently classified to the Mucormycotina, appear as the basal group of the Dikarya.

Key words: ribosomal DNA (rDNA); nuclear 18S (SSU) and 28S (LSU); translation elongation factor 1 alpha; actin; beta tubulin.

Introduction

Zygomycetes are a fascinating and versatile group of true fungi (Mycota) of immense biological importance. Thought to be amongst the first terrestric fungi to colonize the earth, they evolved different ways of interacting with fungi of other groups, plants, animals or man ranging from parasitic towards symbiotic reactions. Morphologically, their mycelia can differentiate either to asexually developing sporangia or sporangiola or, once a compatible mating partner is available and the environmental conditions are favourable, to sexually formed zygospores. This makes them ideally suited as model systems for studying fundamental evolutionary and developmental processes. The Mucorales Fries 1832 represent one of the most prominent orders of the zygomycetes (Fries, 1832). Recently the order Mucorales gained the rank of a division and the term Mucoromycotina was introduced (James *et al*., 2006a). Members of this particular group have biotechnological importance in the biotransformation of steroids, the production of rennin-like cheese-clotting proteases and in the fermentation of food, especially soy protein (e.g. soy souce, tempeh). In addition they produce an array of vitamin-like compounds like carotenes or carotenoids. The amazing versatility of the Mucorales has made them interesting to scientists in recent years. Given that three genome sequences (*Rhizopus oryzae*: http://www.broad.mit.edu/annotation/genome/rhizopus_oryzae/Home.html, *Phycomyces blakesleeanus*: http://genome.jgi-psf.org/Phybl1/Phybl1.home.html, *Mucor circinelloides*: http://mucorgen.um.es/) have been completed and many phylogenetic projects were finalized or are presently underway, the point has been reached where there is an urgent need to summarize and review, but also to revise, the current systematics and phylogeny of these interesting microorganisms.

Table 1. Nucleotide sequences of genes encoding 18S and 28S rRNA, beta-tubulin, translation elongation factor EF1-alpha and actin, which were used for the reconstruction of the 5-gene phylogeny shown in figure 2. GenBank accession numbers in bold were generated during this study.

no.	species and classification	Genebank accession number				
		18S rDNA	28S rDNA	beta-tubulin	EF1-alpha	actin
	Rhodophyta					
1	*Cyanidioschyzon merolae*	AB158485	AB158485	AB095181	AB095182	AB095179
	Viridiplantae					
2	*Arabidopsis thaliana*	X52322	X52322	AY054693	AY133532	NM_001036427
3	*Oryzae sativa*	AF069218	M11585	X79367	AF030517	X15863
	Metazoa					
4	*Caenorhabditis* sp.	X03680	X03680	U55260	NM_076922	X16796
5	*Drosophila melanogaster*	M21017	M21017	M20419	NM_165850	M18830
6	*Homo sapiens*	U13369	U13369	AF141349	NM_001402	NM_005159
7	*Xenopus laevis*	X02995	X02995	L06232	BC043843	BC099316
	Ascomycota					
8	*Neurospora crassa*	X04971	M38154	M13630	D45837	U78026
	Basidiomycota					
9	*Coprinus cinereus*	M92991	AF041494	AB000116	DN593200	AB034637
10	*Schizophyllum commune*	X54865	AF334751	X63372	X94913	AF156157
	Zygomycota, Mortierellales					
11	*Dissophora decumbens*	AF157133	AF157187	AY944863	AF157247	AJ287155
12	*Mortierella verticillata*	AF157145	AF157199	AF162071	AF157262	AJ287170
	Zygomycota, Mucorales					
13	*Absidia glauca*	AF157118	AF157172	AY944776	X54730	AJ287135
14	*Actinomucor elegans*	AF157119	AF157173	AY944783	AF157229	AJ287137
15	*Blakeslea trispora*	AF157124	AF157178	AY944767	AF157235	AJ287143
16	*Chaetocladium brefeldii*	AF157125	AF157179	AY944766	AF157236	AJ287144
17	*Chlamydoabsidia padenii*	AF113415	AF113453	AY944785	AF157238	AJ287146
18	*Choanephora infundibulifera*	AF157127	AF157181	AY937398	AF157239	AJ287147
19	*Cokeromyces recurvatus*	AF113416	AF113454	**AY944812**	AF157242	AJ287150
20	*Ellisomyces anomalus*	AF157134	AF157188	**AY944815**	AF157249	AJ287157
21	*Fennellomyces linderi*	AF157135	AF157189	AY944817	AF157250	AJ287158
22	*Gilbertella persicaria*	AF157136	AF157190	AY937400	AF157251	AJ287159
23	*Halteromyces radiatus*	AF157138	AF157192	AY944788	AF157253	AJ287161
24	*Mucor hiemalis*	AF113428	AF113468	AY937401	AF157265	AJ287174
25	*Mucor mucedo*	X89434	AF113470	AY944791	AF157267	AJ287176
26	*Mucor racemosus*	AF113430	AF113471	AY937402	AF157268	AJ287177
27	*Mycocladus corymbifer*	AF113407	AF113445	**AY944774**	AF157227	AJ287134
28	*Mycotypha africana*	AF157147	AF157201	AY944805	AF157271	AJ287180
29	*Parasitella parasitica*	AF157149	AF157203	AY944793	AF157273	AJ287182
30	*Phycomyces blakesleeanus*	AF157151	AF157205	AY944795	AF157275	AJ287184
31	*Pilaira anomala*	AF157152	AF157206	AY944806	AF157276	AJ287185
32	*Poitrasia circinans*	AF157155	AF157209	AY937399	AF157279	AJ287188
33	*Radiomyces spectabilis*	AF157157	AF157211	AY944808	AF157281	AJ287190
34	*Rhizopus microsporus*	AF157158	AF157212	AF162065	AF157288	AJ287197
35	*Rhizopus oryzae*	AF113440	AF113481	AY944797	AF157289	AJ287198
36	*Spinellus fusiger*	AF157159	AF157213	AY944800	AF157292	AJ287201

Table 1. (Continued)

no.	species and classification	Genebank accession number				
		18S rDNA	**28S rDNA**	**beta-tubulin**	**EF1-alpha**	**actin**
	Zygomycota, Mucorales					
37	*Syncephalastrum racemosum*	X89437	AF113484	AY944810	AF157295	AJ287204
38	*Syzygites megalocarpus*	AF157162	AF157216	AY944802	AF157296	AJ287205
39	*Thamnostylum piriforme*	AF157164	AF157218	AY944819	AF157298	AJ287207
40	*Umbelopsis isabellina*	AF157166	AF157220	AY944824	AF157300	AJ287209
41	*Umbelopsis ramanniana*	X89435	AF113463	AF162073	AF157258	AJ287166
42	*Utharomyces epallocaulus*	AF157168	AF157222	AY944807	AF157302	AJ287211
43	*Zygorhynchus* sp.	AF157170	AF157224	**AY944804**	AF157304	AJ287213

Experimental Strategies and Methods

Fungal Cultivation and Maintenance

The fungal strains were grown on 30.0 g L^{-1} malt extract agar supplemented with 5.0 g L^{-1} yeast extract and maintained in liquid nitrogen using 10-20 [v/v] % glycerol or 20 [v/v] % skimmed milk (Oxoid) aqueous solutions as cryoprotectants. All fungi are deposited at the Fungal Reference Centre (Institute of Microbiology, University Jena; http://www.prz.uni-jena.de) and available upon request.

Micromorphological Investigations were Performed using a Combination of Light and Scanning Electron Microsopy

Micromorphological and ultrastructural apomorphies as shown in figures 3 - 7 were investigated by Light microscopy (Axiophot, Zeiss, Germany) and by scanning electron microscopy (Philips XL 30 ESEM, The Netherlands and Emitech K500 sputter coater, England) in accordance to Voigt and Olsson (2008).

Amplification and DNA Sequencing of Molecular Marker Genes

The purification of genomic DNA, the PCR amplification using forward primer βtub1 5`CAR GCY GGT CAR TGY GGT AAC CA 3`and the reverse primer βtub4r 5`GC CTC AGT RAA YTC CAT YTC RTC CAT 3`, cloning and sequencing of beta-tubulin fragments was previously described by Einax and Voigt (2003). GenBank accession numbers of the sequences generated in this study are listed in table 1 and refer to the accession numbers in bold print.

Table 2. Nucleotide sequences of genes encoding 18S and 28S rRNA, which were in addition to the rDNA sequences of table 1 used for the reconstruction of the phylogeny shown in figure 1

no.	species and classification	Genebank accession number	
		18S rDNA	28S rDNA
	Metazoa		
44	*Ciona intestinalis*	AB013017	AF212177
	Ascomycota		
45	*Candida albicans*	AF114470	U45776
46	*Schizosaccharomyces pombe*	X58056	Z19136
	Basidiomycota		
47	*Hericium erinaceum*	AF146778	AJ406494
48	*Trametes versicolor*	AY705965	AJ406538
	Zygomycota, Endogonales		
49	*Endogone lactiflua*	DQ536471	DQ273788
50	*Endogone pisiformis*	DQ322628	DQ273811
	Zygomycota, Entomophthorales		
51	*Basidiobolus ranarum*	AF113414	AF113452
52	*Conidiobolus coronatus*	AF113417	AF113455
53	*Conidiobolus lamprauges*	AF113420	AF113458
54	*Entomophthora muscae*	AY635820	DQ273772
	Zygomycota, Harpellales		
55	*Capniomyces stellatus*	AF007531	AF031073
56	*Furculomyces boomerangus*	AF277013	AF031074
57	*Smittium culisetae*	D29950	AF031072
	Zygomycota, Kickxellales		
58	*Coemansia reversa*	AF007533	AF031067
59	*Kickxella alabastrina*	AF007537	AF031064
60	*Linderina pennispora*	AF007538	AF031063
	Zygomycota, Mortierellales		
61	*Lobosporangium transversale*	AF113424	AF113462
	Zygomycota, Mucorales		
62	*Gongronella butleri*	AF157137	AF157191
63	*Pilobolus umbonatus*	AF157153	AF157207
64	*Umbelopsis nana*	AF157167	AF157221
	Zygomycota, Zoopagales		
65	*Kuzuhaea moniliformis*	AB016010	DQ273796
66	*Piptocephalis corymbifera*	AB016023	AY546690
	Chydridiomycota		
67	*Chytriomyces* sp.	M59758	AY349065
68	*Gonapodya* sp.	AF164329	AY349059
69	*Monoblepharis* sp.	AY349029	AY349061
70	*Rhizophydium* sp.	AF164266	DQ485549
	Glomeromycota		
71	*Geosiphon pyriformis*	X86686	AM183920
72	*Gigaspora margarita*	AM181143	AF396783
73	*Glomus clarum*	AJ852597	AF396791
74	*Glomus intraradices*	AJ852526	AJ854583
75	*Glomus mosseae*	AJ306438	DQ469126

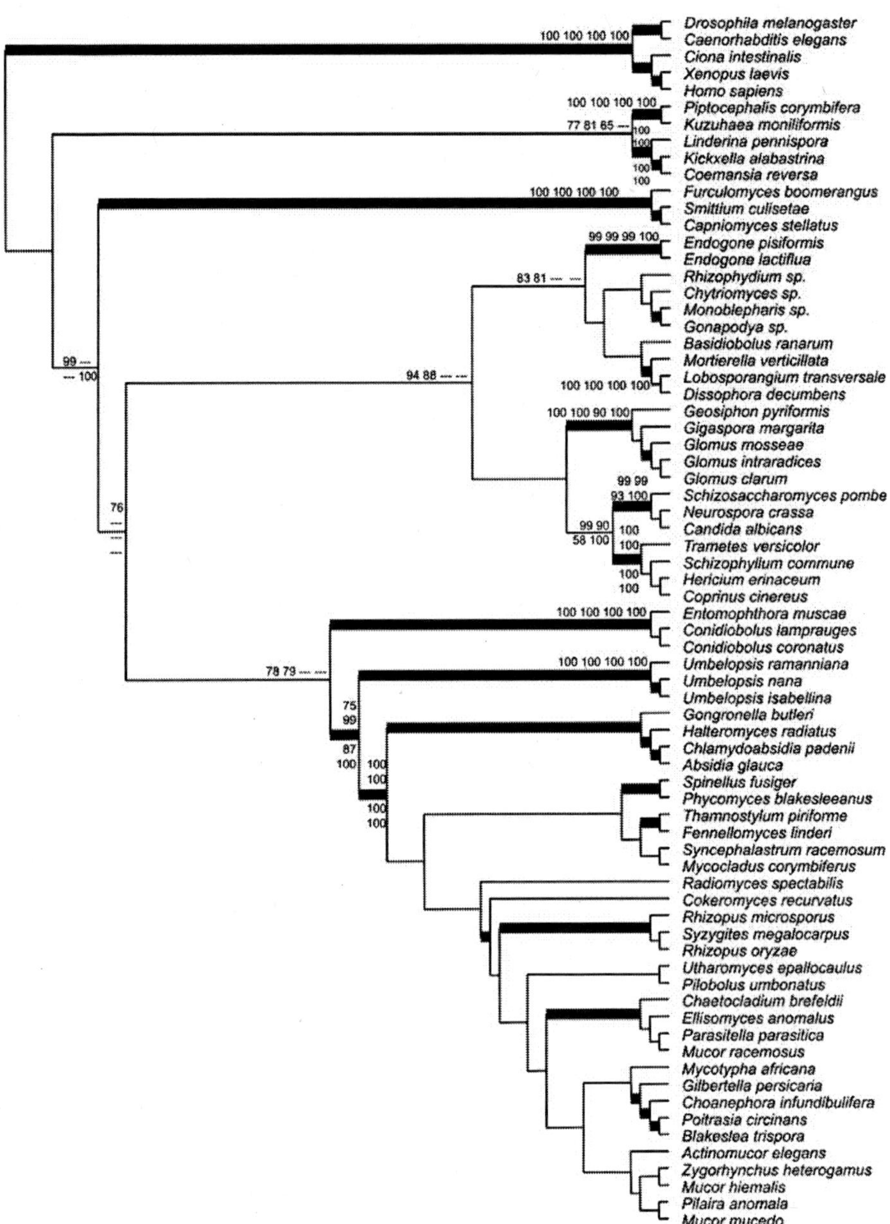

Figure 1. Profile Distance phylogram based on a combined alignment of 18S and 28S rDNA nucleotide sequences (spanning 1431 and 387 characters, respectively) from 72 taxa. Branch supports (bootstrap proportions and posterior probabilities) derived from four different analyses are given at the branches in the following order: Bootstrap proportions (BPs) from Profile Distance, BPs from conventional neighbor-joining, posterior probabilities (PP) from Bayesian inference, and BPs from Maximum Likelihood analyses. Branches in bold print indicate branch supports greater than or equal to 75% in all four calculations. The phylogram was rooted to the Metazoa (*Drosophila melanogaster*, *Caenorhabditis elegans*, *Ciona intestinalis*, *Xenopus laevis* and *Homo sapiens*), which is widely accepted as the monophyletic sistergroup of the chitinous fungi.

Phylogenetic tree Reconstructions Based on Bayesian Inference, Maximum-Likelihood, Profile and Conventional Distance Analyses

Forty-three nucleotide sequences of 18S and 28S rDNA, beta-tubulin (*btub*), translation elongation factor EF1-alpha (*tef*) and actin (*act*) are listed in table 1 and used for the five-gene phylogeny shown in figure 2. Additionally, thirty-two nucleotide sequences of 18S and 28S rDNA (Table 2) were retrieved from GenBank (http://ncbi.nlm.nih.gov) and used for phylogenetic analyses in conjunction with the rDNA sequences of table 1 (except those of the Rhodophyta and the Viridiplantae). Alignments were carried out in ClustalX version 1.83 (Higgins and Sharp, 1988, 1989; Thompson *et al.*, 1997) and were visualized in BioEdit version 7.0.3 (Hall, 1999) for manual control. The reconstruction of the five-gene phylogenies was conducted based on an alignment from 43 taxa, each with 871, 402, 811, 1168 and 1112 characters for 18S and 28S rDNA, *act*, *btub* and *tef*, respectively. The combined dataset of 4364 aligned nucleic acid characters in total was subjected to Bayesian inference (Huelsenbeck and Ronquist, 2001) and distance reconstructions using PAUP* v4.0b10 (Swofford, 1998) and ProfDist version 0.9.6 (Müller *et al.*, 2004; Friedrich *et al.*, 2005). The profiles were generated automatically using Jukes-Cantor as the underlying distance correction model. Minimal bootstrap values of 75 % and 95 % identity of uncorrected p-distances were used as thresholds for profile generation. Conventional distance analyses applying neighbor-joining (Saitou and Nei, 1987) of Jukes-Cantor distances were performed using PAUP* v4.0b10. Negative branch lengths were prohibited in this analysis. All other options were set to factory defaults. Bootstrap analyses (Felsenstein, 1985) with 1000 replicates of neighbor-joining searches on Jukes-Cantor distances were used to elucidate branch stability. Bayesian inference analyses were computed with MrBayes v3.0b4 (Huelsenbeck and Ronquist, 2001). Bootstrap proportions (BPs) enforced to the 50 % majority rule option were obtained by 1000 bootstrap replicates of Jukes-Cantor distances. Bayesian inference was initiated from a random starting tree and four Markov chains were run for 500,000 generations, with tree sampling frequency set to 500, printing frequency set to 1000. Initial log likelihoods were 108838.988204, 109157.490403, 108012.694722, 108839.764651 for chains 1 to 4, respectively. The first 25% of the generated trees were discarded (burn-in). The consensus tree was calculated using the halfcompat option. The node confidence values are presented as posterior probabilities (in percent). The resulting Bayesian inferred tree is shown in figure 2.

An alignment of 72 combined nucleotide sequences of 1431 characters from 18S rDNA and 387 characters from 28S rDNA were subjected to distance analyses using PAUP* v4.0b10 and ProfDist v0.9.6, maximum likelihood analysis using PAUP* and Bayesian inference using MrBayes v3.0b4. Profile distance (PD), neighbor-joining (NJ) and Bayesian Inference (BI) were carried out as described for the five-gene genealogy. The resulting profile neighbor-joining tree is shown in figure 1. To compare tree topologies inferred from different algorithms maximum likelihood (ML), neighbor-joining and Bayesian analyses (with initial log likelihoods of 68665.449956, 69791.602544, 68500.384517, 68563.660987 for Markov chains 1 to 4, respectively) were conducted on the same dataset. The trees resulting form those three additional analyses are not shown. Maximum likelihood was performed using a heuristic search. The number of substitution types was set to 2 corresponding to the HKY85 variant model. Transition/transversion ratio was 2 (kappa = 3.8665967). Assumed nucleotide frequencies were empirical with A=0.39018 C=0.15629 G=0.22199 T=0.23153. The distribution of rates at variable sites was equal. The number of distinct data patterns under this HKY85 model was 1248. A molecular clock was not enforced. Starting branch lengths were obtained using the Rogers-Swofford approximation method (Rogers and Swofford, 1998). Trees with approximate likelihoods 5% or further from the target score were rejected without additional iteration. Branch-length optimization was set to one-dimensional Newton-Raphson with pass limit = 20 and delta=1e-006. The starting tree was obtained *via* stepwise addition choosing as-is for the addition of the sequences. One tree was held at each step. The branch-swapping algorithm was tree-bisection-reconnection (TBR), the 'MulTrees' option was in effect, steepest descent option was not. Initial ´MaxTrees´setting was 1000 and auto-increased by 100. Branches collapsed (creating polytomies) if branch length was less than or equal to 1e-008. 115280 rearrangements were tried on the combined 18S and 28S rDNA data set. The score of the best tree found was 39266.29752. One tree was retained (not shown). Bootstrapping was carried out using 100 maximum likelihood replicates with a fast-heuristic search. All phylograms were displayed and printed in TreeView version 1.6.6. (Page, 1996). The application of different algorithms resulted in similar tree topologies using both datasets. Therefore only one tree is shown. The phylogenetic trees and the aligned data matrices including all accession numbers of the nucleotide sequences used in this phylogenetic analyses are deposited in TreeBase (http://www.treebase.org/treebase: study accession no. S2031, matrix accession nos. M3813 - M3814).

Combined Nuclear rDNA Phylogenies Suggest that the Order Mucorales is Phylogenetically Closely Related to the Order Entomophthorales

In accordance with several other phylogenies (Voigt *et al.*, 1999; Voigt and Olsson, 2008; Voigt and Wöstemeyer, 2001; O´Donnell *et al.*, 2001; Tanabe *et al.*, 2004; James *et al.*, 2006b; Hibbett *et al.*, 2007) we find that the order Mucorales is a monophyletic group. This argument is strengthened by branch supports of 75%, 99%, 87% and 100% for BPs of PD, NJ, posterior probabilities (PPs) of BI and BPs of ML bootstrap analyses, respectively (Figure 1). The order Entomophthorales G. Winter 1880 appear to be the closest phylogenetic neighbour of the Mucorales, which may have evolved from the Entomophthorales in a paraphyletic manner. This relationship is supported exclusively by bootstrap analyses set to distance optimality criteria (BPs of 78% and 79% for PD and NJ bootstrap analyses, respectively), but not by likelihood bootstrap analyses like ML and BI. The mucoralean-entomophthoralean clade forms a monophyletic sister group (BP-PD=76%) to the Endogonales-Mortierellales-*Basidiobolus*-Chytridiomycota-Dikarya clade. That particular clade is supported by 94% and 88% in bootstrap analyses using distance trees. The Endogonales Moreau ex R.K. Benjamin 1979 (BP-PD, NJ and PP-BI = 99%; BP-ML=100%) and Mortierellales Cavalier-Smith 1998 (BP and PP=100%), together with the chytrids including *Basidiobolus*, form a monophyletic sister group to the dikaryan-glomeromycotan clade, a relationship which is however not statistically supported. The Dikarya (BP-PD=99%; BP-NJ=90%; PP-BI=58%; BP-ML=100%) comprise Ascomycota (BP-PD, NJ=99%; PP-BI=93%; BP-ML=100%) and Basidiomycota (BP and PP=100%) as monophyletic sister groups. The presence of a dikaryotic stage within the ontogeny of these fungi seems to be a stable synapomorphy for the Dikarya (as used in James *et al.*, 2006a,b and established in Hibbett *et al.*, 2007). The Glomeromycota, which were traditionally grouped as Glomales J.B. Morton & Benny 1990 to the class Zygomycetes but are now treated as a separate phylum as proposed by Schüßler *et al.* (2001), form a well established monophylum (BP-PD, NJ=100%; PP-BI=90%; BP-ML=100%). The exclusion of the Glomeromycota from the Zygomycetes represented the first step of a higher rank classification, which is adapted to a natural system. Since the phylum Zygomycota including the Trichomycetes, which possess orders grouping among the Mesomycetozoa (Cafaro, 2005) and is here represented by the harpellalean clade (BP and PP=100%), contains a vast variety of polyphyletic groups, efforts have been made to classify the monophyletic groups into separate subphyla, e.g. the Endogonales, the Mortierellales and the Mucorales into the Mucoromycotina and the Entomophthorales into the Entomophthoromycotina (Hibbett *et al.*, 2007). In contradiction to that and the RPB1(DNA dependent RNA polymerase II largest subunit)-based phylogeny of Tanabe *et al.* (2004), where *Mortierella* was resolved as the most basal divergence within the Mucorales, our analyses suggest that the Mucorales and the Entomophthorales should be grouped to the Mucoromycotina but the Mortierellales and the Endogonales should be grouped separately as a lineage basal to the derived fungi like the Dikarya and the Glomeromycota. Here the elimination of the subphylum Entomophthoromycotina and the subphylum Mortierellomycotina subphyl. nov. is proposed. The obligately mycoparasitic Piptocephalidaceae J. Schröter 1886 (BP and PP=100%) and the Kickxellales Kreisel ex R.K. Benjmain 1979 (BP and PP=100%) group together in one monophylum (BP-PD=77%; BP-NJ=81%, PP-BI=65%), indicating a close relationship between Zoopagales Bessey ex R.K. Benjamin 1979 and Kickxellales. All these phylogenetic connections show that the higher rank level grouping among the Zygomycota is far from being solved and that, following the intentions of Cavalier-Smith (1998), the subphyla should be given the rank of phyla: the regularly septated (with lenticular cavity) Kickxellomycota phyl. nov. (incl. the Kickxellales-Harpellales-Dimargaritales clade, Tanabe *et al.*, 2004), the unseptated Mucoromycota phyl. nov. (incl. Mucorales and Entomophthorales), the irregularily septated Mortierellomycota phyl. nov. (incl. Mortierellales and Endogonales), and the monoordinal Zoopagomycota phyl. nov. consisting of the Zoopagales.

Phylogenetic Relationships within the Order Mucorales Support the Umbelopsidaceae, the Absidiaceae *Sensu Stricto*, the Choanephoraceae *Sensu Lato* and the Phycomycetaceae as Natural Groups

Based on the combined LSU and SSU analysis shown in figure 1 and the five-gene phylogeny demonstrated in figure 2 the Umbelopsidaceae (Meyer and Gams, 2003) represent the most basal monophyletic lineage within the Mucorales, a relationship which is supported to 100% in all four branch stability analyses (BP-PD, NJ, ML and PP-BI).

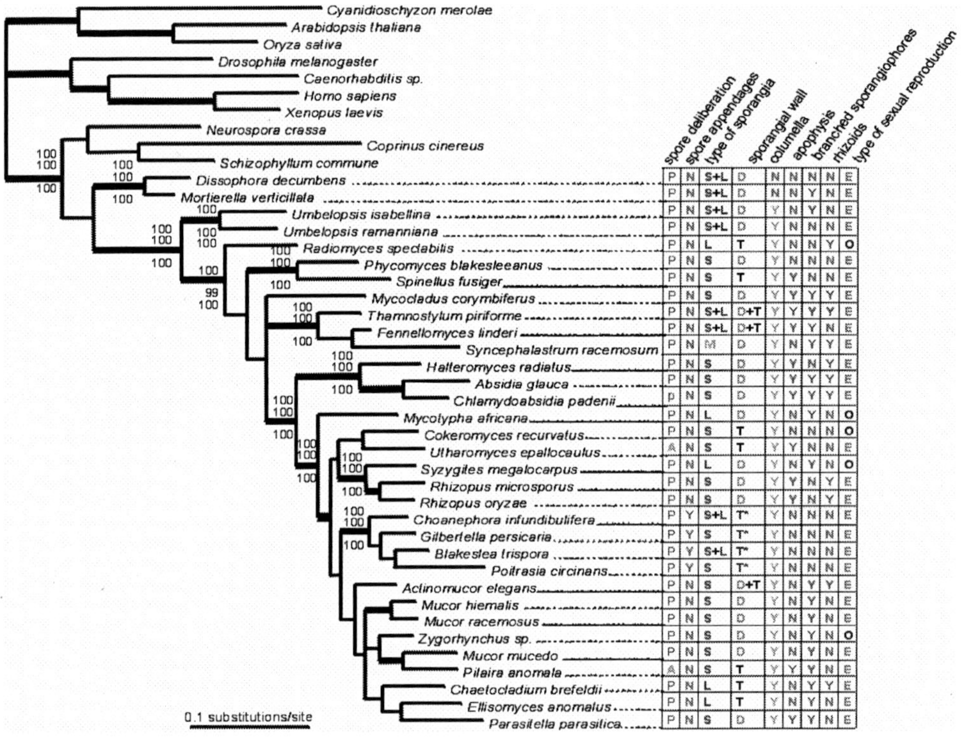

Figure 2. Five-gene phylogeny inferred by Bayesian analysis based on a combined alignment of 18S rDNA, 28S rDNA, actin, beta-tubulin and EF1-alpha nucleotide sequences from 43 taxa with 871, 402, 811, 1168 and 1112 characters, respectively. Posterior probabilities from the Bayesian analysis and Bootstrap proportions (BPs) from Profile Distances and neighbor-joining analyses are given at the branches corresponding to the same order from top to bottom. Bold branches indicate a BP greater than or equal to 75 % in all three analyses. Green plants (*Cyanidioschyzon merolae*, *Arabidopsis thaliana* and *Oryza sativa*) were used as outgroup taxa. For the zygomycetes (Mortierellales and Mucorales) morphological characters are indicated such as the mechanism of spore deliberation (A…active or P…passive), the presence of spore appendages (Y…yes or N…no), the type of sporangia (S…sporangia, L…sporangiola, M…merosporangia), the consistency of the sporangial wall (D…deliquescent, T…persistent, T*…persistent and half splitting), the presence of a columella, the presence of an apophysis, the presence of rhizoids, branching of sporangiophores (Y…yes, N…no) and the type of sexual reproduction (O…homothallic, E…heterothallic, ?...unknown).

Furthermore, the Absidiaceae *sensu stricto* (containing only the mesophilic representatives of the *Absidia* spp.) and the Choanephoraceae *sensu lato* (including the monotypic Gilbertellaceae) always form well-supported monophyletic groups in phylogenetic analyses utilizing two different data sets and four different algorithms. Recently, it became evident that a trichotomous grouping among the *Absidia* spp. and their closest relatives like *Chlamydoabsidia*, *Halteromyces* and *Gongronella* is concordant with the morphology of the zygospores, namely sterile hair–like, mycelial appendages on the suspensors of the zygospores in the mesophilic group, smooth-walled zygospores surrounded by equatorial rings in the thermotolerant group and *Mucor*-like rough-walled zygospores in the mycoparasitic group. Based on these findings the introduction of two new families for separation of the mesophilic Absidiaceae Arx 1982 from the thermotolerant and mycoparasitic *Absidia* spp. was proposed by Hoffmann *et al.* (2007) and a new family the Mycocladiaceae K. Hoffmann, S. Discher & K. Voigt 2007 was erected.

In addition to this surprising relationship, the common monophyletic origin of the Choanephoraceae–Gilbertellaceae clade could be confirmed in both analyses. Sporangia with persistent, sutured walls splitting in half at maturity and longitudinally striated sporangiospores with polar clusters of hair-like (ciliate) appendages represent synapomorphic characters of this group. The ultrastructure of sporangia, sporangiola and sporangiospores show several morphological characters (e.g. longitudinal sutures splitting the sporangial walls, striations and polar ciliate appendages on the sporangiospores), which provide evidence in support of a common monophyletic origin of the Gilbertellaceae Benny 1991 and the Choanephoraceae J. Schröter 1894 supporting the fusion of both families into the Choanephoraceae *sensu lato* (Voigt and Olsson, 2008).

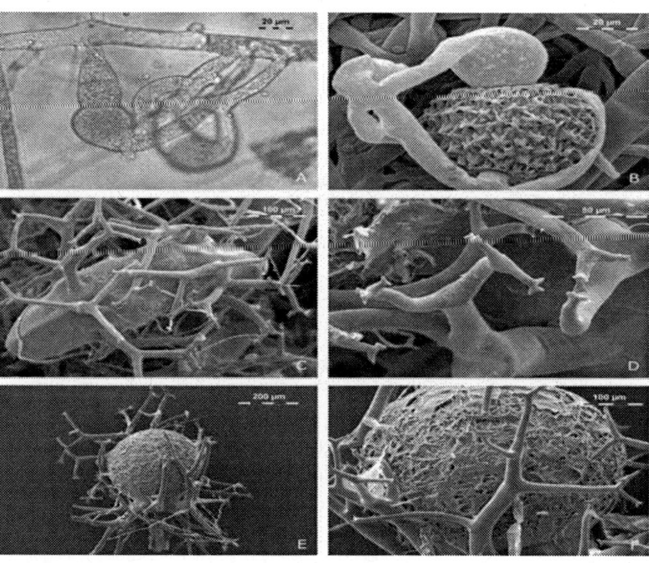

Figure 3. Light (A) and scanning electron (B-F) micrographs of zygospores from *Zygorhynchus heterogamus* (A), *Zygorhynchus moelleri* (B) and *Phycomyces blakesleeanus* (C-F) represented by different images: young zygospore following fusion of progametangia (C), dichotomously branched appendices arising from suspensors (D), mature zygospore (E), surface of the epispore (F). Values on scale bars indicate magnification.

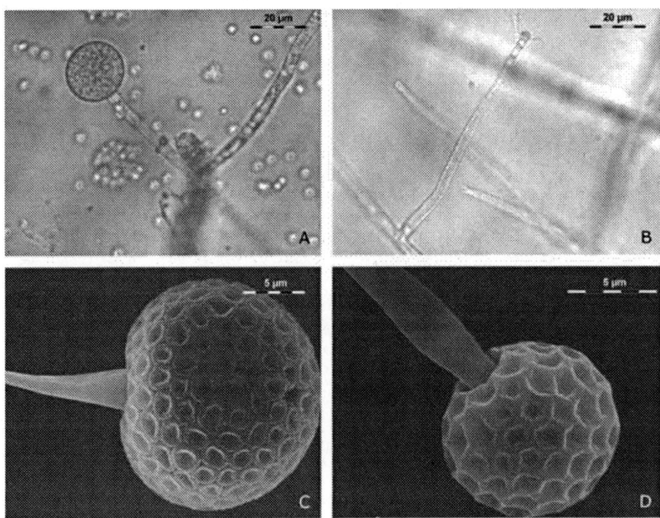

Figure 4. Light (A-B) and scanning electron (C, D) micrographs of multi-spored sporangia from *Umbelopsis* spp. bearing no apophyses (A, C, D) and a tiny, buttonhole-like columella (B). The sporangial wall *Umbelopsis isabellina* FSU 4745 shows regular ornamentations (C, D). The lengths of scale bars indicate 20 and 5 micrometers for light and scanning electron microscopy, respectively.

The Phycomycetaceae Arx 1982, traditionally treated to be monogeneric with three species of the genus *Phycomyces* and micromorphologically characterized by the development of large, unbranched sporangiophores and during zygosporogenesis by the formation of coiled tong-like suspensors bearing branched appendages (Figure 3C-F, Kirk *et al.*, 2001), appears to include the genus *Spinellus*, which is in accordance to Arx (1982). *Spinellus* is widespread in northern temperate regions and encompasses five species, all parasitic on the basidiocarps of agaric fungi, especially *Mycena*. The sporangia of both genera can be distinguished by the presence of a well-developed apophysis, a sterile and funnel-like transition structure between sporangiophore and columella. Only species of the genus *Spinellus* have apophysate sporangia (Figure 2). In contrast, the sporangiophores of *Phycomyces* show strong positive phototropism, while the light-dependence of the sporangiophores of *Spinellus* is less pronounced, but both form large unbranched sporangiophores with large multi-spored sporangia of up to 1 mm in size. The less-pronounced positive phototropism of the apophysate sporangia and the lacking dichotomously branched appendages of the suspensors typic for *Phycomyces* zygospores might be micromorphological arguments why *Spinellus* was classified into the Mucoraceae.Physiological properties aiming at light-dependent morphogenesis, nutrition or thermotolerance like phototropism, coprophilism and thermophilism, respectively, are neither symplesio- nor synapomorphic and can be secondarily lost during evolution. For instance, *Phycomyces blakesleeanus*, *Mycotypha africana* and *Mucor mucedo* belong to phylogenetically distant clades (Figures 1, 2), but develop sporangiophores which bend towards visual light sources and are therefore positively phototropic. Also coprophilism, the ability of the microorganism to live and propagate on herbivore dung, evolved multiple times during the mucoralean radiation. *Utharomyces epallocaulus* and *Pilaira anomala*, both coprophilous, appear on different clades (Figures 1, 2). Moreover, thermotolerance also appears several times in the evolution of the Mucorales, e.g. *Mycocladus corymbifer* growing at temperatures up to 52°C (Hoffmann *et al.*, 2007), *Fennellomyces linderi*, which grows at 37°C, the threshold temperature value permissive for thermotolerant but suppressive for mesophilic mucoralean species (Hoffmann *et al*, 2007), and *Rhizopus microsporus* (especially, var. *oligosporus*) used in soy fermentation (tempeh) as well as *R. oryzae* appear on different well-supported major clades (Figure 2).

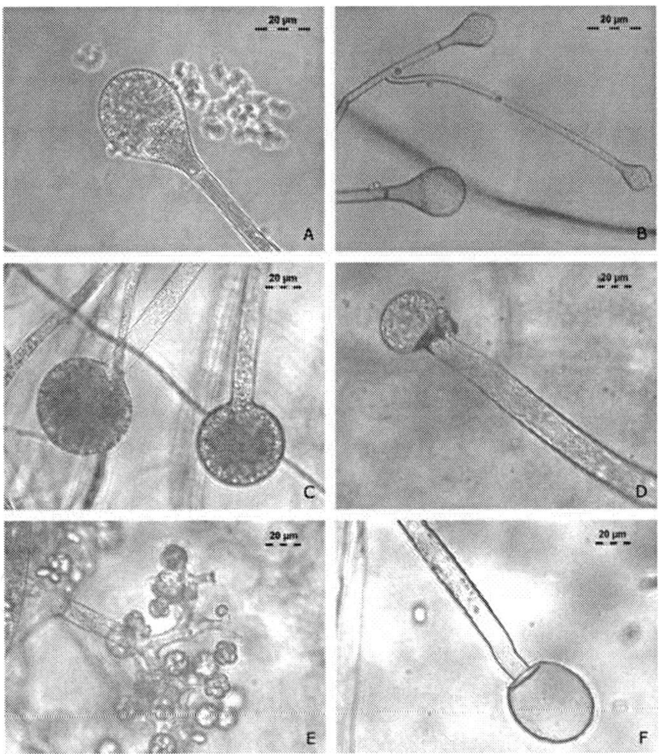

Figure 5. Light micrographs of multi-spored sporangia (A, C) and few-spored sporangiola (E) and columellae of sporangia (B, D, F) from *Absidia glauca* (A, B), *Zygorhynchus heterogamus* (C, D) and *Thamnidium elegans* (E, F). The lengths of scale bars indicate 20 micrometers in each image.

Phylogeny *Versus* Morphology: Evolutionary Implications of a five-gene Genealogy Based on the Elucidation of Morphological Synapomorphies with Phylogenetic Relevance

The five-gene phylogeny shown in figure 2 is an attempt to test the phylogenetic significance and relevance of micromorphological characters like (i) active or passive mechanism of asexually formed mitospore discharge, (ii) the presence or absence of spore appendages, (iii) the presence or absence of multi-spored sporangia (Figures 4A, C, D; 5A, C; 6A, B), few-/uni-spored sporangiola (Figures 5E; 6C, D) or merosporangia (Figure 7D) expressed by the type of sporangia, (iv) the consistency of the mitosporangial wall, (v) the presence or absence of a columella, which is a bulbous vesicle at the sporangiophore apex (Figures 4B; 5B, D, F), (vi) the presence or absence of an apophysis, (vii) the grade of branching of the sporangiophores, (viii) the presence or absence of rhizoids (Figure 6A) and (ix) the type of sexual reproduction which results in the formation of zygospores (homothallically, Figure 3A-B, or heterothallically formed zygospores, Figure 3C-F). Although the taxa included in our analysis are restricted, the use of Bayesian inference, profile and conventional distance algorithms gave similar tree topologies, which are highly suitable for the elucidation of higher-level phylogenetic relationships within the Fungi.

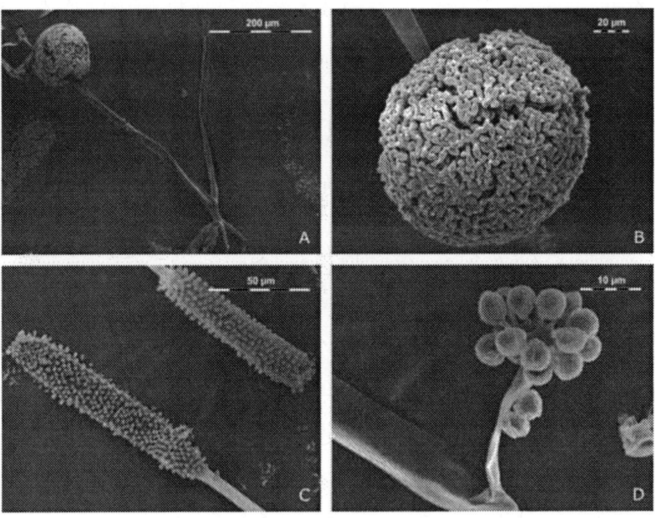

Figure 6. Scanning electron micrographs of multi-spored sporangia from *Rhizopus arrhizus* FSU 2822 (A, B) and uni-spored sporangiola (C, D) from *Mycotypha microspora* FSU 313 (C) and *Cunnighamella elegans* FSU 4755 (E, F). Sporangiophores in *Rhizopus arrhizus* arise from rhizoids (A). The lengths of scale bars indicate 20 micrometers in each image.

Focusing on the properties of asexually formed mitospores and their compartments, spore deliberation, sporangial type (sporangia or sporangiola) and wall (deliquescent or persistent) and appendices of sporangiophores like apophyses, branches/stolons or rhizoids have evolved multiple times during evolution of the Mucorales and are therefore not suitable for the characterization of natural groups (Figure 2). But appendaged mitospores in persistent sporangia bearing longitudinal sutures enabling the sporangia to split into two halves, which characterize the Choanephoraceae *sensu lato* (Voigt and Olsson, 2008), and the presence of a columella characterize monophyletic groups and are therefore suitable synapomorphies. Shape and size of the columella are phylogenetically not relevant. Tiny little columellae (Figure 4B) characterize the monogeneric Umbelopsidaceae (Meyer and Gams, 2003) with *Umbelopsis* as the type genus (Figure 4), a family that belongs to the Mucorales but is basal to the core Mucorales. Mucoralean multi-spored sporangia, so far formed, are always borne on a more or less well-developed columella. The moderately multi-spored sporangiola of the bigeneric family Radiomycetaceae Hesseltine & J.J. Ellis 1974 with *Radiomyces* as type genus and *Radiomyces spectabilis* Embree 1959 as type species are borne on complex ampullae, which are developmentally homologous with columellae. The sporangiola of the bigeneric family Mycotyphaceae Benny & R.K. Benjamin 1985 as well as the merosporangia of the monogeneric Syncephalastraceae Naumov ex R.K. Benjamin 1959 are borne on dehiscent pedicels, which can also be considered to be homologous to a columella. Merosporangia represent a special type of the few-spored sporangiola, in which the arrangement of the mitospores is row-like and surrounded by a sporangiolar wall (Figure 7D). The developmental stages of merosporangia formation is demonstrated in figure 7. Based on the phylogenies shown in figures 1 and 2, the hypothesis that the merosporangia-forming genus *Syncephalastrum* forms a bridge from basal mucoralean towards derived conidiogenous ascomycete fungi cannot be supported, a finding which is in accordance with Voigt *et al.* (1999), Voigt and Wöstemeyer (2001), O'Donnell *et al.* (2001), James *et al.* (2006b) and Voigt and Olsson (2008). Rather the Syncephalastraceae appear in close phylogenetic relationship to *Fennellomyces linderi* and *Thamnostylum piriforme* (BP=100%; Figure 2). Furthermore, it can be speculated that the pedicellate sporangiola of *Chaetocladium* and *Ellisomyces* are precursors of columellate sporangia. The family Thamnidiaceae Fitzpatrick 1930, comprising twelve genera with a total of 22 species, is characterized by diffluent, columellate sporangia and few- to one-spored, persistent walled, columellate sporangiola borne on the same or separate, morphologically identical sporangiophores or exclusively sporangiola-forming species (Kirk *et al.*, 2001). This family, and the family Mucoraceae Dumort. 1882, representing the two largest families of the order Mucorales, are highly polyphyletic (Figure 2). The fact that the majority of derived mucoralean fungi form sporangia or sporangia in combination with sporangiola shows that the amount of mitospores produced within an asexual propagation structure has no taxonomic significance. Mucoralean fungi, which only form sporangiola, are a minority. Subsequently, a tendency towards propagation *via* multi-spored can be observed in the evolution of the Mucorales (Figure 2).

For a long time, the evolution of an active spore deliberation mechanism was believed to be unique for the coprophilous family Pilobolaceae Corda 1842. But the trigeneric Pilobolaceae, containing the *Utharomyces*, *Pilaira* and the type genus *Pilobolus*, appear to be biphyletic (Figures 1, 2). While *Utharomyces epallocaulus* and *Pilobolus umbonatus* are monophyletic, *Pilaira anomala* clusters within a different clade and is closely related to *Mucor mucedo*, a relationship, which is confirmed by Voigt and Olsson (2008). Therefore, it can be concluded that an active spore deliberation device appeared at least twice in mucoralean evolution. The fact that active spore deliberation mechanisms have multiple evolutionary origins can be further substantiated by the presence of forcibly discharged spores of the Entomophthorales, the order phylogenetically closest to the Mucorales.

The formation and type of zygospores in an interaction between self-fertile (homothallic) or self-sterile (heterothallic) compatible mating individuals has been a controversial subject. The type of zygospore (smooth-, rough- or warty-walled episporous) and its suspensors (opposed, coiled, tong-like apposed, bearing branched or unbranched appendages) has been used in taxonomy. The taxonomic discrimination based on zygospore morphology is appropriate and efficient in the case of the discrimination between the appendaged suspensor-forming Absidiaceae *sensu stricto*, the unappendaged suspensor-forming, smooth-walled zygosporous Mycocladiaceae and the unappendaged suspensor-forming, rough-walled mycoparasitic group of *Absidia* that deserves to be included in a separate genus or even family (Hoffmann *et al.*, 2007). But in the case of the Pilobolaceae and the Choanephoraceae, which were originally characterized by (smooth or striate, respectively) zygospore formation between tongs-like or apposed suspensors, zygospore morphology is not phylogenetically significant. *Gilbertella persicaria* (synonymous to *Choanephora persicaria* E.D. Eddy 1925; Mil'ko, 1967) develops non-choanephoraceous *Mucor*-type rough-walled, appendaged zygospores between two opposed suspensors. Therefore Hesseltine (1960) placed *Gilbertella* in the Mucoraceae, a view shared by Zycha *et al.* (1969), Arx (1982) and Kirk (1984). But Hesseltine and Ellis (1973) allied *Gilbertella* again with *Choanephora* and *Blakeslea* in the Choanephoraceae, a view supported by Voigt and Olsson (2008).

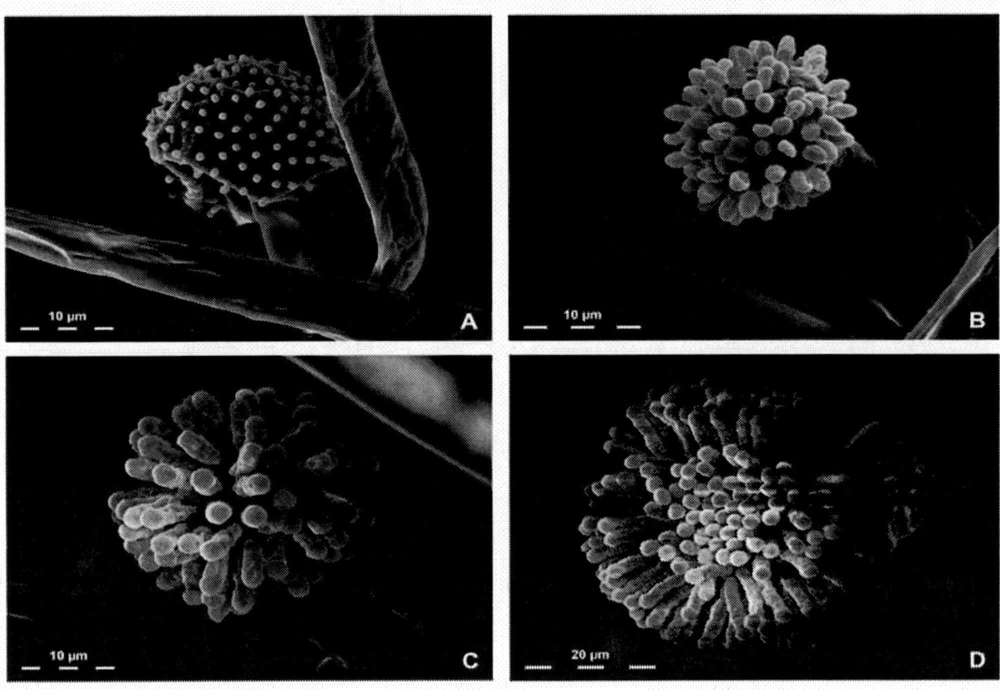

Figure 7. Scanning electron micrographs of merosporangia from *Syncephalastrum racemosum* at four different developmental stages: 1. initial phase of merosporangial development (A), 2. elongation of merosporangia (B), 3. differentiation (C) and maturation (D) of merosporangiospores. The lengths of scale bars indicate magnification.

Conclusions

1. Multi-gene phylogenies combined with micromorphological and ultrastructural characters provide a powerful tool for the delimitation of natural groups within the Mucoromycotina (formerly classified into the Zygomycetes).
2. The order Entomophthorales groups together with the Mucorales, but the orders Endogonales and Mortierellales group apart from the Mucorales, an observation, which suggests the elimination of the subphylum Entomophthoromycotina, the inclusion of the Entomophthorales within the Mucoromycotina and the erection of a

new subphylum, the Mortierellomycotina containing the Mortierellales and the Endogonales.
3. The close phylogenetic relatedness between Entomophthorales and Mucorales can be substantiated by the presence of active spore discharge mechanisms in all of the entomophthoralean and some of the mucoralean fungi.
4. The Mucoraceae, the Thamnidiaceae and the Pilobolaceae are polyphyletic.
5. Absidiaceae, Mycocladiaceae, Umbelopsidaceaee, Choanephoraceae and Phycomycetaceae, both *sensu lato*, are monophyletic and mostly consistent with traditional phenotype-based classification schemes.
6. Phylogenetic evidence was provided that the Phycomycetaceae, formerly monogeneric, can be converted to be bigeneric by the inclusion of the genus *Spinellus* in addition to *Phycomyces*, a finding which was earlier proposed by Arx (1982).
7. Columella, longitudinal sutures in the sporangial wall and mitospore appendages possess phylogenetic relevance.
8. Physiological parameters such as coprophilism, phototropism and thermotolerance as well as micromorphological characters like the development of an active spore deliberation mechanism, the type of sporangia and sporangial wall, the appearance of an apophysis, rhizoids and branched or stoloniferous sporangiophores, and the morphology and type of zygosporogenesis lack phylogenetic relevance.

Future Lines of Research

1. The bigeneric nature of the Phycomycetaceae needs extended testing and substantial exemplification.
2. The polyphyletic families Mucoraceae, Thamnidiaceae and Pilobolaceae deserve more attention to adapt their subgroups to natural clades.
3. Septal pore ultrastructure gains increasing importance in zygomycete phylogeny.
4. Initiatives such as the generation of multi (20-50)-gene genealogies within the Assembling the Fungal Tree of Life (AFTOL) project will provide an immense step towards a natural system of the basal fungal lineages including the Mucorales.

Acknowledgements

This work was supported by the Deutsche Forschungsgemeinschaft (VO 772/4-1,2 & 3) and by the Thüringer Ministerium für Wissenschaft, Forschung und Kunst to KH. The study would not have been realized without the inspiring interaction with colleagues, especially Imke Schmitt (University of Minnesota, Minneapolis, USA), Rytas Vilgalys and Andrej Gryganski (both at Duke University, Durham, NC, USA), from the *Assembling the Fungal Tree of Life* (AFTOL) project. We express our gratitude to Mrs. Gisela Baumbach for excellent assistance in strain maintenance and to Nadine Piekarski for her kind help with scanning electron microscopy.

REFERENCES

Arx, J. A. von (1982). On Mucoraceae s. str. and other families of the Mucorales. *Sydowia 35*: 10-26.

Cafaro, M.J. (2005). Eccrinales (Trichomycetes) are not fungi, but a clade of protists at the early divergence of animals and fungi *Mol. Phyl. Evol. 35*: 21-34.

Cavalier-Smith, T. (1998). A revised six-kingdom system of life. *Biol. Rev. 73*: 203-266.

Einax, E. and Voigt, K. (2003). Oligonucleotide primers for the universal amplification of beta-tubulin genes facilitate phylogenetic analyses in the regnum Fungi, *Org. Divers. Evol. 3*: 185-194.

Felsenstein, J. (1985). Confidence limits of phylogenies: An approach using the bootstrap. *Evolution 39*: 783-791.

Friedrich, J., Dandekar, T., Wolf, M. and Müller, T. (2005). ProfDist: a tool for the construction of large phylogenetic trees based on profile distances. *Bioinformatics 21(9)*: 2108-2109.

Fries, EM. (1832). Systema Mycologicum, Vol. 3. Ernest Mauriti, Greifswald pp. 261-524.

Hall, T.A. (1999). BioEdit: a user-friendly biological sequence alignment editor and analysis program for Windows 95/98/NT. *Nucl. Acids Symp. Ser. 41*: 95-98.

Hesseltine, C. W. (1960) *Gilbertella* gen. nov. (Mucorales). *Bulletin of the Torrey Botanical Club 87*: 21-30.

Hesseltine, C. W, Ellis, J. J. (1973) Mucorales. In: Ainsworth G. C., Sparrow F. K., Sussman A.S. Eds. *The fungi. An advanced treatise*, Vol. 4B. Academic Press, New York, pp. 187-217.

Hibbett, D.S., Binder, M., Bischoff, J.F., Blackwell, M., Cannon, P.F., Eriksson, O.E., Huhndorf, S., James, T., Kirk, P.M., Lücking, R., Thorsten, Lumbsch, H., Lutzoni, F., Matheny, P.B., McLaughlin, D.J., Powell, M.J., Redhead, S., Schoch, C.L., Spatafora, J.W., Stalpers, J.A., Vilgalys, R., Aime, M.C., Aptroot, A., Bauer, R., Begerow, D., Benny, G.L., Castlebury, L.A., Crous, P.W., Dai, Y.C., Gams, W., Geiser, D.M., Griffith, G.W., Gueidan, C., Hawksworth, D.L., Hestmark, G., Hosaka, K., Humber, R.A., Hyde, K.D., Ironside, J.E., Kõljalg, U., Kurtzman, C.P., Larsson, K.H., Lichtwardt, R., Longcore, J., Miadlikowska, J., Miller, A., Moncalvo, J.M., Mozley-Standridge, S., Oberwinkler, F., Parmasto, E., Reeb, V., Rogers, J.D., Roux, C., Ryvarden, L., Sampaio, J.P., Schüssler, A., Sugiyama, J., Thorn, R.G., Tibell, L., Untereiner, W.A., Walker, C., Wang, Z., Weir, A., Weiss, M., White, M.M., Winka, K., Yao, Y.J. and Zhang, N. (2007). A higher-level phylogenetic classification of the Fungi. *Mycol Res. 111*: 509-547.

Higgins, D.G. and Sharp, P.M. (1988). CLUSTAL: a package for performing multiple sequence alignment on a microcomputer. *Gene 73*: 237-244.

Higgins, D.G. and Sharp, P.M. (1989). Fast and sensitive multiple sequence alignments on a microcomputer. *Comput. Appl. Biosci. 5(2)*: 151-153.

Hoffmann, K., Discher, S. and Voigt, K. (2007). Revision of the genus *Absidia* (Mucorales, Zygomycetes): based on physiological, phylogenetic and morphological characters: Thermotolerant *Absidia* spp. form a coherent group, the Mycocladiaceae fam. nov. *Mycol. Res. 111*: 1169-1183.

Huelsenbeck, J.P. and Ronquist, F. (2001). MRBAYES: Bayesian inference of phylogeny. *Bioinformatics 17*: 754-755.

James, T. Y., Kauff, F., Schoch, C.L., Matheny, P.B., Hofsetter, V., Cox, C.J., Celio, G., Gueidan, C., Fraker, E., Maidlikowska, J.H., Lumbsch, T., Rauhut, A., Reeb, V., Arnold, A.E., Amtoft, A., Stajich, J.E., Hosaka, K., Sung, G.-H., Johnson, D., O'Rourke, B., Crockett, M., Binder, M., Curtis, J.M., Slot, J.C., Wang, Z., Wilson, A.W., Schassler, A., Longcore, J.E., O'Donnell, K., Mozley-Standridge, S., Porter, D., Letcher, P.M., Powell, M.J., Taylor, J.W., White, M.M., Griffith, G.W., Davies, D.R., Humber, R.A., Morton, J.B., Sugiyama, J., Rossman, A.Y., Rogers, J.D., Pfister, D.H., Hewitt, D., Hansen, K., Hambleton, S., Shoemaker, R.A., Kohlmeyer, J., Volkmann-Kohlmeyer, B., Spotts, R.A., Serdani, M., Crous, P.W., Hughes, K.W., Matsuura, K., Langer, E., Langer, G., Untereiner, W.A., Lacking, R., Badel, B., Geiser, D.M., Aptroot, A., Diederich, P., Schmitt, I., Schultz, M., Yahr, R., Hibbett, D.S., Lutzoni, F., McLaughlin, D.J., Spatafora, J.W. and Vilgalys, R. (2006a). Reconstructing the early evolution of fungi using a six-gene phylogeny. *Nature 443*: 818-822.

James, T.Y., Letcher, P.M., Longcore, J.E., Mozley-Standridge, S.E., Porter, D., Powell, M.J., Griffith, G.W. and Vilgalys, R. (2006b). A molecular phylogeny of the flagellated fungi (Chytridiomycota) and description of a new phylum (Blastocladiomycota). *Mycologia 98*: 860-871.

Kirk, P. M. (1984) A monograph of the Choanephoraceae. *Mycological Papers 152*: 1-61.

Kirk, P. M., Cannon, P. F., David, J. C. and Stalpers, J. A. (2001). *Ainsworth & Bisby's Dictionary of the fungi*. 9th ed. CABI Publishing, Wallingford.

Meyer, W. and Gams, W. (2003). Delimitation of *Umbelopsis* (Mucorales, Umbelopsidaceae fam. nov.) based on ITS sequence and RFLP data. *Mycol. Res. 107*: 339-350.

Mil'ko, A. A. (1967) Taxonomy and synonyms Mucorales. *Mikol. Fitopath. 1*: 26-35.

Müller, T., Rahmann, S., Dandekar, T. and Wolf, M. (2004). Accurate and robust phylogeny estimation based on profile distances: a study of the Chlorophyceae (Chlorophyta). *BMC Evol. Biol. 4*: 10.1 186/1471-2148-4-20.

O'Donnell, K. L, Lutzoni, F. M., Ward, T. J. and Benny, G. L. (2001) Evolutionary relationships among mucoralean fungi (Zygomycota): evidence for family polyphyly on a large scale. *Mycologia 93*: 286-297.

Page, R.D.M. (1996). TreeView: an application to display phylogenetic trees on personal computers. *Computational and Applied Biosciences 12*: 357-358.

Rogers, J.S. and Swofford, D.L. (1998). A fast method for approximating maximum likelyhood of phylogentic trees from nucleotide sequences. *Systematic Biology 47*: 77-89.

Saitou, N. and Nei, M. (1987). The neighbor-joining method: a new method for reconstructing phylogenetic trees. *Molecular Biology and Evolution 4*: 406-425.

Schüßler, A., Schwarzott, D. and Walker, C. (2001). A new fungal phylum, the Glomeromycota: phylogeny and evolution. *Mycol. Res. 105*: 1413-1421.

Swofford, D.L. (1998). PAUP*: Phylogenetic Analysis using Parsimony (*and Other Methods). Version 4. Sinauer Associates, Sunderland, MA.

Tanabe, Y., Saikawa, M., Watanabe, M.M. and Sugiyama, J. (2004). Molecular phylogeny of Zygomycota based on EF-1α and RPB1 sequences: limitations and utility of alternative markers to rDNA. *Mol. Phyl. Evol. 30*: 438-449.

Thompson, J.D., Gibson, T.J., Plewniak, F., Jeanmougin, F. and Higgins, D.G. (1997). The Clustal X windows interface: flexible strategies for multiple sequence alignment aided by quality analysis tools. *Nucleic Acids Res. 25*: 4876-4882.

Voigt, K., Cigelnik E. and O'Donnell, K. L. (1999). Phylogeny and PCR identification of clinically important zygomycetes based on nuclear ribosomal-DNA sequence data. *J. Clin. Microbiol. 37*: 3957-3964.

Voigt, K. and Olsson, L. (2008). Molecular phylogenetic and scanning electron microscopical analyses places the Choanephoraceae and the Gilbertellaceae in a monophyletic group within the Mucorales (Zygomycetes, Fungi). *Acta Biol. Hung. 59*: in press.

Voigt, K. and Wöstemeyer, J. (2001). Phylogeny and origin of 82 zygomycetes from all 54 genera of the Mucorales and Mortierellales based on combined analysis of actin and translation elongation factor EF-1α genes. *Gene 270*: 113-120.

Zycha, H., Siepmann, R., Linnemann, C. (1969) *Mucorales, eine Beschreibung aller Arten und Gattungen dieser Pilzgruppe*, J. Cramer, Lehre.

In: Current Advances in Molecular Mycology
Eds: Y. Gherbawy, R.L. Mach and M.K. Rai

ISBN 978-1-60456-909-4

Chapter XIII

Molecular Identification and Characterization of Indoor Wood Decay Fungi

O. Schmidt

Department of Wood Biology, University of Hamburg, Leuschnerstr. 91d, 21031 Hamburg, Germany.

Abstract

The chapter describes molecular techniques and results related to the identification and characterization of indoor wood-decay basidiomycetes, namely protein-based techniques such as SDS-PAGE and immunological assays and DNA-based techniques such as RAPD-analysis, use of rDNA for RFPLs, specific-priming PCR, DNA sequencing and analyses of microsatellites and AFLPs.

Key words: Indoor wood-decay fungi, molecular methods, characterization.

Introduction

The indoor wood-decay fungi are the economically most important wood-inhabiting fungi as they deteriorate wood at the end of the series `forestry - timber harvest - storage - woodworking - indoor use´ (Schmidt 2006). In Europe, most damage is caused by brown-rot fungi that degrade conifer wood. About 80 basidiomycetes have been found in 800 North German buildings within seven years by Huckfeldt (Table 1; Huckfeldt and Schmidt 2006).

Their biology and significance for buildings were described by Huckfeldt and Schmidt (2006) and Schmidt (2007). Emphasis is placed on *Serpula lacrymans* (Schmidt 2000), which dominates in most European countries. However, described methods can be applied to any fungus. For refurbishment and for scientific purpose, the identity of species should be known.

Molecular methods to identify and characterize organisms are based on the objective information deriving from the molecules of the target organism. Thus, molecular methods have been established since the 1980s to identify wood-decay fungi and for phylogenetic research (Jellison and Goodell 1988; Palfreyman et al. 1988; Schmidt and Kebernik 1989). This overview describes molecular techniques and results related to indoor basidiomycetes. Advantages and disadvantages of the techniques are mentioned.

Table 1. Species, type of rot, abundance and significance of indoor basidiomycetes found in 800 North German buildings during a seven-year investigation (after Schmidt 2007)

Species	Rot	Abundance	Significance
Serpula lacrymans (Wulfen) J. Schroet.	b	189	from cellar to roof (not on sunlit timber)
Coniophora puteana (Schumach.) P. Karst.	b	109	from cellar to roof
Donkioporia expansa (Desm.) Kotl. & Pouzar	w	75	from cellar to roof, windows
Antrodia spp.	b	27	from cellar to roof
Antrodia vaillantii (DC.) Ryvarden	b	23	from cellar to roof
Coprinus spp., 4 species	w	23	on walls, ceilings with reed-layer
Tapinella panuoides (Batsch) E.-J. Gilbert	b	22	cellar, below bathrooms, windows, roof
Oligoporus spp.	b	19	from cellar to roof
Asterostroma cervicolor (Berk. & Curtis) Massee	w	18	cellar, subfloor, roof, walls, ceilings
Coniophora marmorata Desm.	b	16	from cellar to roof
Gloeophyllum spp.	b	15	windows, doors, bathrooms, roof
Serpula himantioides (Fr.) P. Karst.	b	15	from cellar to roof
Oligoporus placenta (Fr.) Gilb. & Ryvarden	b	13	flooring, staircases, windows
Antrodia sinuosa (Fr.) P. Karst.	b	12	from cellar to roof
Gloeophyllum sepiarium (Wulfen) P. Karst.	b	12	windows, doors, roof
Antrodia xantha (Fr.) Ryvarden	b	11	roof
Gloeophyllum abietinum (Bull.) P. Karst.	b	11	windows, doors, bathrooms, roof
Trechispora spp.	w	9	windows, cellar, below bathrooms, roof
Dacrymyces stillatus Nees	b	8	windows, doors, facades
Leucogyrophana pinastri (Fr.) Ginns & Weresub	b	8	cellar, subfloor, beam-ends, bathrooms
Phanerochaete spp.	w	8	moist wood constructions
Phellinus contiguus (Pers.) Pat.	w	8	windows, half-timbering, shingle roofs
Trechispora farinacea (Pers.) Liberta	w	8	roof, below defect sanitary facilities
Grandinia spp., *Hyphoderma* spp., *Hyphodontia* spp.	w	7	cellar, roof, purlin, beams
Diplomitoporus lindbladii (Berk.) Gilb. & Ryvarden	w	6	subfloor, roof
Gloeophyllum trabeum (Pers.) Murrill	b	5	windows, doors
Leucogyrophana pulverulenta (Sowerby) Ginns	b	5	subfloor, beam-ends, bathrooms, kitchens
Lentinus lepideus (Fr.) Fr.	b	4	cellar, subfloor, beam-ends
Leucogyrophana mollusca (Fr.) Pouzar	b	3	cellar
Resinicium bicolor (Alb. & Schwein.) Parmasto	w	3	indoor chipboard
Trechispora mollusca (Pers.) Liberta	b	3	behind roof insulation
Antrodia serialis (Fr.) Donk	b	2	beam-ends in upper floor
Cerinomyces pallidus G.W. Martin	b	2	windows outside
Coniophora arida (Fr.) P. Karst.	b	2	floorboard
Cylindrobasidium laeve (Pers.) Chamuris	w	2	half-timbering in a mill
Fomitopsis rosea (Alb.: Schwein) P. Karst.	b	2	beams under floor

Table 1. (Continued)

Species	Rot	Abundance	Significance
Hyphoderma praetermissum (P. Karst.) J. Erikss. & Å. Strid	w	2	flooring, roof, windows outside
Leucogyrophana spp.	b	2	cellar
Pleurotus cornucopiae (Paulet) Rolland	w	2	wooden ship
Pleurotus ostreatus (Jacq.) P. Kumm.	w	2	rubbish below living space
Pluteus cervinus P. Kumm.	w	2	?
Tomentella sp. 2 species	-	2	cellar
Trametes hirsuta (Wulfen) Pilát	w	2	windows
Antrodia gossypium (Speg.) Ryv.	b	1	wall
Antrodia malicola (Berk. & M.A. Curtis) Donk	b	1	?
Asterostroma laxum Bres.	w	1	staircase
Bjerkandera adusta (Willd.) P. Karst.	w	1	beam
Botryobasidium spp.	w	1	cellar
Coniophora fusispora (Cooke & Ellis) Cooke	b	1	beam end
Crepidotus spp.	w	1	kitchen flooring
Fomitopsis pinicola (Sw.) P. Karst.	b	1	roof
Grifola frondosa (Dicks.) Gray	w	1	oak half-timbering with soil contact
Heterobasidion annosum (Fr.) Bref.	w	1	introduced with moist timber
Hyphodontia alutaria (Burt) J. Erikss.	w	1	purlin
Hyphodontia breviseta (P. Karst.) J. Erikss.	w	1	roof
Hyphodontia floccosa (Bourdot & Galzin) J. Erikss.	w	1	cellar
Hyphodontia nespori (Bres.) J. Erikss. & Hjortstam	w	1	roof
Leccinum sp.	-	1	cellar
Merulius tremellosus Schrad.	w	1	cellar
Oligoporus rennyi (Berk. & Broome) Donk	b	1	cellar ceiling
Phellinus pini (Brot.) Bondartsev & Singer	w	1	flooring timber
Pleurotus pulmonarius (Fr.) Quél.	w	1	wooden ship
Radulomyces confluens (Fr.) M.P. Christ.	w	1	window
Ramariopsis kunzei (Fr.) Corner	-	1	cellar
Schizophyllum commune Fr.	w	1	roof
Schizopora paradoxa (Schrad.) Donk	w	1	rafter, window
Stereum rugosum Pers.	w	1	oak half-timbering with soil contact
Trametes ochracea (Pers.) Gilb. & Ryvarden	w	1	window
Trametes versicolor (L.) Lloyd	w	1	window
Trechispora invisitata (H.J. Jacks.) Liberta	w	1	?
Trichaptum abietinum (Dicks.) Ryvarden	w	1	introduced with moist timber
Tubaria furfuracea (Pers.) Gillet	-	1	roof in extended stable
Volvariella bombycina (Schaeff.) Singer	w	1	introduced with moist timber

b = brown rot, w = white rot, - = assumably no wood-decay.

Protein-Based Techniques

SDS-PAGE

In SDS-PAGE (sodium dodecyl sulfate polyacrylamide gel electrophoresis), the whole cell protein is extracted from fungal tissue, denatured and negatively charged with mercaptoethanol and sodium dodecyl sulphate (SDS). The proteins are separated according to size on acrylamide gels and visualized by Coomassie blue or other stainings. The banding pattern obtained discriminates organisms at the species level and slightly below. The technique was able to distinguish a number of wood-decay fungi (Vigrow et al. 1991a; Palfreyman et al. 1991; Schmidt and Moreth 1995). For example, the closely related *Serpula lacrymans* and *S. himantioides* could be separated by their respective specific banding pattern (Schmidt and Kebernik 1989; Figure 1). SDS-PAGE is a rapid technique when the sample originates from a pure culture. The technique is advantageous to basic research. Due to the great number of protein bands of a sample, SDS-PAGE did not reach practical application to identify unknown samples of wood-decay fungi. Isozyme analyses are not available on the common indoor wood-decay fungi.

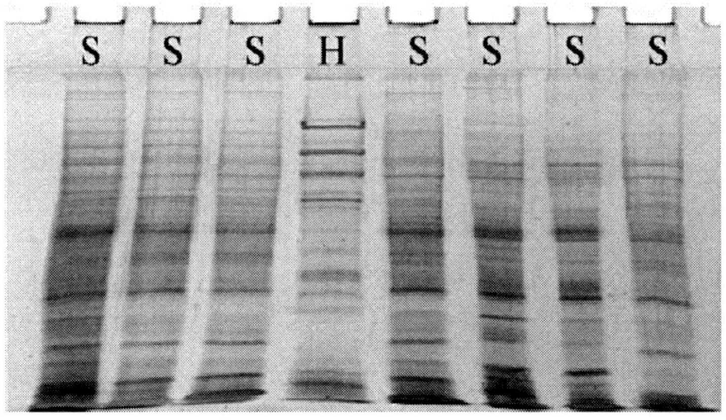

Figure 1. Protein bands of isolates of *Serpula lacrymans* (S) after SDS-PAGE. Culture H was identified by ITS sequencing to be *S. himantioides* (Schmidt and Kebernik 1989).

Immunological Methods

Immunological methods have been performed on several indoor wood-decay fungi, including *Coniophora puteana*, *Gloeophyllum trabeum*, *Lentinus lepideus*, *Oligoporus placenta* and *Serpula lacrymans* (e.g., Jellison and Goodell 1988; Palfreyman et al. 1988; Glancy et al. 1990, Vigrow et al. 1991b; Toft 1993; Clausen 1997). The diagnostic potential lies in the identification of species without the need of a prior isolation and pure culturing and in the detection of fungi at early stages of decay (Clausen and Kartal 2003). However, the experiments may exhibit cross-reactions with non-target organisms. Immunological methods have been also applied to visualize the distribution of enzymes of wood-decay fungi within and around the hypha and in woody tissue (e.g., Kim et al. 1991).

DNA-Based Techniques

Most investigations on wood-decay fungi use the polymerase chain reaction (*PCR*) to amplify the DNA of the sample, either in the traditional form or as multiplex PCR and real-time PCR (e.g., Vainio and Hantula 2000; Hietala et al. 2003; Eikenes et al. 2005). To visualize the PCR products and to assess their quality, agarose gel electrophoresis is commonly used.

RAPD Analysis

The technique of RAPD (randomly amplified polymorphic DNA) analysis uses only one, short (often 10mer) and randomly chosen PCR primer which anneals several times to the genome. Thus, RAPD analysis discriminates fungi by a polymorphic banding pattern at low taxonomical level, namely isolates, intersterility groups and species. Theodore et al. (1995) showed polymorphism among eight isolates of *Serpula lacrymans*. Similarity between other *S. lacrymans* isolates (Figure 2) was found by Schmidt and Moreth (1998). *Serpula himantioides* (Figure 2) and *Coniophora puteana* showed isolate polymorphism. RAPD analysis does not require prior information of the target DNA and is a rapid technique when starting from pure cultures. However, short primers imply an increased sensitivity to contamination. Several primers should be used to avoid spurious results. The technique is unsuited for species identification from unknown samples by comparison, as other, as not yet investigated fungi may by chance share a similar banding pattern. The method, however, is able to identify special isolates of wood-decay fungi, as was shown by Göller and Rudolph (2003): The isolate Eberswalde 15 of *Coniophora puteana* is an obligatory test fungus for testing the efficacy of wood preservatives according to the European standard EN 113. The isolate is known for its variable behaviour in wood decay tests between different test institutions. RAPD analysis showed that some alleged Ebw. 15 cultures held in different test laboratories are actually subcultures from the British facultative test isolate FPRL 11e, thereby explaining the varying test results.

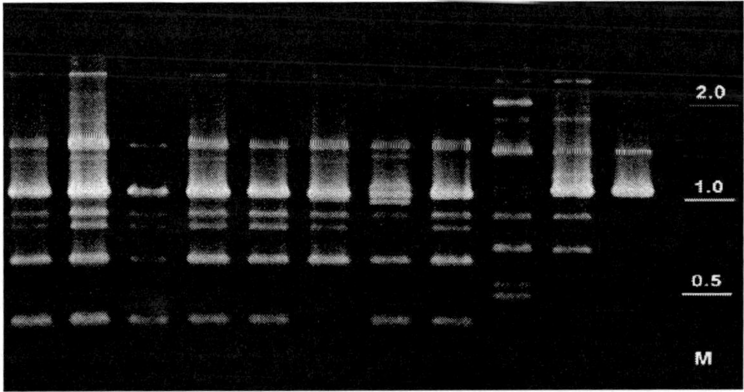

Figure 2. RAPD-pattern of isolates of *Serpula lacrymans* (8 left-most lanes) and *S. himantioides* (3 right-most lanes) obtained with primer GGACTCCACG. M = Marker (kb) (Schmidt and Moreth 1998).

Use of rDNA

The investigation of rDNA (ribosomal DNA) has become popular approach (Figure 3). The conserved rDNA regions 18S and 28S are preferentially used for phylogenetic analyses of genera, families, and higher taxonomic groups. The rapidly evolving internal transcribed spacers (ITS I and II) are used for closely related species. The intergenic spacers (IGS I and II) show the highest intraspecific diversity among all rDNA regions due to length polymorphism. Depending on the intension, the RNA genes or the spacers are used for analyses. These domains are either restricted by endonucleases for RFLP studies or they are sequenced.

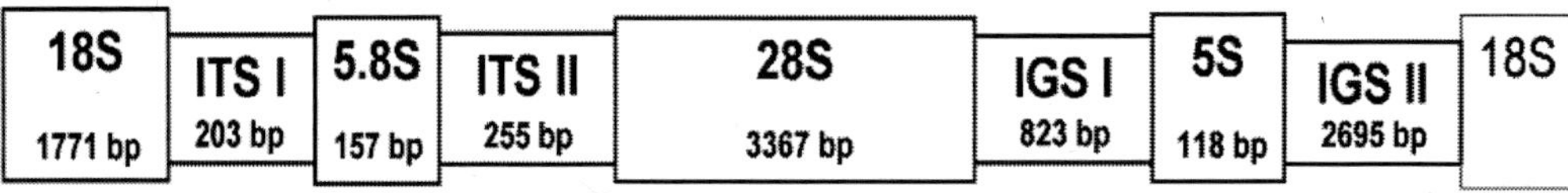

Figure 3. Schematic diagram of one rDNA unit. The number in the boxes is the size in base pairs for *Antrodia vaillantii*.

RFLP Analyses of rDNA

RFLP analyses after restriction of the ITS region with endonucleases differentiated single isolates of *Coniophora puteana*, *Gloeophyllum trabeum* and *Oligoporus placenta* (Zaremski et al. 1999). Various isolates of the closely related *Serpula lacrymans* and *S. himantioides* exhibited distinct fragment profiles after digestion with the endonucleases *Hae*III/*Taq*I (Schmidt and Moreth 1999). *Taq*I differentiated *S. lacrymans*, *S. himantioides*, *Donkioporia expansa*, *C. puteana*, *Antrodia vaillantii*, *O. placenta*, and *Gloeophyllum sepiarium* (Figure 4). ITS-RFLP analysis is currently a favoured database for the identification of wood decay and associated fungi (Zaremski et al. 1999; Adair et al. 2002; Diehl et al. 2004; Råberg et al. 2004).

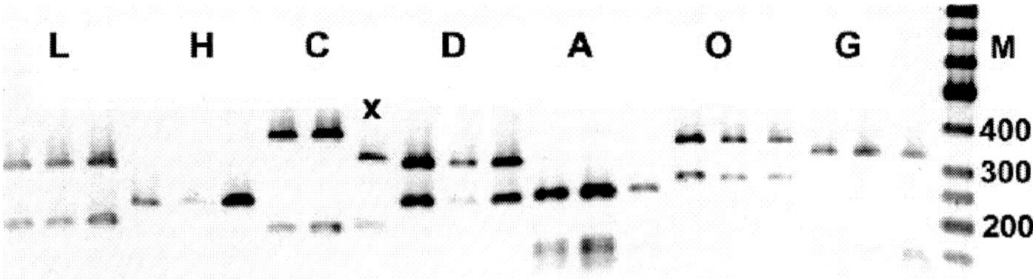

Figure 4. ITS-RFLP-pattern of *Serpula lacrymans* (L), *S. himantioides* (H), *Coniophora puteana* (C), *Donkioporia expansa* (D), *Antrodia vaillantii* (A), *Oligoporus placenta* (O) and *Gloeophyllum sepiarium* (G) after restriction with restriction endonuclease *Taq*I. Culture (x) subsequently identified by sequencing to be *Coniophora olivacea*. M = Marker (bp) (Schmidt 2006).

The rapid and inexpensive technique does not require prior information on the target DNA. However, the limited ITS size of only 600–700 bases prevents a separation of all approximately 80 fungal species that have been found in buildings. An as yet unanalyzed species may feign another fungus by exhibiting similar fragments. Isolate variation in the ITS, removing a recognition site of a restriction enzyme or providing a new restriction site, results in an unspecific fragment pattern.

In T-RFLP (terminal restriction fragment length polymorphism) analysis, one or both primers are connected with a fluorescent dye. When the end fragments with the dye undergo sequence analysis, only these fragments are determined. Råberg et al. (2005) used the technique for *C. puteana* and *O. placenta*.

SSPP

Based on the sequence divergence among species, oligonucleotide sequences may be used to design species-specific PCR primers, which in turn may be used for SSPP (species-specific priming PCR).

Specific oligonucleotide sequences located in the ITS II region of seven indoor basidiomycetes were designed for subsequent identification of unknown samples (Moreth and Schmidt 2000; Schmidt and Moreth 2000). Using the ITS 1 primer of White et al. (1990) as a forward primer the amplified ITS fragments exhibit a distinct and predictable size for the fungi (Figure 5).

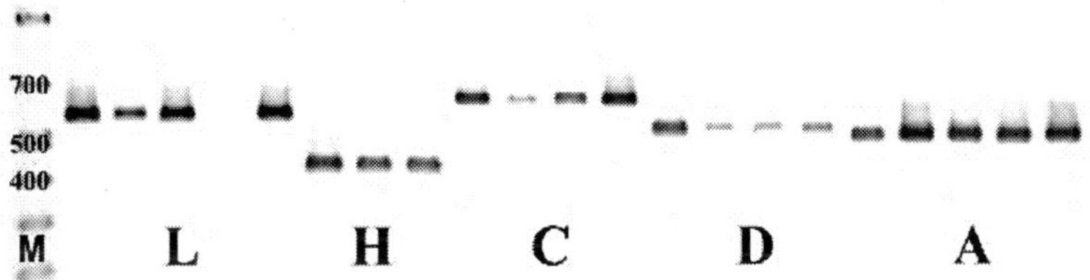

Figure 5. Positive PCR reaction within the ITS region by specific reverse primers. L = *Serpula lacrymans*: ATGTTTCTTGCGACAACGAC, H = *S. himantioides*: TCCCACAACCGAAACAAATC, C = *Coniophora puteana*: AGTAGCAAGTAAGGCATAGA, D = *Donkioporia expansa*: TCGCCAAAACGCTTCACGGT, A = *Antrodia vaillantii*: CACCGATAAGCCGACTCATT. Three to five mycelial samples of these species are shown. M = Marker (bp) (Moreth and Schmidt 2000).

SSPP is a precise and rapid technique. At first sight, the technique seems to be a powerful molecular identification tool for fungi. Subsequent restriction of the PCR amplicon as well as the use of fungal pure cultures, axenically obtained samples, and precautions to exclude DNA from the laboratory or from contaminated field material are not required. The technique is used in Germany for commercial fungal diagnosis. Limitations are: Relative to ITS-RFLPs, the ITS size of only 600–700 nucleotides prevents the design of specific primers for all of the approximately 80 indoor basidiomycetes. An as yet unanalyzed species may react with the primer of another fungus which would indicate a wrong name. The ITS sequences of closely related species from the genera *Antrodia* (Schmidt and Moreth 2003b),

Coniophora (Schmidt et al. 2002a/2003) and *Gloeophyllum* (Schmidt et al. 2002b) are too similar for species-specific primers.

Sequencing of rDNA

Sequencing of ribosomal DNA domains avoids the main limitations of RFLP and SSPP analyses because the complete information of the sequence of the target DNA is obtained. Usually, sequences are deposited in one of the international databases, which, however, exchange all data: EMBL-EBI (European Molecular Biology Laboratory, European Bioinformatics Institute, www.ebi.ac.uk), NCBI-GenBank (National Center for Biotechnology Information, USA, www.ncbi.nl,.nih.gov), DDBJ (DNA data bank of Japan, www.ddbj.nig.ac.jp).

Use of rDNA Sequences for Identification

The ITS sequences of a great number of wood-decay fungi are known. The sequence of isolate S7 was in 1999 the first ITS deposition from *Serpula lacrymans* in the databases (AJ 245948, Schmidt and Moreth 2000). Table 2 shows the accession numbers of ITS sequences of 18 indoor wood-decay species (Schmidt and Moreth 2002/2003a). The list comprises most of the common indoor basidiomycetes (cf. Table 1). The sizes, including the ITS 1 and ITS 4 primers of White et al. (1990), range from 625 bp in *Gloeophyllum sepiarium* to 734 bp in *Leucogyrophana pinastri*. Intraspecific variation is neglectable. There is only one variable nucleotide position and a gap among six isolates of *S. lacrymans*. A maximum of five variations were found among 17 isolates of *C. puteana*. Thus, ITS sequences can be used to identify unknown fungal samples through sequence comparison using the Basic Local Alignment Search Tool (BLAST) (e.g., www.ncbi.nlm.nih.gov/blast). BLAST has revealed the ITS-sequence identity of a `wild′ *S. lacrymans* isolate from the Himalayas by comparing it with indoor isolates (White et al. 2001; also Palfreyman et al. 2003), detected misnamed isolates of *S. lacrymans* (Horisawa et al. 2004), identified *Antrodia* spp. and *Serpula* spp. isolations (Högberg and Land 2004), confirmed *C. puteana* isolates (Råberg et al. 2004), and re-identified fungi from inoculated wood samples (Zaremski et al. 2005).

Limitations of ITS-sequencing are: The identification of unknown samples is time-consuming because each sample must be sequenced. Samples can be only identified by BLAST, if the correct control sequence is deposited in the databases. However, sequencing of the ITS is currently the best molecular tool for species identification because a sample is identified by its unique series of about 650 nucleotides.

Use of rDNA Sequences for Phylogenetics

ITS sequences and those from the 18S, 28S and IGS rDNA are also used for phylogenetic analyses. A phylogenetic tree based on the ITS sequences of the indoor

Coniophoraceae is shown in Figure 5 (Moreth and Schmidt 2005). A phylogenetic ITS tree of *Serpula lacrymans* isolates showed that isolates collected in nature in Czech Republic, India, Pakistan and Russia group in the branch of indoor isolates but differ from wild Californian isolates (Kauserud et al. 2004a).

Partial 28S rDNA sequences were used by Bresinsky et al. (1999) and Jarosch and Besl (2001) for *S. lacrymans*, *S. himantioides*, *Meruliporia incrassata* and for *Coniophora* and *Leucogyrophana* species. Complete 18S and 28S rDNA sequences are known for several indoor wood-decay fungi (Moreth and Schmidt 2005, Table 2). The size of the 18S rDNA for the different species amounts to 1800 bp, and the 28S rDNA is around 3300 bp. Albeit limited to maximum three investigated isolates per species, intraspecific variation was low.

Table 2. Sequenced and deposited rDNA regions of indoor wood-decay fungi. Grey: sequence known; some sequences not yet deposited. 1-28: number of sequenced isolates. Six-digit number: EMBL accession number (supplemented from Schmidt 2006)

	18S	ITS I	5.8S	ITS II	28S	IGS I	5S	IGS II
Serpula lacrymans	3 440945 440946	7 245948 249268 419907 419908 419909 419910			3 440939 440940 440941	3	3	3
	437601 536022 536023							
Serpula himantioides	3 440947 440948	12 245949 419911			3 440942 440943 440944	3	3	3
	437602 536024 536025							
Meruliporia incrassata		2 419912 419913				2	2	2
Leucogyrophana mollusca		6 419914 419915				2	2	2 partial
Leucogyrophana pinastri		4 419916 419917				2	2	2
Coniophora puteana	1 488581	28 249502 249503 344109 344110			1 583426	2	2	2
Coniophora marmorata	2 540306	4 518879 518880			1 583427	2	2	2
Coniophora arida	1 488582	3 345007 344113						
Coniophora olivacea	1 488905	5 344112 345009						
Antrodia vaillantii	1 488583	12 249266 344140 421007 421008			1 583429			
	1 286436							
Antrodia sinuosa	1 488906	5 345011 416068						
Antrodia serialis		8 344139 345010						
Antrodia xantha	1 488584	6 345012 415569			1 583430			
Oligoporus placenta		8 249267 416069						

Table 2. (Continued)

	18S	ITS I	5.8S	ITS II	28S	IGS I	5S	IGS II
Gloeophyllum abietinum	2 560802		5 420947 420948		1 583431			
Gloeophyllum sepiarium	2 540308		5 344141 420946		1 583432			
Gloeophyllum trabeum			6 420949 420950					
Donkioporia expansa	2 540307		2 249500 249501		1 583428			

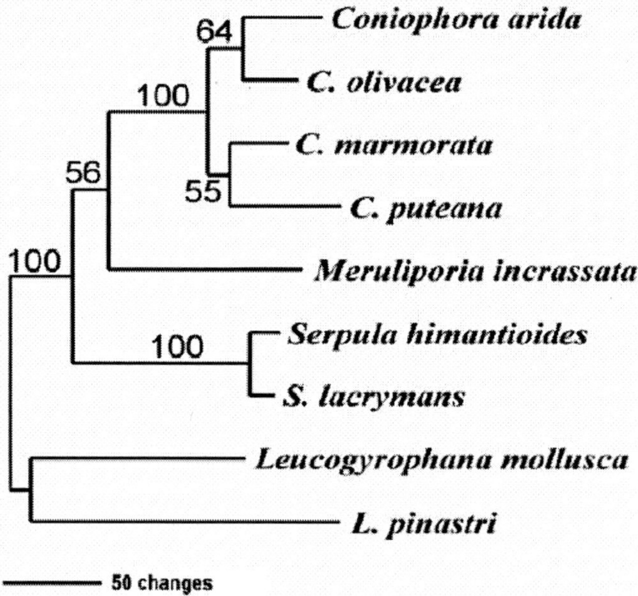

Figure 6. Most parsimonious phylogenetic tree of indoor wood-decay Coniophoraceae based on rDNA-ITS sequences. Bootstrap support values over 50% are shown. (Moreth and Schmidt 2005).

The complete sequence of the intergenic spacer (IGS I and IGS II) was available for few basidiomycetes, namely the pathogen *Filobasidiella neoformans* (Fan et al. 1995) and the ectomycorrhizal fungus *Laccaria bicolor* (Martin et al. 1999). Complete IGS sequences of some common indoor wood-decay fungi (*Serpula lacrymans*, *S. himantioides*, *Meruliporia incrassata*, *Leucogyrophana pinastri*, *Coniophora puteana*, *C. marmorata* and *Antrodia vaillantii*) (Table 2) are meanwhile deposited (Schmidt and Moreth 2008). In *A. vaillantii*, the IGS comprises 3636 bp and consists of the shorter IGS I (823 bp) and the longer IGS II (2695 bp) sequences. The intergenic spacers of *A. vaillantii* are separated by the short 5S rDNA (118 bp), whose transcription occurs in the same direction as the 18S-5.8S-28S genes, which is the case in most basidiomycetes. In the above Coniophoraceae, 5S rDNA sequence occurs in reverted direction. The complete sequence of one rDNA unit was available for the eukaryotic parasite *Encephalitozoon cuniculi* (Peyretaillade et al. 1998). If ongoing whole genome sequencing analyses are not taken into account, the whole rDNA sequence of *A.*

vaillantii is the first basidiomycetous deposition (AM286436). The sizes of the rDNA regions of *A. vaillantii* are shown in Figure 3. By means of the IGS sequences, the complete rDNA unit is now available for five indoor basidiomycetes (Schmidt and Moreth 2008).

AFLP Analysis

AFLP (amplified fragment length polymorphism) analysis is based on (1) total genomic restriction, (2) ligation of primer adapters, and (3) unselective followed by selective PCR of anonymous DNA fragments from the entire genome.

AFLP analysis of 19 European isolates of *S. lacrymans* indicated that the fungus in Europe is genetically extremely homogeneous by observing that only five out of 308 scored AFLP fragments were polymorphic (Kauserud et al. 2004b). A worldwide sample of 91 isolates was divided into two distantly related main groups: isolates from natural environments in California (*S. lacrymans* var. *shastensis*) and isolates from buildings and nature in Japan, Europe, northeast America and Oceania (*S. lacrymans* var. *lacrymans*) (Kauserud et al. 2007).

AFLP markers are more reproducible compared to RAPD and microsatellites (see below) analysis and give a higher resolution.

SSR Analysis

Microsatellites or simple sequence repeats (SSR) are variable genomic regions of tandem repeats of up to seven nucleotides. The variability of the number of repeats at a particular locus and the conservation of the flanking sequences make microsatellites valuable genetic markers, providing information for the identification and on genetic diversity and relationship among genotypes.

Microsatellite markers were developed for *S. lacrymans* by Högberg. et al. (2006). Kauserud et al. (2007) showed with these markers that a worldwide collection of 84 isolates of *S. lacrymans* var. *lacrymans* was divided into three main groups: mainland Asia as presumable origin of the variety, mainly Japan, and a cosmopolitan group, predominantly Europe, North America and Oceania.

DNA-Arrays (DNA-Chips, Microarrays)

DNA-arrays are chips carrying up to 10,000 different DNA probes, e.g. oligonucleotides, which are raster-like bound on their surface. DNA molecules of unknown samples hybridize specifically with the corresponding DNA probe, and the hybridized chip area is detected colorimetrically. Results on indoor wood-decay fungi are not known, but respective results are running.

Additional Molecular Methods

In the technique of *MALDI-TOF MS* (matrix-assisted laser desorption/ionization time-of-flight mass spectrometry), biomolecules or whole cells are embedded in a crystal of matrix molecules, which absorb the energy of a laser. The sample is ionized and transferred to the gas phase. The ions are accelerated in an electric field, and their time of flight is determined in a detector.

Figure 7 shows the first MALDI-TOF MS fingerprints of basidiomycetes, namely, those of the closely related sister taxa *S. lacrymans*, *S. himantioides* and *C. puteana*, *C. marmorata* (Schmidt and Kallow 2005). The obtained spectra may be used for subsequent diagnosis of unknown fungal samples by comparison. The technique is rapid, but it needs expensive equipment.

Currently, the technique of *Py-GC/MS* (Pyrolysis-gas chromatography/mass spectrometry) was used to distinguish mycelial cultures of five indoor wood-decay basidiomycetes (Odermatt and Schmidt, unpubl.).

MVOCs (Microbial Volatile Organic Compounds) such as pinenes, acrolein, and ketones were found in *S. lacrymans*, *C. puteana* and *O. placenta* (Korpi et al. 1999). Blei et al. (2005) distinguished pure cultures of *A. sinuosa*, *C. puteana*, *Donkioporia expansa*, *G. sepiarium*, *S. lacrymans* and *S. himantioides*. Field experiments, however, were influenced by the distance of sampling from the infested and/or destroyed wood and also by the rate of air change in buildings.

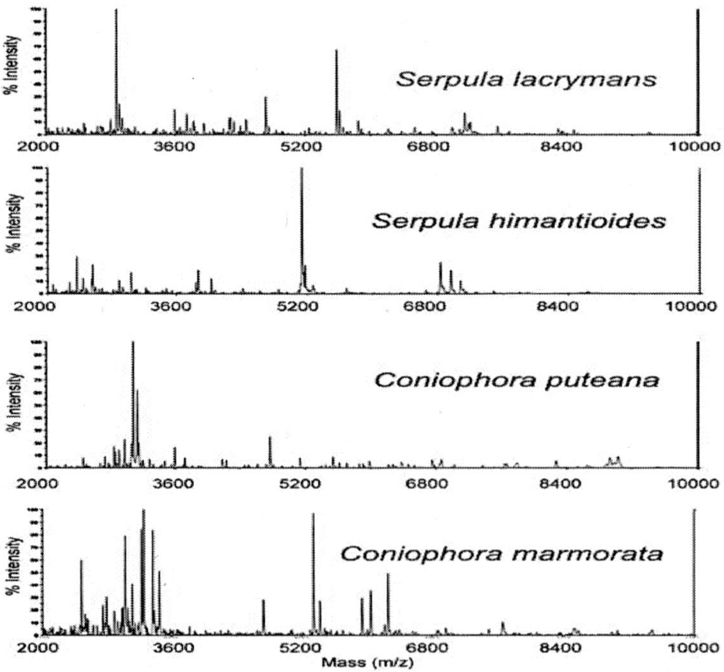

Figure 7. MALDI-TOF mass spectra of mycelia of each two closely related *Serpula* and *Coniophora* species (Schmidt and Kallow 2005).

Future Line of Work on *Serpula Lacrymans*

Little information is available in view of expression and regulation of important wood-decay enzymes such as endoglucanses for cellulose degradation. Knowledge mainly derives from white-rot basidiomycetes. For the important indoor brown-rot basidiomycetes, first studies on the endoglucanases (EC 3.2.1.4) of *S. lacrymans* provided a partial gene sequence of the enzyme (Peters 2007).

The USA Department of Energy Joint Genome Institute is currently sequencing the whole genome of *S. lacrymans*. Two monokaryons from breeding experiments (Schmidt and Moreth-Kebernik 1991) are used.

References

Adair, S., Kim, S.H. and Breuil, C. (2002). A molecular approach for early monitoring of decay basidiomycetes in wood chips. *FEMS Microbiol. Lett. 211*: 117–122.

Blei, M., Fiedler, K., Rüden, H. and Schleibinger, H.W. (2005). Differenzierung von Holz zerstörenden Pilzen mittels ihrer mikrobiellen flüchtigen organischen Verbindungen (MVOC). In: Keller, R., Senkpiel, K., Samson, R.A. and Hoekstra, E.S., (Eds.), *Mikrobielle allergische und toxische Verbindungen*. Schriftenreihe Institut Medizin. Mikrobiol. Hygiene Universität Lübeck 9, pp. 163-178.

Bresinsky, A., Jarosch, M., Fischer, M., Schönberger, I. and Wittmann-Bresinsky, B. (1999). Phylogenetic relationships within *Paxillus* s.l. (Basidiomycetes, Boletales): separation of a southern hemisphere genus. *Plant Biol. 1*: 327–333.

Clausen, C. (1997). Immunological detection of wood decay fungi – an overview of techniques developed from 1986 to present. *Int. Biodet. Biodegrad. 39*: 133–143.

Clausen, C.A. and Kartal, S.N. (2003). Accelerated detection of brown-rot decay: Comparison of soil block test, chemical analysis, mechanical properties, and immunodetection. *Forest Prod. J. 53*: 90–94.

Diaz, M.R., Boekhout, T., Kiesling, T. and Fell, J.W. (2005). Comparative analysis of the intergenic spacer regions and population structure of the species complex of the pathogenic yeast *Cryptococcus neoformans*. *FEMS Yeast Res. 5*: 1129-1140.

Diehl, S.V., McElroy, T.C. and Prewitt, M.L. (2004). Development and implementation of a DNA-RFLP database for wood decay and wood associated fungi. *Int. Res. Group Wood Preserv. Stockholm,* Doc. 10527.

Eikenes, M., Hietala, A., Alfredsen, G., Fossdal, C.G. and Solheim, H. (2005). Comparison of quantitative real-time PCR, chitin and ergosterol assays for monitoring colonization of *Trametes versicolor* in birch wood. *Holzforschung 59*: 568–573.

Fan, M., Chen, L.-C., Ragan, M.A., Gutell, R.R., Warner, J.R., Currie, B.P. and Casadevall, A. (1995). The 5S rRNA and the rRNA intergenic spacer of the two varieties of *Cryptococcus neoformans*. *J. Med. Vet. Mycol. 33*: 215-221.

Glancy, H., Palfreyman, J.W., Button, D., Bruce, A. and King, B. (1990). Use of an immunological method for the detection of *Lentinus lepideus* in distribution poles. *Journal Institute Wood Sci. 12*: 59-64.

Göller, K. and Rudolph, D. (2003). The need for unequivocally defined reference fungi – genomic variation in two strains named as *Coniophora puteana* BAM Ebw. 15. *Holzforschung 57*: 456–458.

Hietela, A., Eikenes, M., Kvaalen, H., Solheim, H. and Fossdal, C. (2003). Multiplex real-time PCR for monitoring *Heterobasidion annosum* colonization in Norway spruce clones that differ in disease resistance. *Appl. Environ. Microbiol. 69*: 4413–4420.

Högberg, N. and Land, C.J. (2004). Identification of *Serpula lacrymans* and other decay fungi in construction timber by sequencing of ribosomal DNA – a practical approach. *Holzforschung 58*: 199–204.

Högberg, N., Svegården, I.B. and Kauserud, H. (2006). Isolation and characterization of fifteen polymorphic microsatellite markers for the devasting dry rot fungus *Serpula lacrymans*. *Molec. Ecol. Notes 6*: 1022-1024.

Horisawa, S., Sakuma, Y., Takata, K. and Doi, S. (2004). Detection of intra- and interspecific variation of the dry rot fungus *Serpula lacrymans* by PCR-RFLP and RAPD analysis. *J. Wood Sci. 50*: 427–432.

Huckfeldt, T. and Schmidt, O. (2006). *Hausfäule- und Bauholzpilze*. Rudolf Müller, Köln.

Jarosch, M. and Besl, H. (2001). *Leucogyrophana*, a polyphyletic genus of the order Boletales (Basidiomycetes). *Plant Biol. 3*: 443–448.

Jellison, J. and Goodell, B. (1988). Immunological detection of decay in wood. *Wood Sci. Technol. 22*: 293–297.

Kasuga, T. and Mitchelson, K.R. (2000). Intersterility group differentiation in *Heterobasidion annosum* using ribosomal IGS1 region polymorphism. *Forest Pathol. 30*: 329–344.

Kauserud, H., Högberg, N., Knudsen, H., Elbornes, S.A. and Schumacher, T. (2004a) Molecular phylogenetics suggest North American link between the anthropogenic dry rot fungus *Serpula lacrymans* and its wild relative *S. himantioides*. *Molec. Ecol. 13*: 3137-3146.

Kauserud, H., Schmidt, O., Elfstrand, M. and Högberg, N. (2004b). Extremely low AFLP variation in the European dry rot fungus (*Serpula lacrymans*): implications for self/nonself-recognition. *Mycol. Res. 108*: 1264–1270.

Kauserud, H., Svegården, I.B., Sætre, G.-P., Knudsen, H., Stensrud, Ø., Schmidt, O., Doi, S., Sugiyama, T. and Högberg, N. (2007). Asian origin and rapid global spread of the destructive dry rot fungus *Serpula lacrymans*. *Molec. Ecol. 16*: 3350-3360.

Kim, Y.S., Goodell, B. and Jellison, J. (1991). Immuno-electron microscopic localization of extracellular metabolites in spruce wood decayed by brown-rot fungus *Postia placenta*. *Holzforschung 45*: 389-393.

Korpi, A., Pasanen, A.L. and Viitanen, H. (1999). Volatile metabolites of *Serpula lacrymans*, *Coniophora puteana*, *Poria placenta*, *Stachybotris chartarum* and *Chaetomium globosum*. *Building Environm. 34*: 205-211.

Martin, F., Selosse, M.-A. and Le Tacon, F. (1999). The nuclear rDNA intergenic spacer of the ectomycorrhizal basidiomycete *Laccaria bicolor*: structural analysis and allelic polymorphism. *Microbiol. 145*: 1605-1611.

Moreth, U. and Schmidt, O. (2000). Identification of indoor rot fungi by taxon-specific priming polymerase chain reaction. *Holzforschung 54*: 1–8.

Moreth, U. and Schmidt, O. (2005). Investigations on ribosomal DNA of indoor wood decay fungi for their characterization and identification. *Holzforschung 59*: 90–93.

Palfreyman, J.W., Glancy, H., Button, D., Bruce, A., Vigrow, A., Score, A. and King, B. (1988). Use of immunoblotting for the analysis of wood decay basidiomycetes. *Int. Res. Group Wood Preserv. Stockholm,* Doc. 2307.

Palfreyman, J.W., Vigrow, A., Button, D., Hegarty, B. and King, B. (1991). The use of molecular methods to identify wood decay organisms. 1. The electrophoretic analysis of *Serpula lacrymans. Wood Protect. 1*: 15–22.

Palfreyman, J.W., Gartland, J.S., Sturrock, C.J., Lester, D., White, N.A., Low, G.A., Bech-Andersen, J. and Cooke, D.E.L. (2003). The relationship between "wild" and "building" isolates of the dry rot fungus *Serpula lacrymans. FEMS Microbiol. Lett. 228*: 281–286.

Peters, S. (2007). *Vorkommen und Aktivität von Cellulasen bei Serpula lacrymans.* Diploma-Thesis, University Hamburg.

Peyretaillade, E., Biderre, C., Peyret, P., Duffieux, F., Méténier, G., Gouy, M., Michot, B. and Vivarès, C.P. (1998). Microsporidian *Encephalitozoon cuniculi*, a unicellular eukaryote with an unusual chromosomal dispersion of ribosomal genes and a LSU rDNA reduced to the universal core. *Nucleic Acids Res. 26*: 3513-3520.

Råberg, U., Högberg, N. and Land, C.J. (2004). Identification of brown-rot fungi on wood in above ground conditions by PCR, T-RFLP and sequencing. *Int. Res. Group Wood Preserv. Stockholm,* Doc. 10512.

Råberg, U., Högberg, N. and Land, C.J. (2005). Detection and species discrimination using rDNA T-RFLP for identification of wood decay fungi. *Holzforschung* 59: 696-702.

Schmidt, O. (2000). Molecular methods for the characterization and identification of the dry rot fungus *Serpula lacrymans. Holzforschung 54*: 221–228.

Schmidt, O. (2006). *Wood and tree fungi. Biology, damage, protection, and use.* Springer, Berlin Heidelberg New York

Schmidt, O. (2007). Indoor wood-decay basidiomycetes: damage, causal fungi, physiology, identification and characterization, prevention and control. *Mycol. Progr. 6*: 261-279.

Schmidt, O. and Kallow, W. (2005). Differentiation of indoor wood decay fungi with MALDI-TOF mass spectrometry. *Holzforschung 59*: 374–377.

Schmidt, O. and Kebernik, U. (1989). Characterization and identification of the dry rot fungus *Serpula lacrymans* by polyacrylamide gel electrophoresis. *Holzforschung 43*: 195–198.

Schmidt, O. and Moreth, U. (1995). Detection and differentiation of *Poria* indoor brown-rot fungi by polyacrylamide gel electrophoresis. *Holzforschung 49*: 11–14.

Schmidt, O. and Moreth, U. (1998). Characterization of indoor rot fungi by RAPD analysis. *Holzforschung 52*: 229–233.

Schmidt, O. and Moreth, U. (1999). Identification of the dry rot fungus, *Serpula lacrymans*, and the wild merulius, *S. himantioides*, by amplified ribosomal DNA restriction analysis (ARDRA). *Holzforschung 53*: 123–128.

Schmidt, O. and Moreth, U. (2000). Species-specific PCR primers in the rDNA-ITS region as a diagnostic tool for *Serpula lacrymans. Mycol. Res. 104*: 69-72.

Schmidt, O. and Moreth, U. (2002/2003a). Data bank of rDNA-ITS sequences from building-rot fungi for their identification. *Wood Sci. Technol. 36*: 429–433, revision in *Wood Sci. Technol. 37*: 161–163.

Schmidt, O. and Moreth, U. (2003b). Molecular identity of species and isolates of internal pore fungi *Antrodia* spp. and *Oligoporus placenta. Holzforschung 57*: 120-126.

Schmidt, O. and Moreth, U. (2008). Ribosomal DNA intergenic spacer of indoor wood-decay fungi. *Holzforschung 62:* DOI 10.1515.

Schmidt, O. and Moreth-Kebernik, U. (1991). Monokaryon pairings of the dry rot fungus *Serpula lacrymans. Mycol. Res. 95*: 1382–1386.

Schmidt, O., Grimm, K. and Moreth, U. (2002a/2003). Molecular identity of species and isolates of the *Coniophora* cellar fungi. *Holzforschung* 56: 563-571, addendum, *Holzforschung 57*: 228.

Schmidt, O., Grimm, K. and Moreth, U. (2002b). Molekulare und biologische Charakterisierung von *Gloeophyllum*-Arten in Gebäuden. *Zeitschr. Mykologie 68*: 141-152.

Theodore, M.L., Stevenson, T.W., Johnson, G.C., Thornton, J.D. and Lawrie, A.C. (1995). Comparison of *Serpula lacrymans* isolates using RAPD PCR. *Mycol. Res. 99*: 447–450.

Toft, L. (1993). Immunological identification in vitro of the dry rot fungus *Serpula lacrymans*. *Mycol. Res.* 97: 290–292.

Vainio, E.J. and Hantula, J. (2000). Direct analysis of wood-inhabiting fungi using denaturing gradient gel electrophoresis of amplified ribosomal DNA. *Mycol. Res. 104*: 927–936.

Vigrow, A., King, B. and Palfreyman, J.W. (1991b). Studies of *Serpula lacrymans* mycelial antigens by Western blotting techniques. *Mycol. Res. 95*: 1423–1428.

Vigrow, A., Palfreyman, J.W. and King, B. (1991a). On the identity of certain isolates of *Serpula lacrymans*. *Holzforschung 45*: 153–154.

White, N.A., Dehal, P.K., Duncan, J.M., Williams, N.A., Gartland, J.S., Palfreyman, J.W. and Cooke, D.E.L. (2001). Molecular analysis of intraspecific variation between building and "wild" isolates of the dry rot fungus *Serpula lacrymans* and their relatedness to *S. himantioides*. *Mycol. Res. 105*: 447–452.

White, T.J., Bruns, T., Lee, S. and Taylor, J. (1990). Amplification and direct sequencing of fungal ribosomal genes for phylogenetics. In: Innis, M.A., Gelfand, D.H., Sninisky, J.J. and White, T.J. (Eds.) *PCR protocols*. Academic Press, San Diego, pp. 315–322.

Zaremski, A., Ducousso, M., Prin, Y. and Fouquet, D. (1999). Molecular characterization of wood-decaying fungi. *Bois Forêts Tropiques, Special Issue*, 76–81.

Zaremski, A., Ducousso, M., Domergue, O., Fardoux, J., Rangin, C., Fouquet, D., Joly, H., Sales, C., Dreyfus, B. and Prin, Y. (2005). In situ molecular detection of some white-rot and brown-rot basidiomycetes infecting temperate and tropical woods. *Can. J. Forest Res. 35*: 1256-1260.

In: Current Advances in Molecular Mycology
Eds: Y. Gherbawy, R.L. Mach and M.K. Rai

ISBN 978-1-60456-909-4

Chapter XIV

MOLECULAR TOOLS FOR IDENTIFICATION AND DIFFERENTIATION OF DIFFERENT HUMAN PATHOGENIC *CANDIDA* SPECIES

***V. V. Tiwari*[1], *M. N. Dudhane*[2] *and M. K. Rai*[1]**
[1]Department of Biotechnology, SGB Amravati University, Amravati-444 602, Maharashtra state, India;
[2]Department of Biochemistry, PDMMC, Amravati -444 602, Maharashtra state, India.

ABSTRACT

Due to the increasing number of immunocompromised hosts and advances in medical technology, there has been a consistent rise in the number of cases of invasive fungal infections. The fungi as simple eukaryote display various emergent properties like development, aging, mating, pathogenicity, transvection and gene silencing. Yet microbial systems provide unique opportunities to develop novel global approaches in genomics and computational biology to explain different phenomena. As candidosis incidence continue to rise, quick laboratory identification of *Candida* species is becoming increasingly important for a growing population of patients at high risk. Molecular tools for integrating genomics information are reported. Several genomic approaches like PCR-based assay, *in situ* hybridization are described here to examine fungal pathogenicity and development for identification. It is hoped that the experimental tractability of fungal systems will lead to a new hypothesis-driven genomics.

Key words: immunocompromised, pathogenecity, *Candida*, candidosis, PCR.

INTRODUCTION

There have been a growing number of fungal infections with a dramatic increase in the population of severely immunocompromised patients. These infections are mainly due to impairments in host defense mechanisms as a consequence of viral infections, especially the human immunodeficiency virus, hematological disorders such as different types of leukemia, organ transplants and more intensive and aggressive medical practices. Many clinical procedures and treatments, such as surgery, the use of catheters, injections, radiation, chemotherapy, antibiotics, and steroids, are risk factors for fungal infections. The new opportunistic pathogens have increased the understanding of medical mycology, and unexpected changes have been observed in the pattern of fungal infections in humans. It is also possible that most of the recently reported taxa have caused infections, which previously passed unnoticed due to inadequate diagnostic expertise. The rapid appearance of range of new pathogens has created a growing interest in fungal systematics. Fungal taxonomy is a dynamic and progressive discipline that requires deep study and nomenclature. For clinicians and clinical microbiologists the identification is difficult. Another difficulty for clinicians and microbiologists is that fungi are mostly classified on the basis of their appearance rather than on the nutritional and biochemical differences. This implies that different concepts have to be applied in fungal taxonomy. Generally, medical mycologists are familiar with only one aspect of pathogenic fungi, i.e., the stage that develops by asexual reproduction. Usually microbiologists ignore or have sparse information about the sexual stages of these organisms. However, the sexual stages are precisely the baseline of fungal taxonomy and nomenclature. It seems evident that in the near future, modern molecular techniques will allow most of the pathogenic and opportunistic fungi to be connected to their corresponding sexual stages and integrated into a more natural taxonomic scheme. The aim is to update our present understanding of the systematics of pathogenic and opportunistic fungi, emphasizing their relationships with the currently accepted taxa of the phyla Ascomycota and Basidiomycota.

Fungi and eukaryotic organisms have approximately 300,000 different species. Of these, about 200 are potential parasites, with only a few of these affecting humans. Fungal diseases of mammals, mycoses range from the common mild cutaneous or subcutaneous skin infections, such as atheletes foot, to be potentially lethal acute or chronic infection of deep tissues that are typically caused by *Candida* species. Of the *Candida* species afflicting humans, *Candida albicans* is by far the most common. *Candida albicans* belongs to the class Ascomycetes and the family, Saccharomycetaceae. This yeast can live as harmless commensal in many different body locations, and is carried in almost half of the population. However, in response a change in the host environment, *C. albicans* can convert from a benign commensal into a disease causing pathogen, causing infections in the oral, gastrointestinal and genital tracts. The infection caused by *C. albicans* can be defined in two broad categories, superficial mucocutaneous and systemic invasive, which involves the spread of *C. albicans* to the blood stream (candidemia) and to the major organs. Systemic candidemia is often fatal. Superficial infections affect the various mucous membrane surfaces of the body such as in oral and vaginal thrush. The incidence of vulvovaginal candidiasis (thrush) has increased approximately 2 fold in the last decade. Approximately 75% of all women experience a clinically significant episode of vulvovaginal cadidiasis (VVC) at least once during the reproductive period. VVC is a relatively benign condition that responds well to anti-fungal treatment. It is proposed that the infection is due to the minor changes in epithelial conditions, such as pH, altered glucose/glycogen concentration or changes in epithelial integrity. During pregnancy, the risk of vaginal thrush increases, possibly due to changes in hormone production, leading to increased glycogen content in the vagina. The pathogenesis of recurrent vaginal thrush involves a defect in the local immunity to candidiasis, possibly through inappropriate prostaglandin E2 (PGE2) production. The role of prostaglandins during the infection is not very clear, however it has been demonstrated that mononuclear cells from the patients suffering from recurrent vaginal cadidiasis produce higher levels of PGE2 as compared with cells from control women, indicating the important role of PGE2 during infection. Recurrent vaginal candidiasis is also common in female patients with acquired immune deficiency syndrome (AIDS), suggesting a role for depressed cell mediated immunity in candidiasis. Factors responsible for recurrent vaginal candidiasis may originate from the microorganism and the host cells, tissues and organs. Therefore, the severity of *Candida* infection depends upon the status of the host's immune system. However, there are differences in the pathogenicity of *C. albicans* strains suggests that strain related virulence factors may play a role in disease severity. Numerous virulence factors have been attributed to the pathogenecity of *C. albicans.* These include dimorphism, phenotypic switching and immune interference.

Dimorphism and Phenotypic Switching:

Candida albicans is a diploid asexual and dimorphic fungus and depending upon environmental conditions can exist as unicellular yeast (blastospores and chlamydospores) as well as in different filamentous forms (hypha, pseudo hyphae). Several studies suggest that the ability of *C. albicans* to switch between the yeast and the mycelial forms is an important virulence factor. Increased adherence to oropharengeal surfaces has been observed for the mycelial form. Decreased adherence has been demonstrated by a non-germ tube producing variant in experimental vaginitis. Moreover, *C. albicans* cannot only change its cellular morphology in response to growth conditions, but can also irreversibly switch its cellular phenotype both *in vitro* and *in vivo*. This switching is most easily observed in the morphology of colonies. These phenotypic differences are a product of differences in surface protein expression. The different protein expressions result in differential adherence characteristics for the switch variants and differential sensitiveness to neutrophil and phagocytic leucocytic killing. This suggests that differences in the phenotype may allow for increased resistance to immune attack or increased invasiveness. The process of morphogenesis and phenotype switching is found to be dependent upon the lipid composition of *C. albicans*. Lipids constitute about 3.8-4.3% of the dry weight of the fungal cell and are important structural and functional molecules in *C. albicans.*

Disease Diagnosis:

Unfortunately, diagnosis of invasive candidiasis is still problematic, because clinical symptoms are non-specific and conventional assays are not satisfactorily precise and may take several days to obtain results. Blood cultures, which are assumed to be the most reliable marker of invasive candidiasis, are commonly negative, whereas positive cultures from other sites may represent colonization (de Marie, 2000). Also, blood culture requires 2–5 days for correct identification. Molecular-genetic approaches, which rely on rapid detection of *Candida* yeasts nucleic acids in clinical samples, offer a promising alternative.

In addition to detection of *Candida* yeasts, species identification is very much in demand in some situations, because susceptibility to antifungal agents, probability of resistance development, and ability to cause disease can vary among different species. The most often identified species has always been *Candida albicans*, however the incidence of non-*albicans Candida* species has been increasing with the changing spectrum of patients and these species have also been associated with higher mortality (Nguyen *et al.*, 1996). *C. glabrata* and *C. krusei* show lower sensitivity to fluconazole in contrast to *C. albicans*, and emergence of secondary resistance in *C. lusitaniae* to amphotericin B has also been observed rarely (Krcmery *et al.*, 2002; Capoor *et al.*, 2005). *C. parapsilosis* persists in hospital environments, thereby enhancing the chance of nosocomial infection. It also commonly colonizes skin of patients and is able to form biofilms on plastics. This often results in candidaemia in patients with indwelling venous catheters.

Also, epidemiological investigations of infection outbreaks in hospitals are highly desirable to identify the source and route of infection to eradicate it. Such investigations require accurate strain typing, because in candidaemia it is problematic to distinguish between endogenous source of infection and infection transmitted exogenously from other infected patients or even health care workers (Pfaller, 1996). Similarly, to the species identification, phenotyping or genotyping can be applied in strain typing. Phenotyping methods characterize products of gene expression (Tenover *et al.*, 1997), thus reflecting genetic diversity. This is also their main shortcoming; because even minor variations in growth conditions can influence gene expression, which leads to poor, inter laboratory reproducibility. Phenotyping systems are also considered cumbersome, time-consuming and provide limited data for differentiation between epidemiologically unrelated isolates. In addition, some *Candida* species are capable of spontaneous switching between numbers of phenotypes (Soll, 1992), so their phenotypic characteristics may be unstable or variable. In contrast to phenotyping, genotyping techniques detect differences in genetic information directly. Therefore, they are less sensitive to variations in growth conditions and also possess several other advantages over phenotyping procedures, e.g. higher discriminatory power, speed and reproducibility.

This chapter focuses on the vigorously fermenting field of molecular genetic approaches to *Candida* yeast identification and typing. Not only are new techniques and their modifications published at a growing pace, but also commercially available kits and other initiatives continually contribute to their standardization and ease of use, thus facilitating implementation of molecular approaches in routine use. It seems that molecular techniques will soon revolutionize the identification of *Candida* species.

A. Detection and Species Identification

Techniques that rely on amplification of target yeast DNA are used almost exclusively for detection purposes, because the amount of yeast DNA available in a clinical sample is typically very low. These techniques include PCR (Polymerase chain reaction) and NASBA (Nucleic acid sequence based amplification). Alternatively, yeast cells can first be multiplied during cultivation in blood culture bottles, and *Candida* DNA can then be detected in positive bottles by FISH with species-specific probes. This approach requires more time and is most probably less sensitive than PCR or NASBA. On the other hand, it may be more economic, because only positive bottles are examined. Also, results of FISH detection can later be verified by phenotyping of subcultured isolates, which are in addition available for strain typing.

1. PCR-Based Methods

The invention of PCR was a landmark in the progress of molecular microbiology and has had a substantial impact on the diagnosis of infectious diseases. The key strongpoint of these techniques consists in the amplification and detection of minute amounts of microbial nucleic acid in the background of host DNA. PCR-based methods can be appreciable especially when conventional methods are not available, are insensitive or slow.

1.1. Target and Primer Selection

Generally, two strategies of PCR target selection can be adopted. If species-specific sequences are selected as primer-annealing sites, PCR will enable highly specific detection of just one pathogenic yeast species. On the other hand, when universal pan fungal sequences are targeted, PCR will result in amplicons in case any fungal DNA is present in samples. *Candida*-genus specific sequences can also be targeted to detect all *Candida* yeast species. If a broader spectrum of species is targeted, post-PCR analysis is necessary for subsequent species identification. To ensure high sensitivity of PCR detection, primers should preferentially target multicopy genes. Also, high specificity should be secured by targeting sequences specifically found only in the pathogen of interest. The ribosomal RNA (rRNA) gene appears to meet both of these criteria. A tandem array of 50 to 100 copies of the rRNA gene can be found in the haploid genome of all fungi. This consists of the small subunit rRNA gene (18S), the 5.8S gene and the large subunit rRNA (25S) gene, separated by the internal transcribed spacer regions, ITS1 and ITS2. While rRNA genes are highly conserved in fungi, ITS regions involve both highly variable and highly conserved areas (Reiss *et al.*, 1998), thus allowing the generation of species, genus or fungus specific primer. The gene family of secreted aspartyl proteinases represents another group of sequences found in several copies in the yeast genome, which can serve as a target for PCR-amplification (Flahaut *et al.*, 1998). In addition, there are many examples of single-copy genes that can be caught by PCR and seem to be highly fungal specific, e.g. the P450 lanosterol-14á demethylase gene (Burgener-Kairuz *et al.*, 1994), actin gene (Kan, 1993) and heat shock protein 90 gene (Crampin *et al.*, 1993).

1.2. Nested PCR

Nested PCR can be used to increase both the sensitivity and specificity of PCR detection. In this approach, two rounds of PCR are performed. In the first round, outer primers target a larger region for amplification. Amplicons from this round are then added as template into the second round reaction mixture, where inner primers target a fragment of the first round amplicon. Specificity of the assay is increased, because four primers have to anneal in an arranged fashion instead of just two in a single PCR. Sensitivity is increased, because addition of fresh reagents and dilution of first round amplicons in the second round mixture enables additional amplification of a fragment of the amplicons from the first round mixture. Two-rounds setting of nested PCR can also be used to combine the advantages of broad-range and species-specific targeting of yeast sequences. The outer primers can target universal sequences resulting in amplicons in a broad range of yeast species, followed by several parallel second round reactions with species-specific inner primer pairs. When second round primers are carefully designed to prevent interference, primer mixes can be used in a common reaction mixture to reduce costs, in an approach called multiplex PCR. Nested PCR approach was adapted for use in *Candida* species detection by Kanbe *et al.* (2002), and Bougnoux *et al.* (1999) On the other hand, the extreme sensitivity of nested PCR results in its major drawback – the occurrence of false-positive results mainly due to the cross-contamination with previously amplified products (Louie *et al.*, 2000), and also due to contamination with environmental microorganisms, or even contaminated commercially available reagents (Loeffler *et al.*, 1999). To avoid this pitfall, laboratories must follow stringent precautions such as establishing separate rooms and equipment for each step of the PCR and other procedures (Kwok *et al.*, 1989).

1.3. Real-Time PCR

The real-time polymerase chain reaction uses fluorescent reporter molecules to visualize the production of amplicons during each cycle of the PCR reaction. This is in contrast to endpoint detection in conventional PCR, where the amplicon is detected after completed amplification only. Real-time monitoring of amplification based on increase of fluorescence of reporter molecules enables quantification of the target DNA, because the time at which the amplicons reach a specific fluorescence level during cycling corresponds with the starting amount of target DNA. This correlation is impossible in the case of conventional PCR, where the final amount of amplicons always reaches a uniform level due to inhibition of further amplification in the plateau phase of the reaction. The process of amplification can be monitored either using labelled probes which specifically hybridise to the newly formed amplicon molecules, or by staining newly formed double-stranded DNA molecules with non-specific dsDNA binding dyes (e.g. SYBR Green I, BEBO, LC Green or BOXTO). The use of probes increases the specificity of PCR, because an additional sequence homology between the amplicon and probe is necessary for successful reporting of amplification. When a dsDNA binding dye is used instead of a specific probe, melting analysis of the amplicon has to be performed subsequently to verify the identity of the amplicon. Sometimes, unambiguous differentiation between specific and non-specific products can be problematic. On the other hand, melting analysis can provide additional useful information about the amplified sequence. Traditionally, one has to choose between the use of a sequence-specific probe and a non-specific dsDNA binding dye, because of spectra overlap in fluorescent dyes used for labelling of probes and for dsDNA staining. Recently, however, the use of BOXTO dsDNA-binding dye has been reported as compatible with the use of probes labelled by FAM (Lind *et al.*, 2006).

The use of an integrated thermocycler/fluorimeter with highly efficient heat exchange mechanism has significantly shortened the turnaround time of real-time PCR. Both amplification and detection take place in the same closed vessel, reducing post-amplification manipulation steps and dramatically decreasing the risk of false-positive results. Despite greater start-up expense and the lack of standardization, the oncoming explosion of new chemistries and instrumentation, sensitivity, reproducibility and potential for high-throughput, have made the real-time PCR attractive and indispensable for diagnostic mycology.

Several studies have reported the identification of *Candida* species by targeting the rRNA gene complex using real-time PCR (Guiver *et al.*, 2001; Pryce *et al.*, 2003; Hsu *et al.*, 2003; White *et al.*, 2003; Selvarangan *et al.*, 2003). The wide variety of fluorescent dyes available makes it possible to amplify multiple templates in a single tube, as fluorescent dyes with different emission spectra may be attached to the different probes, providing the dyes are compatible with the excitation and detection criteria of the real-time cycler used (Mackay, 2004). Bu *et al.* (2005) described the detection and quantification of five fungal species in a multiplex real-time PCR assay.

1.4. Post-PCR Analysis

Whether conventional or real-time PCR is used, several options for post-PCR analysis are available to characterize the amplicon and conclusions can be drawn on its species specificity, especially if universal sequences are targeted for amplification. Obviously, the only ultimate and most accurate way of post-PCR analysis is direct sequencing (Jung *et al.*, 1992). Although commercial systems are available (Hall *et al.*, 2003), this option is still too expensive and laborious for routine use. However, alternative sequencing techniques, e.g. pyrosequencing, are under continuous development and promise further reduction of costs in future. All the other techniques of amplicon post-PCR analysis rely in some way on characterization of its sequence-related variability. The length of the amplicon can be roughly estimated by agarose gel electrophoresis, which represents the most simple and traditional technique of post-PCR analysis, also called Amplified product length polymorphism (APLP). More accurate length characterization of amplicons can be achieved by polyacrylamide gel electrophoresis, which can be automated in a capillary-based analyzer (Marino *et al.*, 1994; Chen *et al.*, 2000). Restriction analysis of amplicons represents a rather cheap and elegant but laborious and more time-consuming technique (Morace *et al.*, 1997; Dendis *et al.*, 2003). Similarly, single-strand conformational polymorphism (SSCP) can be employed to evaluate sequence-based characteristics of amplicons (Jaeckel *et al.*, 1998; Hui *et al.*, 2000), but it is not widely used because of special expertise and labour needed for correct performance.

To avoid the time-consuming and laborious electrophoresis step, used traditionally in the above-mentioned techniques, two alternative approaches can be applied. Microtitration plate enzyme immunoassay (PCR-EIA) can be utilized as a user-friendly alternative, which also improves detection of sensitivity (Fujita *et al.*, 1995). Amplicons hybridise to two oligonucleotide probes, a biotin-labelled genus-specific probe and a digoxigenin-labelled species-specific probe, and the hybridization complex is added into streptavidine-coated wells of a microtiter plate. The positive reaction is caught by peroxidase-conjugated anti-digoxigenin antibodies. Melting analysis of amplicons represents another recent promising choice, because of its simple, rapid and economic performance. It relies on staining of double stranded DNA (amplicon) with a fluorescent dye, which is released from the structure once DNA strands are separated during denaturation (melting) with increasing temperature. The concomitant decrease of fluorescence can be followed by a fluorimeter, either embedded into a real-time PCR instrument or used as a separate melter instrument (Ririe *et al.*, 1997). Characteristic melting curves are recorded, because the process of melting depends both upon the length of the amplicon and upon its sequence, where AT rich regions melt at lower temperatures, whereas GC-rich ones at higher temperatures. Use of melting analysis in post-PCR analysis of pathogenic yeasts amplicons has not been reported yet. However, there is no doubt about its potential in this area, particularly since high resolution melting analysis (HRMA) and saturating dyes are now available (Herrmann *et al.*, 2006). The first use of HRMA in diagnostic microbiology was reported in mycobacteria recently (Odell *et al.*, 2005).

2. Nucleic Acid Sequence Based Amplification (NASBA)

NASBA is a specific and very sensitive RNA amplification technique, which exploits the action of three enzymes, i.e. reverse transcriptase, RNase H and T7 RNA polymerase, in an isothermal amplification process with cDNA as an intermediate (Compton, 1991; Kievits *et al.*, 1991) (Figure 1). In medically important fungi, conserved regions of the 18S rRNA gene can be targeted by the amplification (Widjojoatmodjo *et al.*, 1999). Labelled oligonucleotide probes are then hybridized to an internal specific sequence of the *Candida* yeast species. Amplification and detection can be completed within few hours and the analysis has shown a detection limit of 1 CFU. NASBA has been evaluated to detect six various *Candida* species (Loeffler *et al.*, 2003). The main benefits of NASBA compared to PCR are no need of thermal cycling instrument and specific detection of living yeast cells, because RNA unlike DNA is rapidly degraded outside cells. The main disadvantage, which prevents more widespread use of NASBA, is the high price of the three enzymes mixture.

3. Identification by Fluorescence in situ Hybridisation (FISH)

Fluorescent *in situ* hybridisation (FISH) with fluorescein labelled oligonucleotide probes is a convenient way to detect yeasts without the need of pure culture. The employment of novel PNA (peptide nucleic acid) probes combines their high-affinity with advantages of targeting highly structured rRNA region, which has extended the potential of this method. Briefly, probes are hybridized to smears made directly from the contents of blood culture bottles on a slide, non-hybridized probes are washed out and slides are examined by fluorescence microscopy to reveal the presence of the organism. The sensitivity of the method has been estimated at least as similar to most results obtained by PCR-based assays (Lischewski *et al.*, 1997). Due to a simple technical protocol with the exclusion of DNA extraction, the entire PNA FISH requires only 2.5 hours after a blood culture is designated positive by an automated blood culture system, the whole procedure is suitable for automation. FISH including probes specific for *Candida* species has been demonstrated to be a reasonable diagnostic tool for species identification (Kempf *et al.*, 2000). PNA FISH has been developed to differentiate *C. albicans* from non-*albicans Candida* species (Rigby *et al.*, 2002), evaluated in a multicenter study (Wilson *et al.*, 2005) and its implementation in hospital reduced antifungal drug expenses (Alexander *et al.*, 2006). The same group also conducted a PNA FISH assay to differentiate *C. albicans* from *C. dubliniensis* (Oliviera *et al.*, 2001).

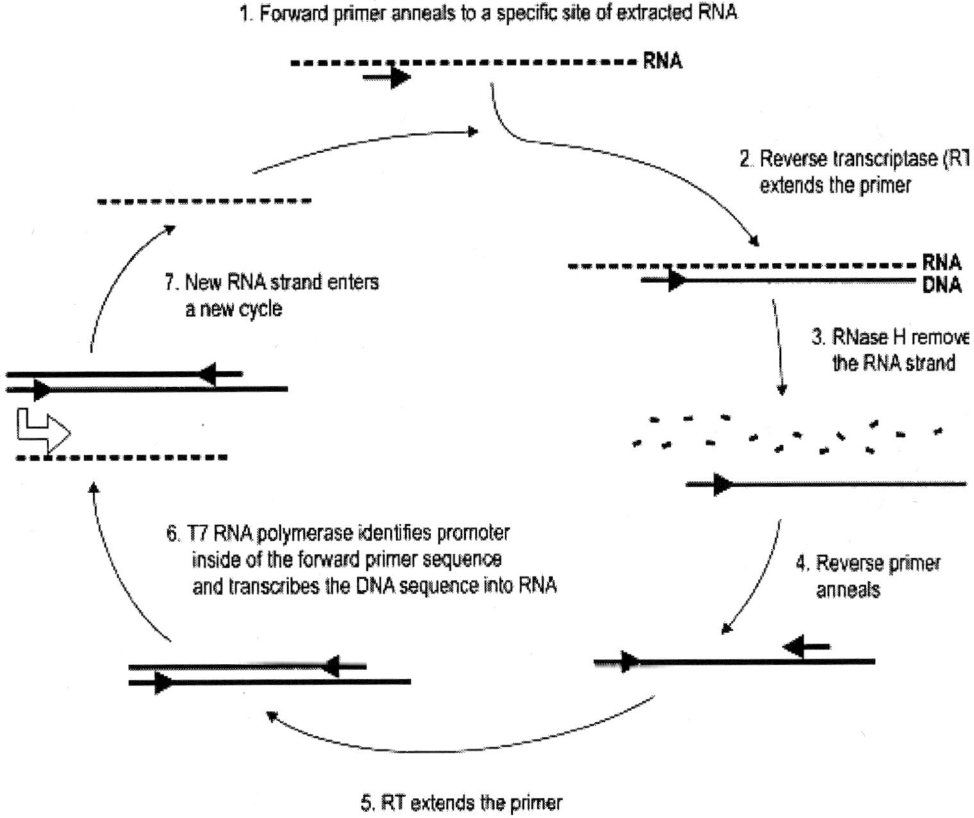

Figure 1. Nucleic Acid Sequence Based Amplification (NASBA) reaction steps.

B. Strain Typing

In contrast to detection techniques, which target universal or species-specific sequences, strain typing aims to differentiate specific strains or clones of a given species among clinical isolates. Therefore, pure culture of each isolate has to be available. Again, a number of techniques seek to address sequence polymorphisms of strains or clones. As more such polymorphisms are included into the comparison, as higher discriminatory power can be achieved. Therefore, it is sometimes recommended to apply at least two independent techniques in a given array of isolates to verify the results and increase the discriminatory power.

1. Electrophoretic Karyotyping (EK)

Most fungal species display chromosome-length polymorphism, which results from unequal chromosomal rearrangements or from chromosome breakage and healing (Zolan, 1995). Sequences that cause these unequal rearrangements are mainly transposable elements and other dispersed repeats. Chromosome-length polymorphisms can be assessed by electrophoretic karyotyping, exploiting a modified type of electrophoresis. During conventional electrophoresis, fragments longer than 50 kb show the same mobility and remain unseparated. Pulse Field Gel Electrophoresis (PFGE) allows separation of extremely large DNA fragments or even whole chromosomes in agarose gel using an alternating electric field. In order to avoid mechanical breakage of DNA molecules during common extraction procedures, intact cells are first embedded into an agarose block and lysis is carried out in a plug of agarose excised from the initial block. The agarose matrix keeps the DNA molecules intact, in place, while allowing the reagents to diffuse freely. A plug can be incubated with detergents and enzymes to remove cellular components from the DNA. Then the plug is inserted into the gel and intact chromosomes migrate from the plug into the gel during electrophoresis. After each change of polarity of the current during pulsed-field gel electrophoresis, DNA coils reorient and move in a different direction. This helps them to pass through the gel. With each reorientation of the electric field, smaller sized DNA molecules realign and move in the new direction more quickly than the larger ones. Thus, the larger DNA molecules lag behind which contributes to size-separation of individual chromosomes.

Chromosome sizes can vary greatly between different *Candida* strains, thus resulting in various banding patterns known as electrophoretic karyotypes. Evaluation of differences in banding patterns can be performed visually or can be computer-assisted. If chromosomal length polymorphism is not satisfactorily discriminatory, chromosomes can be digested by rare-cutting restriction endonucleases prior to electrophoresis, to get several variable large DNA fragments. Although the equipment needed for PFGE is rather expensive and sample preparation is labour-intensive, tedious and not suitable for analysis of large numbers of samples, EK is a well-established method in *Candida* spp. typing (Monod *et al.*, 1990). It has been successfully used to differentiate strains of *C. albicans*, *C. lusitaniae*, *C. parapsilosis*, *C. tropicalis* and *C. glabrata* (Magee *et al.*, 1987; Merz *et al.*, 1988; Asakura *et al.*, 1991; Carruba *et al.*, 1991; Espinel-Ingro *et al.*, 1999), and to distinguish *C. albicans* from the phenotypically close species *C. dubliniensis* (Sullivan *et al.*, 1998). It also performed well when compared to other typing techniques (Lopez-Ribot *et al.*, 2000).

2. Restriction Analysis (REA)

In this technique, genomic DNA is typically cleaved by a frequently cutting restriction endonuclease to result in sequence-dependent restriction fragment length polymorphism (RFLP). This can be visualized by separating the fragments yielded using common agarose gel electrophoresis. The resulting banding patterns show interstrain variations as a result of the polymorphic nature of restriction site sequences or as a result of deletions and insertions in the DNA stretches between cleaving sites. Restriction analysis is rapid, easy, inexpensive, but restriction patterns are very complex and therefore difficult to compare. The majority of intense bands in an RFLP pattern represent rDNA sequences and mitochondrial DNA sequences; these fragments do not provide enough information to assess the relatedness of moderately related isolates. For easier comparison, separated fragments can be transferred onto a membrane and hybridized with a labeled fingerprinting probe that will recognize relatively few fragments of restricted DNA. Multi-copy probes designed to bind repeat sequences dispersed throughout the genome (e.g. RNA genes, mitochondrial DNA sequences and repetitive sequences) were applied (Soll, 2000). In contrast to bacteria, endonuclease digestion of ribosomal cistrons generates fragments of similar relative size in yeasts, resulting in a simple Southern blot hybridisation pattern, which is low in resolution for strain discrimination (Gil-Lamaignere *et al.*, 2003). Fungal rRNA and mitochondrial probes therefore have not been generally used in broad epidemiological studies (Soll, 2000). Use of fragments containing repetitive genomic sequences as probes, e.g. the Ca3 probe, has been more successful (Anderson *et al.*, 1993). Synthetic sequences derived from microsatellites have also been used to hybridize to hypervariable loci of fungal DNA cut with restriction endonuclease (Sullivan *et al.*, 1993). The locus of interest can also be amplified using gene-specific primers and then subjected to REA (Olive *et al.*, 1999). Trost *et al.* (2004) evaluated identification and strain characterization of clinically relevant *Candida* species by amplification of intergenic spacers ITS1 andITS2 followed by evaluation of RFLP of PCR products after sequence-specific enzymatic cleavage.

3. Random Amplified Polymorphic DNA (RAPD) or Arbitrarily primed PCR (AP-PCR)

Two groups originally developed this technique independently (Williams *et al.*, 1990; Welsh *et al.*, 1991). Although the term AP-PCR may better fit the principle of the technique, the term RAPD (pronounced *rapid*) is more widely used because of its simplicity. RAPD is a PCR-fingerprinting technique, which employs a single short primer (typically 10 bases in length), sequence of which is chosen arbitrarily rather than based on knowledge of the targeted genomic sequence. Therefore, RAPD requires no prior knowledge of sequence of the examined organism and can be applied universally. At the beginning, a set of oligonucleotide primers of random sequences has to be screened; those with optimal performance are chosen for further analysis. Because of its short sequence, a RAPD primer is able to anneal too many locuses throughout the genome.

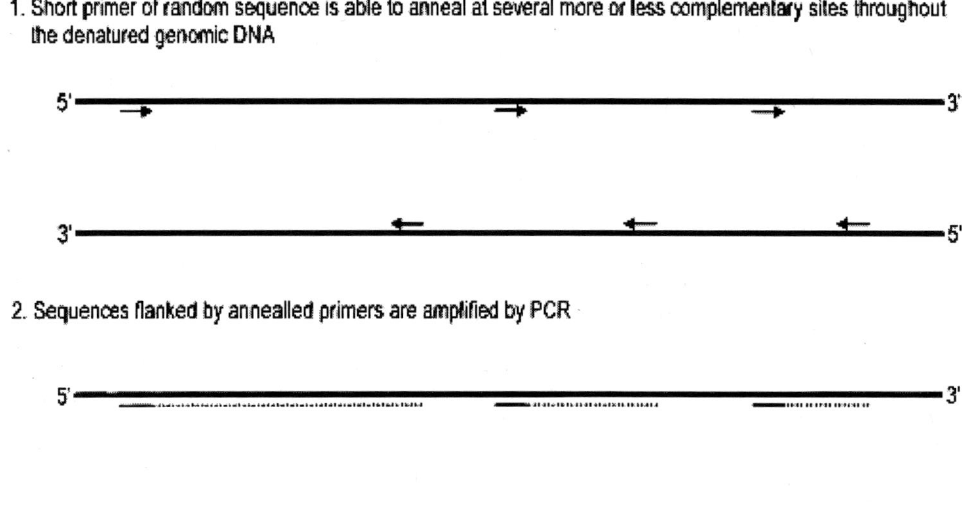

Figure 2. Essential steps of Random Amplified Polymorphic DNA (RAPD) technique.

In addition, when RAPD is performed for typing purposes, annealing occurs under low stringency conditions (typically 35–40°C, 2.5 mM MgCl2). Then, a primer likewise hybridizes with many additional imperfectly matched sequences with sufficient affinity. If two molecules of the same primer anneal in a proper orientation close enough for the PCR to proceed efficiently, an amplicon is generated. Several amplicons are typically generated in a complex genomic DNA, which can be visualized as a banding pattern by gel electrophoresis. Because of competition for annealing of the primer, minor inter-strain polymorphisms in annealing sites result in strain-specific variations of banding patterns. For schematic overview of the technique see Figure 2. The majority of amplified fragments originate from unique sequences rather than from repetitive elements (Welsh *et al.*, 1995). When targeting the entire genome, RAPD generally results in more complex patterns than standard PCR or RFLP and may increase the likelihood of detecting interstrain differences (Swaminathan *et al.*, 1995). RAPD is easily designed, technically simple, economic and also faster than other typing methods. However, special emphasis should be laid on the performance of this technique, because minor differences in experimental conditions can result in different profiles, which compromises intra- and interlaboratory reproducibility of RAPD. There are many factors, that can influence the appearance or disappearance of bands, including Mg2+ concentration, primer/template concentration ratio, *Taq* polymerase concentration and source, the model of thermal cycler etc. (Penner *et al.*, 1993; Meunier *et al.*, 1993; Tyler *et al.*, 1997) RAPD applied to eukaryotes seems to be more reproducible due to the higher stability of their genomes (Tyler *et al.*, 1997). To overcome reproducibility obstacles, PCR can be also targeted to specific sequences, for example microsatellite sequences, and performed at higher stringency (Shemer *et al.*, 2001). RAPD combines the advantage of operational simplicity with no need of prior sequence data, which makes it suitable especially for typing of a collection of isolates of less studied microorganisms, when interlaboratory comparison is not required. Therefore, RAPD is the most widely used typing technique in clinical mycology. It has been repeatedly shown, that RAPD can be employed as a convenient tool for species identification as well as strain typing (Lehmann *et al.*, 1992; Liu *et al.*, 1996; Stefan *et al.*, 1997; Bautista-Munoz *et al.*, 2003). On the other hand, problems with possible bias during complicated comparison of banding patterns consisting of bands of variable intensity should not be denied. To overcome this, a new McRAPD approach, which takes advantage of melting analysis of RAPD amplicons, has been introduced recently (Plachy *et al.*, 2005). It omits both gel electrophoresis and tedious analysis of banding patterns and directly yields numerical data, which can be subjected to automated unbiased analysis. It was first employed for rapid and accurate species identification in five pathogenic yeast species. However, when high-resolution melting device and saturating dye are used to increase its discriminatory power, it seems to be well suitable for strain typing as well.

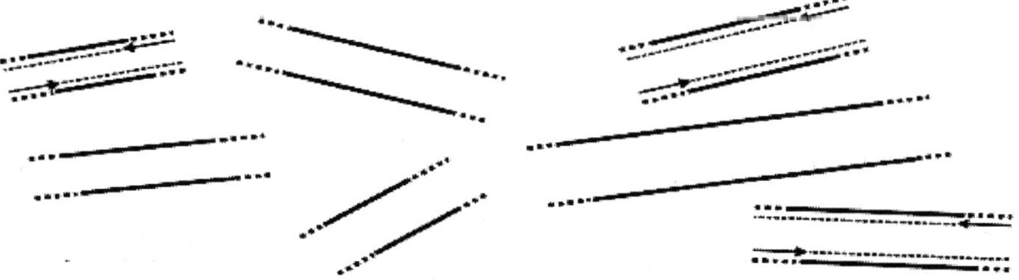

Figure 3. Essential steps of Amplified Fragment Length Polymorphism (AFLP) technique.

4. Amplified Fragment Length Polymorphism (AFLP)

Amplified fragment length polymorphism is a PCR based DNA fingerprinting technique (Vos *et al.*, 1995). In short, the AFLP procedure usually starts by digestion of genomic DNA using two restriction enzymes (generally a hex-cutter and a tetra-cutter) followed by legation of double-stranded adapters to the ends of the restriction fragments. Afterwards, amplification of the restriction fragments by PCR using two primers complementary to the adapter- and restriction site sequences takes place. To reduce the amount of resulting amplicons, one to three selective nucleotides are usually added to the 3'-ends of primers. Then, a subset of restriction fragments, harboring complementary sequence to the primer extension adjacent to the restriction site, is amplified. Finally, amplified restriction fragments are separated by gel electrophoresis and analyzed. Primers may be labeled by fluorescent tags enabling a computer-based automated sequence analyzer to read the polyacrylamide gel electrophoretic patterns (Fluorescent Amplified Fragment Length Polymorphism, FAFLP). For schematic overview of the technique see Figure 3. A typical AFLP fingerprint represents between 50-100 bands which greatly increase the discriminatory power of AFLP and this facilitates its use in epidemiological studies. On the other hand, the expense and expertise needed discourage its routine use. Like RAPD and RFLP, AFLP requires no prior sequence information. However, compared to RAPD, specific primers and stringent annealing temperature ensure that AFLP is a highly reproducible and robust method, whereas compared to RFLP, fingerprints obtained with AFLP are more informative, easier to read and reading can be automated. At first, AFLP was used for bacteria typing in microbiology (Lin *et al.*, 1996). More recently, pathogenic *Candida* species identification (Borst *et al.*, 2003) and typing (Ball *et al.*, 2004) have also been reported.

5. DNA Sequencing

Obviously, the most accurate way to compare two individuals, strains or clones is to sequence their entire genomes. Of course, such an approach is impracticable for routine use. Instead, parts of a selected fungal gene can be amplified and resulting sequences can be compared (Chen *et al.*, 2002). However, focusing on just one variable fragment can hardly provide enough data for strain typing and is usually suitable for species identification only. Coleman *et al.* (1997) studied a fragment of the V3 variable region of the large ribosomal subunit genes from *C. dubliniensis* isolates and found out that it was significantly different from the other species analysed. Including more locuses into sequencing and consecutive comparison can increase the discriminatory power. Botterel *et al.* (2001) used sequencing of three polymorphic microsatellite markers to *C. albicans* typing. To date, the most complex and promising use of sequencing for typing purposes is represented by the Multi Locus Sequence Typing (MLST) approach. MLST focuses on the nucleotide polymorphism of internal fragments of several housekeeping genes, where each unique allele combination determines a sequence type of strain. Fragments of the housekeeping genes are first amplified by PCR and then sequenced. This approach has many advantages – it provides unambiguous, portable and easy to standardize results. Absolute interlaboratory reproducibility can be achieved, enabling global epidemiological studies. The MLST approach is developed continuously as an open platform (Aanensen *et al.*, 2005). Bougnoux *et al.* (2002) investigated the potential value of MLST for characterization of clinical isolates of *C. albicans* and found it a highly resolving and stable method. MLST was also compared to other fingerprinting methods already established for *C. albicans* by Robles *et al.* (2004) MLST system for strain differentiation has been already set up for *C. tropicalis*, *C. glabrata* and *C. albicans* (Tavanti *et al.*, 2005). Although the technique is currently not feasible for most laboratories, increasingly user-friendly automation together with cutting of costs due to newly emerging technologies (e.g. pyrosequencing) promise broader availability in the future.

6. DNA-Microarrays

Microarray-based systems offer an attractive outlook not only for the future of strain typing. They offer high level of sensitivity, specificity and throughput capacity, without requiring *a priori* knowledge of specific sequences. Chips or microarrays are high-density microscopic sets of oligonucleotide probes immobilized on solid surface, to which nucleic acid samples are hybridised. Perfectly matched sequences from the sample hybridize more efficiently to the corresponding oligomers on the array and give stronger signal than mismatched bound sequences. The final signal is detected by high-resolution fluorescent scanning and analyzed by computer software, thus enabling automation and standardization (Kurella *et al.*, 2001). Easier management of the vast data generated and reduction of the costs of DNA-chips are only a matter of time. Then, microarrays surely will move from the research area to clinical practice. For typing purposes, microarrays can be directed to identify the presence and quantity of different sequence variants of specific genes or regions, e.g. rRNA genes, internal transcribed spacers (ITSs) in particular. Ongoing sequencing projects in pathogenic yeasts will also soon enable quite straightforward designing of whole-genome DNA microarrays (Cummings *et al.*, 2000). The use of microarrays for microbial fingerprinting has been already reported for *Salmonella enterica* isolates (Willse *et al.*, 2004), for closely related *Xanthomonas* pathovars (Kingsley *et al.*, 2002) and *Mycobacterium* species (Troesch *et al.*, 1999). Rapid automated performance of tens of thousands of hybridization assays on a tiny chip represents the strongest point of this technology.

Conclusion

To conclude, both yeast species identification and strain typing mainly relies on several modifications of two basic technologies – amplification of a DNA fragment followed by its analysis by different means, or, hybridization of the total genomic DNA to a set of probes. It is not easy to foresee, which of these approaches will ultimately prevail. Most probably, the repertoire of techniques applied will be reduced only partly, because different techniques can best match different particular needs. Both the costs, easy operation and the reproducibility and resolving power will be considered for particular applications of different techniques.

Also, both a trend towards better performance of less reproducible or less discriminatory but cost-effective techniques and a trend towards lower costs of highly reproducible and discriminatory but rather expensive techniques can be clearly observed. There is no doubt that PCR will be continued as the core of rapid detection and identification techniques. Due to its low cost and potential for automation, high resolution melting analysis (HRMA) promises to bring outstanding progress in post-PCR analysis in the near future. Melting analysis has already been applied in yeast species identification (Plachy *et al.*, 2005) and our preliminary data show, that it is also suitable for improving the potential of RAPD typing. Furthermore, the potential of HRMA to at least partly substitute for sequencing in MLST can be envisaged, possibly establishing a new Multi Locus Melting Typing (MLMT) approach. This should bring the merits of these technologies even closer to all routine laboratories in near future. In addition, the DNA-microarrays technology surely has the potential to revolutionize DNA-based diagnostics also in the field of clinical mycology in the more distant future.

REFERENCES

Aanensen, D. M. and Spratt, B. G. (2005). The multilocus sequence typing network: mlst.net. *Nucleic Acids Res 33*: W728–W733.

Alexander, B. D., Ashley, E. D., Reller, L. B., Reed, S. D. (2006). Cost savings with implementation of PNA FISH testing for identification of *Candida albicans* in blood cultures. *Diagn Microbiol Infect Dis* (in press).

Anderson, J., Srikantha, T., Morrow, B., Miyasaki, S. H., White, T. C., Agabian, N., Schmid, J., Soll, D. R. (1993). Characterization and partial nucleotide sequence of the DNA fingerprinting probe Ca3 of *Candida albicans*. *J Clin Microbiol 31*: 1472–80.

Asakura, K., Iwaguchi, S. I., Homma, M., Sukai, T., Higashide, K., Tanaka, K. (1991). Electrophoretic karyotypes of clinically isolated yeasts of *Candida albicans* and *C. glabrata*. *J Gen Microbiol 137*: 2531–8.

Ball, L. M., Bes, M. A., Theelen, B., Boekhout, T., Egeler, R. M., Kuijper, E. J. (2004). Significance of amplified fragment length polymorphism in identification and epidemiological examination of *Candida* species colonization in children undergoing allogenic stem cell transplantation. *J Clin Microbiol 42*: 1673–9.

Bautista-Munoz, C., Boldo, X. M., Villa-Tanaca, L., Hernandez- Rodriguez, C. (2003). Identification of *Candida* spp. by randomly amplified polymorphic DNA analysis and differentiation between *Candida albicans* and *Candida dubliniensis* by direct PCR methods. *J Clin Microbiol 41*: 414–20.

Borst, A., Theelen, B., Reinders, E., Boekhout, T., Fluit, A. C., Savelkoul, P. H. M. (2003). Use of amplified fragment length polymorphism analysis to identify medically important *Candida* spp., including *C. dubliniensis*. *J Clin Microbiol 41*: 1357–62.

Botterel, F., Desterke, C., Costa, C., Bretagne, S. (2001). Analysis of microsatellite markers of *Candida albicans* used for rapid typing. *J Clin Microbiol 39*: 4076–81.

Bougnoux, M. E., Dupont, C., Mateo, J., Saulnier, P., Faivre, V., Payen, D., NicolasChanoine, M. H. (1999). Serum is more suitable than whole blood for diagnosis of systemic candidiasis by nested PCR. *J Clin Microbiol 37*: 925–30.

Bougnoux, M. E., Morand, S., d'Enfert, C. (2002). Usefulness of multilocus sequence typing for characterization of clinical isolates of *Candida albicans*. *J Clin Microbiol 40*: 1290–7.

Bu, R., Sathiapalan, R. K., Ibrahim, M. M., Al-Mohsen, I., Almodovar, E., Gutierrez, M. I., Bhatia, K. (2005). Monochrome LightCycler PCR assay for detection and quantification of five common species of *Candida* and *Aspergillus*. *J Med Microbiol 54*: 243–8.

Burgener-Kairuz, P., Zuber, J. P., Jaunin, P., Buchman, T. G., Bille, J., Rossier, M. (1994). Rapid detection and identification of *Candida albicans* and *Torulopsis (Candida) glabrata* in clinical specimens by species specific nested PCR amplification of a cytochrome P450 lanosterol-alpha-demethylase (L1A1) gene fragment. *J Clin Microbiol 32*: 1902–7.

Capoor, M. R., Nair, D., Deb, M., Verma, P. K., Srivastava, L., Aggarwal, P. (2005). Emergence of non-*albicans Candida* species and antifungal resistance in a tertiary care hospital. *Jpn J Infect Dis 58*: 344–8.

Carruba, G., Pontieri, E., De Bernardis, F., Martino, P., Cassone, A. (1991). DNA fingerprinting and electrophoretic karyotypes of environmental and clinical isolates of *Candida parapsilosis*. *J Clin Microbiol 29*: 916–22.

Chen, S. C. A., Halliday, C. L., Meyer, W. (2002). A review of nucleic acid based diagnostic tests for systemic mycoses with an emphasis on polymerase chain reaction-based assays. *Med Mycol 40*: 333–57.

Chen, Y. C., Eisner, J. D., Kattar, M. M., Rassoulian-Barrett, S. L., LaFe, K., Yarfitz, S. L., Limaye, A. P., Cookson, B. T. (2000). Identification of medically important yeasts using PCR-based detection of DNA sequence polymorphisms in the internal transcribed spacer 2 region of the rRNA genes. *J Clin Microbiol 38*: 2302–10.

Coleman, D., Sullivan, D., Harrington, B., Haynes, K., Henman, M., Shanley, D., Bennet, D., Moran, G., McCreary, C., O'Neill, L. (1997). Molecular and phenotypic analysis of *Candida dubliniensis*: a recently identified species linked with oral candidasis in HIV-infected patients. *Oral Dis 3*: 96–101.

Compton, J. (1991). Nucleic acid sequence-based amplification. *Nature 350*: 91–2.

Crampin, A. C., Matthews, R. C. (1993). Application of the polymerase chain reaction to the diagnosis of candidosis by amplification of an HSP 90 gene fragment. *J Med Microbiol 39*: 233–8.

Cummings, C. A., Relman, D. A. (2000). Using DNA microarrays to study host-microbe interactions. *Emerg Infect Dis 6*: 513–25.

de Marie, S. (2000). New developments in the diagnosis and management of invasive fungal infections. *Haematologica 85*: 88–93.

Dendis, M., Horváth, R., Michálek, J., Rùžièka, F., Grijalva, M., Bartoš, M., Benedík, J. (2003). PCR-RFLP detection and species identification of fungal pathogens in patients with febrile neutropenia. *Clin Microbiol Infect 9*: 1191–202.

Espinel-Ingro, A., Vazquez, J. A., Boikov, D., Pfaller, M. A. (1999). Evaluation of DNA-based typing procedures for strain categorization of *Candida* spp. *Diagn Microbiol Infect Dis 33*: 231–9.

Flahaut, M., Sanglard, D., Monod, M., Bille, J., Rossier, M. (1998). Rapid detection of *Candida albicans* in clinical samples by DNA amplification of common regions from *C. albicans*-secreted aspartic proteinase genes. *J Clin Microbiol 36*: 395–401.

Fujita, S., Lasker, B. A., Lott, T. J., Reiss, E., Morrison, C. J. (1995). Microtitration plate enzyme immunoassay to detect PCR-amplified DNA from *Candida* species in blood. *J Clin Microbiol 33*: 962–7.

Gil-Lamaignere, C., Roilides, E., Hacker, J., Müller, F. M. C. (2003). Molecular typing of fungi-a critical review of the possibilities and limitations of currently and future methods. *Clin Microbiol Infect 9*: 172–85.

Guiver, M., Levi, K., Oppenheim, B.A. (2001). Rapid identification of *Candida* species by TaqMan PCR. *J Clin Pathol 54*:362–6.

Hall, L., Wohlel, S., Roberts, G. D. (2003). Experience with the MicroSeq D2 large-subunit ribosomal DNA sequencing kit for identification of commonly encountered, clinically important yeast species. *J Clin Microbiol 41*: 5099–102.

Herrmann, M. G., Durtschi, J. D., Bromley, L. K., Wittwer, C. T., Voelkerding, K. V. (2006). Amplicon DNA melting analysis for mutation scanning and genotyping: cross-platform comparison of instruments and dyes. *Clin Chem 52*: 494–503.

Hsu, M. C., Chen, K. W., Lo, H. J., Chen, Y. C., Liao, M. H., Lin, Y. H., Li, S. Y. (2003). Species identification of medically important fungi by use of real-time LightCycler PCR. *J Med Microbiol 52*: 1071–6.

Hui, M., Ip, M., Chan, P. K., Chin, M. L., Cheng, A. F. (2000). Rapid identification of medically important *Candida* to species level by polymerase chain reaction and single-strand conformational polymorphism. *Diagn Microbiol Infect Dis 38*: 95–9.

Jaeckel, S., Epplen, J. T., Kauth, M., Miterski, B., Tschentscher, F., Epplen, C. (1998). Polymerase-chain reaction-single strand conformation polymorphism or how to detect reliably and efficiently each sequence variation in many samples and many genes. *Electrophoresis 19*: 3055–61.

Jung, M., Dritschilo, A., Kasid, U. (1992). Reliable and efficient direct sequencing of PCR-amplified double-stranded genomic DNA template. *PCR Methods Appl 1*: 171–4.

Kan, V. L. (1993). Polymerase chain reaction for the diagnosis of candidaemia. *J Infect Dis 168*: 779–83.

Kanbe, T., Horii, T., Arishima, T., Ozeki, M., Kikuchi, A. (2002). PCR based identification of pathogenic *Candida* species using primer mixes specific to *Candida* DNA topoisomerase II genes. *Yeast 19*:973–89.

Kempf, V. A. J., Trebesius, K., Autenrieth, I. B. (2000). Fluorescent *in situ* hybridization allows rapid identification of microorganisms in blood cultures. *J Clin Microbiol 38*: 830–8.

Kievits, T., Van Gemen, B., Van Strijp, D., Schukking, R., Dircks, M., Andriaanse, H., Malek, L., Sooknanan, R., Lens, P. (1991). NASBA isothermal enzymatic *in vitro* nucleic acid amplification optimized for the diagnosis of HIV-1 infection. *J Virol Methods 35*: 273–86.

Kingsley, M. T., Straub, T. M., Call, D. R., Daly, D. S., Wunschel, S. C., Chandler, D. P. (2002). Fingerprinting closely related *xanthomonas* pathovars with random nanomer oligonucleotide microarrays. *Appl Environ Microbiol 68*: 6361–70.

Krcmery, V., Barnes, A. J. (2002). Non-*albicans Candida* spp. causing fungaemia: pathogenicity and antifungal resistance. *J Hosp Infect. 50*: 243–60.

Kurella, M., Hsiao, L. L., Yoshida, T., Randall, J. D., Chow, G., Sarang, S. S., Jensen, R. V., Gullans, S. R. (2001). DNA microarray analysis of complex biologic processes. *J Am Soc Nephrol 12*: 1072–8.

Kwok, S. and Higuchi, R. (1989). Avoiding false positives with PCR. *Nature 339*: 237–8.

Lehmann, P. F., Lin, D., Lasker, B. A. (1992). Genotypic identification and characterization of species and strains within the genus *Candida* by using random amplified polymorphic DNA. *J Clin Microbiol 30*: 3249–54.

Lin, J. J., Kuo, J., Ma, J. (1996). A PCR-based DNA fingerprinting technique: AFLP for molecular typing of bacteria. *Nucleic Acids Res 24*: 3649–50.

Lind, K., Stahlberg, A., Zoric, N., Kubista, M. (2006). Combining sequence-specific probes and DNA binding dyes in real-time PCR for specific nucleic acid quantification and melting curve analysis. *BioTechniques 40*: 315–319.

Lischewski, A., Kretschmar, M., Hof, H., Amann, R., Hacker, J., Morschhäuser, J. (1997). Detection and identification of *Candida* species in experimentally infected tissue and human blood by rRNA-specific fluorescent *in situ* hybridization. *J Clin Microbiol 35*: 2943–8.

Liu, D., Coloe, S., Jones, S. L., Baird, R., Pedersen, J. (1996). Genetic speciation of *Candida* isolates by arbitrarily primed polymerase chain reaction. *FEMS Microbiol Lett 145*: 23–6.

Loeffler, J., Dorn, C., Hebart, H., Cox, P., Magga, S., Einsele, H. (2003). Development and evaluation of the Nuclisens Basic Kit NASBA for the detection of RNA from *Candida* species frequently resistant to antifungal drugs. *Diagn Microbiol Infect Dis 45*: 217–20.

Loeffler, J., Hebart, H., Bialek, R., Hagmeyer, L., Schmidt, D., Serey, F. P., Hartmann, M., Eucker, J., Einsele, H. (1999). Contaminations occurring in fungal PCR assays. *J Clin Microbiol 37*: 1200-2.

Lopez-Ribot, J. L., McAtee, R. K., Kirkpatrick, W. R., Perea, S., Patterson, T. F. (2000). Comparison of DNA-based typing methods to assess genetic diversity and relatedness among *Candida albicans* clinical isolates. *Rev Iberoam Micol 17*: 49–54.

Louie, M., Louie, L., Simor, A. E. (2000). The role of DNA amplification technology in the diagnosis of infectious diseases. *CMAJ 163*: 301–9.

Mackay, I. M. (2004). Real-time PCR in the microbiology laboratory. *Clin Microbiol Infect 10*: 190–212.

Magee, B. B. and Magee P. T. (1987). Electrophoretic karyotypes and chromosome numbers in *Candida* species. *J Gen Microbiol 133*: 425–30.

Marino, M. A., Turni, L. A., Del Rio, S. A., Williams, P. E. (1994). Molecular size determinations of DNA restriction fragments and polymerase chain reaction products using capillary gel electrophoresis. *J Chromatogr A 676*: 185–9.

Merz, W. G., Connelly, C., Hieter, P. (1988). Variation of electrophoretic karyotype among clinical isolates of *Candida albicans*. *J Clin Microbiol 26*: 842–5.

Meunier, J. R., Grimont, P. A. (1993). Factors affecting reproducibility of random amplified polymorphic DNA fingerprinting. *Res Microbiol 144*: 373–9.

Monod, M., Porchet, S., Baudraz-Rosselet, F., Frenk, E. (1990). The identification of pathogenic yeast strains by electrophoretic analysis of their chromosomes. *J Med Microbiol 29*:123–9.

Morace, G., Sanguinetti, M., Posteraro, B., Lo Cascio, G., Fadda, G. (1997). Identification of various medically important *Candida* species in clinical specimens by PCR-restriction enzyme analysis. *J Clin Microbiol 35*: 667–72.

Nguyen, M. H., Peacock, J. E., Morris, A. J., Tanner, D. C., Nguyen, M. L., Snydman, D. R., Wagener, M. M., Rinaldi, M. G., Yu, V. L. (1996). The changing face of candidemia: emergence of non-*Candida albicans* species and antifungal resistance. *Am J Med 100*: 617–23.

Odell, I. D., Cloud, J. L., Seipp, M., Wittwer, C. T. (2005). Rapid species identification within the *Mycobacterium chelonae*-abscessus group by high-resolution melting analysis of hsp65 PCR products. *Am J Clin Pathol 123*: 96–101.

Olive, D. M. and Bean, P. (1999). Principles and applications of methods for DNA-based typing of microbial organisms. *J Clin Microb 37*: 1661–9.

Oliviera, K., Haase, G., Kurtzman, C., Hyldig-Nielsen, J. J., Stender, H. (2001). Differentiation of *Candida albicans* and *Candida dubliniensis* by fluorescent *in situ* hybridization with peptide nucleic acid probes. *J Clin Microbiol 39*: 4138–41.

Penner, G. A., Bush, A., Wise, R., Kim, W., Domier, L., Kasha, K., Laroche, A., Scoles, G., Molnar, S. J., Fedak, G. (1993). Reproducibility of random amplified polymorphic DNA (RAPD) analysis among laboratories. *PCR Methods Appl 2*: 341–5.

Pfaller, M. A. (1996) Nosocomial candidiasis: emerging species, reservoirs and modes of transmission. *Clin Infect Dis 22*: 89–94.

Plachý, R., Hamal, P., Raclavský, V. (2005). McRAPD as a new approach to rapid and accurate identification of pathogenic yeasts. *J Microbiol Methods 60*: 107–13.

Pryce, T. M., Kay, I. D., Palladino, S., Heath, C. H. (2003). Real-time automated polymerase chain reaction (PCR) to detect *Candida albicans* and *Aspergillus fumigatus* DNA in whole blood from high-risk patients. *Mycology 47*: 487–96.

Reiss, E., Tanaka, K., Bruker, G., Chazalet, V., Coleman, D., Debeaupuis, J. P., Hanazawa, R., Latge, J. P, Lortholary, J., Makimura, K., Morrison, C. J., Murayama, S. Y., Naoe S, Paris S, Sarfati J, Shibuya K, Sullivan D, Uchida K, Yamaguchi H. (1998). Molecular diagnosis and epidemiology of fungal infections. *Med Mycol 36*: 249–57.

Rigby, S., Procop, G. W., Haase, G., Wilson, D., Hall, G., Kurtzman, C., Oliveira, K., Oy, S. V., Hyldig-Nielsen, J. J., Coull, J., Stender, H. (2002). Fluorescence *in situ* hybridization with peptide nucleic acid probes for rapid identification of *Candida albicans* directly from blood culture bottles. *J Clin Microbiol 40*: 2182–6.

Ririe, K. M., Rasmussen, R. P., Wittwer, C. T. (1997). Product differentiation by analysis of DNA melting curves during the polymerase chain reaction. *Anal Biochem 245*:154–60.

Robles, J. C., Koreen, L., Park, S., Perlin, D. S. (2004). Multilocus sequenced typing is a reliable alternative method to DNA fingerprinting for discriminating among strains of *C. albicans*. *J Clin Microbiol 42*: 2480–8.

Selvarangan, R., Bui, U., Limaye, A. P., Cookson. (2003). Rapid identification of commonly encountered *Candida* species directly from blood culture bottles. *J Clin Microbiol 41*:5660–4.

Shemer, R., Weissman, Z., Hashman, N., Kornitzer, D. (2001). A highly polymorphic degenerate microsatellite for molecular strain typing of *Candida krusei*. *Microbiology 147*: 2021–8.

Soll, D. R. (1992) High-frequency switching in *Candida albicans*. *Clin Microbiol Rev 5:* 183–203.

Soll, D. R. (2000). The ins and outs of DNA fingerprinting of infectious fungi. *Clin Microbiol Rev 13*: 332–70.

Stefan, P., Vazquez, J. A., Boikov, D., Xu, C., Sobel, J. D., Atkins, R. A. (1997). Identification of *Candida* species by randomly amplified polymorphic DNA fingerprinting of colony lysates. *J Clin Microbiol 35*: 2031–9.

Sullivan, D. and Coleman, D. (1998). *Candida dubliniensis*: characteristics and identification. *J Clin Microbiol 36*: 329–34.

Sullivan, D. J., Bennett, D., Henman, M., Harwood, P., Flint, S., Mulkahy, F., Shanley, D., Coleman, D. (1993). Oligonucleotide fingerprinting of isolates of *Candida* species other than *C. albicans* and of atypical *Candida* species from human immunodefficiency virus-positive and AIDS patients. *J Clin Microbiol 31*: 2124–33.

Swaminathan, B. and Barrett, T. J. (1995). Amplification methods for epidemiologic investigations of infectious diseases. *J Microbiol Methods 23*: 129–39.

Tavanti, A., Davidson. A. D., Johnson, E. M., Maiden, M. C., Shaw, D. J., Gow, N. A., Odds, F. C. (2005). Multilocus sequence typing for differentiation of strains of *Candida tropicalis*. *J Clin Microbiol 43*: 5593–600.

Tenover, F. C., Arbeit, R. D., Goering, R. V. (1997). How to select and interpret molecular strain typing methods for epidemiological studies of bacterial infections: a review for healthcare epidemiologists. *Infect Control Hosp Epidemiol 18*: 426–39.

Troesch, A., Nguyen, H., Miyada, C. G., Desvarenne, S., Gingeras, T. R., Kaplan, P. M., Cros, P., Mabilat, C. (1999). *Mycobacterium* species identification and rifampicin resistance testing with high-density DNA probe arrays. *J Clin Microbiol 37:* 49–55.

Trost, A., Graf, B., Eucker, J., Sezer, O., Possinger, K., Gobel, U. B., Adam, K. (2004). Identification of clinically relevant yeasts by PCR/RFLP. *J Microbiol Methods 56*: 201–11.

Tyler, K. D., Wang, G., Tyler, S. D., Johnson, W. M. (1997). Factors affecting reliability and reproducibility of amplification-based DNA fingerprint of representative bacterial pathogens. *J Clin Microbiol 35*: 339–46.

Vos, P., Hogers R, Bleeker, M., Reijans, M, van der Lee, T., Hornes, M., Frijters, A., Pot, J., Peleman, J., Kuiper, M., Zabeau, M. (1995). AFLP: a new technique for DNA fingerprinting. *Nucleic Acids Res 23*: 4407–14.

Welsh, J. and McClelland, M. (1991). Genomic fingerprinting with APPCR using pairwise combinations of primers: Application to genetic mapping of the mouse. *Nucleic Acids Res 19*: 861–6.

Welsh, J., Ralph, D., McClelland. (1995). DNA and RNA fingerprinting using arbitrarily primed PCR. In: Innis MA, Gelfand DH, Sninsky JJ, editors. *PCR strategies*. Academic Press, San Diego/New York p. 249–76.

White, P. L., Shetty, A., Barnes, R.A. (2003). Detection of seven *Candida* species using the LightCycler system. *J Med Micobiol 52*: 229–38.

Widjojoatmodjo, M. N., Borst, A., Schukking, R. A. F., Box, A. T. A., Tacken, N. M. M., Gemen, B. V., Verhoef, J., Top, B., Fluit, A.C. (1999). Nucleic acid sequence-based amplification (NASBA) detection of medically important *Candida* species. *J Microbiol Methods 38*: 81–90.

Williams, J. G. K., Kubelik, A. R., Livak, K. J., Rafalski, J. A., Tingey, S. V. (1990). DNA polymorphisms amplified by arbitrary primers are useful as genetic markers. *Nucleic Acids Res 18*: 1631–5.

Willse, A., Straub, T. M., Wunschel, S. C., Small, J. A., Call, D. R., Daly D. S., Chandler, D. P. (2004). Quantitative oligonucleotide microarray fingerprinting of Salmonella enterica isolates. *Nucleic Acids Res 32*: 1848–56.

Wilson, D. A., Joyce, M. J., Hall, L. S., Reller, L. B., Roberts, G. D., Hall, G. S., Alexander, B. D., Procop, G. W. (2005). Multicenter evaluation of a *Candida albicans* peptide nucleic acid fluorescent *in situ* hybridization probe for characterization of yeast isolates from blood cultures. *J Clin Microbiol 43*: 2909–12.

Zolan, M.E. (1995). Chromosome-length polymorphism in fungi. *Microbiol Rev 59*: 686–98.

INDEX

2

2D, 50, 51

A

aberrant, 227
abortion, 96
abundance, 168, 334, 335
access, 65, 66, 67, 68, 74, 76, 78, 80, 91, 229, 265, 282, 302
accessibility, 66, 78, 264
accuracy, 302
ACEI, 218
acetate, 100, 234, 237, 239
achievement, 86
acid, 9, 28, 45, 97, 99, 100, 159, 183, 186, 206, 225, 227, 234, 236, 237, 242, 246, 353, 365, 370
acidic, 182, 223, 224, 225, 227, 230, 231, 232, 233, 234, 243, 245
acidity, 225
actin, 23, 26, 27, 30, 32, 203, 285, 302, 311, 314, 315, 316, 319, 322, 332, 353
activation, 50, 68, 205, 212, 223, 225, 228, 229, 230, 242, 243
activators, 49
active site, 221
acute, 181, 350
adaptation, 28, 40, 62, 231, 242
adhesion, 32
adjustment, 3
aerobic, 92
Africa, 180, 199
Ag, 238
agar, 101, 103, 174, 175, 176, 177, 178, 179, 180, 181, 182, 183, 252, 254, 267, 316
Agaricus bisporus, 3, 34, 36
age, 34, 92
agent, 4, 5, 6, 7, 46, 74, 101, 139, 150, 159, 165, 195
agents, 1, 4, 5, 6, 7, 85, 103, 173, 197, 266, 284, 352
aggregation, 20, 21, 231, 242
aggressiveness, 110, 164
aging, 349
agrarian, 86
agricultural, xiii, 40, 43, 52, 63, 64, 66, 67, 80, 82, 233, 240, 245
agricultural crop, 40
agriculture, xiii, xiv, 42, 43, 67, 74, 78, 86, 101
aid, 305
AIDS, 94, 173, 195, 231, 351, 369
air, 171, 172, 180, 200, 344
alanine, 225, 236
alcohol, 218
alfalfa, 61, 162
algae, 199
algorithm, 6, 320
alkaline, 195, 201, 223, 224, 225, 226, 227, 228, 229, 230, 231, 232, 233, 234, 236, 242, 243, 244, 246
alkaline phosphatase, 224
allele, 42, 152, 183, 362
alleles, 42, 61, 143, 148, 151
alpha, 22, 24, 25, 26, 28, 29, 72, 153, 185, 197, 198, 199, 220, 222, 284, 314, 315, 316, 319, 322, 365
alternative, 42, 81, 203, 213, 227, 236, 301, 332, 352, 355, 368
alternatives, 3
amino, 45, 69, 99, 155, 183, 186, 214, 225, 226, 227, 236, 237, 238

amino acid, 45, 69, 155, 183, 186, 214, 225, 226, 227, 236, 237, 238
amino acids, 155, 225, 226, 227, 236, 238
amorphous, 232, 233
Amsterdam, 36, 102, 134, 137, 259
AMT, 250, 258
analog, 98
analysis of variance, 239
analytical techniques, 50
androgen, 220
anemia, 100
animal feed, 207
animal health, 85
animals, 1, 3, 7, 8, 32, 33, 34, 65, 86, 92, 94, 95, 97, 98, 108, 136, 171, 186, 195, 207, 264, 266, 301, 310, 314, 330
annealing, 88, 147, 151, 187, 353, 360, 362
annotation, 9, 12, 16, 24, 26, 32, 43, 44, 47, 53, 72, 73, 74, 76, 77, 132, 301, 314
Anopheles gambiae, 22, 24
anorexia, 85
ANOVA, 239
antagonist, 202, 214
antagonistic, 64, 73
antagonists, 220
antecedents, 140
anthropogenic, 346
antibiotic, 173, 255, 258
antibiotic resistance, 258
antibiotics, 86, 92, 195, 223, 224, 240, 350
antibody, 108, 208
antigen, 16, 17, 24, 30
antimicrobial, 45, 61, 284
antisense, 49, 54, 56, 58, 62, 233
anti-sense, 49
antisense RNA, 56, 62
appetite, 96
application, xiii, xiv, 4, 5, 6, 33, 36, 71, 80, 141, 147, 148, 149, 162, 163, 177, 179, 197, 200, 308, 309, 320, 331, 336
APS, 84, 136, 168
aquatic, 195, 201
aqueous solution, 316
aqueous solutions, 316
Arabidopsis thaliana, 7, 9, 10, 11, 12, 13, 14, 15, 16, 17, 18, 19, 20, 21, 22, 23, 24, 25, 31, 47, 315, 322
arbuscular mycorrhizal fungi, 51
arbuscular mycorrhizal fungus, 104
Archaea, 312
Argentina, 60, 106
argument, 3, 320
arithmetic, 146
arrest, 26, 34
ARS, xii, 81, 83, 249, 267, 295, 297
arthritis, 95
arthropods, 167
artificial, 89, 114, 118, 131, 172, 208, 266
asexual, 2, 219, 281, 300, 304, 327, 350, 351
Asia, 108, 133, 343
Asian, 346
aspergillosis, 227
Aspergillus niger, 210, 219, 222, 227, 244, 246
assessment, 66
assignment, 56, 266
associations, 40, 42, 47, 67, 133, 264
assumptions, 43
asymptomatic, 139, 140
ATP, 20, 26, 32
ATPase, 224, 245
atrophy, 97
attachment, 258
attention, xiii, 7, 153, 185, 195, 208, 249, 250, 329
atypical, 266, 311, 369
Australia, 167, 188, 189
Austria, vii, ix, x, xiv, 130, 205, 223, 263, 295, 305, 308
automation, 357, 363, 364
autonomous, 33, 155, 163
autophagy, 21, 22, 24
availability, 44, 45, 47, 50, 51, 282, 363

B

BAC, 17, 22, 76, 83
backcross, 145
bacteria, 47, 90, 106, 148, 252, 359, 362, 367
bacterial, 25, 40, 44, 56, 58, 75, 82, 89, 151, 260, 369
bacterial infection, 369
bacterial strains, 40
baking, 95
banks, 65
barley, 42, 46, 53, 61, 76, 86, 93, 96, 99, 107, 108, 112, 114, 115, 116, 117, 118, 120, 121, 122, 125, 126, 127, 128, 129, 130, 131, 134, 135, 136, 137, 150, 166
barrier, 151, 224
base pair, 89, 149, 227, 256, 268, 338
basic research, 336

basidiomycetes, 3, 60, 264, 333, 334, 339, 340, 342, 344, 345, 347, 348
Bayesian, 198, 314, 318, 319, 322, 326, 331
Bayesian analysis, 322
B-cell, 31
BEA, 111, 112, 116, 117, 120, 130, 131
beams, 334
behavior, 211
benefits, 356
benign, 350
beta, 4, 15, 17, 22, 24, 31, 58, 153, 195, 197, 198, 199, 218, 220, 221, 232, 240, 245, 254, 266, 284, 285, 302, 305, 314, 315, 316, 319, 322, 330
bias, 289, 304, 361
binding, 8, 14, 15, 17, 18, 19, 20, 21, 22, 23, 26, 27, 28, 29, 30, 32, 33, 45, 50, 144, 147, 185, 211, 212, 213, 214, 215, 216, 217, 218, 220, 226, 227, 228, 229, 231, 233, 236, 242, 245, 246, 354, 367
binuclear, 220
bioassay, 259
bioassays, 141
biocatalysts, 266
biochemical, 46, 172, 174, 183, 206, 239, 281, 304, 305, 350
biocontrol, 82, 139, 161, 197
bioconversion, 240
biodegradable, 101
biodiversity, 197, 289, 304, 306
biofilm formation, 48, 232
biofilms, 352
biofuels, 240
bioinformatics, xiii, xiv, 4, 52, 65, 74, 80, 81
biologic, 367
biological, 4, 5, 35, 51, 64, 65, 67, 69, 73, 80, 82, 101, 103, 109, 137, 157, 195, 203, 224, 232, 241, 250, 258, 264, 287, 301, 305, 314, 330
biological control, 82, 103, 157, 195
biological control agents, 82
biological processes, 224
biologically, 85, 100, 173
biology, xiii, 2, 3, 8, 36, 43, 48, 52, 64, 74, 80, 82, 90, 105, 141, 154, 157, 161, 166, 304, 308, 309, 333, 349
biomass, 232
biomolecules, 344
biopolymer, 210
biosciences, 78
biosensors, 48
biosynthesis, 53, 111, 231, 246, 251, 260
biosynthetic pathways, 111
biotechnological, 217, 300, 314
biotechnology, xiii, xiv, 61, 64, 80
biotic, 48
biotin, 356
biotransformation, 314
birth, 96, 97
black, 121, 122, 124, 125, 126, 175, 176, 177, 178, 179, 180, 181, 182, 210, 255
bleaching, 207
blindness, 98
blocks, 45, 150
blood, 96, 100, 350, 352, 353, 356, 364, 365, 366, 367, 368, 369, 370
blood cultures, 364, 366, 370
blood stream, 350
body weight, 100
boiling, 95
BOLD, 301, 303, 309
bonds, 232, 233
bone, 96, 100, 136
bone marrow, 96, 100, 136
bootstrap, 153, 187, 191, 192, 194, 299, 305, 318, 319, 320, 330
Boston, 83
brain, 98
branching, 2, 281, 322, 326
Brazil, xi, 60, 147, 223, 235, 240, 245
breakdown, 58, 98, 151, 207, 233
breast, 95, 101
breeding, 40, 42, 43, 80, 89, 345
British, 105, 337
browser, 7, 301
buffer, 144, 234, 254
buildings, 333, 334, 339, 343, 344
burn, 319
burning, 100
bust, 43

C

Ca^{2+}, 220
cache, 7
Caenorhabditis elegans, 26, 32, 47, 53, 55, 57, 318
calcium, 17, 35
California, vii, 39, 47, 141, 144, 159, 160, 163, 308, 343
calmodulin, 153, 220
calreticulin, 17
cAMP, 26, 58
Canada, 95, 108, 304

cancer, 169
Candida, vi, 12, 19, 20, 21, 26, 27, 31, 32, 46, 53, 54, 58, 59, 60, 61, 173, 230, 240, 241, 242, 243, 244, 245, 246, 311, 317, 349, 350, 351, 352, 353, 354, 355, 356, 357, 358, 359, 362, 364, 365, 366, 367, 368, 369, 370
candidates, 51, 142
candidiasis, 350, 352, 365, 368
capacity, 100, 206, 228, 230, 240, 363
capillary, 355, 367
carbohydrate, 21, 45, 212, 309
carbon, 57, 209, 211, 213, 216, 218, 219, 220, 221, 232, 234
Carica papaya, 87
carotenoids, 314
carrier, 25
cartilage, 30
case study, 53, 83, 163, 301
casein, 17
cassettes, 258
catalysts, 218
categorization, 141, 366
catheters, 350, 352
cations, 259
cattle, 94, 97
CBS, 102, 188, 189, 190, 193, 267, 279, 280, 281, 283, 297, 299, 300
cDNA, 18, 19, 45, 46, 47, 58, 256, 356
celery, 167
cell, 2, 3, 11, 16, 18, 20, 23, 24, 26, 27, 28, 30, 32, 33, 35, 36, 45, 47, 48, 50, 53, 59, 80, 109, 151, 152, 158, 159, 168, 174, 207, 208, 211, 213, 216, 217, 228, 230, 231, 233, 240, 242, 243, 250, 257, 261, 281, 304, 310, 335, 351
cell cycle, 24, 30, 228
cell death, 28, 33, 36
cell differentiation, 11, 16, 33, 243
cell growth, 35
cell metabolism, 33, 240
cell organelles, 257
cell surface, 20, 26, 27, 32, 230
cellulose, 151, 206, 208, 209, 210, 211, 213, 219, 220, 221, 232, 233, 241, 243, 245, 345
cellulosic, 232
central nervous system, 30
centralized, 303
centromere, 21
cephalosporin, 246
cereals, 85, 86, 93, 104, 105, 108, 111, 113, 131, 132, 134, 137, 138, 166
chelators, 231
chemical, 2, 67, 101, 111, 130, 232, 345
chemicals, 2, 218
chemotherapy, 54, 350
chicken, 20
children, 364
chimpanzee, 31
China, 60, 130
Chinese, 16, 26
chitin, 345
chloride, 231
chloroform, 109, 115, 254
chloroplast, 20, 22, 25
cholesterol, 174, 196
chromatin, 19, 29
chromatograms, 112
chromatography, 101, 216
chromosome, 9, 16, 17, 22, 48, 55, 258, 358, 367
chromosomes, 358, 368
chronic, 95, 350
ciliate, 323
Ciona, 317, 318
cis, 208, 212
cistron, 167
citrus, 153, 167
classes, 91, 154, 232, 265, 282
classical, 52
classification, 35, 76, 135, 141, 158, 159, 173, 198, 199, 265, 266, 281, 300, 305, 306, 315, 316, 317, 321, 329, 330
classified, 149, 177, 314, 324, 328, 350
clean-up, 252
cleavage, 89, 225, 228, 229, 230, 231, 247, 359
clients, 63
clinical, 105, 111, 149, 162, 232, 284, 295, 306, 308, 311, 350, 352, 353, 357, 361, 363, 364, 365, 366, 367, 368
clinical symptoms, 352
clinical syndrome, 111
clinically significant, 350
clinicians, 350
clonality, 156
clone, 9, 16, 17, 18, 19, 22, 257, 258
clones, 42, 45, 76, 250, 251, 254, 257, 258, 346, 357, 362
cloning, xiii, 147, 154, 165, 168, 227, 240, 241, 244, 261, 268, 309, 316
cluster analysis, 146, 151
clustering, 46, 75
clusters, 147, 148, 162, 176, 184, 323, 327

Co, 144, 255, 258
coagulation, 31
coagulation factor, 31
cocoa, 46
codes, 227, 230, 231
coding, 43, 45, 47, 88, 91, 142, 153, 183, 184, 189, 215, 216, 228, 235, 258, 282, 284, 302
codominant, 145, 151
codon, 236
codons, 236, 237
coffee, 85, 86
colic, 98
collaboration, 66, 113
colleges, xiii
colonisation, 47
colonization, 47, 56, 58, 110, 151, 161, 308, 345, 346, 352, 364
colonizers, 140
colony-stimulating factor, 31
commercial, 109, 339, 355
commodities, 86
common symptoms, 92
communication, 300
communities, 287
community, 78, 300, 303, 307
compaction, 57
compatibility, 3, 41, 60, 135, 139, 141, 142, 143, 146, 156, 160, 163, 168
competence, 140
competition, 213, 214, 217, 220, 286, 360
complement, 50, 258, 301
complementary, 55, 61, 150, 280, 362
complementary DNA, 61
complex interactions, 250
complexity, 5, 43, 45, 65, 267
components, 3, 33, 46, 50, 64, 68, 88, 174, 206, 207, 229, 241, 358
composite, 136, 187
composition, 138, 183, 231, 281, 304, 351
compounds, 85, 95, 109, 131, 205, 207, 208, 314
comprehension, 208
computer, 51, 64, 65, 80, 131, 358, 362, 363
computer software, 51, 363
computing, 6, 73
concentration, 2, 7, 123, 351, 360
concordance, 135
conditioning, 207
confidence, 319
configuration, 219
conformational, 148, 149, 157, 221, 355, 366
conformational states, 221
confusion, xiii, 179
Congress, 133, 135, 138, 179, 198
congruence, 200
conidiophore, 139, 140
conifer, 333
conjugation, 28
consensus, 75, 213, 216, 228, 233, 265, 266, 319
conservation, 52, 91, 162, 239, 306, 343
construction, 4, 147, 216, 330, 346
consumption, 97
contaminants, 45
contaminated food, 94
contamination, 64, 94, 95, 113, 116, 118, 119, 120, 129, 130, 131, 133, 237, 337, 354
context-dependent, 231
control, 1, 2, 3, 8, 34, 39, 46, 49, 52, 53, 63, 64, 66, 73, 82, 86, 92, 101, 103, 106, 116, 117, 118, 173, 203, 208, 211, 224, 225, 236, 240, 243, 244, 253, 267, 309, 319, 340, 347, 351
controlled, 3, 33, 242, 258, 268
convergence, 156
conversion, 207
conversion rate, 207
cooking, 93
cooling, 88
corepressor, 31
corn, 52, 86, 95, 96, 97, 98, 99, 100, 101, 111, 207, 257
correlation, 43, 107, 110, 113, 120, 121, 122, 123, 124, 125, 130, 131, 143, 145, 147, 148, 156, 205, 354
correlations, 34, 147, 148
cortex, 27, 140
cost-effective, 364
costs, 210, 354, 355, 363
cotton, 85, 86, 93, 94, 102, 157, 203
cotyledon, 13
coverage, 76, 254, 256, 257, 303
covering, 3, 76
crack, 98
cracking, 98
CRC, 137, 311
crop production, 80
crops, 42, 46, 85, 86, 96, 98, 104, 108, 133, 134, 139, 152, 161, 249, 250, 260, 307, 309
cross-linking, 50
Cryptococcus, 12, 13, 14, 15, 16, 18, 21, 22, 23, 27, 28, 29, 30, 31, 32, 44, 49, 56, 74, 165, 311, 345
crystal, 173, 344

crystalline, 232, 233
crystals, 180, 181, 200
CTAB, 109
C-terminal, 226, 230
C-terminus, 225, 227, 229, 230
culm, 116, 117, 118, 119, 120
cultivation, 133, 234, 235, 237, 239, 252, 255, 266, 282, 287, 316, 353
cultural, 164, 171, 172, 173, 176, 182, 189, 200, 201
cultural character, 164, 173, 176, 182, 189, 200, 201
culture, 2, 65, 66, 82, 101, 151, 171, 172, 173, 174, 175, 181, 189, 196, 201, 208, 224, 225, 228, 234, 260, 267, 336, 352, 353, 356, 357, 368, 369
culture conditions, 208
culture media, 171, 224
cycles, 43, 88, 96, 97, 145, 187, 239
cycling, 144, 187, 354, 356
cysteine, 229, 241, 242
cytochrome, 301, 365
cytoplasm, 12, 53, 174
cytosine, 24, 31
cytoskeletal, 23
cytosolic, 185
cytotoxic, 98
cytotoxicity, 134, 137
Czech Republic, 341

D

D. melanogaster, 27, 29
dairy, 95, 212
dairy products, 212
data base, 284, 294, 301, 303
data mining, 4, 64
data set, 184, 320, 323
data structure, 4, 5
database, 1, 4, 5, 6, 7, 8, 9, 12, 16, 24, 26, 32, 34, 35, 46, 50, 63, 64, 65, 66, 67, 68, 69, 71, 72, 74, 75, 76, 77, 78, 81, 82, 83, 84, 252, 255, 258, 284, 289, 294, 295, 338, 345
data-mining, 1, 36
death, 20, 36, 59, 98
decay, 232, 266, 333, 334, 335, 336, 337, 338, 340, 341, 342, 343, 344, 345, 346, 347, 348
decisions, 3, 101
decomposition, 64
defects, 96, 97, 230, 231
defense, 60, 62, 68, 90, 152, 158, 350
defense mechanisms, 68, 152, 350
defibrillation, 207
deficiency, 8, 351
degenerate, 369
degradation, 21, 49, 206, 207, 209, 210, 219, 229, 232, 233, 240, 241, 243, 345
degrading, 44, 59, 83, 151, 159, 163, 206, 250
degree, 92, 98, 146, 147, 154, 239, 255
dehydrogenase, 12, 23, 28, 31, 220
delays, 6
delivery, 64, 258
delta, 20, 27, 320
demand, 352
denaturation, 88, 149, 356
denatured, 149, 335
denaturing gradient gel electrophoresis, 348
density, 46, 147, 363, 369
dentin, 17
deoxynivalenol, 85, 94, 96, 97, 107, 110, 134, 135
deoxynucleotide, 144
Department of Agriculture, 78
Department of Energy, 5, 218, 345
deposition, 340, 343
deposits, 298
depressed, 351
derivatives, 110, 173, 208
dermatophytes, 197
desire, 63
desorption, 50, 344
destruction, 217
detection, 65, 66, 68, 71, 80, 85, 88, 90, 100, 102, 104, 108, 109, 131, 132, 133, 134, 136, 137, 138, 144, 147, 148, 149, 157, 159, 160, 162, 164, 165, 166, 195, 228, 252, 254, 336, 345, 346, 348, 352, 353, 354, 355, 356, 357, 364, 365, 366, 367, 370
detection techniques, 357
detergents, 358
developing countries, 63, 81, 94
developmental process, 1, 3, 4, 8, 33, 34, 83, 314
deviation, 286
diacylglycerol, 20
diagnostic, 63, 64, 74, 139, 150, 154, 156, 160, 174, 176, 177, 183, 336, 347, 350, 355, 356, 357, 365
diagnostic markers, 150
dialysis, 252
diarrhea, 85
diet, 97
differentiation, v, vi, 2, 4, 22, 25, 35, 47, 60, 85, 90, 137, 141, 142, 160, 162, 168, 171, 172, 173, 174, 175, 178, 182, 195, 231, 284, 305, 306, 311, 328, 346, 347, 349, 352, 354, 363, 364, 368, 369
digestion, 50, 76, 92, 143, 149, 254, 338, 359, 362

digestive process, 95
digestive tract, 96
dimorphism, 228, 351
diploid, 20, 40, 231, 242, 351
disaster, 133
discipline, 78, 350
discrimination, 91, 151, 153, 156, 164, 183, 327, 347, 359
discriminatory, 352, 358, 361, 362, 364
disease gene, 50
diseases, 46, 63, 64, 69, 79, 86, 93, 94, 104, 111, 140, 195, 200, 202, 250, 260, 307, 308, 309, 350
dispersion, 347
displacement, 22
disposition, 140
dissociation, 22, 25
distal, 16
distributed computing, 6
distribution, 2, 23, 30, 67, 68, 78, 140, 142, 150, 152, 154, 161, 228, 320, 336, 345
divergence, 44, 155, 321, 330, 339
diversification, 154, 155
diversity, 3, 72, 77, 91, 133, 138, 140, 142, 143, 144, 147, 151, 152, 157, 159, 164, 167, 168, 211, 232, 305, 306, 338
division, 22, 314
DNA polymerase, 88, 105, 144
DNA sequencing, 52, 198, 282, 287, 306, 316, 333, 362, 366
DNase, 220
doctor, 63
dogs, 94
doors, 334
dosage, 230
download, 4, 63, 72
downregulating, 230
drainage, 207
Drosophila, 7, 20, 21, 23, 24, 27, 28, 29, 34, 35, 47, 202, 315, 318
drought, 98
drug discovery, 65, 83, 284
drug targets, 52
drug-resistant, 86
drugs, 65, 218, 367
dry, 86, 128, 172, 346, 347, 348, 351
dsRNA, 49
dung, 266, 324
duplication, 302
duration, 6, 95
dyes, 354, 355, 356, 366, 367
dynamin, 19

E

E. coli, 251, 254, 260
ears, 96, 98
earth, 86, 314
ecological, vi, 42, 64, 67, 103, 140, 263, 264
ecologists, 48
ecology, 40, 48, 58, 104, 300
economic, 67, 86, 94, 101, 353, 356, 360
economic losses, 94, 101
economic performance, 356
economy, 85, 86, 101
ecosystems, 64
ectoderm, 27
education, 63
efficacy, 49, 337
egg, 100
eggs, 100
Egypt, ix, xiv, 63, 169
Egyptian, 305
electric field, 344, 358
electrochemical, 224
electron, 174, 311, 316, 323, 324, 332, 346
electronic, 65, 66, 78
electrophoresis, 50, 89, 90, 109, 149, 152, 166, 183, 187, 235, 254, 335, 337, 347, 355, 358, 359, 360, 362, 367
ELISA, 118, 132
elongation, 19, 72, 139, 153, 156, 159, 165, 184, 185, 189, 199, 201, 284, 311, 314, 315, 319, 328, 332
embryo, 23
embryonic, 15, 26, 32
embryos, 9
emission, 355
employment, 39, 356
encoding, 18, 43, 53, 151, 152, 155, 157, 160, 161, 163, 185, 196, 199, 205, 206, 207, 209, 210, 211, 212, 214, 215, 216, 218, 219, 220, 221, 222, 228, 230, 231, 233, 244, 245, 246, 253, 284, 285, 301, 302, 315, 317
endocytosis, 260
endogenous, 49, 352
endonuclease, 89, 338, 359
endothelial, 22
end-users, 80
energy, 344
England, 95, 108, 130, 134, 268, 294, 316

English, 54, 132, 309
enlargement, 97
ENN, 119
environment, xiii, 2, 63, 195, 201, 224, 230, 231, 239, 350
environmental, 93, 95, 101, 110, 229, 230, 241, 243, 244, 245, 267, 289, 294, 300, 314, 351, 354, 365
environmental conditions, 93, 95, 300, 314, 351
environmental factors, 110
enzymatic, 105, 151, 210, 211, 232, 309, 359, 366
enzyme, xiii, 89, 151, 205, 207, 208, 209, 210, 211, 215, 217, 221, 251, 254, 260, 345, 355, 366
enzyme immunoassay, 355, 366
enzymes, 33, 45, 89, 92, 142, 143, 150, 151, 159, 163, 205, 206, 207, 208, 210, 211, 218, 219, 221, 222, 223, 224, 227, 228, 232, 233, 240, 250, 258, 261, 336, 345, 356, 358
epidemic, 42, 166
epidemics, 42, 65, 93
epidemiological, 43, 58, 352, 359, 362, 363, 364, 369
epidemiology, 67, 154, 368
epistasis, 49
epithelial cell, 21, 28
epithelial cells, 21
equipment, 131, 145, 287, 344, 354, 358
ergosterol, 345
erythroid, 22, 23
Escherichia coli, 53, 173, 233, 251
esophagus, 100
EST, 45, 55, 61, 75, 77, 84
ester, 99
esterase, 199, 220
estimating, 91, 289, 304
estrogen, 97
ethanol, 207, 220
Ethiopia, 60
ethylene, 22
etiologic agent, 235
eukaryote, 1, 44, 57, 305, 347, 349
eukaryotes, 1, 3, 7, 34, 91, 185, 195, 202, 306, 311, 360
eukaryotic, 1, 16, 23, 33, 34, 43, 54, 58, 75, 82, 90, 141, 142, 150, 185, 198, 240, 342, 350
eukaryotic cell, 33, 90, 141, 185
Eurasia, 180
Europe, 42, 61, 108, 110, 111, 130, 132, 133, 137, 138, 167, 295, 299, 333, 343
European, 53, 58, 60, 61, 64, 66, 79, 82, 83, 111, 132, 133, 134, 136, 137, 138, 158, 165, 168, 183, 203, 218, 220, 261, 267, 303, 333, 337, 340, 343, 346
European Union, 66
evidence, 2, 42, 61, 69, 137, 146, 161, 166, 209, 211, 212, 213, 215, 219, 232, 323, 329, 331
evolution, 3, 40, 42, 43, 45, 61, 62, 91, 142, 155, 156, 162, 163, 195, 197, 198, 199, 200, 202, 220, 282, 301, 304, 305, 306, 307, 310, 311, 324, 326, 327, 331
evolutionary, 1, 4, 34, 42, 43, 90, 137, 155, 156, 161, 162, 166, 171, 184, 185, 197, 198, 202, 282, 298, 311, 314, 325, 327
evolutionary process, 43
excision, 156, 254
excitation, 355
exclusion, 321, 357
execution, 6
exercise, 1
exons, 153
experimental condition, 360
expert, 64
expertise, 350, 355, 362
exponential, 9, 12, 16, 24, 26, 32, 239
exponential functions, 9, 12, 16, 24, 26, 32
exposure, 95
expressed sequence tag, 52, 57, 60, 61
external environment, 224
external influences, 266
extinction, 43
extracellular, 153, 160, 206, 218, 221, 241, 243, 258, 346
extraction, 101, 109, 115, 131, 132, 146, 207, 254, 268, 309, 357, 358
exudate, 176
eye, 23, 95, 106, 198, 287

F

F. solani, 72, 86, 87, 93, 94, 95, 101, 150, 153
F. vasinfectum, 86
failure, 94
false, 50, 121, 122, 123, 124, 289, 301, 354, 355, 367
false positive, 50, 121, 122, 123, 124, 367
family, vi, 8, 9, 10, 11, 12, 13, 14, 15, 16, 20, 23, 24, 25, 29, 30, 32, 40, 42, 43, 45, 47, 95, 161, 162, 165, 173, 197, 198, 200, 245, 246, 282, 294, 298, 313, 323, 326, 327, 328, 331, 350, 353
family members, 246
family structure, vi, 282, 313, 314

family system, 313
farmers, 106, 114
fat, 53
fatty acid, 23
fee, 80
fermentation, 101, 206, 266, 314, 325
fertilization, 25, 133
fertilizers, 240
fiber, 93
fibers, 207, 232
fibrils, 151
Filobasidiella neoformans, 342
filters, 5
filtration, 186, 207
fine tuning, 215
fingerprinting, 42, 88, 89, 102, 106, 137, 144, 148, 150, 154, 158, 160, 163, 164, 165, 166, 167, 168, 311, 359, 362, 363, 364, 365, 367, 368, 369, 370
Finland, vii, x, xi, xiv, 107, 108, 110, 112, 113, 114, 119, 130, 132, 133, 135, 137, 138, 219
first responders, 67
fish, 25, 27, 28, 30, 94
FISH, 353, 356, 364
fitness, 36
flight, 50, 344
flooring, 334, 335
flow, 4, 42, 43, 80
fluorescence, 354, 356
fluorimeter, 355, 356
fluorogenic, 109, 131
focusing, 152, 224, 362
food, 64, 65, 83, 88, 92, 103, 104, 110, 114, 206, 219, 250, 260, 261, 266, 284, 305, 314
food industry, 65
food production, 284
forebrain, 27
forensic, 66
forestry, 333
forests, 203
fossil, 281
France, 158
freeze-dried, 186
freezing, 93
fruits, 85, 86, 308
functional activation, 223, 226, 228, 229, 239
functional analysis, 57
fungal infection, 52, 74, 93, 95, 282, 309, 349, 350, 365, 368
fungal metabolite, 173, 195, 200, 202
fungal spores, 93
fungicide, 108, 113, 114, 127, 128, 129
fungicides, 133
Fusarium oxysporum, xiv, 41, 48, 58, 59, 60, 72, 73, 101, 102, 103, 104, 105, 106, 135, 139, 140, 142, 143, 144, 145, 146, 147, 148, 150, 151, 152, 154, 157, 158, 159, 160, 161, 162, 163, 164, 165, 166, 167, 168, 185, 202, 227, 241, 304
fusion, 28, 29, 48, 323

G

GABA, 225, 226
gas, 344
gas chromatograph, 344
gas phase, 344
gastrointestinal, 85, 350
GDP, 22, 25, 185
gel, 50, 53, 60, 89, 90, 101, 109, 149, 152, 166, 187, 235, 254, 335, 337, 347, 355, 358, 359, 360, 362, 367
gels, 88, 149, 152, 252, 335
GenBank, 64, 71, 78, 109, 189, 190, 192, 235, 236, 237, 238, 252, 255, 256, 257, 258, 265, 267, 268, 283, 284, 291, 294, 295, 298, 300, 303, 315, 316, 319, 340
gene amplification, 287
gene expression, 26, 45, 46, 47, 48, 49, 54, 56, 57, 61, 62, 205, 208, 211, 212, 213, 214, 215, 216, 217, 219, 220, 221, 223, 224, 241, 244, 245, 352
gene silencing, 49, 54, 56, 60, 349
gene transfer, 59
genealogy, 156, 319, 325
generalization, 265
generation, 46, 48, 65, 251, 319, 329, 353
genetic, xiii, 2, 3, 34, 40, 42, 43, 46, 48, 51, 52, 53, 54, 55, 63, 64, 65, 67, 69, 76, 80, 83, 89, 92, 106, 111, 134, 141, 143, 144, 145, 146, 147, 148, 149, 150, 153, 154, 156, 158, 162, 163, 164, 167, 168, 178, 185, 199, 203, 224, 239, 250, 258, 261, 282, 301, 311, 312, 343, 352, 367, 369, 370
genetic diversity, 46, 67, 141, 143, 145, 146, 147, 148, 149, 156, 162, 163, 261, 312, 343, 352, 367
genetic drift, 43
genetic factors, 258
genetic information, 89, 145, 352
genetic linkage, 65
genetic marker, 76, 106, 144, 153, 168, 343, 370
genetics, xiii, 2, 39, 48, 52, 56, 86, 103, 165, 197, 223
genome sequences, 6, 47, 51, 74, 314

genome sequencing, 40, 43, 51, 74, 77, 250, 342
genomes, 1, 3, 4, 5, 6, 7, 8, 9, 12, 23, 24, 25, 26, 31, 32, 33, 36, 40, 44, 52, 65, 74, 77, 82, 90, 106, 150, 155, 162, 168, 208, 215, 253, 360, 362
genomic, xiii, 4, 13, 15, 16, 17, 22, 23, 27, 28, 29, 30, 31, 32, 36, 44, 45, 46, 47, 51, 52, 55, 58, 61, 64, 65, 67, 68, 74, 76, 77, 80, 84, 88, 92, 105, 115, 142, 144, 146, 152, 156, 162, 186, 235, 237, 246, 251, 252, 254, 256, 257, 258, 259, 268, 301, 316, 343, 346, 349, 359, 360, 362, 363, 366
genomic regions, 142, 156, 251, 343
genomics, v, xiii, xiv, 39, 45, 52, 54, 61, 65, 68, 73, 74, 75, 81, 82, 83, 84, 86, 260, 349
genotype, 41, 67, 183, 297, 299, 301
genotypes, 41, 42, 46, 54, 136, 183, 303, 343
genre, 132
genus Candida, 367
Germany, viii, ix, x, xi, xiv, 63, 145, 187, 188, 263, 264, 268, 279, 283, 295, 297, 299, 300, 304, 313, 316, 333, 339
germination, 51, 174, 260
Gibberella, 18, 19, 26, 32, 52, 72, 74, 95, 111, 112, 164, 166, 169, 200
gibberellin, 22
glioma, 30
Glomus intraradices, 317
glucocorticoid receptor, 220
glucose, 100, 101, 206, 209, 213, 216, 219, 220, 232, 233, 234, 351
glucosidases, 232, 234
glutamate, 28
glutamine, 35, 225, 236
glutathione, 12, 27
glycerol, 316
glycine, 25, 236
glycogen, 34, 351
glycoprotein, 20, 23, 28
glycoproteins, 151, 230
glycosylation, 50
glycosylphosphatidylinositol, 53
goals, 4, 63
gonadotropin, 25, 30
GPI, 20, 51
G-protein, 45, 230
graduate students, xiii, xiv
grafting, 259
grain, vi, 93, 94, 95, 96, 97, 98, 100, 104, 107, 108, 109, 113, 114, 115, 119, 120, 121, 122, 123, 127, 130, 131, 133, 138, 249, 250, 257, 259
grain dust, 133
grains, 93, 103, 107, 108, 110, 111, 120, 121, 122, 123, 124, 125, 126, 130, 131, 133, 134, 136, 138, 305
grants, 132
granulocyte, 31
grasses, 5, 86, 164, 169
gravity, 2
Greece, 168
green fluorescent protein, 58
grouping, 89, 147, 153, 321, 323
groups, 5, 6, 42, 69, 92, 112, 132, 135, 141, 142, 143, 144, 145, 146, 149, 151, 160, 163, 165, 173, 174, 175, 186, 225, 265, 266, 281, 299, 300, 314, 321, 322, 323, 326, 328, 337, 338, 343, 359
growth, 2, 3, 20, 21, 28, 33, 47, 49, 51, 54, 64, 77, 80, 86, 96, 97, 99, 100, 103, 105, 152, 158, 173, 178, 179, 181, 211, 216, 219, 221, 224, 227, 229, 230, 233, 234, 236, 241, 242, 243, 251, 255, 257, 266, 281, 286, 289, 294, 351, 352
growth factor, 28
growth inhibition, 227
growth rate, 219, 255
growth temperature, 290, 294
guidelines, 302
Guinea, 17
gums, 100

H

habitat, 2, 86, 99, 101, 245
haemopoiesis, 30
haploid, 3, 20, 353
haplotype, 142, 143, 144
haplotypes, 142, 143, 148, 156
harbour, 302
harmful, 108
harvest, 333
harvesting, 108, 112, 113, 114, 121, 127, 128, 129, 131
Hawaii, 105
hazards, 95
head, 93, 95, 98, 105, 106, 108, 111, 132, 133, 135, 137, 142, 160
healing, 358
health, 40, 52, 63, 85, 94, 95, 97, 98, 352
health care, 352
health care workers, 352
health effects, 98
health problems, 97
healthcare, 369

heart, 87
heat, 32, 353, 355
heat shock protein, 353
helix, 225
hematological, 350
hemicellulose, 151, 232, 233
hemisphere, 264, 267, 345
hemorrhage, 96
heterogeneity, 156, 232
heterologous protein production, 208, 217
heterozygosity, 148
heuristic, 187, 319
high resolution, 356, 364
high risk, 349
high temperature, 88
high-risk, 67, 68, 368
Hilbert, 62
histogenesis, 36
histone, 19, 153
HIV, 94, 173, 202, 365, 366
HIV-1, 202, 366
homogeneous, 343
homolog, 18, 19, 20, 23, 24, 26, 27, 28, 30, 31, 32, 202, 229, 230, 231, 236, 243, 245, 247, 251
homologous genes, 57
homology, 1, 8, 9, 13, 15, 16, 17, 22, 23, 26, 27, 28, 29, 30, 31, 32, 49, 145, 152, 231, 354
homopolymers, 232
Honduras, 60
honey, 176, 309
horizontal gene transfer, 48, 154, 301
hormone, 18, 25, 30, 97, 351
horse, 267
horses, 94, 98
hospital, 352, 357, 365
hospitals, 352
host, 39, 40, 41, 42, 43, 45, 47, 50, 51, 52, 53, 54, 59, 60, 61, 62, 67, 68, 69, 84, 86, 87, 92, 94, 98, 101, 102, 139, 140, 141, 151, 152, 156, 158, 161, 171, 172, 183, 198, 203, 208, 230, 241, 249, 250, 254, 258, 260, 286, 350, 353, 365
host population, 54
HPLC, 112
human, vi, 4, 6, 7, 17, 19, 20, 21, 22, 23, 24, 25, 26, 27, 28, 29, 30, 31, 32, 35, 44, 52, 53, 55, 63, 74, 85, 86, 94, 95, 100, 106, 142, 149, 165, 166, 171, 195, 200, 202, 223, 224, 228, 230, 231, 235, 236, 240, 242, 246, 266, 309, 349, 350, 367, 369
human development, 19
human genome, 4, 52, 165
human immunodeficiency virus, 350
humans, 92, 98, 108, 111, 223, 224, 239, 266, 282, 310, 350
humidity, 2, 93
Hungarian, 198
Hungary, viii, ix, xi, xiv, 171, 187, 188, 189, 197, 198, 263, 313
hyaline, 172, 175, 176, 177, 178, 179, 180, 181, 182, 183, 199
hybrid, 50, 54, 62
hybridization, 62, 67, 76, 137, 145, 149, 203, 252, 254, 255, 257, 258, 282, 356, 363
hydro, 229, 230
hydrogen, 201, 232
hydrogen bonds, 232
hydrogenation, 218
hydrolases, 205, 208, 210, 216, 219, 229
hydrolysis, 185, 206, 207, 218, 233
hydrophilic, 229, 230
hydrophobic, 228
hydroxyproline, 23
hypersensitive, 220
hypothesis, 40, 51, 52, 143, 197, 213, 215, 236, 301, 327, 349

I

id, 20, 21, 64, 79, 80
Idaho, vii, viii, 263, 284
identity, 8, 67, 187, 188, 189, 201, 236, 256, 257, 267, 279, 280, 281, 282, 284, 286, 289, 294, 295, 297, 298, 299, 300, 302, 319, 333, 340, 348, 354
Illinois, 295, 297
illumination, 187
images, 69, 76, 286, 287, 288, 289, 290, 291, 303, 323
immune system, 96, 98, 351
immunity, 351
immunocompromised, 86, 94, 230, 349, 350
immunological, 333, 345
immunosuppression, 85
impairments, 350
implementation, 67, 80, 345, 352, 357, 364
in situ, 48, 58, 349, 356, 366, 367, 368, 370
in situ hybridization, 349, 366, 367, 368, 370
in vitro, 2, 33, 35, 64, 152, 173, 212, 214, 216, 217, 228, 242, 348, 351, 366
in vivo, 64, 134, 196, 200, 201, 214, 216, 254, 257, 260, 351
inactivation, 151, 154, 164

inactive, 155, 156, 229
incidence, 98, 349, 350, 352
inclusion, 6, 184, 328, 329
incubation, 88, 100, 234, 236, 267
independence, 86
India, vii, viii, ix, x, xi, xiv, 39, 85, 86, 92, 94, 102, 103, 138, 139, 143, 150, 158, 171, 172, 180, 182, 183, 197, 198, 200, 201, 202, 245, 341, 349
Indian, 86, 101, 150, 172, 173, 175, 195, 198, 199, 200, 201, 202, 203
indication, 68
indicators, 113, 306
indices, 302
indigenous, 305
inducer, 208, 209, 210, 211, 213, 215, 220, 231
induction, 30, 152, 208, 209, 210, 211, 212, 214, 215, 217, 218, 219, 221, 226, 230, 233, 242
industrial, 205, 206, 208, 217, 219, 220, 224, 239, 240, 245, 266, 284
industrial application, 208, 217, 245
industry, 74, 205, 206, 207, 218
inefficiency, 48
infection, 45, 46, 50, 53, 58, 60, 61, 69, 85, 86, 90, 93, 94, 101, 104, 120, 130, 132, 133, 135, 136, 140, 151, 159, 160, 161, 162, 164, 167, 200, 227, 230, 231, 232, 235, 240, 250, 251, 308, 350, 352, 366
infections, 85, 86, 94, 95, 98, 106, 145, 195, 230, 350
infectious, 353, 367, 369
infectious disease, 353, 367, 369
infectious diseases, 353, 367, 369
inferences, 184
infertility, 85
information retrieval, 35
ingestion, 92
inhalation, 235
inheritance, 150
inhibition, 35, 58, 131, 211, 354
inhibitor, 22, 25, 173, 196, 202, 230
inhibitors, 109, 173, 202
inhibitory, 109, 131
initiation, 26, 46
injection, 49, 232
injections, 350
innovation, 80
inoculation, 114, 118, 131
inoculum, 255
insects, 264
insertion, 152, 154, 155, 208, 217, 249, 251, 252, 254, 255, 256, 257, 258, 294, 295, 297
insight, 44, 45, 48, 65, 153, 205
institutions, 337
instruments, 366
insulation, 334
insulin, 26, 29
insulin-like growth factor I, 29
integration, 48, 50, 251, 252, 258, 260
integrin, 31
integrity, 231, 351
intensity, 361
intentions, 321
interaction, 2, 23, 41, 42, 48, 50, 51, 52, 57, 62, 69, 73, 161, 214, 215, 216, 231, 246, 259, 285, 286, 303, 327, 329
interactions, 6, 34, 39, 40, 42, 43, 45, 46, 48, 49, 50, 51, 52, 54, 57, 58, 59, 60, 61, 62, 69, 80, 81, 84, 158, 159, 163, 212, 246, 249, 250, 260, 313, 365
interface, 4, 5, 203, 234, 303, 310, 332
interference, 39, 49, 54, 55, 351, 354
interferon, 30
internalization, 246
international, 80, 82, 267, 309, 340
internet, 4, 63, 64, 65, 66, 80, 82
interpretation, 103, 148, 295, 303
interrelationships, xiii
intervening sequence, 227
intrinsic, 49
intron, 153, 185, 187, 189, 227, 236, 237
introns, 153, 227, 236, 237, 284
invasive, 20, 21, 229, 230, 243, 286, 349, 350, 352, 365
inversions, 44, 144
inverted repeats, 144, 212
investigations, 60, 112, 211, 347
ion channels, 229
ion transport, 25
ionic, 224, 228
ionization, 344
ions, 23, 35, 344
Ireland, 35
iron, 231, 308
irritation, 85
island, 106
isoelectric point, 50
isolation, 92, 100, 101, 115, 135, 143, 206, 219, 235, 260, 267, 287, 303, 309, 336
isothermal, 356, 366
isozyme, 135, 183, 199

isozymes, 152
Israel, 147
Italy, 165, 218, 220
iteration, 320

J

Japan, 64, 95, 152, 311, 340, 343
Japanese, 25, 30, 260
Java, 4
judge, 3
Jung, 355, 366
justification, 298

K

K-12, 53
kappa, 218, 320
kappa B, 218
karyotype, 143, 154, 367
karyotypes, 259, 358, 364, 365, 367
karyotyping, 358
kernel, 74
ketones, 344
kidney, 180
kidneys, 98
killing, 351
kinase, 17, 19, 20, 26, 29, 58, 256
knockout, 48, 155, 258
Korea, 143
Korean, 307

L

labour, 148, 266, 287, 355, 358
lactams, 224
lactose, 210, 212, 216, 221
lambda, 254, 261
lamellae, 34
language, 78
large-scale, 42, 45, 46, 48, 50, 62, 74, 132, 200
laser, 50, 58, 344
lead, 52, 64, 69, 82, 141, 208, 218, 225, 301, 349
legume, 60, 249, 250, 259, 260
legumes, vi, 201, 249, 250, 257, 259, 260, 261
Leguminosae, 181
lens, 28
Lepidoptera, 153, 165
leptin, 29
lesions, 100
lettuce, 161, 165, 167
leukemia, 31, 350
lichen, 264
life cycle, 69
life sciences, 64
life span, 36
ligand, 15, 29
ligands, 23
lignin, 232
lignocellulose, 44, 59, 83, 232
likelihood, 8, 319, 320, 360
limitations, 72, 145, 203, 267, 332, 340, 366
linear, 232, 233
linkage, 42, 58, 156, 218
links, 5, 6, 65, 67, 78, 98
lipase, 23
lipid, 12, 136, 351
lipid peroxidation, 136
lipopolysaccharide, 61
liquefaction, 207
liquid chromatography, 134
liquid nitrogen, 316
liquor, 100
literature, 48, 69, 78, 140, 203, 224, 236
lithium, 231
liver, 98
localization, 63, 221, 225, 229, 236, 244, 346
location, 184, 185, 186
locus, 9, 12, 19, 20, 21, 22, 26, 40, 41, 53, 135, 152, 153, 160, 224, 227, 295, 298, 343, 359
London, 34, 36, 81, 105, 161, 202, 218, 219, 225, 304, 305, 306
long distance, 42
long period, 287
long-term, 67, 95
losses, 93
low molecular weight, 209
low-level, 159
LSU, 197, 264, 268, 282, 293, 295, 302, 303, 314, 322, 347
lung, 308
lungs, 235
Lycopersicon esculentum, 87
lymphoid, 18
lysine, 242, 311
lysis, 358
lysosome, 229

M

machine-readable, 78
machinery, 153, 241, 247
machines, 73, 207
macrophage, 20, 27, 29
macrophages, 246
maintenance, 154, 227, 231, 232, 236, 240, 303, 316, 329
maize, 5, 46, 74, 86, 93, 94, 98, 102, 105, 106, 135, 305
malt extract, 267, 316
Mammalian, 23
mammals, 350
management, 7, 33, 43, 53, 63, 64, 65, 67, 104, 164, 363, 365
management practices, 64
manganese, 35
mango, 46
manipulation, xiii, 48, 62, 355
manufacturer, 186
manure, 267
mapping, 36, 50, 54, 62, 76, 144, 168, 311, 369
marine environment, 172, 176
marker genes, 264, 268, 316
Markov, 319
Markov chain, 319
Maryland, vii, 39
mass spectrometry, 50, 344, 347
mathematical, 40, 43
matrix, 27, 50, 151, 180, 182, 216, 268, 295, 297, 299, 300, 320, 344, 358
maturation, 4, 12, 15, 33, 227, 232, 328
Maximum Likelihood, 318
measures, 67, 250, 259
mechanical, iv, 345, 358
mechanical properties, 345
media, 100, 101, 103, 172, 174, 201, 208, 213, 234, 236, 255, 266, 267, 287
Medicago truncatula, 55
medicinal, 202
medicinal plants, 202
medicine, 63, 74
Mediterranean, 79, 203
meiosis, 2, 228, 229, 230, 241
melanin, 251, 259, 260, 261
melt, 356
melter, 356
melting, 254, 354, 356, 361, 364, 366, 367, 368
membranes, 26, 252, 254
meristem, 8, 9, 10, 11, 12, 13, 14, 15, 16, 23, 33
mesenchyme, 23, 30
mesoderm, 27, 30
metabolic, 33, 220, 224, 281
metabolic pathways, 33, 224
metabolism, 30, 43, 45, 210, 213, 216, 218, 221, 231
metabolite, 67, 112, 136, 174, 307
metabolites, 84, 85, 92, 101, 103, 136, 137, 138, 173, 175, 197, 223, 224, 239, 284, 308, 346
metabolizing, 213
metamorphosis, 4
Metazoa, 1, 4, 6, 7, 9, 12, 32, 202, 315, 317, 318
metazoan, 25
Mexican, 161
Mg^{2+}, 360
mice, 47, 136
microarray, 39, 45, 46, 47, 53, 54, 55, 58, 61, 132, 250, 367, 370
Microarrays, 61
microbes, 151, 173, 243, 305, 310
Microbes, 246
microbial, 47, 48, 56, 58, 66, 83, 221, 282, 304, 305, 349, 353, 363, 368
Microbial, 46, 62, 66, 75, 245, 251, 344
microbial community, 282, 304
microorganism, 230, 240, 266, 324, 351
microorganisms, 64, 164, 173, 202, 224, 227, 232, 239, 314, 354, 361, 366
microphotographs, 285
microsatellites, 109, 150, 158, 160, 162, 164, 165, 168, 333, 343, 359
microscope, 48, 281
microscopy, 202, 316, 356
microtubule, 16, 260
microtubules, 185
migration, 31, 43, 88, 149
milk, 212, 316
mimicking, 7, 225
mining, 46
Minnesota, 295, 329
minority, 327
misidentification, 301
misidentified, 289, 294, 295
misinterpretation, 148
misleading, 302
mitochondrial, 25, 90, 106, 142, 156, 166, 197, 198, 301, 307, 359
mitochondrial DNA, 90, 106, 142, 307, 359
mitogen, 17, 23
mitogen-activated protein kinase, 17, 23

mitosis, 306
mitotic, 156, 308
mobility, 17, 25, 148, 183, 213, 358
model system, 314
modeling, 58, 80
models, 42, 43, 51, 74, 158
modernization, 265
modulation, 32, 224
modules, 4
moisture, 93, 95, 96, 97, 98, 100, 105
moisture content, 97, 100
mold, 96, 98, 305
molecular biology, 83, 172, 250
molecular markers, xiii, 85, 89, 134, 139, 145, 160, 161, 167, 173, 183, 264, 266, 267, 301
molecular mechanisms, 39, 40, 154, 240
molecular structure, 76, 158
molecular weight, 50, 148, 231, 254
molecules, 49, 89, 102, 185, 208, 209, 213, 215, 224, 334, 343, 344, 351, 354, 358, 360
monograph, 173, 197, 312, 331
monomer, 56, 212, 232
monomeric, 29, 210
monomers, 206
mononuclear cell, 351
mononuclear cells, 351
Montenegro, 52, 56
Morocco, 168
morphogenesis, 2, 3, 4, 33, 35, 46, 51, 61, 228, 230, 231, 324, 351
morphological, 69, 70, 72, 101, 108, 109, 110, 112, 114, 131, 138, 139, 140, 171, 172, 173, 178, 183, 186, 189, 191, 192, 194, 195, 245, 264, 265, 266, 267, 279, 280, 281, 282, 284, 287, 291, 294, 295, 296, 297, 298, 299, 301, 302, 303, 307, 322, 323, 325, 330
morphology, xiii, 42, 141, 159, 162, 172, 173, 176, 182, 200, 201, 228, 255, 281, 282, 287, 294, 297, 300, 304, 323, 325, 327, 329, 351
mortality, 98, 100, 111, 352
mortality rate, 98, 100, 111
mosaic, 45
Moscow, vii, viii, 136, 263
moths, 159
motives, 212
mouse, 16, 17, 18, 22, 23, 24, 25, 26, 27, 28, 29, 30, 31, 32, 230, 231, 232, 369
mouse model, 230
mouth, 100
movement, 66
mRNA, 16, 17, 18, 19, 24, 26, 27, 28, 32, 45, 46, 47, 49, 56, 58, 185, 228, 236, 237, 246
mtDNA, 142, 156
mucin, 28
mucous membrane, 350
multidrug resistance, 28, 51
muscle, 19
mushrooms, 36
mutagenesis, 39, 45, 47, 48, 52, 55, 58, 59, 155, 231, 241, 250, 251, 257, 258
mutant, 47, 62, 200, 211, 216, 219, 227, 230, 231, 246, 251
mutants, 47, 48, 49, 224, 228, 230, 231, 232, 241, 243, 245, 249, 251, 257, 258
mutation, 3, 27, 43, 145, 154, 212, 217, 229, 258, 366
mutation rate, 154
mutations, 162, 214, 224, 227, 251
mycelium, xiii, 2, 36, 91, 108, 176, 177, 178, 179, 180, 181, 182, 183, 186, 255, 260
mycobacteria, 356
Mycobacterium, 363, 368, 369
mycology, xiii, xiv, 63, 65, 78, 79, 81, 102, 249, 250, 350, 355, 361, 364
myo-inositol, 14

N

N-acety, 21, 24
NAM, 8, 9, 10, 11, 12, 13, 14, 15, 16, 33
naming, 66
NAS, 353, 356, 357, 366, 367, 370
Nash, 35
National Academy of Sciences, 135
natural, 42, 43, 54, 57, 89, 120, 130, 131, 141, 149, 168, 208, 209, 265, 267, 312, 314, 321, 322, 326, 328, 329, 343, 350
natural environment, 343
natural selection, 42
neck, 176, 179
necrosis, 22, 96, 202
needles, 305
negative selection, 56
nematode, 228, 243
Nepal, 102, 158
nested PCR, 354, 365
Netherlands, ix, 137, 138, 145, 188, 189, 259, 260, 263, 267, 268, 279, 283, 297, 299, 300, 316
network, 76, 78, 113, 364
neutropenia, 365

neutrophil, 351
New York, 36, 104, 157, 197, 203, 304, 305, 307, 308, 310, 330, 347, 370
New Zealand, 188
Newton, 320
Ni, 35, 166
nickel, 218
Nicotiana tabacum, 24
Nielsen, 55, 82, 136, 368
Nigeria, 60, 93, 102
nitrate, 154
nitrogen, 133
non-random, 33
normal, 3, 33, 86, 98, 116
normal development, 33
normal distribution, 116
normalization, 239
North America, 111, 130, 137, 167, 343, 346
North Carolina, 73, 95
Norway, 130, 134, 137, 346
Norway spruce, 346
novelty, 67
nptII, 255, 258
N-terminal, 226
NTS, 142, 184
nuclear, vi, 21, 22, 23, 26, 27, 28, 32, 54, 90, 91, 104, 135, 141, 143, 159, 165, 166, 184, 195, 196, 197, 198, 225, 229, 236, 242, 244, 263, 267, 282, 284, 285, 286, 287, 288, 292, 293, 295, 302, 305, 307, 311, 314, 320, 332, 346
nuclease, 49
nuclei, 3
nucleic acid, 5, 6, 88, 165, 319, 352, 353, 356, 363, 365, 366, 367, 368, 370
nucleosome, 19
nucleotide sequence, 71, 88, 89, 152, 154, 156, 183, 214, 216, 264, 267, 268, 282, 285, 288, 292, 293, 310, 318, 319, 322, 331, 364
nucleotides, 49, 91, 227, 237, 339, 340, 343, 362
nucleus, 24, 226
nutrient, 2, 101, 232, 267
nutrients, 207
nutrition, 86, 228, 250, 324
nylon, 53, 252, 254

O

oat, 86, 108, 120, 121, 175, 176, 177, 178, 179, 180, 181, 182, 183
obesity, 29
obligate, 266
observations, 33, 140, 230, 266, 281
Oceania, 343
oedema, 85
oil, 86, 165, 207
oil production, 207
oilseed, 53, 159
oligodeoxynucleotides, 106
oligomers, 363
oligonucleotides, 144, 147, 150, 216, 343
oligosaccharides, 206, 208, 209, 233
olive, 46
oncogene, 19, 29, 30
onion, 86, 94
on-line, 35, 64, 66, 67, 71, 74, 75, 76, 78, 79, 81, 265, 267, 268, 291, 303
onychomycosis, 95
operator, 7
operon, 142
optical, 239
optimal performance, 359
optimization, 239, 320
oral, 350, 365
organ, 2, 64, 350
organic, 101, 103, 133, 266
organic matter, 101, 103
organism, 6, 7, 43, 46, 47, 50, 69, 76, 90, 154, 201, 224, 298, 301, 334, 356, 359
organization, xiii, 1, 34, 58, 76, 162, 246, 260, 301, 303
organizations, 64
orientation, 360
Ottawa, 304
ovaries, 97
overproduction, 208
oxalic, 227
oxalic acid, 227
oxidative, 240
oxidative stress, 240
oxygen, 173, 232

P

p53, 16, 17, 24, 26, 30, 31, 33, 35
Pacific, 82, 84, 218, 308
packaging, 254
Pakistan, 341
Panama, 87
Pap, 263
paper, 53, 78, 86, 112, 113, 206, 207, 218, 234

paracoccidioidomycosis, 235
parasite, 57, 58, 61, 183, 342
parasites, 61, 67, 140, 266, 350
Paris, 146, 155, 158, 166, 179, 368
Parkinson, 65, 83
partnership, 67
passive, 322, 325
pathogenesis, 39, 43, 45, 46, 47, 51, 68, 69, 84, 140, 151, 158, 159, 161, 231, 241, 257, 258, 351
pathogenic, vi, 39, 41, 44, 46, 47, 49, 51, 52, 60, 63, 65, 69, 71, 74, 75, 77, 86, 87, 92, 93, 109, 111, 114, 135, 136, 138, 139, 140, 141, 143, 145, 146, 147, 148, 151, 153, 155, 156, 157, 158, 160, 163, 164, 166, 167, 168, 199, 202, 203, 224, 227, 249, 250, 251, 257, 259, 260, 261, 308, 310, 345, 349, 350, 353, 356, 361, 362, 363, 366, 368
pathogens, 39, 40, 41, 42, 43, 44, 45, 51, 54, 57, 59, 63, 64, 66, 67, 68, 69, 74, 76, 78, 83, 84, 93, 137, 140, 141, 142, 143, 148, 149, 151, 155, 164, 167, 231, 243, 249, 250, 251, 258, 259, 260, 266, 284, 350, 365, 369
pathologist, 78, 80
pathologists, 63, 64, 65, 66, 141
pathology, 63, 64, 65, 78, 79, 80, 82, 200, 250
pathways, 46, 50, 68, 80, 232, 234, 241, 311
patients, 86, 94, 231, 349, 350, 351, 352, 365, 368, 369
patterning, 26, 32
peat, 203
pectin, 151, 152, 228, 246
pectins, 207
penalty, 7
penicillin, 242, 246
Penicillium brevicompactum, 280
Pennsylvania, 104, 157, 166
peptide, 14, 25, 50, 356, 368, 370
peptides, 61
performance, 355, 360, 363, 364
permit, 68
Peroxisome, 18
personal, 110, 111, 200, 300, 309, 331
personal communication, 110, 111
personal computers, 200, 309, 331
Peru, 166
pest management, 64
pests, 240
pH, vi, 100, 160, 223, 224, 225, 226, 227, 228, 229, 230, 231, 232, 233, 234, 235, 236, 237, 239, 240, 241, 242, 243, 244, 245, 246, 351
pH values, 224, 234, 235, 236
phage, 249, 251, 254, 256, 257, 258, 261
phagocytic, 351
pharmaceutical, xiii
phenotype, 42, 49, 183, 225, 227, 230, 251, 297, 299, 301, 305, 329, 351
phenotypes, 42, 49, 67, 251, 258, 352
phenotypic, 42, 65, 66, 67, 77, 228, 264, 266, 289, 300, 351, 352, 365
phenylketonuria, 160
pheromone, 3, 28
Phoenix, 87
phosphatases, 242, 246
phosphorylation, 50, 215, 218, 229
photoreceptor, 23
phototropism, 324, 329
phylogenetic, vi, 1, 3, 34, 63, 68, 69, 72, 77, 88, 91, 101, 103, 109, 113, 135, 138, 139, 142, 153, 162, 183, 184, 185, 186, 189, 190, 192, 193, 194, 198, 200, 201, 202, 263, 265, 268, 282, 283, 287, 288, 298, 299, 300, 301, 305, 306, 307, 309, 310, 311, 314, 319, 320, 323, 325, 327, 329, 330, 331, 332, 334, 338, 340, 342
phylogenetic tree, 63, 103, 142, 190, 193, 194, 200, 268, 288, 298, 299, 309, 320, 330, 331, 340, 342
phylogeny, xiii, xiv, 69, 91, 146, 153, 157, 166, 184, 195, 199, 200, 203, 264, 266, 282, 298, 299, 300, 304, 307, 308, 310, 311, 314, 315, 317, 319, 321, 322, 325, 329, 331, 332
phylum, 265, 307, 310, 321, 331
physical interaction, 50
physical properties, 158
physiological, 34, 39, 40, 140, 141, 147, 172, 186, 227, 236, 264, 266, 281, 287, 291, 305, 307, 330
physiology, 48, 284, 300, 347
phytopathogens, 64
Phytotoxicity, 260
pig, 17, 26
pigments, 174, 179, 180, 196
pigs, 94
Pisum sativum, 18, 87, 105, 187, 202
pitch, 141, 159, 207
placenta, 334, 336, 338, 339, 341, 344, 346, 348
plants, 1, 3, 7, 9, 32, 33, 39, 40, 41, 42, 43, 46, 48, 49, 52, 53, 54, 57, 59, 60, 61, 64, 65, 66, 68, 69, 85, 86, 87, 89, 93, 94, 98, 102, 108, 109, 114, 137, 139, 140, 141, 142, 143, 149, 151, 152, 158, 159, 160, 166, 167, 171, 172, 181, 183, 186, 195, 196, 199, 223, 224, 239, 245, 250, 253, 260, 264, 282, 286, 301, 311, 314, 322
plaques, 254, 257, 258

plasma, 223, 224, 228, 229, 231, 242, 245
plasma membrane, 223, 224, 228, 229, 245
plasmid, 48, 252, 254
plasmids, 251
plasticity, 2, 45
plastics, 352
plastid, 25
platelet, 28
platforms, 7
play, 152, 155, 232, 250, 257, 351
Pleurotus ostreatus, 335
ploughing, 129
PNA, 356, 364
pneumonia, 231
point mutation, 144, 148, 166, 217
poisoning, 92, 96
Poland, 188
polarity, 358
polarized, 257
polyacrylamide, 50, 89, 149, 152, 235, 335, 347, 355, 362
polymerase, xiii, 85, 88, 91, 92, 101, 103, 137, 138, 144, 162, 164, 166, 169, 196, 268, 282, 304, 321, 336, 346, 354, 356, 360, 365, 366, 367, 368
polymerase chain reaction, xiii, 85, 88, 91, 92, 101, 103, 137, 138, 144, 162, 164, 166, 169, 196, 268, 282, 304, 336, 346, 354, 365, 366, 367, 368
polymers, 208
polymorphism, 92, 105, 109, 133, 142, 143, 145, 147, 148, 149, 150, 151, 153, 155, 157, 160, 161, 164, 165, 166, 167, 168, 337, 338, 343, 346, 355, 358, 362, 364, 366, 370
polymorphisms, 102, 106, 141, 142, 143, 144, 151, 156, 162, 165, 166, 168, 169, 183, 358, 360, 365, 370
polypeptide, 13, 20, 28, 226, 231
polyphenols, 207
polysaccharide, 163, 228
polysaccharides, 151, 219
polyunsaturated fat, 305
polyunsaturated fatty acid, 305
poor, 44, 80, 130, 227, 267, 352
population, 42, 43, 57, 60, 65, 68, 86, 90, 111, 141, 145, 147, 148, 151, 154, 155, 165, 295, 345, 349, 350
pore, 174, 329, 348
Porifera, 202
postoperative, 308
post-translational modifications, 50
potato, 42, 51, 64, 94, 118, 132, 167
poultry, 94, 95, 100
power, 67, 153, 291, 352, 358, 361, 362, 363
precipitation, 50
prediction, 80
pregnancy, 351
preparation, 101, 131, 358
preservatives, 337
pressure, 140, 183
prevention, 347
priming, 333, 339, 346
printing, 319
prior knowledge, 359
probability, 8, 33, 352
probe, 102, 103, 115, 136, 149, 150, 160, 174, 235, 252, 254, 257, 343, 354, 356, 359, 364, 369, 370
procedures, 139, 350, 352, 354, 358, 366
producers, 72, 93, 101, 130
production, 3, 52, 63, 64, 67, 80, 85, 86, 89, 95, 96, 97, 100, 102, 105, 107, 132, 134, 135, 136, 138, 141, 152, 160, 164, 173, 178, 179, 181, 182, 183, 195, 206, 207, 208, 217, 219, 223, 224, 227, 232, 233, 239, 240, 250, 255, 258, 259, 260, 266, 305, 314, 351, 354
profit, 78
program, 5, 6, 116, 187, 305, 330
programming, 4, 36
progressive, 350
prokaryotes, 40
proliferation, 2, 59
promote, 185, 232
promoter, 16, 48, 208, 212, 213, 214, 215, 216, 217, 221, 226, 228, 233, 240, 246, 255, 258
promoter region, 48, 240
propagation, 327
prophase, 3, 33
prophylaxis, 308
proposition, 225
prostaglandins, 351
proteases, 151, 228, 314
protection, 63, 81, 216, 286, 347
protein binding, 229
protein family, 9, 23
protein sequence, 5, 8, 26, 227
protein structure, 80
protein synthesis, 185, 199
proteinase, 29, 366
protein-protein interactions, 50, 54, 62
proteins, 8, 25, 28, 30, 33, 45, 50, 51, 53, 57, 58, 152, 185, 208, 210, 212, 214, 215, 216, 220, 225, 229, 234, 235, 238, 243, 246, 255, 308, 335

proteolysis, 23, 229, 244
proteome, 23, 50, 57, 82
proteomics, 39, 50, 51, 56, 57
prothrombin, 31
protocol, 146, 163, 186, 187, 249, 250, 251, 257, 357
protocols, 168, 203, 268, 311, 348
protons, 224
protoplasts, 48
pseudo, 351
Pseudomonas, 61, 102
public, 3, 51, 63, 74, 267, 284, 302
publishers, 78
PubMed, 284
pulp, 206, 207, 218
pulse, 94
pulses, 85, 86
purification, 50, 109, 160, 221, 222, 316

Q

quality control, 67
quantitative estimation, 47
query, 4, 5, 6, 7, 44, 69, 72, 80, 258

R

R. oryzae, 325
race, 42, 56, 89, 103, 105, 141, 143, 145, 146, 147, 150, 153, 155, 159, 162, 165, 167
radiation, 40, 324, 350
radiolabeled, 100, 103
rain, 112
rainfall, 113
random, 55, 88, 89, 103, 109, 144, 146, 147, 157, 159, 162, 163, 165, 187, 251, 257, 258, 308, 319, 359, 367, 368
random amplified polymorphic DNA, 88, 103, 109, 146, 147, 157, 159, 162, 163, 308, 367, 368
randomly amplified polymorphic DNA, 282, 337, 364, 369
randomness, 252
range, 2, 3, 7, 23, 60, 67, 91, 96, 97, 99, 101, 139, 140, 141, 145, 181, 182, 224, 228, 239, 255, 266, 267, 281, 301, 340, 350, 354
RAPD, 88, 89, 92, 102, 103, 104, 105, 109, 115, 135, 137, 139, 144, 145, 146, 147, 148, 151, 157, 158, 159, 161, 162, 163, 164, 166, 168, 171, 195, 282, 333, 337, 343, 346, 347, 348, 359, 360, 362, 364, 368
rape, 53, 159
rat, 24, 30, 31
rating scale, 253, 255, 256
rats, 230
raw material, 83
raw materials, 83
reading, xiii, 132, 155, 227, 236, 252, 362
reagents, 354, 358
real time, 233
real-time, 48, 108, 109, 131, 132, 134, 136, 138, 336, 345, 346, 354, 355, 356, 366, 367
receptors, 3, 33, 45, 229
recognition, 3, 34, 42, 45, 72, 89, 135, 140, 208, 209, 210, 220, 226, 227, 229, 242, 301, 310, 339, 346
recombinant DNA, xiii
recombination, 148, 258, 259
reconstruction, 264, 288, 300, 301, 315, 317, 319
recruiting, 215
recycling, 229
red light, 43
reduction, 3, 355, 363
redundancy, 49
regeneration, 33, 35
regional, 30, 33, 57, 267
regression, 116, 122, 137
regression analysis, 137
regular, 324
regulation, xiii, xiv, 16, 21, 23, 30, 46, 55, 59, 80, 205, 208, 211, 212, 213, 214, 215, 216, 217, 218, 221, 222, 224, 226, 228, 229, 230, 231, 233, 234, 239, 240, 241, 242, 244, 246, 305, 345
regulators, 3, 32, 33, 46, 58, 61, 213, 217, 232
relational database, 46, 69, 76
relationship, 47, 54, 135, 137, 146, 171, 243, 314, 320, 322, 323, 327, 343, 347
relationships, 1, 34, 60, 77, 85, 90, 91, 101, 148, 150, 151, 153, 156, 161, 167, 183, 184, 185, 186, 190, 191, 192, 193, 194, 197, 198, 202, 203, 260, 265, 282, 288, 299, 300, 301, 308, 310, 322, 326, 331, 345, 350
relative size, 359
relatives, 167, 308, 323
relevance, 217, 231, 239, 284, 325, 329
reliability, 264, 267, 301, 369
repeatability, 148
repetitions, 101
repression, 56, 213, 216, 217, 220, 226, 229, 243
repressor, 30, 213, 216, 217, 218, 219, 221, 230, 231, 240
reproduction, 281, 350

reproductive organs, 97
research, 5, 7, 8, 35, 50, 65, 66, 75, 78, 80, 90, 92, 98, 100, 106, 113, 115, 130, 141, 147, 155, 162, 168, 172, 183, 195, 205, 208, 250, 251, 257, 259, 265, 266, 284, 295, 334, 363
researchers, xiii, xiv, 47, 64, 76, 78, 139, 154
reservoirs, 368
residues, 96, 225, 227, 232, 233, 236, 238, 242, 289
resistance, 23, 24, 39, 40, 41, 42, 43, 45, 49, 52, 53, 54, 56, 57, 58, 60, 67, 81, 86, 95, 105, 160, 255, 258, 259, 346, 351, 352, 365, 367, 368, 369
resolution, 89, 90, 108, 141, 153, 184, 307, 343, 359, 361, 363, 368
resources, 2, 37, 61, 63, 74, 76, 77, 80, 82, 83, 303
respiratory, 26, 27, 32, 98
respiratory problems, 98
response format, 5
restriction enzyme, 89, 92, 143, 149, 150, 162, 164, 254, 339, 362, 368
restriction fragment length polymorphis, 102, 104, 142, 143, 149, 282, 339, 359
retina, 27
retinoblastoma, 19
retinoic acid, 27
revaluation, 297
reverse transcriptase, 163, 237, 356
RFLP, 88, 89, 90, 101, 103, 134, 135, 139, 141, 143, 148, 157, 163, 166, 282, 331, 338, 339, 340, 345, 346, 347, 359, 360, 362, 365, 369
rhizobia, 104
rhizoid, 289
rhizosphere, 92, 142, 166
Rho, 22, 25, 29, 60
rhodium, 218
Rhodophyta, 315, 319
ribosomal, vi, 18, 31, 78, 90, 91, 92, 102, 103, 104, 106, 110, 141, 142, 143, 149, 153, 155, 156, 157, 161, 167, 168, 183, 184, 189, 195, 196, 197, 202, 203, 263, 264, 267, 282, 287, 288, 302, 306, 307, 310, 311, 312, 314, 332, 338, 340, 346, 347, 348, 353, 359, 362, 366
ribosomal RNA, 90, 91, 92, 103, 142, 156, 168, 184, 189, 196, 203, 264, 311, 314, 353
ribosome, 185
ribosomes, 90
rice, 5, 11, 17, 24, 25, 30, 44, 46, 47, 50, 52, 54, 56, 57, 58, 60, 62, 64, 74, 76, 81, 83, 86, 99, 100, 134, 161, 168, 175, 176, 177, 180, 181, 259
rings, 179, 181, 183, 323
RISC, 49
risk, 66, 131, 156, 350, 351, 355
risk factors, 350
RNA, 5, 20, 22, 39, 49, 53, 54, 55, 56, 57, 90, 91, 102, 103, 109, 155, 183, 184, 189, 197, 236, 237, 239, 305, 306, 308, 321, 338, 353, 356, 359, 367, 370
RNAi, 39, 45, 49, 53, 57
RNAs, 46, 54
Romania, 198
room temperature, 186, 267
root cap, 12
routines, 6
Russia, 110, 133, 136, 138, 341
Russian, 136
rust, 54, 55, 60
ruthenium, 218
Rutherford, 53, 62
rye, 56, 112, 120, 121, 166

S

Saccharomyces cerevisiae, 17, 18, 20, 21, 26, 43, 54, 62, 82, 220, 241, 243, 259
safety, 113
saline, 176
salmon, 30, 180, 182
Salmonella, 363, 370
sample, 64, 72, 101, 116, 117, 118, 123, 124, 149, 173, 244, 336, 340, 343, 344, 353, 358, 363
sampling, 101, 106, 303, 308, 319, 344
Sao Paulo, 203
saprophyte, 183
Saudi Arabia, 85
Saudia Arabia, vii, viii, xi, xii
savings, 364
scaffold, 233
Scanning electron, 285, 290, 291, 326, 328
scanning electron microscopy, 268, 282, 285, 303, 314, 316, 324, 329
schema, 52
schizophrenia, 22
Schmid, 364
science, 65, 78
scientific, 36, 67, 78, 80, 88, 224, 236, 258, 333
scientific community, 80, 258
scientists, xiii, xiv, 78, 99, 300, 314
Sclerotinia sclerotiorum, 72, 227, 238, 245
scores, 6, 255
scripts, 4
SDS, 235, 333, 335, 336

search, 1, 3, 4, 5, 6, 7, 9, 34, 36, 50, 63, 64, 69, 81, 264, 284, 288, 302, 319
searches, 1, 4, 5, 6, 7, 8, 9, 12, 24, 26, 32, 78, 187, 255, 256, 267, 268, 282, 288, 289, 302, 319
searching, 6, 77, 256, 284
secrete, 224
secretion, 62, 151, 157, 218, 220, 228, 257, 258
security, 64, 82, 83
sediments, 173
seed, 12, 15, 23, 94, 115, 131, 136, 159, 183
seeds, 85, 86, 93, 305
segregation, 258
selecting, 156, 255
senescence, 36
sensation, 100
sensing, 229
sensitivity, 91, 123, 149, 232, 337, 352, 353, 354, 355, 356, 363
separation, 50, 149, 174, 323, 339, 345, 358
septum, 174
sequencing, 45, 52, 55, 59, 73, 74, 76, 90, 91, 141, 149, 157, 168, 195, 203, 240, 254, 268, 282, 302, 311, 316, 336, 338, 340, 345, 346, 347, 348, 355, 362, 363, 364, 366
series, 82, 174, 267, 333, 340
serine, 17, 29, 62, 220, 228, 243
serum, 196
services, iv, 6, 75
severity, 98, 110, 253, 255, 351
sex, 21, 92, 97
sexual development, 16
sexual reproduction, 143, 322, 326
shape, 2, 33, 36, 139, 140, 171, 182, 247
shares, 235
sharing, 67
sheep, 94, 95
Sheep, 24
shock, 25, 32
shoot, 22
short period, 96
Siberia, 180
sickle cell, 105
sickle cell anemia, 105
side effects, 95
signal transduction, 45, 50, 68, 223, 225, 228, 229, 239, 241, 242, 244, 246
signaling, 29, 242, 243, 244
signalling, 50, 62, 225, 227, 228, 229, 240, 241
signals, 2, 199, 213, 215
signs, 95, 98
silica, 101
silico methods, 47
silver, 89, 162, 235
similarity, 1, 4, 5, 6, 7, 8, 9, 10, 12, 21, 25, 31, 32, 46, 60, 69, 77, 88, 89, 92, 109, 142, 143, 145, 146, 150, 156, 230, 246, 255, 256, 258, 288, 289, 294, 295, 302
simulation, 64, 80
single nucleotide polymorphism, 109, 132
siRNA, 49
sites, 8, 28, 46, 66, 78, 89, 143, 144, 147, 150, 153, 154, 161, 190, 191, 192, 212, 213, 216, 227, 228, 233, 246, 249, 251, 252, 254, 255, 256, 257, 258, 320, 352, 353, 359, 360
skills, 36
skin, 85, 95, 96, 106, 350, 352
society, 75
sodium, 100, 152, 212, 237, 239, 335
software, 7, 9, 12, 16, 24, 26, 32, 71, 75, 76, 80
soil, xiii, 52, 86, 92, 96, 101, 103, 109, 153, 159, 164, 165, 171, 172, 176, 178, 180, 182, 183, 195, 197, 200, 201, 202, 266, 267, 295, 304, 335, 345
soils, 140, 157, 168, 195
Sorghum, 41, 87
sorting, 229
South Africa, 95
South America, 108, 235
Southern blot, 359
Southern Hemisphere, 98
Soviet Union, 100, 111
soy, 207, 314, 325
soybean, 77, 86, 94, 199
soybeans, 305
spacers, 91, 142, 183, 184, 195, 338, 342, 359, 363
Spain, 157, 168
spatial, 43, 250
specialists, 69
specialization, 133, 141
speciation, 40, 171, 175, 260, 308, 367
specificity, 23, 41, 91, 141, 152, 172, 197, 198, 260, 353, 354, 355, 363
spectra, 46, 48, 344, 355
spectrophotometry, 109, 134
spectrum, 132, 352, 353
speed, xiii, 7, 352
sphingosine, 98
spleen, 96
sporangium, 286, 287, 289, 291, 299
spore, 49, 61, 109, 260, 294, 322, 325, 326, 327, 329
stability, 64, 207, 319, 322, 360

stages, 3, 36, 40, 45, 46, 60, 94, 96, 100, 161, 327, 328, 336, 350
standard deviation, 256
standardization, 352, 355, 363
standards, 131
starch, 21, 207
stars, 121, 122, 124, 125, 126
starvation, 231
statistics, 6
stem cell transplantation, 364
sterile, 81, 323, 324, 327
steroid, 22
steroids, 266, 314, 350
stochastic, 164
stock, 254, 257
stomach, 100
storage, 97, 266, 333
strain, 17, 48, 59, 66, 67, 74, 83, 88, 89, 100, 113, 114, 145, 149, 152, 155, 181, 211, 212, 213, 215, 216, 217, 218, 231, 246, 251, 252, 253, 254, 255, 267, 279, 280, 281, 284, 286, 294, 295, 299, 303, 311, 329, 351, 352, 353, 357, 359, 360, 362, 363, 366, 369
strain improvement, 211
strains, 58, 63, 65, 66, 69, 86, 88, 89, 91, 94, 98, 99, 100, 102, 105, 109, 112, 114, 134, 139, 140, 141, 142, 143, 145, 148, 153, 155, 157, 160, 164, 165, 167, 178, 179, 191, 192, 194, 205, 208, 211, 216, 218, 219, 251, 253, 258, 260, 264, 267, 279, 280, 288, 295, 298, 299, 302, 303, 305, 306, 309, 316, 346, 351, 357, 358, 362, 367, 368, 369
strategies, 1, 4, 34, 36, 42, 48, 67, 82, 85, 101, 203, 208, 210, 214, 267, 310, 316, 332, 353, 370
strength, 224
stress, 93, 98, 228, 243
students, 303
subgroups, 145, 329
subjective, 109, 300
subsidy, 78
substances, 208, 209, 212, 218
substitutes, 206
substitution, 165, 319
substrates, 172, 206, 267, 303
sucrose, 101, 118
suffering, 95, 351
sugar, 45, 132, 146, 159, 234
sugar beet, 132, 146, 159
sugar cane, 234
sugars, 210, 218, 310
sulfate, 152, 235, 335
sulphate, 335
sunflower, 196, 201
supernatant, 234
supply, 65, 69, 76
suppression, 35, 49, 96
suppressor, 16, 17, 24
surgery, 350
surprise, 9
survival, 140
susceptibility, 41, 352
swelling, 97
switching, 7, 208, 351, 352, 369
symbiosis, 47, 56, 62
symbiotic, 43, 46, 47, 51, 61, 84, 285, 314
symptom, 152
symptoms, 92, 96, 98, 150
syndrome, 19, 30, 351
synergistic, 206
synteny, 44, 45
synthesis, 73, 224, 240, 243, 250, 251, 261, 304
synthetic, 101, 106, 267
Syria, 189
systematic, 55
systematics, 101, 166, 199, 200, 266, 282, 303, 305, 306, 307, 308, 311, 314, 350
systems, 33, 39, 48, 49, 51, 55, 58, 73, 221, 240, 300, 301, 349, 352, 355, 363

T

T-2 toxin, 85, 93, 94, 95, 99, 100, 101, 102, 103, 105, 110, 111, 130, 136
Taif, vii, viii, xi, xii, 85, 139
tandem mass spectrometry, 50
tandem repeats, 91, 150, 184, 343
targets, 46, 49
taxa, 52, 67, 68, 69, 89, 91, 142, 173, 184, 190, 191, 194, 197, 266, 268, 292, 293, 295, 301, 304, 305, 314, 318, 319, 322, 326, 344, 350
taxonomic, 6, 66, 69, 90, 91, 101, 102, 136, 139, 141, 174, 178, 183, 186, 194, 198, 200, 265, 288, 300, 327, 338, 350
taxonomy, xiii, 2, 4, 5, 6, 7, 67, 69, 86, 89, 101, 104, 136, 157, 166, 172, 174, 184, 197, 264, 265, 300, 301, 303, 304, 327, 350
tea, 179
technological, 80
technology, xiii, 49, 62, 147, 300, 349, 363, 364, 367
temperature, 2, 47, 88, 93, 96, 97, 98, 100, 105, 144, 160, 201, 232, 290, 325, 356, 362

test data, 99
testes, 97
Texas, 60
textile, 206, 219
textile industry, 206, 219
theoretical, 236
theory, xiii, 35, 39, 42
therapeutics, 46, 51
therapy, 282
thermal, 356, 360
threat, 83
threonine, 17, 29
threshold, 290, 297, 298, 302, 325
thresholds, 319
thrush, 350
thyroid, 18
timber, 333, 334, 335, 346
time consuming, 139, 141, 195
tissue, 2, 45, 94, 97, 110, 140, 151, 157, 159, 258, 266, 335, 336, 367
tobacco, 61, 94
tolerance, 167, 228, 232, 243, 260
tomato, 58, 82, 86, 94, 151, 158, 160, 161, 163, 167
topology, 197
total product, 256
totipotent, 2
toxic, 95, 100, 103, 111, 134
toxicity, 92, 98, 99
toxin, 67, 68, 72, 92, 95, 97, 98, 100, 101, 110, 116, 164, 259
toxins, 92, 94, 95, 96, 97, 104, 110, 112, 132, 133, 136, 138
trade, 54
tradition, 265
traffic, 229
traits, 67, 74, 149, 199, 264
transcript, 45, 46, 47, 57, 77, 209, 213, 215, 216, 217, 256
transcription, 3, 9, 11, 12, 15, 16, 18, 19, 20, 21, 22, 23, 24, 25, 26, 27, 28, 29, 30, 31, 32, 33, 47, 50, 199, 205, 209, 211, 212, 213, 214, 215, 216, 217, 218, 221, 223, 224, 225, 226, 228, 229, 231, 233, 236, 239, 241, 242, 243, 244, 245, 246, 342
transcription factor, 3, 9, 11, 12, 15, 18, 19, 20, 21, 22, 23, 24, 25, 26, 27, 28, 29, 30, 31, 32, 205, 211, 212, 213, 215, 218, 221, 223, 225, 228, 236, 239, 241, 242, 243, 244, 245, 246
transcription factors, 3, 29, 205, 211, 212, 213, 215, 225, 236
transcriptional, 11, 22, 23, 24, 25, 27, 31, 46, 49, 50, 52, 56, 59, 61, 205, 209, 211, 212, 213, 215, 216, 217, 218, 219, 222, 226, 230, 233, 234, 239, 240, 244, 246
transcriptomics, 56
transcripts, 45, 46, 236, 244
transducer, 32
transducin, 15, 30
transduction, 69
transfer, 12, 155, 161, 163, 167, 252
transformation, 48, 59, 60, 151, 242, 250, 251, 252, 255, 258, 260, 261
transforming growth factor, 17
transgenic, 54, 80
transgenic plants, 54
transition, 235, 324
transitions, 144
translation, 50, 72, 153, 156, 184, 189, 212, 214, 235, 236, 284, 311, 314, 315, 319, 332
translocations, 44
transmembrane, 25, 30, 228, 229
transmission, 164, 368
transparent, 174
transplant, 308
transport, 211, 229, 257
transposases, 152
transposon, 39, 45, 47, 48, 55, 58, 154, 155, 159, 160, 166, 256
transposons, 154, 155, 161, 162, 163
trees, 77, 86, 187, 193, 268, 298, 304, 319, 320, 331
trend, xiii, 171, 184, 364
triage, 266, 311
trial, 78
tribes, 260
Trichoderma viride, 220
trifolii, 190, 193
triggers, 239
trypsin, 50
TSP1, 29
tumor, 16, 22, 25
tumor necrosis factor, 22, 25
tumour, 16, 17, 22, 24, 30
Tunisia, 144, 162
Turkey, 147, 158
Turku, xi, 107, 132, 135
two-dimensional, 50
tyrosine, 18

U

ubiquitin, 22, 23, 25, 26
ubiquitous, 154, 171, 196, 197, 199
ultrastructure, 281, 305, 307, 323, 329
ultraviolet, 144
undergraduate, 303
uniform, 179, 354
uniformity, 53
United Kingdom, 159, 304, 306, 307, 308, 309
United Nations, 66
United States, 68, 93, 95, 98
universities, xiii
urease, 25
URL, 6, 66, 79, 80
USDA, xii, 73, 81, 83
user-defined, 4
users, 7, 68, 78

V

vacuum, 186
vagina, 351
vaginal, 230, 350
vaginal thrush, 350
vaginitis, 230, 351
Valencia, 242
values, 1, 5, 8, 9, 12, 16, 24, 25, 26, 31, 32, 33, 108, 110, 116, 125, 191, 192, 194, 224, 235, 256, 289, 297, 299, 302, 319, 342
van der Waals, 232
van der Waals forces, 232
variability, 45, 63, 86, 92, 101, 105, 147, 150, 154, 155, 167, 183, 266, 309, 343, 355
variable, 42, 45, 91, 110, 143, 150, 151, 160, 179, 181, 182, 227, 267, 302, 320, 337, 340, 343, 352, 353, 358, 361, 362
variation, 42, 49, 54, 57, 90, 91, 92, 101, 105, 110, 113, 134, 138, 140, 141, 142, 144, 146, 147, 149, 153, 154, 157, 158, 162, 164, 166, 167, 183, 197, 203, 224, 226, 255, 261, 282, 312, 339, 340, 341, 346, 348, 366
vascular, 68, 140, 152, 160, 161, 167
vector, 45, 60, 254, 258
vegetables, 85, 86, 94, 104
vegetation, 173
vein, 232
velo-cardio-facial syndrome, 25
Venezuela, 295
vermiculite, 100
versatility, 140, 314
vesicle, 326
Victoria, 106
village, 82
viral, 49, 60, 350
viral infection, 350
virulence, 40, 41, 42, 45, 46, 52, 53, 56, 58, 60, 62, 67, 69, 93, 135, 136, 142, 146, 151, 152, 156, 165, 167, 168, 224, 227, 228, 230, 231, 232, 240, 241, 245, 251, 253, 260, 351
virus, 11, 31, 54, 173, 369
viruses, 49, 90
viscosity, 207
visible, 94, 172, 252
vision, 67
visual, 324
visualization, 35, 68
Vitamin D, 25
vocabulary, 3
vomiting, 85, 96

W

Wales, 134
walking, 98
warrants, 195
Washington, viii, xii, 64, 73, 104, 249, 252
wastes, 233, 240, 245
water, 94, 172, 200, 201, 206
wealth, 52
web, 1, 4, 5, 6, 7, 35, 63, 69, 70, 71, 75, 80, 171, 265
web browser, 7
web sites, 63
web-based, 35, 70, 71
websites, 44, 65
wells, 356
wheat, 52, 60, 76, 81, 86, 93, 95, 96, 97, 99, 105, 107, 108, 111, 112, 114, 116, 117, 118, 120, 121, 122, 123, 125, 126, 130, 131, 132, 133, 134, 135, 136, 137, 150, 160, 162, 164, 165, 166, 308, 309
whole grain, 98
wild type, 228, 232, 245, 251, 253, 255
windows, 7, 203, 310, 332, 334, 335
wine, 207
winter, 100, 111, 113, 121, 133, 303
Wisconsin, 84
women, 350
wood, vi, 333, 334, 335, 336, 337, 338, 340, 341, 342, 343, 344, 345, 346, 347, 348

woods, 348
workers, 172, 216
workflow, 7
World War II, 111
world wide web (www), 5, 78
writing, 5, 78
WWW, 72, 73

X

X chromosome, 29
X-linked, 18
Xylan, 233
xylem, 94, 140

Y

yeast, 3, 21, 46, 47, 49, 50, 51, 54, 56, 57, 62, 82, 91, 101, 200, 207, 220, 229, 230, 232, 235, 236, 237, 239, 241, 243, 244, 246, 257, 264, 267, 304, 306, 310, 316, 345, 350, 351, 352, 353, 354, 356, 361, 363, 366, 368, 370
yield, 35, 86, 93, 108, 151, 207, 294
yield loss, 93

Z

Zea mays, 15, 16, 24, 26, 87, 163
zebrafish, 27
zinc, 11, 15, 21, 23, 25, 27, 28, 31, 223, 225, 227, 235, 236, 238, 239, 242, 243, 246
Zinc, 25, 30, 237
Zn, 214
zygomycosis, 307
zygospore, 287, 323, 327